SPECTROSCOPY AND STRUCTURE OF METAL CHELATE COMPOUNDS

SPECTROSCOPY AND STRUCTURE OF METAL CHELATE COMPOUNDS

KAZUO NAKAMOTO
Illinois Institute of Technology

PAUL J. McCARTHY, S.J.
Canisius College

JOHN WILEY & SONS, INC.

New York · London · Sydney

Library of Congress Catalog Card Number: 67-30634
GB 471 62981X
Printed in the United States of America

CONTRIBUTORS

PROFESSOR YOSHIHIKO SAITO, *The Institute for Solid State Physics, The University of Tokyo, Tokyo, Japan.*

PROFESSOR CURTIS R. HARE, *Department of Chemistry, State University of New York, Buffalo, New York.*

PROFESSOR JUNNOSUKE FUJITA, *Department of Chemistry, Tohoku University, Sendai, Japan.*

PROFESSOR YOICHI SHIMURA, *Department of Chemistry, Osaka University, Osaka, Japan.*

PROFESSOR KAZUO NAKAMOTO, *Department of Chemistry, Illinois Institute of Technology, Chicago, Illinois.*

PROFESSOR SHIZUO FUJIWARA, *Department of Chemistry, The University of Tokyo, Tokyo, Japan.*

PROFESSOR PAUL J. MCCARTHY, S.J., *Department of Chemistry, Canisius College, Buffalo, New York.*

CONTRIBUTORS

PREFACE

In recent years, several excellent volumes have appeared which review either the whole field of coordination chemistry or some specialized topic in the field. None, however, has concentrated on the structure and spectra of metal chelate compounds. In the present volume our aim is to show how spectroscopic and diffraction techniques have been or can be applied to the elucidation of the structure and bonding of metal chelate compounds.

The physical techniques here discussed are x-ray and neutron diffraction; visible and ultraviolet spectroscopy; optical rotatory dispersion (ORD) and circular dichroism (CD); infrared (IR), electron spin resonance (ESR), and nuclear magnetic resonance (NMR) spectroscopy. Microwave, Mössbauer, and nuclear quadrupole resonance spectroscopy, as well as electron diffraction, have been omitted, since they have not yet been extensively applied in studies of metal chelate compounds.

Each chapter consists of an introduction, a section on theory, a section on applications, and a bibliography. The outline of the theory of each technique is meant to provide enough background to make the applications that follow intelligible. These outlines are not intended to be complete treatments; for these the reader must consult the reference books cited at the end of the chapters. Rather, the emphasis of this book is placed on applications of the theories. Each chapter attempts to show the types of structural and bonding information that can be obtained by using a particular technique. Thus, the various references have been classified according to the nature of the information they provide, rather than according to the metal or ligand involved. In each section one or more typical examples have been discussed in detail, and other similar studies have been cited in the bibliography, which also lists books and reviews of general coverage that may prove useful to one who is looking for a diverse, or more detailed, treatment of a specific technique.

Although the restriction of the material in this book to metal chelate compounds is rather arbitrary, it was done mainly to make the volume of material more tractable. As a result, we have in general omitted discussion of nonchelated coordination compounds, metal carbonyls, metal sandwich compounds, and other similar compounds, unless these also contain a chelate ring.

We would like to thank the following people for their helpful comments and discussions: Professor K. J. Watson (Chapter 1); Professors J. R. Perumareddi and W. Schneider (Chapter 2); Dr. T. Bürer (Chapter 3); Professor I. Nakagawa (Chapter 4); Professor C. A. Trapp (Chapter 5); and Professor H. A. Szymanski and Dr. G. N. LaMar (Chapter 6). We are also grateful to Taylor and Francis, Ltd., and to the American Institute of Physics for permission to quote from references 217 and 282, respectively, in Chapter 6.

March 1968

Kazuo Nakamoto
Paul J. McCarthy, S.J.

CONTENTS

SPECTROSCOPY AND STRUCTURE OF METAL CHELATE COMPOUNDS

1

X-RAY AND NEUTRON DIFFRACTION

YOSHIHIKO SAITO

The University of Tokyo

The most striking feature of chemical science since about 1940 has been the increasing importance of physical methods of investigation. The physical methods used for elucidating molecular structure fall into two classes: those that yield detailed information about the whole structure of the molecule, and those that yield fragmentary information concerning individual bonds or particular groups of atoms in a molecule. The first class includes x-ray, neutron, and electron diffraction, the second, various kinds of spectroscopy of the region ranging from microwave to ultraviolet, the measurement of electric and magnetic moments, optical rotatory dispersion, and circular dichroism. Most of these subjects are treated in the following chapters. In this chapter the principles of the techniques of x-ray and neutron diffraction are considered first. The electron diffraction technique is not mentioned here because most of the structures of metal chelates have been elucidated by x-ray and neutron diffraction techniques. Then various metal chelates are discussed to show how diffraction techniques are applied to each of them.

Although chemical methods of investigation leave no doubt about the existence of chelate rings and the general features of the structure of metal chelates, confirmation of the correctness of the structures by diffraction techniques has greatly strengthened our knowledge of chelation. Such techniques have introduced a metrical element into the understanding of chelate rings by revealing the lengths of chemical bonds, the angles between them, and other structural details that could not be gained by purely chemical methods. The determination of precise location of all the atoms in the unit cell of a crystal by structural analysis provides the strongest and most detailed evidence for the existence of chelate rings. Moreover, the distances between the atoms of neighboring molecules and ions can be measured with the same precision as those between the atoms in the molecule itself. The quantitative knowledge of intermolecular distances gained in this way is of greatest importance, especially in structures involving special interactions, such as hydrogen bonds or direct metal-metal interactions. If the wavelength of x-rays is properly chosen, it is possible to determine the absolute configuration of optically active complexes. The absolute configuration of one optically active complex determined by means of x-rays (in conjunction with rotatory dispersion studies) generally allows the configuration of many related coordination compounds to be settled with reasonable certainty.

When the x-ray method was first applied to metal chelates, the analysis was often taken no further than a determination of the general shape and symmetry of the ion or molecule. Now, with the help of greatly improved techniques including the use of electronic computers and automatic single crystal diffractometers, structures as complex as that of vitamin B_{12}, which consists of 105 atoms, have been successfully solved. When the method is carried to its limit, it is capable of revealing, at least in principle, the precise distribution of all the bonding electrons in a molecule. In practice, however, there are several fundamental difficulties which conspire to prevent the attainment of such a complete solution.

1-1 THEORY

A. Principles of X-ray Diffraction

1. General Theory. A crystal is a periodic three-dimensional array of atoms in which the interatomic distances are about 2×10^{-8} cm on the average. This fact reveals the possibility, first realized by von Laue in 1912, of obtaining diffraction effects if radiation with a wavelength in this range (1–2 A) is passed through a crystal. When the incident beam passes over the atoms in a crystal, each atom may be regarded as scattering a wave. A crystal therefore acts as a three-dimensional diffraction grating for x-rays, and three equations (the Laue equations) must be satisfied simultaneously if there is to be constructive interference of monochromatic x-rays:

$$a \cdot (s - s_0) = h\lambda, \qquad b \cdot (s - s_0) = k\lambda, \qquad c \cdot (s - s_0) = l\lambda, \qquad (1\text{-}1)$$

where a, b, and c are the three axial vectors by which the crystal is described, s and s_0 are the unit vectors of the diffracted and incident beams, respectively, λ is the wavelength, and h, k, and l are integers that define the orders of a particular diffracted beam. Bragg showed that the diffraction effect expressed by (1-1) can be regarded as a reflection of the x-rays by the plane with indices *hkl*:

$$\lambda = 2d(hkl) \sin \theta, \tag{1-2}$$

θ being the angle between the incident (or reflected) beam and the plane *hkl*, and $d(hkl)$ the interplanar spacing between successive planes. It follows from the Bragg equation that a particular reflection *hkl* can occur only at angle θ.

In general, reflection takes place from all possible planes (hkl) out to the limiting values of $d(hkl) = \lambda/2$, which occurs when θ is 90°. As the wavelength employed is usually much less than the maximum spacing, it is possible to observe reflections from a very large number of planes. In general, only a few planes are in a position to reflect if the crystal is kept stationary, and it is therefore necessary to rotate the crystal if all the possible reflections are to be observed. A single crystal, when rotated in a collimated beam of monochromatic x-rays, accordingly reflects a large number of discrete beams in directions determined by the geometry of the crystal lattice. The diffracted beams may be recorded photographically (in a variety of goniometers†) or by proportional or other types of counters. The intensity of each reflection is usually found to vary abruptly from plane to plane. The intensity is a function of θ and also, more especially, of the particular distribution of scattering matter across the plane in question.

Since λ is known, the interplanar spacing $d(hkl)$ can be calculated by means of (1-2) if the direction of the diffracted beam is known. The unit cell dimensions can be derived from the values of $d(hkl)$ and the geometrical relations between the diffracted beams. From the volume of the unit cell and the density of the crystal the number of molecules, Z, in the unit cell can be calculated:

$$Z = \frac{\rho V \times 0.6023 \times 10^{24}}{M}, \tag{1-3}$$

where ρ is the density (g cm^{-3}), V the volume (cm^3) of the unit cell, and M the molecular weight.

The theory of space groups tells us that there are 230 essentially different ways of arranging asymmetric but identical bodies in crystal lattice systems. Every crystal must conform to one of these 230 possible systems. Information about the space group is sometimes obtainable from a simple study of reflections, the systematically absent ones being noted. Most of the commonly occurring space groups can be identified uniquely from such systematic absences. Sometimes there is ambiguity, and rigorous determination of the space group can be made only by a much more detailed study of the reflections and their intensities. At present, however, the determination of the space

† The Weissenberg goniometer is most widely used.

group and of the cell dimensions is largely a matter of routine. When the space group is determined, the number of asymmetric units necessary to build up the symmetry of the space group and complete the identically repeating unit cell can be found easily.

The intensity of each reflection, $I(hkl)$, is proportional to the quantity $|F(hkl)|^2$, where $F(hkl)$ is given by

$$F(hkl) = \sum_{j=1}^{N} f_j(hkl) \exp\,[2\pi i(hx_j + ky_j + lz_j)]. \tag{1-4}$$

In this equation $f_j(hkl)$ is called the atomic scattering factor and represents the scattering power of the jth atom, and x_j, y_j, and z_j are the fractional coordinates of the jth atom.† $F(hkl)$ is usually a complex number called the structure factor. It gives the amplitude and the phase of the diffracted wave from one unit cell. Summation must be taken over all the atoms in the unit cell. Some of the atoms are, however, related to others by a symmetry operation of the space group; therefore (1-4) can be simplified to some extent. For instance, if the space group is centrosymmetric and the origin of coordinates is taken at one of the centers of symmetry, the same kind of atoms can always be found at x_j, y_j, z_j and at $-x_j$, $-y_j$, $-z_j$. Consequently the structure factor becomes real and can be written

$$F(hkl) = 2 \sum_{j=1}^{N/2} f_j(hkl) \cos\,[2\pi(hx_j + ky_j + lz_j)]. \tag{1-5}$$

This equation means that the phase angle is either 0° or 180° and that it can be determined by the sign of $F(hkl)$. Values for $f_j(hkl)$ have been calculated for most of the atoms and have been determined experimentally for some atoms. Consequently our problem is to find by means of (1-4) a set of coordinates from the observed values of $F(hkl)$. The magnitude of $F(hkl)$ can be ascertained from the observed intensities, $I(hkl)$.

If an approximate trial structure can be derived, for example, from packing considerations and other clues to the arrangement of molecules, such as the anisotropy of optical properties of single crystals, values of $F(hkl)$ can be evaluated from (1-4) and compared with the observed magnitudes of $F(hkl)$. If the agreement is not satisfactory, the coordinates are adjusted to give better agreement. The final stage of this refinement of the structure is often carried

† X-rays are scattered by extranuclear electrons, and destructive interference between the wavelets scattered by various parts of the electron cloud takes place to an increasing extent as θ increases; $f(hkl)$ therefore falls off from its initial value, $f = Z$, with increasing θ (Z = atomic number).

out on a computer using the least-squares method. The estimate of the agreement is often given by the discrepancy factor (reliability index):

$$R = \frac{\sum \left| |F_o(hkl)| - |F_c(hkl)| \right|}{\sum |F_o(hkl)|}, \tag{1-6}$$

where $F_o(hkl)$ and $F_c(hkl)$ are the observed and calculated structure factors, respectively. Most of the refined structures have a reliability index of less than 0.20. In an accurate determination this index may be reduced to less than 0.10.

2. Electron Density. Instead of finding the atomic positions by trial and error, a more direct approach to the problem can be made by the use of Fourier synthesis. The electrons are distributed between the atomic nuclei, and the electron density varies continuously from point to point within the unit cell according to the position and the nature of the atoms therein. This distribution of electron density varies periodically from one unit cell to the other throughout the crystal, since the crystal is a periodic three-dimensional array of atoms. Because of this periodicity the electron density distribution can be resolved into its Fourier components. Thus the distribution function of the electron density $\rho(xyz)$ can be expanded as a three-dimensional Fourier series whose coefficients are the structure factors:

$$\rho(xyz) = \frac{1}{V} \sum_{-\infty}^{\infty} \sum_{-\infty}^{\infty} \sum_{-\infty}^{\infty} |F(hkl)| \left[\cos 2\pi(hx + ky + lz) - \alpha(hkl)\right], \tag{1-7}$$

where $\alpha(hkl)$ is the phase of $F(hkl)$.

The sum of the Fourier series represented by (1-7) over a sufficient number of terms would then provide a complete and accurate solution of the problem. Unfortunately, the matter is not so straightforward, because the magnitude of $F(hkl)$ can be evaluated easily from the measured intensity of reflections, but the phase constant cannot be directly measured. This is the well-known phase problem of x-ray analysis and, although it cannot be solved directly, methods of overcoming it are known. The most important of these methods is called the "heavy atom method," which depends on the existence in the structure of an atom of considerable scattering power. In metal chelate compounds the central metal atom acts as a heavy atom. The position of the heavy atom can often be found easily by trial or by Patterson syntheses. The Patterson function is written

$$P(uvw) = \frac{1}{V} \sum_{-\infty}^{\infty} \sum_{-\infty}^{\infty} \sum_{-\infty}^{\infty} |F(hkl)|^2 \cos\left[2\pi(hu + kv + lw)\right]. \tag{1-8}$$

This equation resembles a Fourier summation (1-7) in form but uses intensities directly and does not require a knowledge of phases. This distribution, which when plotted out has the same unit cell size as the true unit cell, bears some resemblance to a Fourier summation of electron density insofar as peaks are concerned. But these peaks are no longer identified as atoms. They are, in fact, the ends of interatomic vectors in the true structure with one end situated at the origin. And the peak heights are proportional to the product of the atomic numbers of the atoms concerned. Evidently, heavy atoms in a structure give very high Patterson peaks. It is this effect which helps in finding the positions of heavy atoms. Positions of heavy atoms can usually be fixed without ambiguity from the positions of Patterson peaks and the knowledge of the space group symmetry.

Approximate phase constants can then be calculated on the basis of the position of the heavy atom. These constants are used with the observed F values to evaluate the electron density. The result shows peaks of heavy atoms with faint outlines of lighter atoms. The phase constants are recalculated on the basis of the atomic coordinates thus obtained. The solution is seldom complete in one step and generally requires a great deal of refinement by iterative processes and a forbidding amount of calculation. The latter obstacle, however, has now been effectively removed by the development of electronic digital computers. This whole procedure is called Fourier refinement. At the final stage the refinement of the structure is carried out by the least-squares method. The method requires the finding of a set of atomic coordinates which minimizes $\sum w(|F_o(hkl)| - |F_c(hkl)|)^2$, w being an appropriate weighting function.†

3. Estimation of the Accuracy of Bond Lengths and Angles. Accurate bond lengths and angles are greatly needed to test and to provide additional data for theoretical treatments of metal chelates. The bond lengths and angles are fixed by the cell dimensions and atomic coordinates. Cell dimensions are much more accurate than atomic coordinates. Therefore errors in bond lengths and angles arise primarily from errors in atomic coordinates. There

† *Electron density projection and Patterson projection.* The electron density in a unit cell can be projected onto one face. The electron density projected along the y axis is

$$\rho(xz) = \int \rho(xyz)\, dy. \tag{1-9}$$

From (1-7) we have

$$\rho(xz) = \frac{1}{A_b}\left\{F(000) + \sum_{-\infty}^{\infty}\sum_{-\infty}^{\infty} |F(h0l)| \cos\left[2\pi(hx + lz) - \alpha(h0l)\right]\right\}, \tag{1-10}$$

where A_b is the area of the parallelogram perpendicular to the y axis. For this summation only $F(h0l)$ terms or, analogously for projections along the x and z axes, only $F(0kl)$ or $F(hk0)$ terms are needed. Computation of this series is much easier than for the three-dimensional series, (1-7), but the series is less powerful. A Patterson projection can be evaluated in the same way.

has been much discussion of the probable error of atomic coordinates in crystal structures. Cruickshank [63] has shown that the relation between the standard deviation $\sigma(x)$ of a coordinate of an atom and the difference between observed and calculated structure factors $(F_o - F_c)$ is, for a centrosymmetric structure,

$$\sigma(x) = \frac{2\pi[\sum h^2(F_o - F_c)^2]^{1/2}}{aV(\partial^2\rho/\partial x^2)}, \tag{1-11}$$

where h is the Miller index, a the length of the unit cell edge, V the volume of the unit cell, and ρ the electron density. For a non-centrosymmetric structure $\sigma(x)$ is doubled. For the other coordinates, y and z, corresponding expressions hold. The standard deviation of the length of a bond between two atoms (d_{12}) is given by

$$\sigma^2(d_{12}) = \sigma^2(x_1) + \sigma^2(x_2) + \sigma^2(y_1) + \sigma^2(y_2) + \sigma^2(z_1) + \sigma^2(z_2), \tag{1-12}$$

if the atoms are in general positions not related by symmetry operations.

4. Limitations of the X-ray Method. The scattering power of an atom depends on the number of its extranuclear electrons. It is difficult to locate accurately the positions of very light atoms in the presence of heavy atoms because their contribution to the diffracted beams is small. Thus in the presence of heavy central atoms it is not easy to ascertain the details of the chelate rings. Hydrogen atoms cannot be located accurately by x-rays except under favorable conditions. Their positions may, however, sometimes be inferred from considerations of symmetry (knowing the number to be located in the unit cell) or from the distances between pairs of atoms between which the hydrogen atoms lie. It is equally difficult to distinguish between atoms of rather similar scattering power, as, for example, N and O, or to determine the state of ionization of atoms in a crystal.

Finally, the shape and the size of molecules (or ions) in crystals are sometimes slightly different from those found in gases, liquids, or solution, because in crystals they are stabilized by specific intermolecular forces.

5. Determination of the Absolute Configuration. By the procedures outlined above the precise shape and size of a molecule can be determined, even if the procedure is rather complicated. It is not possible, however, to determine by these ordinary procedures the absolute configuration of optically active isomers, that is, to decide whether the given model corresponds to a dextro- or to a levorotatory isomer. In order to determine the absolute configuration we have to use the anomalous dispersion of x-rays by heavy atoms in the crystal. First we consider the reason why we cannot determine the absolute configuration under ordinary conditions. Consider a unit cell

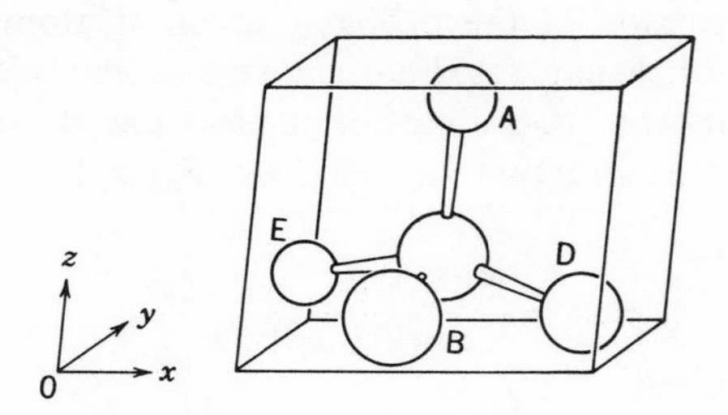

Fig. 1-1. Asymmetric molecule in a unit cell.

containing one asymmetric molecule (Fig. 1-1). The structure factors for the hkl and $\bar{h}\bar{k}\bar{l}$ reflections can be written

$$F_D(hkl) = \sum_j f_j \exp [2\pi i(hx_j + ky_j + lz_j)] = A(hkl) + iB(hkl), \quad (1\text{-}13)$$

$$F_D(\bar{h}\bar{k}\bar{l}) = \sum_j f_j \exp [-2\pi i(hx_j + ky_j + lz_j)] = A(hkl) - iB(hkl), \quad (1\text{-}14)$$

where

$$A(hkl) = \sum_j f_j \cos [2\pi(hx_j + ky_j + lz_j)], \quad (1\text{-}15)$$

$$B(hkl) = \sum_j f_j \sin [2\pi(hx_j + ky_j + lz_j)]. \quad (1\text{-}16)$$

Therefore the intensities of these two reflections are equal:

$$I_D(hkl) = A^2(hkl) + B^2(hkl) = I_D(\bar{h}\bar{k}\bar{l}). \quad (1\text{-}17)$$

This is called Friedel's rule. If the structure shown in Fig. 1-1 is inverted, the enantiomorphous structure illustrated in Fig. 1-2 is obtained. If the atomic

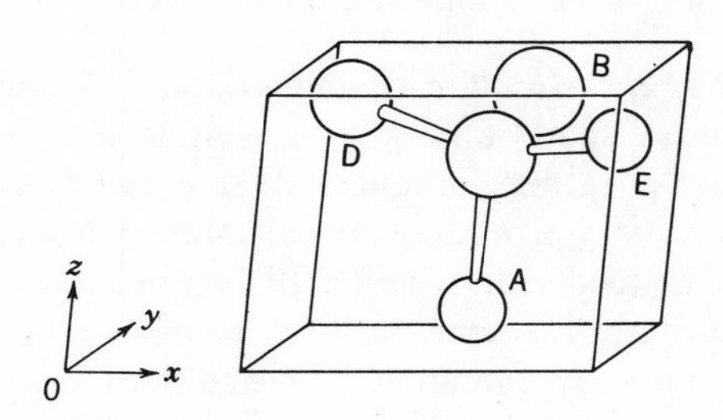

Fig. 1-2. Structure in Fig. 1-1 inverted.

coordinates of the structure shown in Fig. 1-1 are x_j, y_j, z_j, then those in the inverted one are $-x_j$, $-y_j$, $-z_j$. The structure factor $F_L(hkl)$ for the inverted structure can be written

$$F_L(hkl) = \sum f_j \exp\left[-2\pi i(hx_j + ky_j + lz_j)\right]$$

$$= A(hkl) - iB(hkl) = F_D(\bar{h}\bar{k}\bar{l}). \tag{1-18}$$

This is called Bijvoet's relation. Therefore we get

$$I_D(hkl) = I_L(hkl).$$

Thus it is not possible to tell whether the atoms are arranged as in Fig. 1-1 or as in Fig. 1-2. The key to the determination of absolute configuration lies in so arranging the experimental conditions that they give appreciable deviation from Friedel's law. This can be accomplished by ensuring that there is a phase advance between incident and scattered radiation for at least one of the atoms in the crystal. If the wavelength of incident x-rays is chosen to lie close to (but necessarily shorter than) an absorption edge wavelength of the atom in question, which we denote M, x-rays are scattered anomalously, and the scattering factor f_M of the atom M is given by a complex quantity:

$$f_M = f_{oM} + \Delta f'_M + i\,\Delta f''_M = \sqrt{(f_{oM} + \Delta f'_M)^2 + \Delta f''^2_M}\,\exp(i\beta),$$

$$\beta = \tan^{-1}\frac{\Delta f''_M}{f_{oM} + \Delta f'_M}, \tag{1-19}$$

where f_{oM} is the scattering factor when x-rays are scattered normally, and $\Delta f'_M$ and $\Delta f''_M$ are the small correction factors applied to the real and imaginary parts of the scattering factor, respectively. The way in which $\Delta f'_M$ and $\Delta f''_M$ depend on the incident wavelength (in relation to the absorption edge) is illustrated in Fig. 1-3. According to James [126], $\Delta f'_M$ and $\Delta f''_M$ are independent of the angle of diffraction for K absorption edges. As $\Delta f'_M$ is usually negative, $f_{oM} + \Delta f'_M$ is usually smaller than f_{oM}. $\Delta f''_M$ is always positive when the wavelength of the incident radiation is shorter than the absorption edge. Thus a phase advance occurs between the incident and the scattered radiation and it is due to the modification of the scattering power.

That the intensities of hkl and $\bar{h}\bar{k}\bar{l}$ reflections are unequal was first demonstrated on zinc blende crystals by Nishikawa and Matsukawa [166], and by Koster, Knol, and Prins [54, 55]. The first use of this procedure for the determination of the absolute configuration of an organic molecule was made in 1951 by Bijvoet and co-workers [37]. Saito and his co-workers [190–192] determined the absolute configurations of cobalt(III) chelates by the same method.

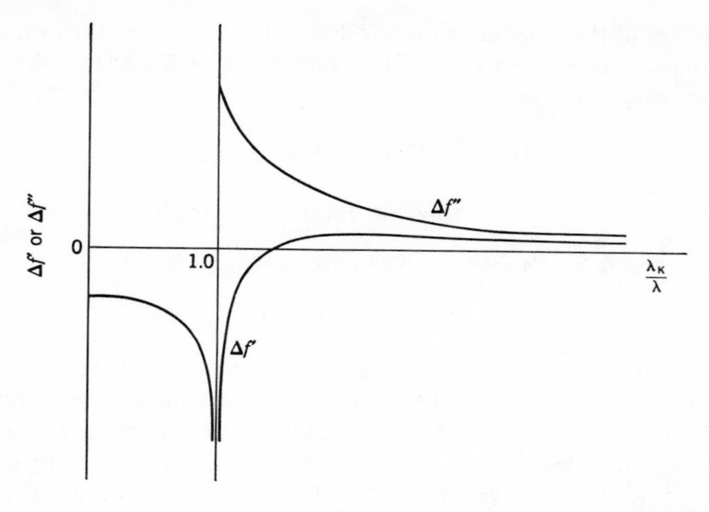

Fig. 1-3. Variation of $\Delta f'$ and $\Delta f''$ with wavelength (λ_K = K absorption edge wavelength).

On the supposition that the unit cell contains only one atom which scatters anomalously, one obtains by inserting (1-19) into (1-4)

$$\begin{aligned}F(hkl) &= \sum_j f_j \exp\,[2\pi i(hx_j + ky_j + lz_j)] \\ &= (f_{o\mathrm{M}} + \Delta f'_{\mathrm{M}} + i\,\Delta f''_{\mathrm{M}}) \exp\,[2\pi i(hx_{\mathrm{M}} + ky_{\mathrm{M}} + lz_{\mathrm{M}})] \\ &\quad + \sum_j{}' f_j \exp\,[2\pi i(hx_j + ky_j + lz_j)], \end{aligned} \tag{1-20}$$

where $\sum'$ means that the summations are taken over all the atoms except M. Equation 1-20 can be written

$$\begin{aligned}F(hkl) &= A(hkl) + iB(hkl), \\ A(hkl) &= (f_{o\mathrm{M}} + \Delta f'_{\mathrm{M}}) \cos\,[2\pi(hx_{\mathrm{M}} + ky_{\mathrm{M}} + lz_{\mathrm{M}})] \\ &\quad - \Delta f''_{\mathrm{M}} \sin\,[2\pi(hx_{\mathrm{M}} + ky_{\mathrm{M}} + lz_{\mathrm{M}})] \\ &\quad + \sum_j{}' f_j \cos\,[2\pi(hx_j + ky_j + lz_j)], \\ B(hkl) &= (f_{o\mathrm{M}} + \Delta f'_{\mathrm{M}}) \sin\,[2\pi(hx_{\mathrm{M}} + ky_{\mathrm{M}} + lz_{\mathrm{M}})] \\ &\quad + \Delta f''_{\mathrm{M}} \cos\,[2\pi(hx_{\mathrm{M}} + ky_{\mathrm{M}} + lz_{\mathrm{M}})] \\ &\quad + \sum_j{}' f_j \sin\,[2\pi(hx_j + ky_j + lz_j)]. \end{aligned}$$

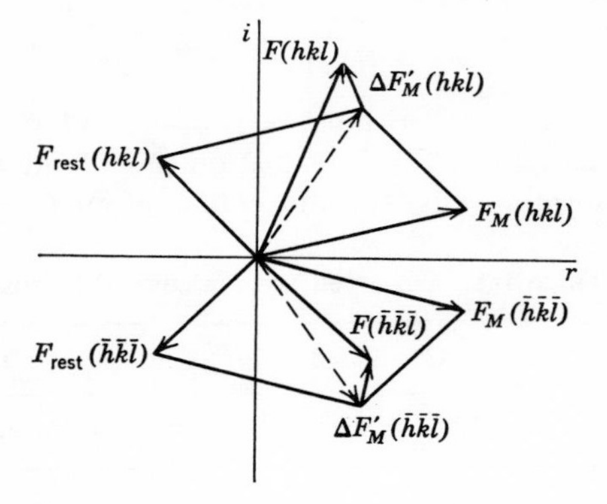

Fig. 1-4. Vector representation of $F(hkl)$ and $F(\bar{h}\bar{k}\bar{l})$.

Therefore we have

$$|F(hkl)| \neq |F(\bar{h}\bar{k}\bar{l})| \tag{1-21}$$

The vector representation of this relation is shown in Fig. 1-4, where the first term of both $A(hkl)$ and $B(hkl)$ is indicated as a vector, F_{M}, the second term as F'_{M}, and the third term as F_{rest}. If the structure is fully known except for the absolute configuration, the actual intensities $I(hkl)$ and $I(\bar{h}\bar{k}\bar{l})$ can be calculated on the basis of a final set of atomic coordinates determined in the usual way, taking the anomalous dispersion into account. If the observed inequality relations are the reverse of those calculated, the set of real atomic parameters must be the inverse of the initial set.

From (1-18) it follows immediately that, if

$$F_{\mathrm{D}}(hkl) \gtreqless F_{\mathrm{D}}(\bar{h}\bar{k}\bar{l}),$$

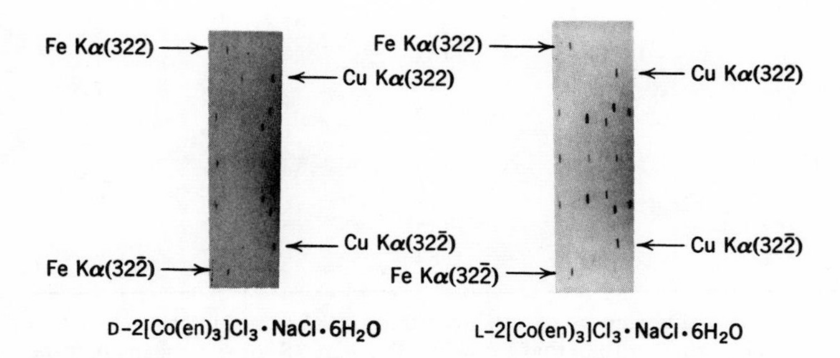

Fig. 1-5. Parts of oscillation photographs of D- and L-$Na[Co(en)_3]_2Cl_7 \cdot 6H_2O$, showing the effect of anomalous dispersion.

then

$$F_{\mathrm{L}}(hkl) \gtreqless F_{\mathrm{L}}(\bar{h}\bar{k}\bar{l});$$

that is, the inequality relations must be reversed in the D- and L-forms.

Since the wavelength of Cu Kα radiation ($\lambda = 1.542$ A) is a little shorter than the K absorption edge of the cobalt atom ($\lambda_{\mathrm{K}} = 1.608$ A), Cu Kα

Table 1-1 Observed and Calculated Intensities

Plane	Calculated		Observed[a]	
hkl	$I(hkl)$	$I(hk\bar{l})$	D-Form	L-Form
202	155	277	$<$	$>$
302	74	162	$<$	$>$
402	1110	1010	?$>$	?$<$
502	396	382	?	?
702	209	201	?	?
112	1	133	$<$	$>$
212	3700	3450	$>$	$<$
312	323	440	$<$	$>$
412	18	31	$<$	$>$
612	42	39	$>$	$<$
122	71	150	$<$	$>$
222	38	157	$<$	$>$
322	142	87	$>$	$<$
422	308	329	?	?
522	0.2	14	$<$	$>$
622	90	81	?$<$	$>$
132	946	893	?$>$	?$<$
232	465	521	?$<$	?$>$
336	121	147	?$<$	?$>$
532	84	75	?$>$	?$<$
632	189	170	?$>$	?
142	775	701	?$>$	?$<$
242	110	98	?$>$	?$<$
342	34	48	?$<$	?$>$
442	3	27	$<$	$>$
542	98	106	?$<$	?$>$
152	315	297	$>$	$<$
252	16	18	?$<$	?
352	143	132	?$>$	?$<$
272	72	64	?$>$	?

[a] The inequality sign $>$ or $<$ shows that the observed intensity for (hkl) is larger or smaller than that for $(hk\bar{l})$. The sign ?$>$ or ?$<$ means that the inequality is not so definite as that without the symbol ?. The symbol ? shows that the inequality between $I(hkl)$ and $I(hk\bar{l})$ could not be observed or was different according to different observers.

radiation will be scattered anomalously by the cobalt atom. The crystal structures of D- and L-2[Co(en)$_3$]Cl$_3$·NaCl·6H$_2$O are fully known except their chirality or "handedness" [162]. Therefore the absolute configuration of these compounds can be determined without ambiguity if the difference between $I(hkl)$ and $I(\bar{h}\bar{k}\bar{l})$ can be observed when Cu Kα radiation is used.

The effect of anomalous dispersion is shown in Fig. 1-5. Here reflections of Fe Kα and Cu Kα from planes (322) and (32$\bar{2}$) are recorded on one film, the latter being equivalent to ($\bar{3}\bar{2}\bar{2}$) because of symmetry. For the D-crystal the intensity of the reflection of Cu Kα from plane (322) is clearly stronger than that from plane (32$\bar{2}$), whereas the reflections from both planes are equal in intensity in the case of Fe Kα radiation. The inequality relation is reversed for the L-crystal. There are many such pairs of reflections for which slight but discernible differences in intensities are observed. Also, the intensity relations observed for one crystal are the reverse of those found for its enantiomorph. In Table 1-1 observed and calculated relations between $I(hkl)$ and $I(\bar{h}\bar{k}\bar{l})$ are compared for all the observed pairs. The complex ion [Co(en)$_3$]$^{3+}$, shown in Fig. 1-18 (p. 29), represents the absolute configuration of $(+)_D$-[Co(en)$_3$]$^{3+}$ found by this investigation.

B. Neutron Diffraction

It is not difficult now to produce a beam of neutrons of uniform velocity, and such neutrons can be diffracted by crystals. The first structure analysis by neutron diffraction was published in 1947, just after nuclear reactors became available as neutron sources. At present, apparatus for using beams of neutrons in crystal structure determinations has been set up in most of the countries where nuclear reactors have been installed. The number of papers on structural studies published has been increasing year by year. Actually, for a number of problems, neutron diffraction can provide information that is not obtainable by x-ray methods. Neutron diffraction has proved particularly valuable for studying the details of structures whose main features have already been established by x-ray diffraction, though it is not necessary to do an x-ray study first.

The neutron beam from the reactor impinges on a crystal which is large enough to receive the entire beam. At any particular angle of incidence, θ, neutrons will be reflected if their wavelength λ satisfies the Bragg equation

$$\lambda = 2d \sin \theta \tag{1-22}$$

where d is the interplanar spacing parallel to the crystal surface. By a suitable choice of the angle θ, a beam of neutrons of wavelength comparable to the interatomic distance (1–2 A) in crystals can be obtained. For this purpose single crystals of lead, aluminum, or copper are employed. The neutron beam from a collimating tube in the reactor falls on a monochromating crystal, which is surrounded by a massive shield attached to the reactor face. A specimen, usually a single crystal set on a goniometer head, is rotated in the neutron beam. The diffracted neutrons are received in a counter, which can be rotated about the center of the spectrometer. The counter is a proportional counter filled with B^{10}-enriched boron trifluoride.

The process of structure analysis by neutron diffraction is very much like that by x-ray diffraction except for the great differences in the size of the apparatus. For example, a BF_3 counter is half a meter long and is placed at about the same distance from the specimen, and the size of the crystal is usually $5 \times 5 \times 5$ mm, much larger than that used for x-ray study. There are also a number of other differences between x-ray and neutron diffraction which have a direct bearing on the type of information obtained by the two methods. In contrast to x-ray scattering, the scattering of neutrons by diamagnetic atoms is purely nuclear. The scattering amplitude of the nucleus varies quite irregularly from element to element, whereas the atomic scattering factors for x-rays increase progressively as the atomic number increases. With one or two exceptions the values of the neutron scattering amplitude† b for all the atoms are about the same within a factor of 3 or 4. Because of the small radius of the nucleus (10^{-13} cm) the neutron scattering factor does not drop off with θ. This is again different from the atomic scattering factor f. When x-rays are scattered by an atom there is a phase change of π, and this is true for the scattering of neutrons by most nuclei. In a few cases, however, there is no phase change at all, and then there is a negative scattering factor. Furthermore, the scattering amplitude differs for each isotope of the same element, and for hydrogen and deuterium the amplitudes are even different in sign. There is another difference from x-ray scattering in the case of atoms which have a resultant magnetic moment due to the presence of unpaired electrons. Such magnetic atoms cause additional scattering of a neutron beam, since neutrons possess a magnetic moment (1.91 nuclear magnetons). This extra scattering is electronic in origin. In a ferromagnetic crystal in the presence of a magnetic field the orientations of the atomic magnetic moment are all parallel. In an antiferromagnetic crystal at temperatures below the transition point all the magnetic moments are antiparallel. In such crystals the ordering of the magnetic moments results in coherent scattering in addition to the normal nuclear scattering. This provides information about the orientation of the magnetic moment in the crystal lattice, information that cannot be obtained by the x-ray method.

In structural studies on chelate compounds, neutron diffraction is used mainly to ascertain the location of light atoms in the presence of heavier ones, and to distinguish between atoms with similar atomic numbers. Not many studies have been made in this field. Some examples are discussed later. Like

† See Table 1-2 (p. 23).

other physical methods, neutron diffraction has its limitations. First, it is not possible to work with very small crystals because they require strong neutron sources, which are difficult to obtain. Secondly, rigorous monochromatization leads to a great loss of intensity, and this in turn leads to poor resolution and a loss of precision in intensity measurements.

C. Examples

Three examples are given here to explain and illustrate the theory. Two are concerned with x-ray diffraction, and the other with neutron diffraction. The compound in example 1 is a typical nickel chelate, and the space group to which it belongs is one that occurs most frequently. The crystal structure is rather simple and the analysis was performed in quite a straightforward fashion. The compound in example 2 is an optically active cobalt chelate. The analysis is more laborious than that in example 1, since the structure is non-centrosymmetric and anomalous dispersion is involved. The determination of the absolute configuration is also illustrated. An analysis by neutron diffraction is presented in example 3.

1. X-ray Diffraction of trans-Diaquobis(pyridine-2-carboxamido)nickel(II) Chloride, **$[Ni(H_2O)_2(piaH)_2]Cl_2$** [149]. The crystals of the compound are blue needles and have a magnetic moment, $\mu_M = 3.25$ B.M. Pyridine-2-carboxamide, abbreviated piaH, may be coordinated as a bidentate ligand in two ways: by means of two nitrogen atoms (I) or one nitrogen and one oxygen atom (II). If coordination occurs in the latter fashion, there are eight

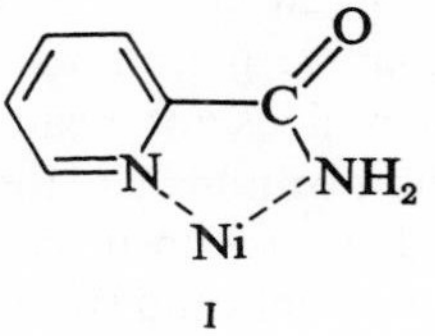

O C N NH₂ Ni I

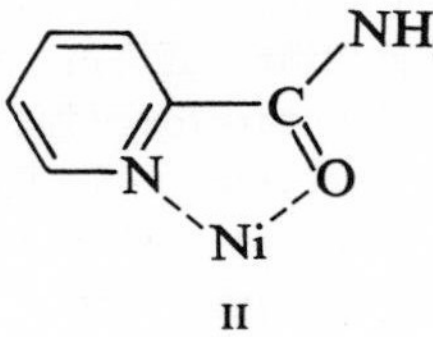

NH₂ C N O Ni II

possible octahedral isomers, as shown in Fig. 1-6. In the case of N,N coordination the same number of isomers is possible. To establish the structure of the chelate the crystals were subjected to x-ray crystal analysis. The unit cell dimensions were determined from Weissenberg photographs taken with Cu Kα radiation ($\lambda = 1.5418$ A). The crystals belong to the monoclinic system and the cell dimensions are

$$a = 13.567 \pm 0.005 \text{ A},$$
$$b = 10.067 \pm 0.005 \text{ A},$$
$$c = 6.355 \pm 0.005 \text{ A},$$
$$\beta = 113.7 \pm 0.1^\circ.$$

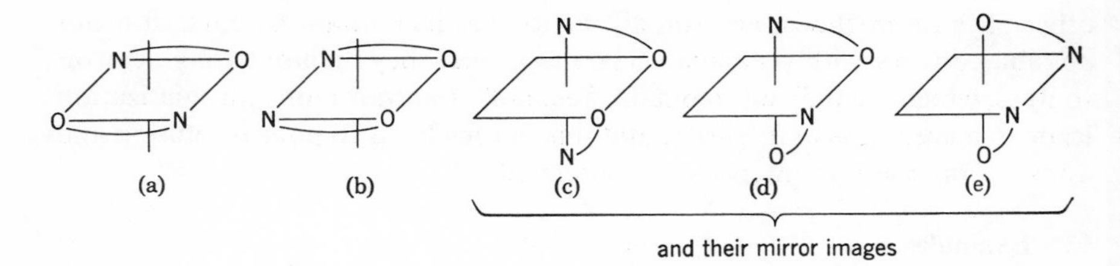

Fig. 1-6. Possible forms of the complex ion $[Ni(H_2O)_2(piaH)_2]^{2+}$. The two unlabeled positions are occupied by H_2O molecules.

The observed density of the crystal is 1.72 g cm^{-3}. From (1-3) the number of formula units in the unit cell is readily found to be two. Small crystal specimens were used, and the diffraction patterns were recorded on Weissenberg photographs. The intensities were estimated visually and were corrected for various factors; 1600 independent reflections were collected.

On examining the reflections it was found that $h0l$ reflections appeared only with l even, and $0k0$ only with k even. One may easily find that the space group is $P2_1/a$.† A "space group" is a group of symmetry elements. If an atom is placed in a quite general position in the unit cell, it is inevitably multiplied by performance of the symmetry operations, and thus other atoms, exactly equivalent to the first one, are found at other positions that are related in a precise way to those of the first. Each space-group has its own characteristic number of equivalent positions. For instance, the space group $P2_1/a$ has four equivalent positions. $P2_1/a$ is a group of twofold screw axis 2_1, glide plane a,‡ and center of symmetry.§ Suppose, however, that we place an atom on one of the centers of symmetry. This symmetry element does not create any atom, but merely makes one half of the atom. Thus positions on the center of symmetry are twofold positions; they are called special positions. $P2_1/a$ has four kinds of special positions. Coordinates of equivalent and special positions and their site symmetry are

$$
\begin{array}{llll}
C_1\text{–}1 & 4 & e & x, y, z;\ \bar{x}, \bar{y}, \bar{z};\ \tfrac{1}{2} - x, \tfrac{1}{2} + y, \bar{z};\ \tfrac{1}{2} + x, \tfrac{1}{2} - y, z \\
C_i\text{–}\bar{1} & 2 & d & \tfrac{1}{2}\tfrac{1}{2}0;\ \tfrac{1}{2}0\tfrac{1}{2} \\
C_i\text{–}\bar{1} & 2 & c & 0\tfrac{1}{2}0;\ 00\tfrac{1}{2} \\
C_i\text{–}\bar{1} & 2 & b & \tfrac{1}{2}00;\ \tfrac{1}{2}\tfrac{1}{2}\tfrac{1}{2} \\
C_i\text{–}\bar{1} & 2 & a & 000;\ 0\tfrac{1}{2}\tfrac{1}{2}
\end{array}
$$

† For details of the symbols see ref. 11, Vol. I, pp. 45–54.

‡ A combination of a mirror reflection followed by a translation of half of the unit cell length. a indicates that the translation is along the a axis.

§ For the arrangement of the symmetry elements the reader is referred to ref. 11, Vol. I, pp. 47–54, 98.

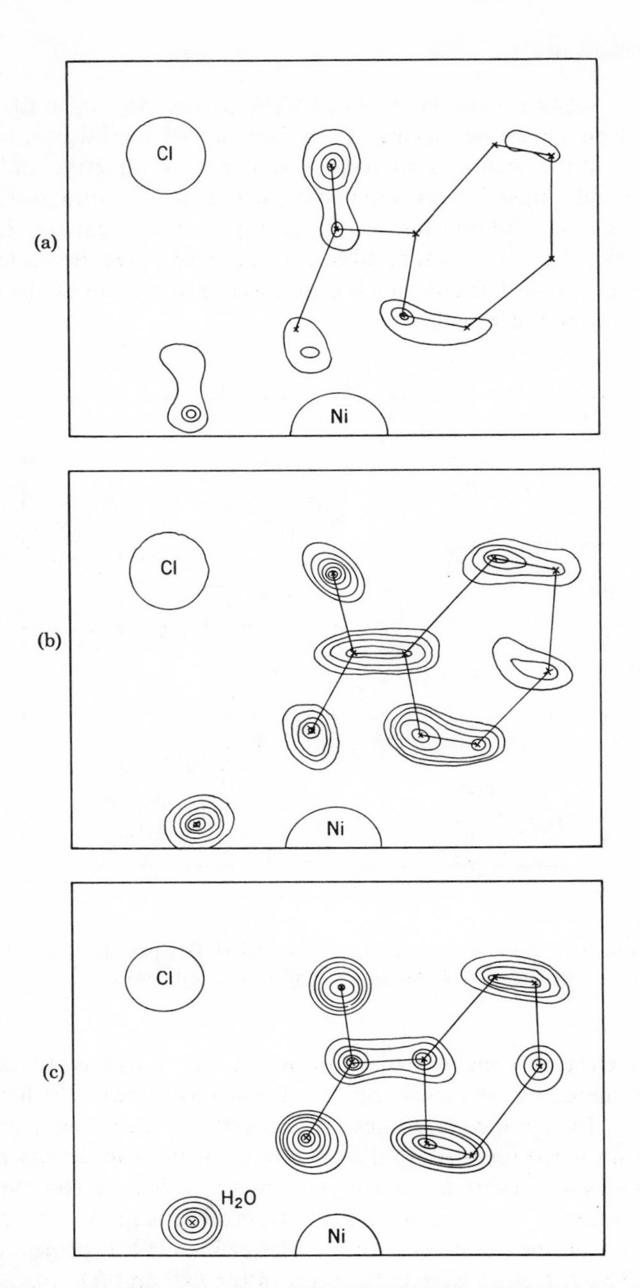

Fig. 1-7. Electron density projections of $[Ni(H_2O)_2(piaH)_2]Cl_2$, showing the process of Fourier refinement. (a) First Fourier map, with phases calculated from the positions of Ni and Cl atoms. (b) Second Fourier map, contribution of lighter atoms included. (c) Final Fourier projection.

Since there are two formula units of $[Ni(H_2O)_2(piaH)_2]Cl_2$ in the unit cell, two nickel atoms must be on one of the four sets of special positions. Thus the complex ion is required to have a center of symmetry, and the two chelating ligands must be coordinated with the nickel atom *trans* to each other (Fig. 1-6a). Without loss of generality one can choose $2a$ for the position of nickel atoms, because all four of these sets give the same arrangement of nickel atoms. The chlorine position was easily found by the Patterson projections, $P(uv)$ and $P(uw)$.

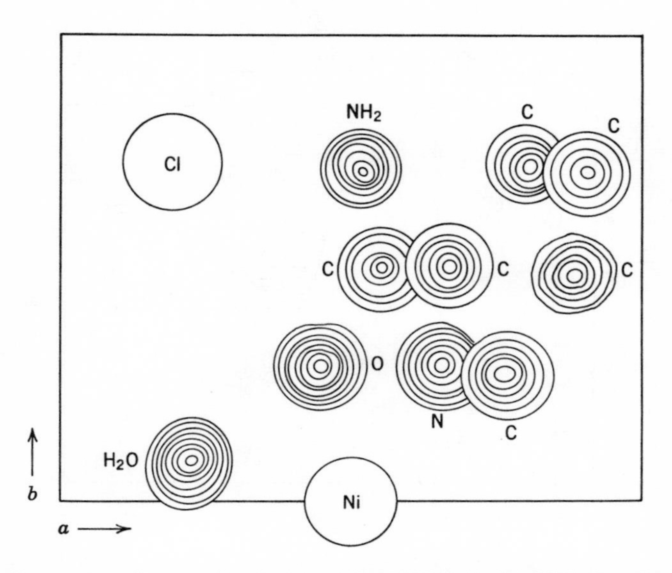

Fig. 1-8. Composite electron density map for $[Ni(H_2O)_2(piaH)_2]Cl_2$. Contours are drawn at intervals of 1 e A^{-3}. Those for Ni and Cl are not drawn.

First, the electron density projection, $\rho(xy)$, which was synthesized using phases determined by the nickel and chlorine atoms, showed the heavy nickel and chlorine atoms quite clearly resolved, together with rather faint outlines of other atoms in the unit cell. In the same way the electron density projection along the b axis was computed. From these projections approximate positions of the lighter atoms were found, and the structure was refined in projections. Figure 1-7 illustrates the process of Fourier refinement in projections along the c axis. The R factors were 0.22 and 0.18 for $hk0$ and $h0l$, respectively.

At this stage of the refinement it was not possible to distinguish between oxygen atoms and nitrogen atoms of the amide groups, and the same atomic scattering factor was given to these atoms. Next, the structure factors of all *hkl* reflections were calculated on the basis of the atomic coordinates deduced from the two-dimensional analysis. All structure factors were calculated for models corresponding to the two linkage isomers I and II. After one cycle of least-squares refinement, R values were reduced to 0.116 for N,O coordination and 0.125 for N,N coordination. At this point it was found by calculation of the bond lengths and angles that the carboxyl oxygen atom was bonded to the nickel atom. Thus the distance between the carbon atom and the atom coordinated with nickel was found to be 1.24 A, and that between the carbon

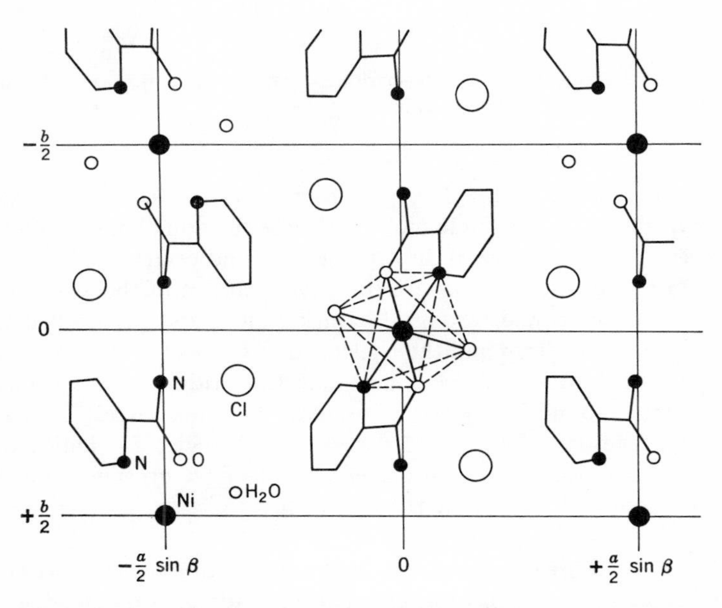

Fig. 1-9. Projection of the structure of $[Ni(H_2O)_2(piaH)_2]Cl_2$ along the c axis.

and the other atom attached to it was 1.31 A. These distances correspond, respectively, to the C=O and C—N bond distances found in the related molecules. One cycle of least-squares refinement was then carried out only for the model having oxygen atoms of piaH as the ligand atoms. The R was 0.100 for all the observed reflections. Figure 1-8 shows the final electron density distribution. This diagram is composed of the sections of the three-dimensional Fourier synthesis through each atomic center. Figure 1-9 shows the projection of the structure along the c axis. The structure is essentially ionic and consists of the positively charged complex ions and chloride ions. The nickel atom is surrounded by a distorted octahedron of six ligand atoms, four

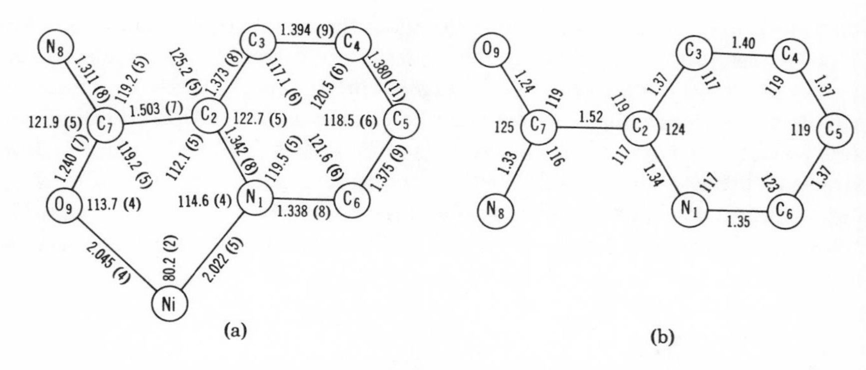

Fig. 1-10. (a) Parameters of the Ni-pyridine-2-carboxamide ring. (b) Parameters of pyridine-2-carboxamide in its crystalline state.

of which are the two carboxyl oxygen atoms and the two ring nitrogen atoms of the pyridine-2-carboxamide molecules. The chelating ligand molecules are planar and lie approximately in a plane with the central nickel atom. The other two coordinating atoms are the oxygen atoms of the water molecules which lie on the normal through the nickel atom and perpendicular to the plane of the nickel atom and the chelate rings (Ni—O = 2.14 A).

Figure 1-10 shows the shape and size of the pyridine-2-carboxamidenickel ring and those found in crystals of pyridine-2-carboxamide [214] itself. The standard deviations of the bond distances and angles are indicated in the parentheses. Coordination with nickel seems to have no significant effect on the dimensions of the pyridine ring.

2. The Crystal Structure and the Absolute Configuration of Tris(l-propylenediamine)cobalt(III) Bromide [124], **[Co(l-pn)$_3$]Br$_3$,** ***Determined by X-ray Diffraction.*** In the course of studies on the crystal structure of tris(bidentate) complexes of cobalt(III), the crystal structure and the absolute configuration of tris(*l*-propylenediamine)cobalt(III) bromide have been determined. The compound exists as orange-red hexagonal needles which are stable in air at room temperature. The shape and size of a unit cell were determined from Weissenberg photographs taken with Cu Kα radiation (λ = 1.542 A). The crystals belong to the hexagonal system, and the cell dimensions are

$$a = 11.08 \pm 0.03 \text{ A},$$
$$c = 8.59 \pm 0.02 \text{ A}.$$

The observed density of the crystal is 1.91 g cm^{-3}. By means of (1-3) a unit cell was found to contain two formula units of [Co(*l*-pn)$_3$]Br$_3$.

The specimen used for the collection of the intensity data was polished by emery paper into a cylindrical form with a diameter of 0.012 cm. The purpose was to reduce the absorption of x-rays by the specimen and to render the correction for the absorption more feasible. Intensities recorded on Weissenberg photographs were estimated visually with a standard intensity scale; 541 independent reflections were observed. Various corrections were then applied and values of $F(hkl)$ were calculated.

On examining the reflections it was found that $00l$ reflections appeared only with l even. No other systematically absent reflections were noted. The space group is accordingly either $P6_3$ or $P6_3/m$. Space group $P6_3/m$ has mirror planes perpendicular to the c axis. Since the crystal contains the optically active complex ion $[Co(l\text{-pn})_3]^{3+}$, the corresponding space group must not possess mirror planes; otherwise the optical antipode will be generated by the operation of reflection, and the crystal structure will be racemic. Consequently the space group was determined as $P6_3$. The space group $P6_3$ has six equivalent positions. It is a group of a threefold axis, 3, a twofold screw axis, 2_1, and a sixfold screw axis, 6_3. $P6_3$ has two kinds of special twofold positions. Coordinates of equivalent and special positions, with their site symmetry, are

C_1–1 6 c	$x, y, z;\ \bar{y}, x-y, z;\ y-x, \bar{x}, z;$
	$\bar{x}, \bar{y}, \frac{1}{2}+z;\ y, y-x, \frac{1}{2}+z;\ x-y, x, \frac{1}{2}+z,$
C_3–3 2 b	$\frac{1}{3}, \frac{2}{3}, z;\ \frac{2}{3}, \frac{1}{3}, \frac{1}{2}+z,$
C_3–3 2 a	$0, 0, z;\ 0, 0, \frac{1}{2}+z.$

Since a unit cell contains only two cobalt atoms, they must be located at one of the two kinds of twofold special positions (a and b). The former was decidedly ruled out on the basis of the packing consideration, because the complex ion is too thick for two ions to be accommodated on the 6_3 axis. Thus the cobalt atoms are located on the threefold axis, and the complex ion must possess the symmetry $3(C_3)$. The positions of the bromine atoms were easily deduced from the Patterson functions projected along the a and c axes. By making use of the results of the structure analyses of related complex compounds [161, 162] a structural model of the complex ion $[Co(l\text{-pn})_3]^{3+}$ could be easily set up. Since the positions of the cobalt and bromine atoms were already fixed, only the azimuthal orientation of the complex ion around the threefold axis was left to be determined. After several trials, the optimum condition was obtained in which the discrepancy factor became 0.26 for $hk0$ data. An a axis projection was also worked out at the same time. Using the results of the c axis projection and the structural model mentioned above, the R factor for the $0kl$ reflections was only 0.25 for the first trial z parameters.

Several three-dimensional Fourier refinements were then made. The final refinement of the atomic parameters was performed on an IBM 7090 computer by the least-squares method. After three cycles the R value dropped and

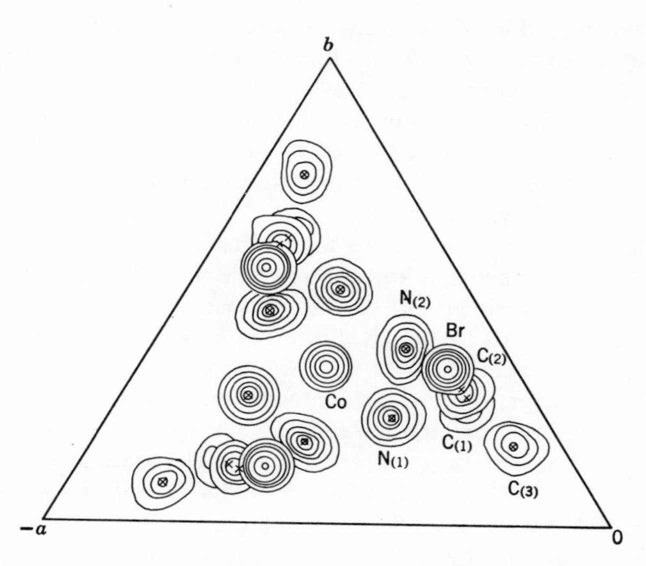

Fig. 1-11. Composite three-dimensional Fourier diagram of $[Co(l\text{-pn})_3]Br_3$. Contours around the cobalt and bromine atoms are drawn at intervals of 10 e A^{-3}, and those for the light atoms at intervals of 2 e A^{-3}.

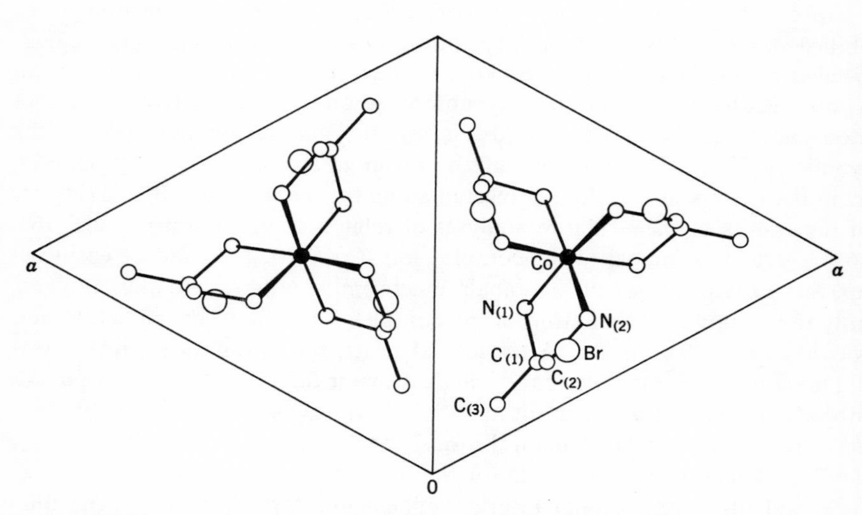

Fig. 1-12. Projections of the structure of $[Co(l\text{-pn})_3]Br_3$ along the c axis.

converged to 0.100 for all the observed reflections. In the calculation of the structure factors the effect of anomalous dispersion was considered. The correction factors were assumed to be

$$\text{for Co,} \quad \Delta f' = -2.2, \quad \Delta f'' = 3.9,$$
$$\text{for Br,} \quad \Delta f' = -0.9, \quad \Delta f'' = 1.5.$$

Since the crystal is non-centrosymmetric, the effect of anomalous dispersion was easily observed, and the absolute configuration of the complex ion was automatically fixed during the course of structure analysis.

A composite electron density diagram made from the sections of the three-dimensional Fourier synthesis is shown in Fig. 1-11. The structure projected along the c axis is illustrated in Fig. 1-12. The structure and absolute configuration are discussed in Sec. 1-2A and Sec. 1-2B.

3. Neutron Diffraction of Uranyl Nitrate Hexahydrate, $\mathbf{UO_2(NO_3)_2 \cdot 6H_2O}$ **[216].** Uranyl nitrate hexahydrate was first studied by Pauling and Dickinson [176] by x-ray diffraction. They determined the unit cell dimensions and believed the space group to be *Cmcm*. Sasvari [195] reported, however, that the compound has a positive piezoelectric effect, which suggested that the crystal has the alternative, non-centrosymmetric space group, $Cmc2_1$. In a two-dimensional x-ray study [79] the structure

$$[UO_2(NO_3)_2(H_2O)_2] \cdot 4H_2O$$

was proposed, but the positions of the crystal water molecules could not be located. Because the uranium atom does not have toward neutrons the dominant scattering power that it has toward x-rays (Table 1-2), a neutron diffraction study was undertaken in order to solve the structure completely.

Table 1-2 Neutron and X-ray Scattering Data

Element	Atomic Number	Neutron Scattering Amplitude[a] b	X-ray Scattering Powers f_x (10^{-12} cm)	
			$\theta = 0°$	$\sin\theta/\lambda = 0.5\ A^{-1}$
H	1	−0.378	0.28	0.02
N	7	0.940	1.97	0.53
O	8	0.577	2.25	0.62
U	92	0.85	25.9	15.3

[a] b is the coherent scattering amplitude in units of 10^{-12} cm.

Large single crystals of $UO_2(NO_3)_2 \cdot 6H_2O$ weighing about 0.5 g were prepared from a slightly acidified aqueous solution; the crystals are orthorhombic. The cell dimensions were determined with a neutron diffractometer using a neutron beam of wavelength 1.065 A. They are

$$a = 13.191 \pm 0.003 \text{ A},$$
$$b = 8.035 \pm 0.002 \text{ A},$$
$$c = 11.467 \pm 0.003 \text{ A}.$$

For collection of intensity data a crystal was shaped to a sphere (weight 0.540 g) with a piece of damp filter paper. This crystal was enclosed in a thin-walled vanadium can and used for intensity measurements. The relative intensities of a total of 1136 independent reflections were measured. The observed F^2 and F values were placed on a nearly absolute scale by comparison with the intensities measured for a magnesium oxide crystal. The structure was solved from the three-dimensional neutron Patterson and Fourier syntheses and refined by the least-squares method. The final R factor was 0.039.

The configuration about the uranium atom is shown in Fig. 1-13. The uranyl group is perpendicular to the plane of the paper and is surrounded equatorially by an irregular hexagon of six oxygen atoms, four from two bidentate nitrate groups and two water oxygens. The uranium-oxygen distances in the uranyl group are not quite equivalent: 1.770 ± 0.007 A and 1.749 ± 0.007 A. The uranyl group is nearly linear, the O-U-O angle being 179.1 ± 0.5°.

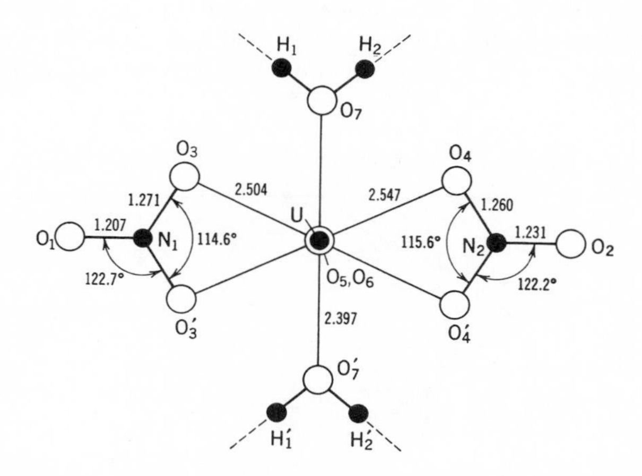

Fig. 1-13. Configuration of uranium atom and nitrate groups in $UO_2(NO_3)_2 \cdot 6H_2O$. Uranyl bonds: $U—O_5 = 1.770$; $U—O_6 = 1.749$; $\angle O_5UO_6 = 179.1°$ [216].

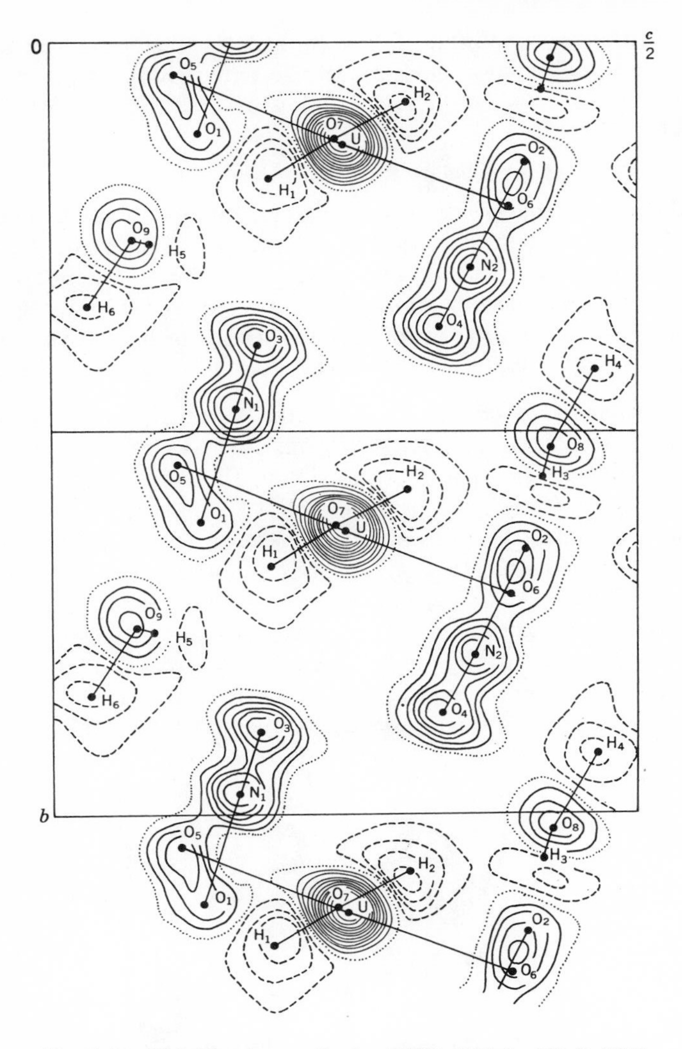

Fig. 1-14. (0*kl*) Fourier synthesis of $UO_2(NO_3)_2 \cdot 6H_2O$ [216].

In Fig. 1-14 is shown the result of an (0*kl*) Fourier synthesis which has some overlap but illustrates the structure well. A two-dimensional projection was not used in the solution of the crystal structure. It is, however, worthwhile to calculate one after completing the determination. Note the relative peak heights of the oxygen and nitrogen atoms as compared with those of uranium. Hydrogen peaks are indicated by broken lines to show their negative scattering amplitudes. A complete three-dimensional structure analysis by x-ray diffraction has also been reported [104].

1-2 APPLICATIONS

A. Conformation of the Chelate Ring

Crystal structure analysis can provide information about the conformation of chelate rings, as well as about other structural details that cannot be ascertained by other chemical and physical methods. One of the most interesting topics concerning the conformation of the chelate ring is the rotational isomerism possible in coordinated ligands such as ethylenediamine. This fact greatly complicates the stereochemistry of these complexes.

1. Ethylenediamine Complexes. That the five-membered ring in coordinated ethylenediamine is puckered was first suggested by the optical rotatory dispersion (ORD) study of *d*-$[Co(en)_3]Br_3$ [133]. This was verified by an x-ray study of $[Cu(en)_2]^{2+}$ by Scouloudi [197], and then of $[CoCl_2(en)_2]^+$ by Nakahara et al. [160]. The existence of puckering in chelate rings has also been established by structural analyses of a number of other metal chelates. These investigations definitely indicate that the ethylenediamine molecule takes the *gauche* configuration in metal chelates. Since there are two possible enantiomorphic *gauche* configurations for ethylenediamine, as shown in Fig. 1-15, two conformations of coordinated ethylenediamine rings are possible. They are mirror images of each other and are designated the δ and λ forms.†

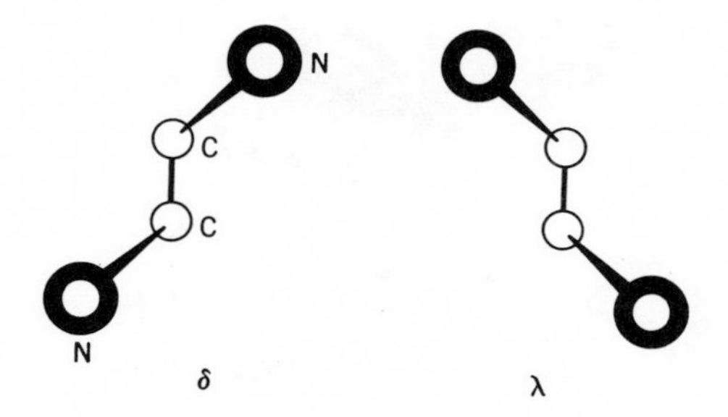

Fig. 1-15. Two possible forms of an ethylenediamine molecule.

† When one looks at the molecule along the axis through the two nitrogen atoms, the sequence of the bonds N—C—C—N defines a static screw. If it is right(left)-handed, the configuration is called $\delta(\lambda)$. This nomenclature was proposed at the Meeting of the Commission on the Nomenclature of Inorganic Compounds of IUPAC, held at Beckenried, Switzerland, August 29 to September 2, 1966.

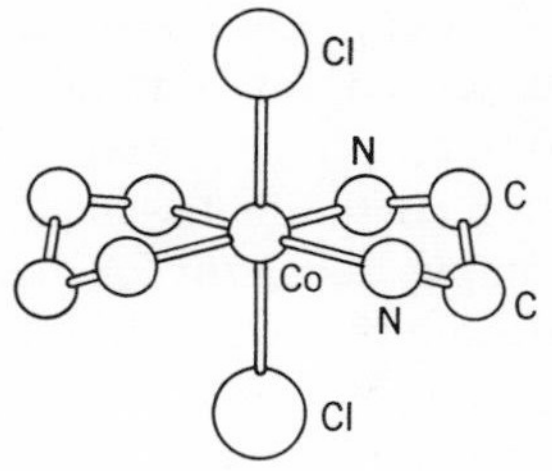

Fig. 1-16. A perspective drawing of the complex ion *trans*-$[CoCl_2(en)_2]^+$.

In the *trans*-dihalogeno-bis(ethylenediamine)cobalt(III) ion, the following three arrangements of two cobalt-ethylenediamine rings can occur:

$$(\delta\delta), \quad (\lambda\lambda), \quad (\delta\lambda).$$

$(\delta\delta)$ and $(\lambda\lambda)$ are mirror images and have the same relative potential energy. The relative potential energies of $(\delta\delta)$ and $(\delta\lambda)$ must, however, be different. This energy difference has been calculated by Corey and Bailar [53], who found that the $(\delta\delta)$ configuration is more stable by about 1 kcal/mole. The configuration of the *trans*-dihalogeno-bis(ethylenediamine)cobalt(III) ion in two isotype crystals† of the composition $[CoX_2(en)_2]X \cdot HX \cdot 2H_2O$ (X = Cl, [160] Br [168]) was found, however, to be centrosymmetric and to have the $(\delta\lambda)$ configuration (Fig. 1-16). In this case the $(\delta\lambda)$ form must be favored by specific intermolecular forces in the solid state. $[CrCl_2(en)_2]Cl \cdot HCl \cdot 2H_2O$ [169] is also an isotype of the Co analogs.

When three bidentate ligands are coordinated octahedrally with a central metal atom, two optically active isomers can occur. This is shown in Fig. 1-17

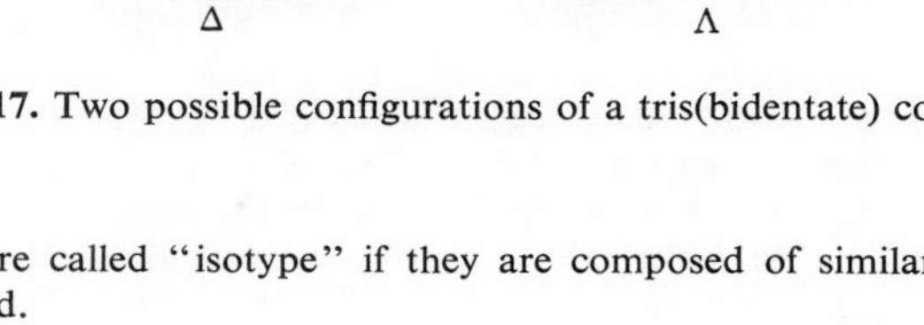

Fig. 1-17. Two possible configurations of a tris(bidentate) complex.

† Two crystals are called "isotype" if they are composed of similar structural units similarly arranged.

in which the isomers are viewed along a threefold axis of the octahedron.† This isomerism has been proved for a number of compounds, the first investigation by x-ray diffraction having been performed by Niekerk et al. [219] for the $[Cr(ox)_3]^{3-}$ ion. If the metal chelate rings are not planar, the situation becomes more complicated. Ethylenediamine again provides one of the best examples. If we denote the two enantiomorphous conformations of the metal-ethylenediamine ring by δ and λ and represent the two possible configurations of a tris(bidentate) complex ion as Δ and Λ, then there are eight possible configurations for an octahedral complex ion, $[M(en)_3]$. They are

$$1\begin{cases}\Delta(\delta\delta\delta)\\ \Lambda(\lambda\lambda\lambda)\end{cases} \quad 2\begin{cases}\Delta(\delta\delta\lambda)\\ \Lambda(\lambda\lambda\delta)\end{cases} \quad 3\begin{cases}\Delta(\delta\lambda\lambda)\\ \Lambda(\lambda\delta\delta)\end{cases} \quad 4\begin{cases}\Delta(\lambda\lambda\lambda)\\ \Lambda(\delta\delta\delta)\end{cases}$$

In ordinary x-ray analysis only the four forms, 1–4, can be identified. In each pair the two isomers are enantiomorphic and indistinguishable by x-rays, unless the absorption edge technique is used. In crystals of $[Co(en)_3]Cl_3 \cdot 3H_2O$, it was found that the complex ion has a threefold axis of symmetry [161]. This finding means that the most stable form must be either 1 or 4. For the Δ-type compound, for example, the difference between structures 1 and 4 can be stated as follows: The direction of the central C—C bond in one en molecule is approximately parallel to the threefold axis of the complex ion in structure 1, whereas it is largely inclined in structure 4. Corey and Bailar [53] label these two isomers the "lel" and "ob" forms, respectively. The x-ray analysis showed that the $[Co(en)_3]^{3+}$ ion has the "lel" form (Fig. 1-18). This finding was confirmed also for

D- and L-$Na[Co(en)_3]_2Cl_7 \cdot 6H_2O$ [162],

D-$[Co(en)_3]Br_3 \cdot H_2O$ [163], and $[Ni(en)_3](NO_3)_2$ [213].

In these studies the shape and size of the $[Co(en)_3]^{3+}$ ion was completely established, and it was concluded that the ion has D_3–32 symmetry within the limits of error of the experiments. This conclusion also agrees with the calculation which found that the "lel" form is more stable by about 2 kcal/mole than the "ob" form [53].

Shape and Size of the Metal-Ethylenediamine Ring. The shape and size of the metal-ethylenediamine ring have been determined by a rather complete x-ray analysis of a tris-ethylenediamine chelate, $Na[Co(en)_3]_2Cl_7 \cdot 6H_2O$ [162].

† The central metal atom and the two ligand atoms of a bidentate ligand define a plane. The two such planes of a given optical isomer of a tris(bidentate) complex define a two-fold axis and a static screw. If this screw is right(left)-handed, it is proposed that the enantiomer be called $\Delta(\Lambda)$. This is the reverse of the nomenclature proposed by Piper [178].

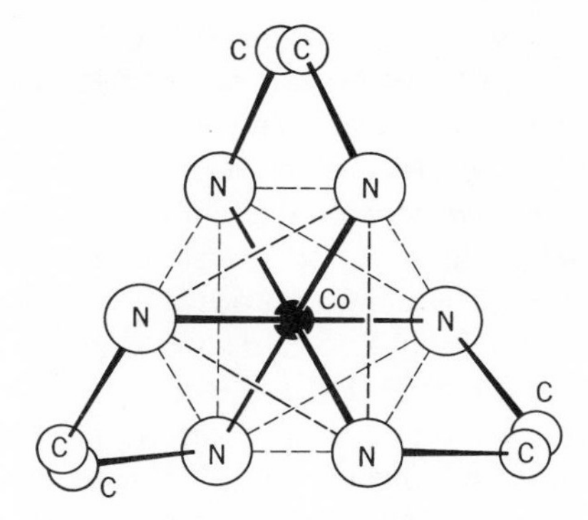

Fig. 1-18. $(+)_D$-$[Co(en)_3]^{3+}$.

In summary form the results are

$$\begin{aligned} C—C &= 1.54 \text{ A}, & \angle CoNC &= 109.5°, \\ C—N &= 1.47 \text{ A}, & \angle NCC &= 109.6°, \\ Co—N &= 2.00 \text{ A}, & \angle NCoN &= 87.4°. \end{aligned}$$

The carbon atoms in the ring are arranged symmetrically 0.26 A above and below the plane determined by the cobalt and nitrogen atoms.

The first complete three-dimensional analysis of a tris-ethylenediamine complex was carried out for $[Ni(en)_3](NO_3)_2$ [213]. The parameters of the Ni-en ring are:

$$\begin{aligned} Ni—N &= 2.120 \pm 0.013 \text{ A}, & \angle NiNC &= 109.7 \pm 1.2°, \\ C—N &= 1.500 \pm 0.025 \text{ A}, & \angle NCC &= 111.1 \pm 2.3°, \\ C—C &= 1.498 \pm 0.028 \text{ A}, & \angle NNiN &= 82.3 \pm 1.0°. \end{aligned}$$

This complex ion has the carbon atoms arranged symmetrically 0.29 A above and below the plane determined by the nickel and nitrogen atoms.

Unlike the symmetrical *gauche* form observed in the nickel and cobalt ethylenediamine complexes described above and in similar compounds [42], the spatial configuration of the ethylenediamine ring in the copper compound $[Cu(en)_2](SCN)_2$ has an unsymmetrical *gauche* form [43]. One carbon atom is 0.16 A and the other carbon atom -0.53 A from the N-Cu-N plane. The same kind of asymmetry has been observed in other copper(II) complexes: $[Cu(en)_2](NO_3)_2$ [135], $[Cu(en)_2][Hg(SCN)_4]$ [197], $[Cu(en)_2(H_2O)X]X$ (X = Cl, Br) [151].

Bis(propylenediamine)cobalt(III) Complexes. When a propylenediamine is introduced in place of ethylenediamine, the stereochemistry is further complicated, since the ligand molecule itself is optically active. From a preliminary x-ray examination of *trans*-$[CoCl_2(dl\text{-}pn)_2]Cl \cdot HCl \cdot 2H_2O$ [136], it is clear that the general features of this crystal structure resemble those of the en analog. Therefore it can be presumed that the complex ion is centrosymmetric and that its structure is closely related to that of *trans*-$[CoCl_2(en)_2]^+$. But, if one uses optically active propylenediamine, the complex is clearly no longer centrosymmetric. There are six possible stereoisomers of the *trans*-$[CoX_2(l\text{-}pn)_2]^+$ ion. First, there are two possible directions of each C—CH_3 bond with respect to the X-Co-X axis. The C—CH_3 bond can lie approximately parallel to the "plane" of the five-membered ring, that is, almost perpendicular to the Co—X bond, or it can stand approximately perpendicular to the plane of the chelate ring. These conformations are called "equatorial" and "axial," respectively. Secondly, there are two possible positions of the C—CH_3 groups with respect to the cobalt atom; these are shown in Fig. 1-19. If we denote these two cases "*cis*" (III) and "*trans*" (IV), referring to the positions of the methyl groups, the six possible isomers may be represented as follows:

"*cis*"(*ax*,*ax*), "*cis*"(*ax*,*eq*), "*cis*"(*eq*,*eq*),
"*trans*"(*ax*,*ax*), "*trans*"(*ax*,*eq*), "*trans*"(*eq*,*eq*).

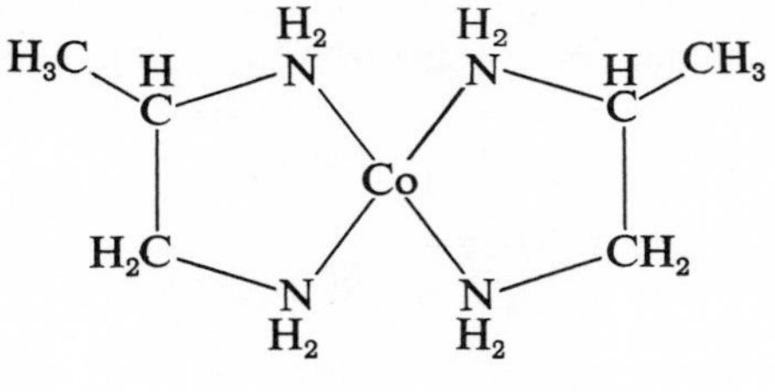

III *cis*

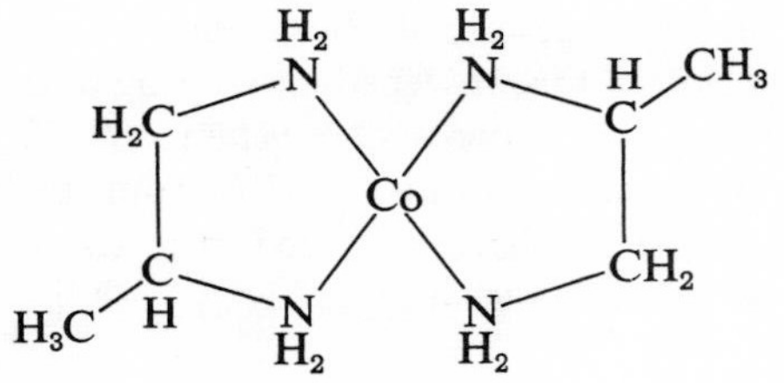

IV *trans*

Fig. 1-19. Possible "*cis*" and "*trans*" configurations in the $[CoCl_2(l\text{-}pn)_2]^+$ ion.

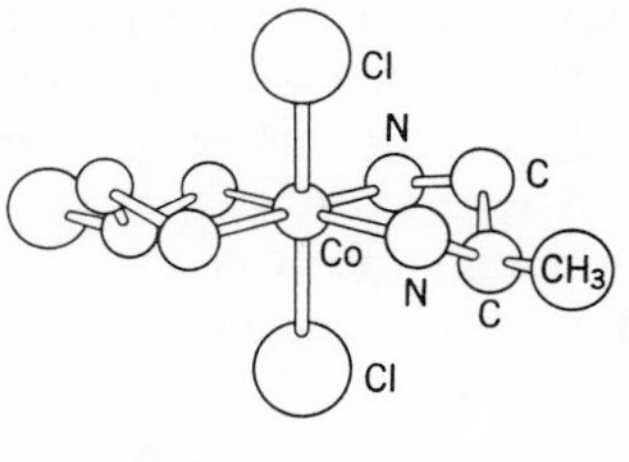

Fig. 1-20. A perspective drawing of the complex ion $[CoCl_2(l\text{-pn})_2]^+$.

These six forms are distinguishable from each other by ordinary x-ray techniques. In fact, the *trans*-$[CoCl_2(l\text{-pn})_2]^+$ ion has been shown [191] to have the "*trans*"(*eq,eq*) form (Fig. 1-20). The complex ion has an approximately twofold axis of symmetry through the central cobalt atom. Accordingly, the most stable form present in the solid state corresponds to the $(\lambda\lambda)$ form with respect to the conformation of the metal chelate rings (see Sec. 1-2B).

Tris(l-propylenediamine)cobalt(III) Ion. The crystal structure of tris(*l*-propylenediamine)cobalt(III) bromide has been determined by three-dimensional Fourier methods [124]. In Fig. 1-21 is shown a drawing of the complex ion, $[Co(l\text{-pn})_3]^{3+}$. It has a threefold axis through the central cobalt atom as a requirement of its space group. The three chelate rings are therefore completely identical. The central C—C bond is nearly parallel to the threefold axis, and the complex ion takes the "lel" form. The parameters of the five-membered chelate ring are very similar to those of the Co-en ring. Methyl

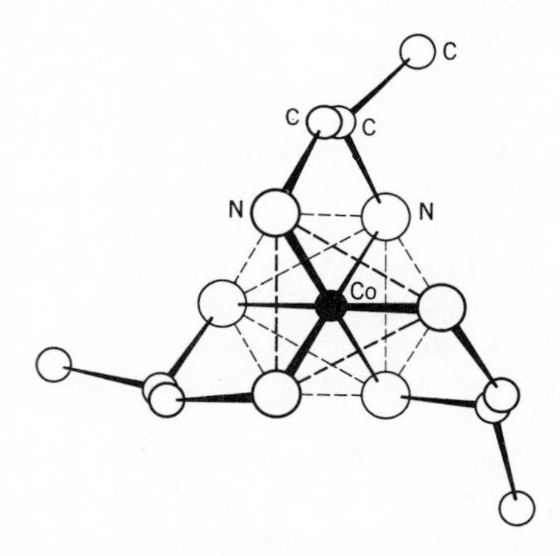

Fig. 1-21. The absolute configuration of the $(-)_D$-$[Co(l\text{-pn})_3]^{3+}$ ion.

substitution on the chelate rings does not seem to disturb the overall features of the rings. Each C—CH_3 bond lies in an equatorial position relative to the plane of the five-membered ring, in agreement with the prediction of Corey and Bailar [53]. The Co-N separation is found to be 2.000 A, which is compared with 1.94–2.03 A found in other Co(III) complexes. Six nitrogen atoms form a slightly distorted octahedron around the Co atom, the angle N-Co-N being 86.5°.

From a number of results hitherto obtained from structure analysis, it seems likely, as Jaeger [125] has pointed out, that the most stable form of the complex ion occurring in the crystals is the one with the highest attainable symmetry. Thus in $[Co(en)_3]^{3+}$ the complex ion has symmetry D_3–32. This is true even when such a symmetry is not required by the space group. For instance, the complex ion in tetragonal crystals of $[Co(en)_3]Br_3 \cdot H_2O$ is also found to have D_3 symmetry [163]. In the case of $[Co(l\text{-}pn)_3]^{3+}$ the highest attainable symmetry is C_3–3.

In addition to those discussed above, the crystal structures of the following complexes involving ethylenediamine have been determined by x-rays: *trans*-$[CoCl_2(en)_2]_2S_2O_6 \cdot H_2O$ [80], *trans*-$[CoCl_2(en)_2]Cl$ [35], $[Pt(en)Br_3]$ [188], *cis*-$[NiCl_2(en)_2]$ and *cis*-$[NiBr_2(en)_2]$ [28], *trans*-$[CoCl_2(en)_2]NO_3$ [170], $[NiXY(en)_2]$ (X, Y = Cl, Br, NO_3, NCS, dimer) [202], and *trans*-$[Ni(NO_3)_2(en)_2]$ [179].

2. Ethylenediaminetetraacetic Acid Complexes. That ethylenediaminetetraacetic acid (H_4EDTA) can function as a hexadentate ligand was confirmed by Weakliem and Hoard [222] by means of x-rays. This had been first suggested by Brintzinger et al. [41] and had been given a more substantial experimental basis by Busch and Bailar [47] and Dwyer et al. [66]. Structure analyses were made for the isomorphous series: $A[M(EDTA)] \cdot 2H_2O$ [A = NH_4^+, Rb^+, M = Co(III); A = NH_4^+, M = Cr(III), Al(III)]. The shape of the complex ion $[Co(EDTA)]^-$ is shown in Fig. 1-22. The cobalt atom of the complex anion is bonded octahedrally to the two nitrogen atoms and to one oxygen atom from each of the four carboxylate groups of EDTA. Three of the five chelate rings (including the ethylenediamine ring) form a girdle about the cobalt. At right angles to the girdle and at right angles to one another are two almost planar rings. A line which, within experimental error, passes through the cobalt atom and the midpoint of the C—C bond of the ethylenediamine ring serves effectively as an axis of twofold symmetry for the entire complex.

Ethylenediaminetetraacetic acid does not always function as a hexadentate ligand. In crystals of $Ni(H_2O)(H_2EDTA)$ and $Cu(H_2O)(H_2EDTA)$ [208], the ligand is pentadentate; a water molecule completes the octahedron in place of a strained chelate ring (Fig. 1-23).

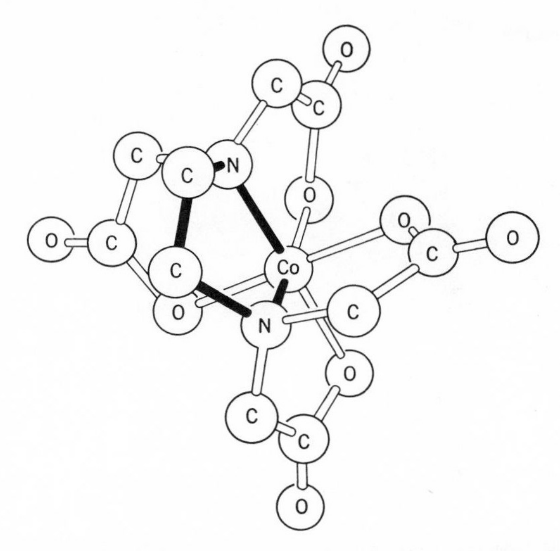

Fig. 1-22. A perspective drawing of the complex ion $[Co(EDTA)]^-$ [222].

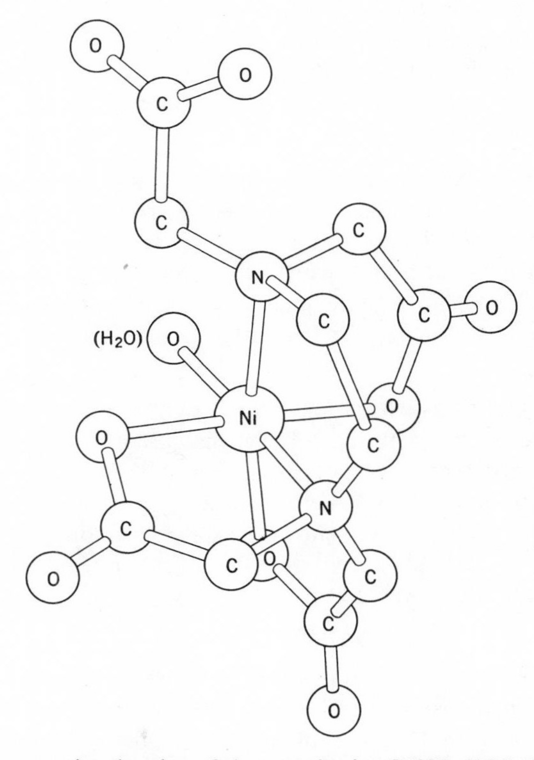

Fig. 1-23. A perspective drawing of the complex ion $[Ni(H_2O)(H_2EDTA)]$ [208].

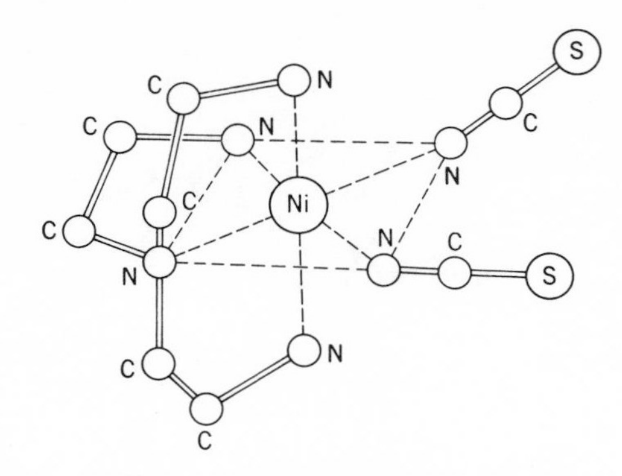

Fig. 1-24. A molecule of $[Ni(tren)(NCS)_2]$.

3. Complexes of 2,2′,2″-Triaminotriethylamine (tren). This amine is tetradentate, and in crystals of $[Ni(tren)(NCS)_2]$ [180, 100] all four nitrogen atoms of tren are coordinated with the same nickel atom. The thiocyanate ions are also coordinated through the nitrogen atoms in *cis* positions, the nickel atom thus being surrounded by six nitrogen atoms arranged in a slightly distorted octahedron (Fig. 1-24).

4. Complexes of Diethylenetriamine (den). Only the crystal structure of tri-oxo-(den)-Mo(VI) has been determined with high accuracy [56]. The molybdenum is six-coordinate and exhibits octahedral coordination, but the distortion from an ideal polyhedron is very considerable. Each of the five-membered rings is puckered in the manner expected both from theory and by analogy with results obtained for $[Co(en)_3]^{3+}$, etc.

5. Acetylacetone Complexes. Through their ability to enolize, β-diketones form stable chelate rings with a large number of metals. Many of the compounds so obtained are neutral, volatile, insoluble in water, but soluble in non-polar solvents. Acetylacetone has received the most attention in this regard, and many interesting structures of acetylacetone chelates have been revealed. X-ray studies have shown bis(acetylacetono)copper(II) to be square-planar [61, 200, 137], and electron diffraction measurements of the vapor confirm the square-planar structure [198]. The complex

$$Co(acac)_2 \cdot 2H_2O \; [45]$$

has a tetragonally distorted octahedral configuration with the water molecules in the *trans* positions. The two chelate rings are almost planar, but the Co atom is out of the plane. The analogous Ni(II) complex is isomorphous and isostructural with the Co(II) complex [157].

The structures of tris-acetylacetonate complexes M(acac)$_3$, M = Fe(III) [185], Cr(III) [159], Mn(III) [158], have been determined accurately by three-dimensional Fourier methods. Three acetylacetonate radicals, $(C_5H_7O_2)^-$, surround the central metal atom with the oxygen atoms in octahedral co-ordination. The formation of the chelate ring results in the resonance structures V and VI, with a planar configuration for the ring system. The

V VI

average intramolecular bond lengths of the chromium, iron, and manganese complexes are shown in Fig. 1-25. They are in acceptable agreement with each other and with the values expected from delocalization of the electrons and the resulting partial bond character in this type of chelate ring. In the Fe(III) and Cr(III) complexes regular octahedra of six ligand oxygens are found, while in the manganese complex the octahedron is distorted. Figure 1-26 shows a view of the distorted octahedron down the one threefold axis which remains after the removal of cubic symmetry. The amount of twist of the upper triangle relative to the lower is approximately 7.5°. This distortion may not be the result of a Jahn-Teller mechanism. The shorter Mn—O bond distances, as compared with the Fe—O bond distances where theoretically and experimentally an octahedral environment is found, suggest that it is energetically more favorable to retain a wider "bite" (O—O distance of 2.80 A) of the bidentate ring than to preserve cubic symmetry as the metal-oxygen distance is decreased. The structures of Al(III) [201] and Co(III) [173] acetylacetonates have also been reported.

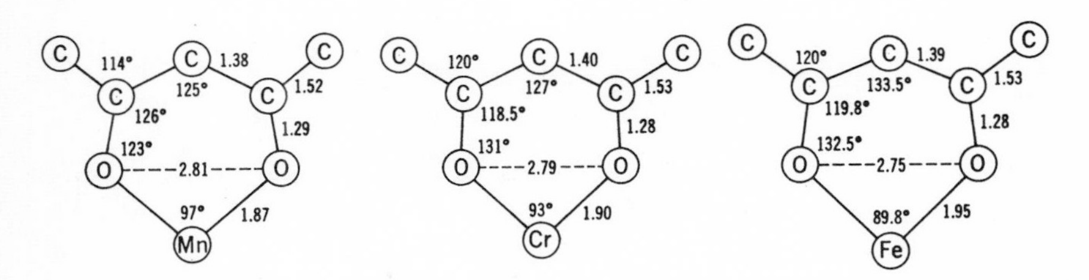

Fig. 1-25. Average intramolecular bond lengths of metal-acetylacetonate rings.

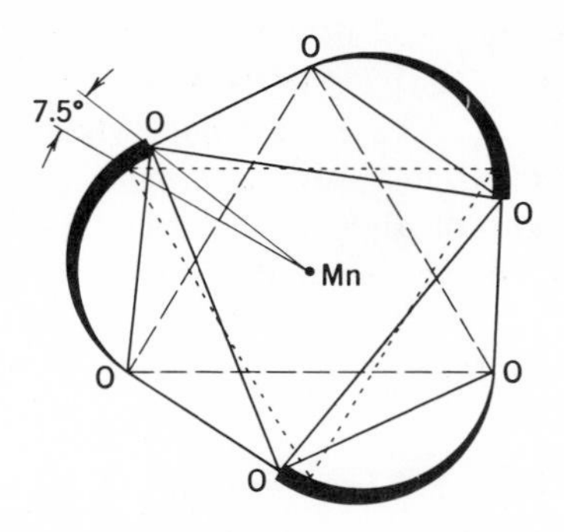

Fig. 1-26. The distorted octahedron around the Mn(III) atom in $[Mn(acac)_3]$.

Thorium acetylacetonate is dimorphic. The α-modification is isomorphous with the Ce(IV) and U(IV) [97, 150] analogs; the β-modification is isomorphous with those of Zr(IV) and Pu(IV) [50]. The occurrence of two crystallographic modifications has been explained by different packing arrangements of otherwise identically built molecules [98]. In $Ce(acac)_4$ a cerium atom is surrounded by eight oxygen atoms in a nearly regular Archimedean antiprism. The complex has D_2–222 symmetry, in which the square edges of an antiprism are bridged by acetylacetonate rings (Fig. 1-27). Average bond lengths and angles in one of the chelate rings of this compound are shown in Fig. 1-28. It is interesting that the O—O distance of 2.8 A is retained when the M—O distance is increased to 2.40 A. (cf. Fig. 1-28). Molecules of $Zr(acac)_4$ [204] are analogous to those of $Ce(acac)_4$; the zirconium atom is coordinated with eight oxygen atoms in an Archimedean antiprism.

The structure of bis(acetylacetono)oxovanadium consists of discrete molecules of $VO(acac)_2$ [64]. Five oxygens lie at the corners of a rectangular

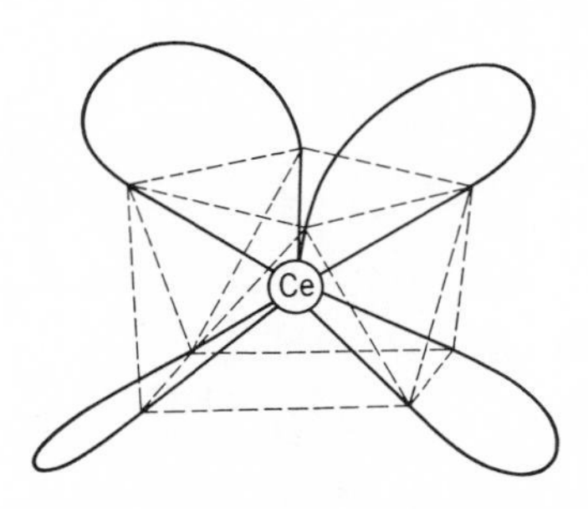

Fig. 1-27. The Archimedean antiprism around the Ce atom in $[Ce(acac)_4]$.

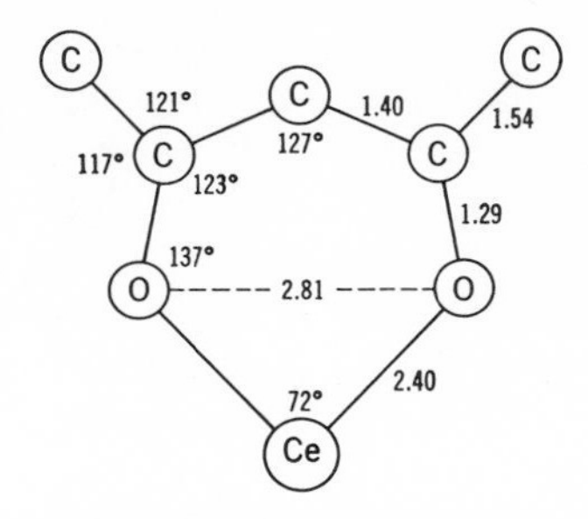

Fig. 1-28. Average intramolecular bond lengths of the Ce-acetylacetonate ring.

pyramid, with the vanadium atom near its center of gravity. The molecule has symmetry C_{2v}–*mm*. Each acac ring is accurately planar, but not quite coplanar with the other acac ring. The planes of these two parts of the molecule meet at an angle of 163°.

The structures of $Be(acac)_2$ [27] and $Zn(acac)_2 \cdot H_2O$ [156] have also been determined. A closely related compound, bis(dipivaloylmethanido)zinc(II),† has also been investigated [57].

B. Determination of Absolute Configuration

Saito and his collaborators found by x-ray analysis that the complex ion $[Co(en)_3]^{3+}$, shown in Fig. 1-18, represents the absolute configuration of $(+)$-$[Co(en)_3]^{3+}$ as explained in Sec. 1-1A. Its conformation can be represented by $\Delta(\delta\delta\delta)$, according to the notation defined in Sec. 1-2A,1. A classical calculation of the absolute configuration of $(+)$-$[Co(en)_3]Cl_3$ was made by Kuhn [141], but his result does not agree with that of the x-ray study and so must be rejected. A theoretical reappraisal of the problem has also been made by Moffitt [154], but this has been criticized by Sugano [211]. The "harmonic oscillator" model of Jones and Eyring [128] predicts the correct configuration for the $(+)$-$[Co(en)_3]^{3+}$ ion.

The absolute configuration of the complex ion $(-)$-$[Co(l\text{-pn})_3]^{3+}$ has been determined in the same way (Sec. 1-1C), and the complex ion shown in Fig. 1-21 was found to represent the absolute configuration of $(-)$-$[Co(l\text{-pn})_3]^{3+}$, which corresponds to the $\Lambda(\lambda\lambda\lambda)$ configuration. If the three methyl groups of $(-)$-$[Co(l\text{-pn})_3]^{3+}$ are replaced by hydrogen atoms, the resulting complex ion

† The dipivaloylmethanido ion is

$$
\begin{array}{l}
(H_3C)_3C\text{—}C\text{=}O \\
\quad\quad\quad\;\; / \\
\quad\quad HC \\
\quad\quad\quad\;\; \backslash\backslash \\
(H_3C)_3C\text{—}C\text{—}O^-
\end{array}
$$

is (−)-$[Co(en)_3]^{3+}$. This is precisely the mirror image of (+)-$[Co(en)_3]^{3+}$ illustrated in Fig. 1-18. In fact Smirnoff [205] found that the ORD curves of $(+)_D$-$[Co(en)_3]^{3+}$ and $(+)_D$-$[Co(d\text{-pn})_3]^{3+}$ correspond well with each other. Consequently the conventional method of correlating the absolute configuration with the ORD curves is proved, for the first time, to be useful and reliable. If the absolute configuration of $\Delta[Co(en)_3]^{3+}$ is taken as a standard and combined with rotatory dispersion studies, it is possible to settle with reasonable certainty the absolute configuration of the enantiomers of many coordination compounds. For example, the similarity of the rotatory dispersion curves for the cations (+)-$[Co(en)_3]^{3+}$, (+)-$[Co(en)_2(NH_3)_2]^{3+}$, (+)-$[Co(en)_2(l\text{-pn})]^{3+}$, (+)-$[Co(en)(l\text{-pn})_2]^{3+}$, and (+)-$[Co(l\text{-pn})_3]^{3+}$ shown in Fig. 1-29 indicates that all have the same configuration as the $\Delta[Co(en)_3]^{3+}$ ion [194].

The absolute configuration of the propylenediamine is of course in agreement with that of $[Co(l\text{-pn})_2Cl_2]^+$ [191]. The absolute configuration of

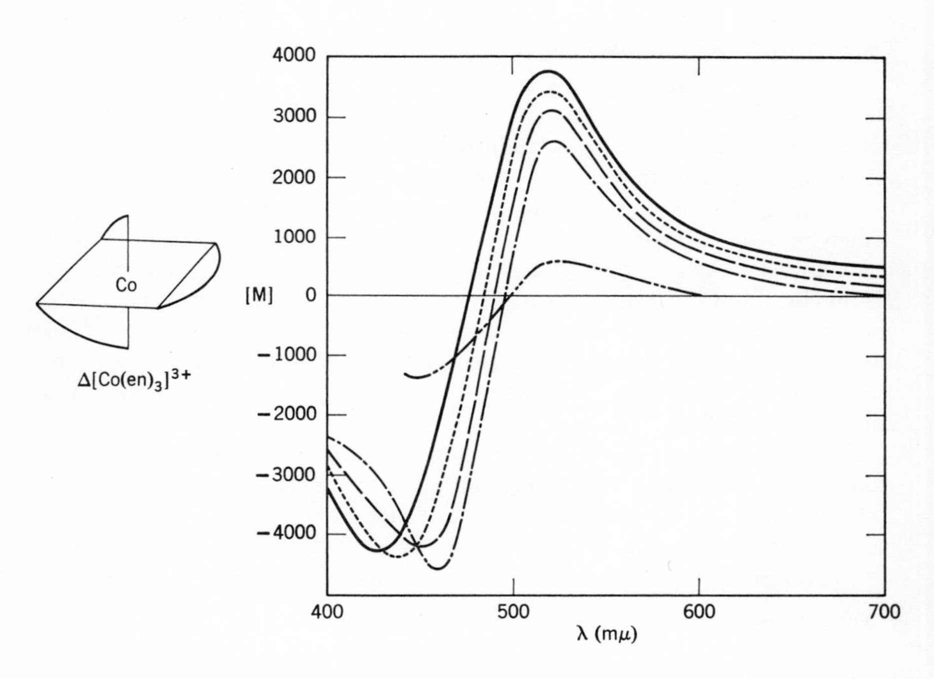

Fig. 1-29. The rotatory dispersion curves of the Δ isomers of the ions $[Co(en)_3]^{3+}$, −·−·−; $[Co(en)_2(NH_3)_2]^{3+}$, −··−··−; $[Co(en)_2(l\text{-pn})]^{3+}$, −−−−−; $[Co(en)(l\text{-pn})_2]^{3+}$, ······; $[Co(l\text{-pn})_3]^{3+}$, ——— [194].

$(-)_D$-propylenediamine (= *l*-pn) is shown in structure VII; heavy wedged lines represent bonds projecting out of the plane of the paper, and dotted

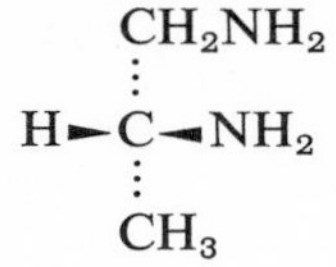

CH_2NH_2
⋮
H—C—NH_2
⋮
CH_3

VII

lines represent bonds projecting behind the plane of the paper. Thus the complex ion $[Co(en)_3]^{3+}$, which is the simplest and most important of these complexes, has been correlated with the *l*-pn and consequently with the whole system of the organic compounds through the absolute configuration of $[Co(l\text{-pn})_3]^{3+}$. When three bidentate ligands, en or pn, are coordinated with the cobalt atom, the central C—C bond prefers the "lel" to the "ob" configuration. In the five-membered cobalt-propylenediamine ring, the C—CH_3 bond has a tendency to take the equatorial configuration rather than the axial one. Furthermore the complex ion seems to favor the highest symmetry, so that the conformations of the three bidentate ligands become identical with one other. If one accepts these three specifications, there is only one way in which three molecules of *l*-propylenediamine will be coordinated with the cobalt atom. The resulting complex assumes the Λ form. These results can be applied to whole series of complexes of this type, such as $[Co(cptn)_3]^{3+}$ and $[Co(chxn)_3]^{3+}$ [193].

C. Determination of *cis-trans* Isomerism

When a structure analysis has been completed, the configuration around the central metal atom is unequivocally fixed; thus the problem of *cis-trans* isomerism is automatically solved. In square-planar metal chelates, *cis-trans* isomerism occurs if the ligand is unsymmetrical (structures VIII and IX).

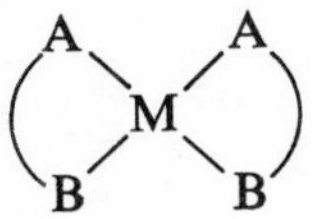

A A
M
B B

VIII *cis*

A B
M
B A

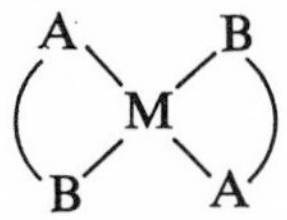

IX *trans*

In octahedral complexes many pairs of *cis-trans* isomers, such as the "violeo" and "praseo" salts (structures X, XI, and XII) are known.

The *cis* isomer is optically active and there are two enantiomorphic forms. The *trans* isomer is generally the more stable complex. The four examples below suffice to explain this point. The reader may find many other examples throughout this chapter.

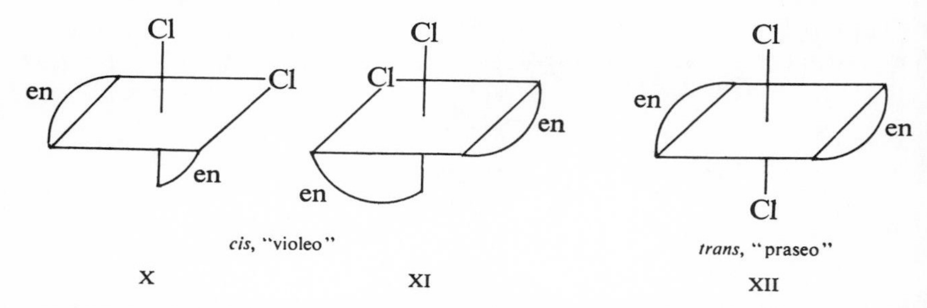

1. Bis(glycino)copper(II) Monohydrate, **$[Cu(NH_2CH_2COO)_2]H_2O$.** The structure analysis of this historically interesting molecule was not performed until quite recently [217]. The two planar glycine rings are attached to the copper atom in the *cis* configuration. Figure 1-30 shows the shape and size of the chelate rings. The

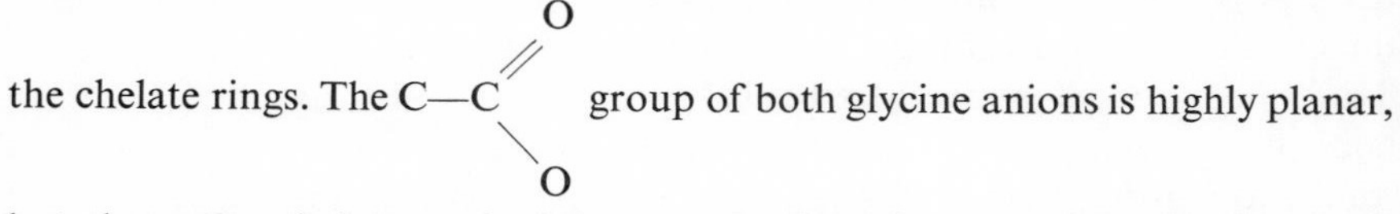

group of both glycine anions is highly planar, but the entire chelate part of the complex is only approximately planar. A weak interaction (Cu···O, 2.74 A) with an adjacent carboxyl oxygen completes the distorted octahedral coordination of the Cu(II) ion. The average bond lengths and angles of the chelate are compared in Fig. 1-30 with those obtained by Marsh [148] for α-glycine. The bond lengths and angles, except those of the carboxyl group, are in good agreement.

2. Bis(methylthiohydroxamato)nickel(II), **$Ni[CH_3CS(NOH)]_2$** [196]. This molecule was also found to have a *cis* configuration. The dimensions of the chelate ring are shown in Fig. 1-31.

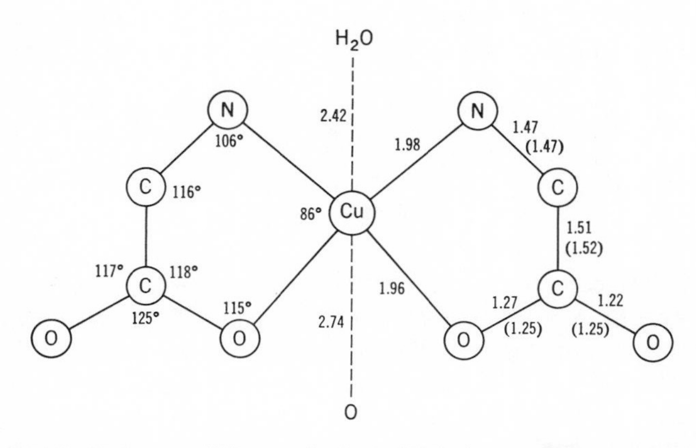

Fig. 1-30. Bis(glycino)copper(II) monohydrate. Data in parentheses are for α-glycine.

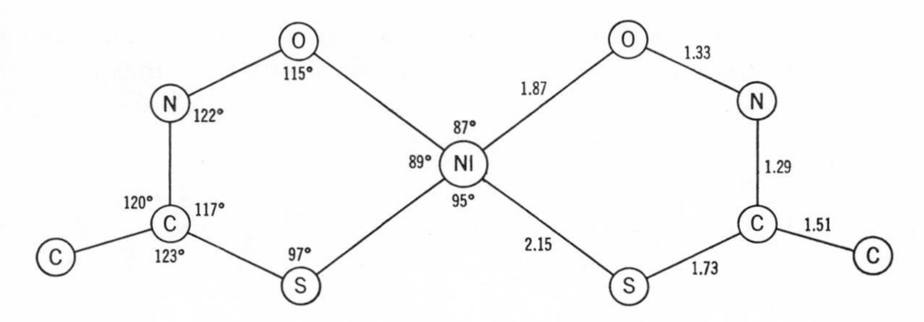

Fig. 1-31. A molecule of Ni[$CH_3CS(NOH)]_2$.

3. ***Bis(pyridine-2-carboxamido)nickel(II) Dihydrate.*** In contrast to the blue paramagnetic complex described in Sec. 1-1C,1, orange-red, diamagnetic crystals of composition Ni(pia)$_2 \cdot 2H_2O$ are obtained when an aqueous solution of nickel(II) chloride is added to an alkaline solution of the ligand. This complex is molecular, and x-ray structure analysis [165] has revealed that the complex molecule has a *trans*, planar structure, and that the two ligands are coordinated with the nickel atom through four nitrogen atoms. The two water molecules are not coordinated directly with the nickel atom, but are located between the complex molecules; they are hydrogen-bonded with each other and are linked to the amide oxygen atoms of the complex molecule (structure XIII).

XIII

4. ***Trans-Diaquobis(pyridine-2-carboxamido)nickel(II) Chloride.*** See Sec. 1-1C, 1.

D. Determination of Linkage Isomerism

Several examples have already been given of chelating molecules for which there are alternative ways of forming rings with metal atoms. In determining linkage isomers it is often necessary to distinguish between two atoms of rather similar scattering power, such as nitrogen and oxygen atoms. In the presence of a very heavy central metal atom it is not easy to distinguish these two atoms by means of peak heights of electron distribution. From the bond distances and angles, however, it is possible to identify such linkage isomers, as the following examples reveal. Sometimes knowledge of the position of hydrogen atoms is useful.

1. Biuret. Biuret, $HN(CONH_2)_2$, may combine with a metal atom by means of two nitrogen atoms, two oxygen atoms, or one nitrogen and one oxygen atom, since the molecule is flexible and rotation around the C—N bonds is possible.

Biuret, hereafter abbreviated bu, forms two kinds of coordination compounds with divalent metals. Of those which can be obtained in an alkaline medium, the potassium bis(biureto)cuprate(II) tetrahydrate,

$$K_2[Cu(bu^{2-})_2]\cdot 4H_2O,$$

has been studied by Freeman, Smith, and Taylor [87]. Two biuret anions are coordinated as bidentate ligands with the copper through the nitrogen atoms of the amido groups. The four nitrogen atoms are located at the corners of an almost perfect square.

In a neutral medium, coordination compounds of the type $M(II)X_2\cdot 2bu$, in which $M(II)X_2 = CuCl_2$, $CuSO_4$, $Cu(NO_3)_2$, $NiCl_2$, $NiSO_4$, $CdCl_2$ are formed. The crystal structure of $CuCl_2\cdot 2bu$ has also been determined [93]. In this compound the biuret is also a bidentate ligand, but coordination occurs through two oxygen atoms. These conclusions were drawn primarily from the dimensions of the chelate rings found in the two copper complexes XIV and XV. At every stage of the refinement all the structure factors for the

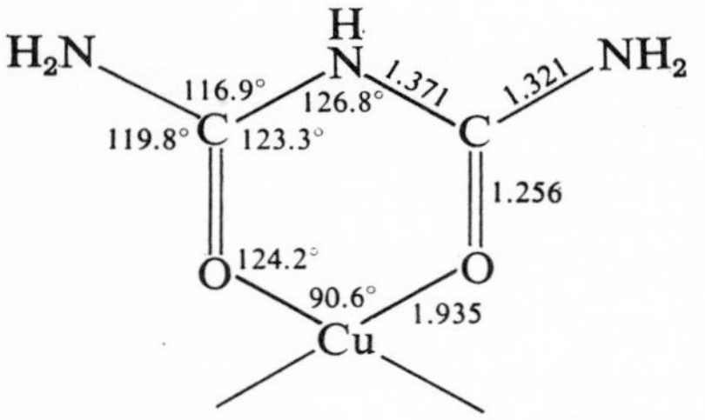

XIV

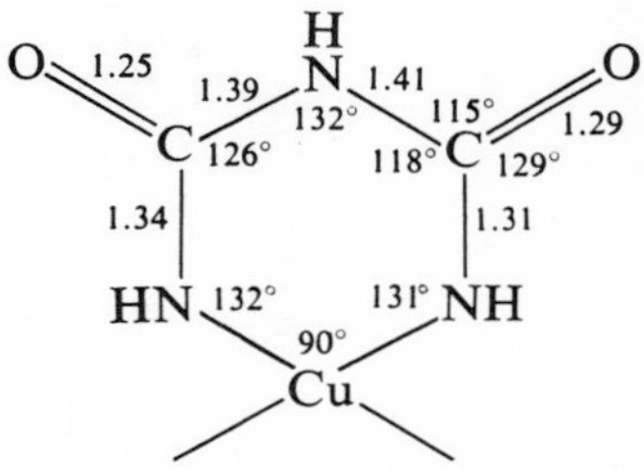

XV

probable models (coordination through four oxygen atoms, two oxygen and two nitrogen atoms, and four nitrogen atoms) were calculated, and the agreement with the observed structure factors was closely examined. An important point established by this analysis is that, within the limits of experimental error, chelation produces no significant change in the linear and angular dimensions of the ligand. This implies that in this case weak chelation does not seriously disturb the bond orders and resonance of the ligand.

In $CdCl_2 \cdot 2bu$, which is isomorphous to $HgCl_2 \cdot 2bu$ [48], the biuret is a unidentate ligand and the configuration (XVI) is *trans*. Coordination occurs

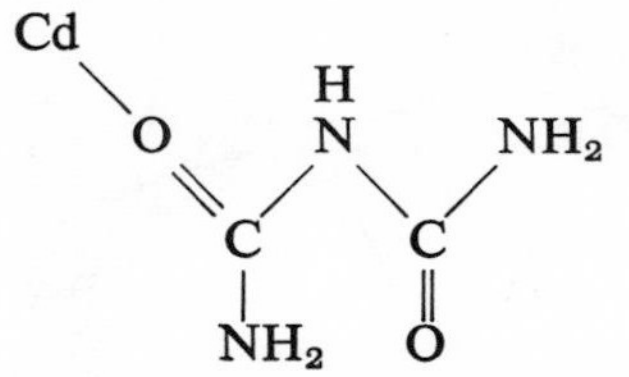

Cd
H
O N NH_2
C C
NH_2 O

XVI

through one of the ligand oxygen atoms. The structure consists of infinite chains of the type shown in XVII, and the coordination around the cadmium atom is octahedral.

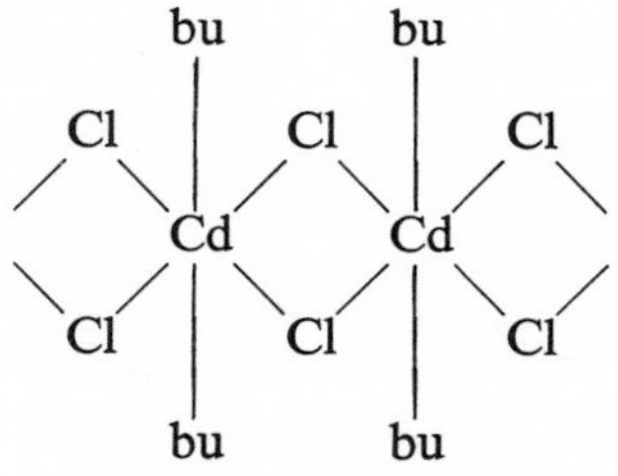

bu bu
Cl Cl Cl
Cd Cd
Cl Cl Cl
bu bu

XVII

***2. Trans-Diaquobis(pyridine-2-carboxamido)nickel(II) Chloride,* $[Ni(H_2O)_2(piaH)_2]Cl_2$.** See Sec. 1-1C, 1.

E. Unidentate and Bidentate Coordination

1. Unidentate Coordination of Acetylacetone. Acetylacetone usually acts as a bidentate ligand, as explained in Sec. 1-2A, 5. But unusual coordination of acetylacetone through the γ-carbon atom has been observed in platinum complexes. In crystals of potassium bis(acetylacetono)chloroplatinate(II), $K[Pt(acac)_2Cl]$, the coordination of the platinum is square-planar with one acetylacetonate ion coordinating through its two oxygens, the other bonding through the γ-carbon atom [74]. A view of the complex ion is shown in Fig. 1-32 with interatomic distances and some of the bond angles. The lengths of the platinum-oxygen bonds *cis* and *trans* to the chlorine are significantly different. The reason may be that the *trans* effect of the C-bonded acac is stronger than that of the Cl atom. In trimethyl(acetylacetono)2,2′-bipyridyl-platinum(IV) the coordination around the platinum is octahedral, and acetylacetone acts as a substituted methyl group [212] (structure XVIII). It is also interesting that the two C═O groups of acac are not parallel.

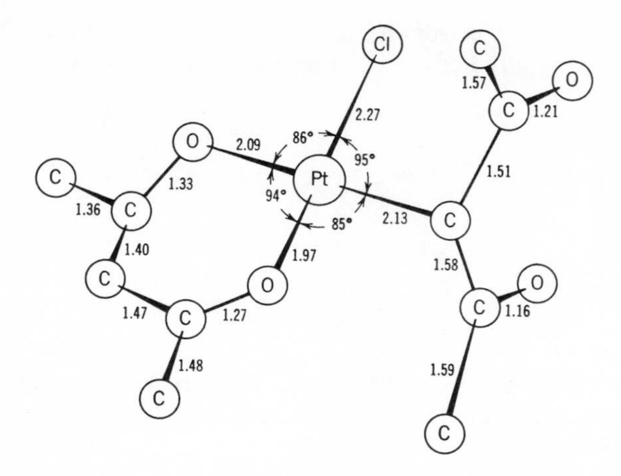

Fig. 1-32. A view of the bis(acetylacetono)chloroplatinate(II) ion [74].

XVIII

2. ***Unidentate Coordination of Biuret.*** See Sec. 1-2D, 1.

3. ***Other examples,*** $\mathbf{CO_3^{2-}}$, $\mathbf{NO_3^-}$, ***etc.*** Although few chemists doubted that the carbonate ion acts as a bidentate ligand in the complex ion $[Co(NH_3)_4CO_3]^+$ [31, 99], it was not until quite recently that this fact was verified by means of x-ray diffraction. The four-membered ring is, as would be expected, a highly strained one, the angle at the cobalt atom being 70.5° instead of 90° as expected for regular octahedral coordination. Similar coordination of CO_3^{2-} ion was found in $[Cu(NH_3)_2CO_3]$ [107, 108]. The carbonate ion acts as a unidentate ligand as in the $[Co(NH_3)_5CO_3]^+$ ion, but its structure has not yet been verified by x-ray diffraction techniques. Bidentate coordination of the nitrate group was found, for example, in gaseous copper(II) nitrate by electron diffraction [142]. Bidentate coordination is also reported for some ClO_4^- and SO_4^{2-} complexes (see Chapter 4).

F. Polymerization of Metal Chelates

1. Acetylacetone-mono(o-hydroxyanil)copper(II) [33]. $CuC_{11}H_{11}NO_2$, the empirical formula of this compound, apparently corresponds to three-coordination around the Cu atom. X-ray structural analysis has shown, however, that a dimer is formed with the two copper atoms bridged by the phenolic oxygen atoms of the ligand molecules, the copper-copper distance being 2.99 A (XIX). Association between pairs of molecules takes place

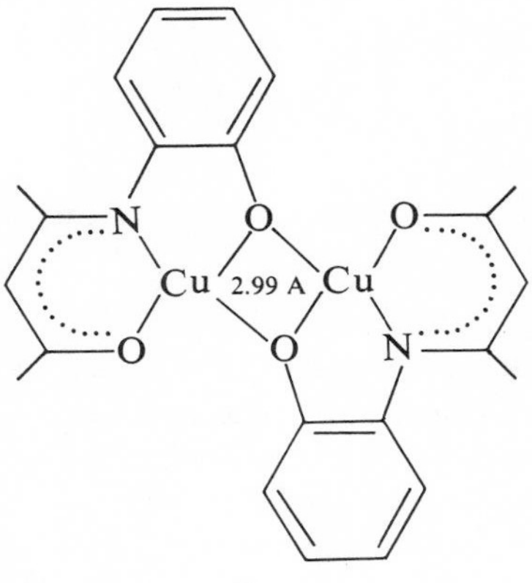

XIX

through a weak bond between the copper atom of one molecule and a hydroxyl oxygen atom of the other (Cu-O, 2.64 A). Although the copper atoms are covalently bonded to an identical group of atoms (one N and three O) at the corners of a somewhat distorted square, they have different environments; one copper is in a distorted square-pyramidal environment, the fifth ligand being an hydroxyl oxygen atom of an adjacent pair of molecules. The other is in a square-planar environment. Structure XIX does not provide an unambiguous explanation of the low magnetic moment observed in the compound (1.37 B.M. at room temperature).

2. Trimethyl(salicylaldehydato)platinum and trimethyl(8-quinolinato)platinum [147]. Both of these two compounds have empirical formulas corresponding to five-coordination. Three-dimensional x-ray crystal structure analyses were carried out at 110 ~ 120°K. Both were found to be dimeric molecules containing a symmetrical $\mathrm{Pt}\overset{\mathrm{O}}{\underset{\mathrm{O}}{\diamond}}\mathrm{Pt}$ bridge (Fig. 1-33). The octahedral coordination of platinum is maintained by the sharing of oxygen atoms.

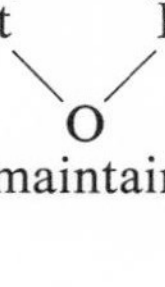

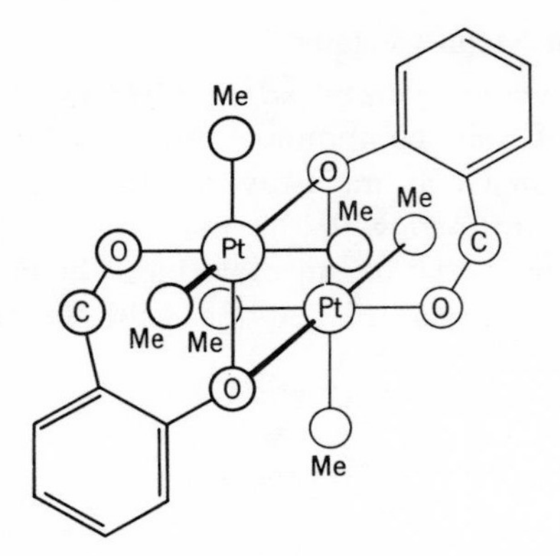

Fig. 1-33. The structure of a trimethyl(salicylaldehydato)platinum molecule [147].

3. Anhydrous Acetylacetonate Complexes [46, 58]. The crystalline anhydrous acetylacetonate complexes of Ni(II) and Co(II) are found to be trimeric and tetrameric, respectively. The structures of trimeric $Ni(acac)_2$ and tetrameric $Co(acac)_2$ are schematically shown in Fig. 1-34. They are similar

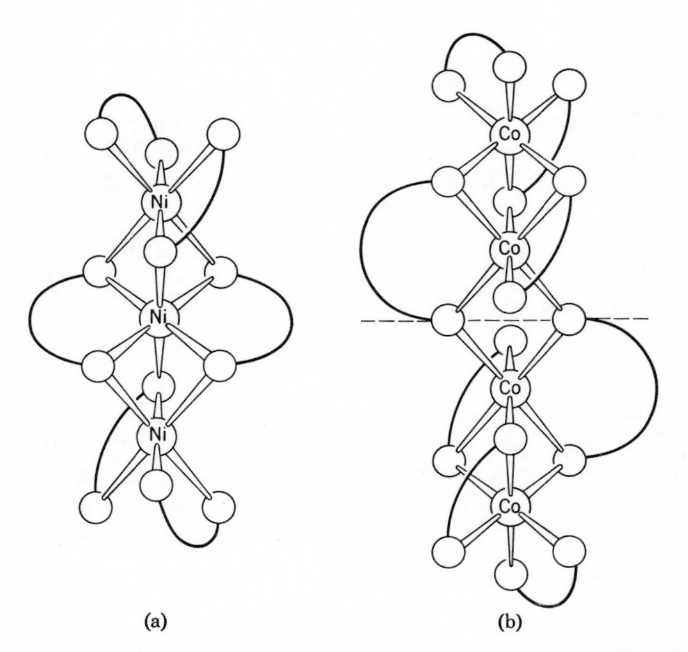

Fig. 1-34. A schematic drawing of the ring arrangements in (a) $[Ni(acac)_2]_3$ and (b) $[Co(acac)_2]_4$ [58].

in that both compounds achieve six-coordination through sharing of oxygen atoms between adjacent metal atoms. Trimerization occurs in $[Ni(acac)_2]_3$ through sharing of triangular faces of octahedra. A trimer is the only polymer that can result if all the units are bridged by sharing a common triangular face, and if all the metal atoms are to be six-coordinate. The structure shown in Fig. 1-34a is one of the nine possible trimeric structures; it has a center of symmetry.

The tetramer $[Co(acac)_2]_4$ (Fig. 1-34b) may be considered to be composed of two diastereoisomeric fragments which are joined along an edge common to two octahedra. Each of the fragments is formed by sharing a common octahedral face between two cobalt atoms. The shape and size of the metal chelate ring are the same as those found in monomeric acetylacetonate complexes.

The Ni(II) compound is also trimeric in benzene [96], but an electron diffraction study of the compound in the vapor state indicates that it is a monomer with an approximately square-planar configuration [199].

4. Dinitrilo Complexes of Copper(I). The crystal structures of the copper(I) derivatives of nitriles of aliphatic dibasic acids, which are salts of the type $Cu[NC—(CH_2)_n—CN]NO_3$, are of quite unusual interest. The Cu(I) atoms are joined into an infinite system by the $—NC—(CH_2)_n—CN—$ groups, which are linked at each end to Cu atoms. There are twice as many cyano groups as Cu atoms, and each Cu forms four tetrahedral bonds. Structure XX shows the succinonitrilo compound ($n = 2$) [130]. In this structure

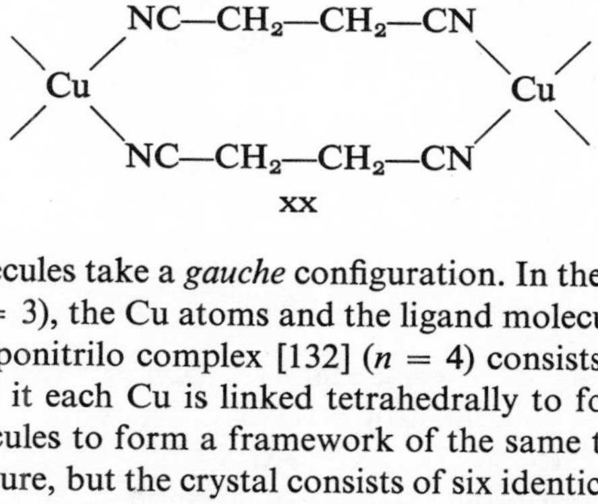

```
    NC—CH2—CH2—CN
 \ /             \ /
 Cu               Cu
 / \             / \
    NC—CH2—CH2—CN
```

XX

succinonitrile molecules take a *gauche* configuration. In the bis(glutaronitrilo) complex [131] ($n = 3$), the Cu atoms and the ligand molecules form puckered layers, but the adiponitrilo complex [132] ($n = 4$) consists of a three-dimensional network. In it each Cu is linked tetrahedrally to four others through linear nitrile molecules to form a framework of the same topological type as the diamond structure, but the crystal consists of six identical interpenetrating networks which are not interconnected by any primary chemical bonds. The nitrate ions are located in the interstices of the networks.

***5. The Cyclic Transition Metal Complex,* $[Ni(SC_2H_5)_2]_6$ [226].** Molecular weight measurements of a nickel mercaptide of empirical formula $Ni(SC_2H_5)_2$ indicates the existence of a hexameric species in solution. Three-dimensional

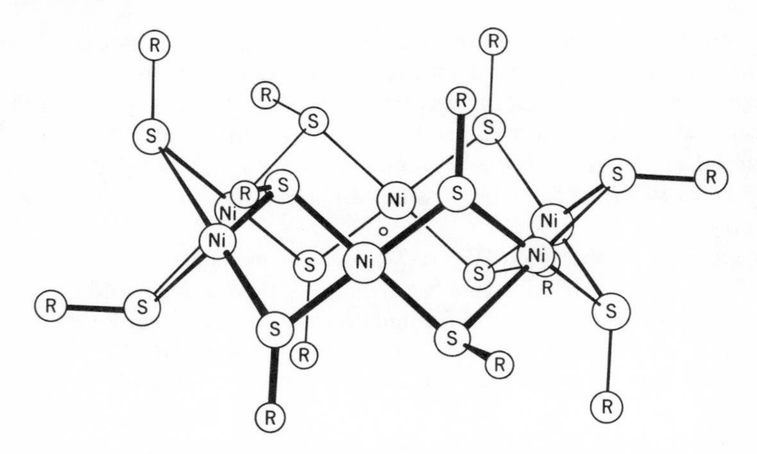

Fig. 1-35. Molecular structure of $[Ni(SC_2H_5)_2]_6$ [226].

analysis has revealed that the crystal consists of cyclic hexamers with the molecular configuration shown in Fig. 1-35. The six nickel atoms form a regular planar hexagon and are connected via mercaptan groups. Pairs of sulfur atoms are uniformly situated above and below the plane of the nickel ring, but equidistant from each pair of adjacent nickel atoms so as to give each similarly coordinated nickel atom an approximately square-planar environment of four sulfur atoms. The hexanuclear nickel complex can be considered to arise from the intersection of these planes of sulfur atoms at the non-bonded S-S edges with a dihedral angle of 120°. The presumably non-bonded Ni-Ni distances around the hexagon are 2.92 A, and the Ni-S distance has an average value of 2.20 A. The Ni-S-Ni and S-Ni-S angles in the

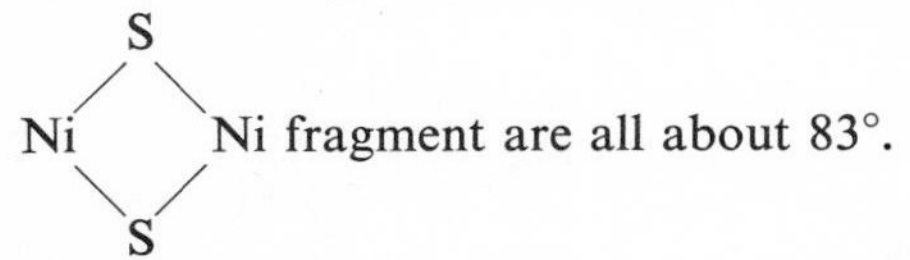

fragment are all about 83°.

6. Other Examples of Polymerization. *Dimers:* Bis(salicylaldehydato)-Cu(II), bis(8-hydroxyquinolinato)Cu(II) [36], bis(*cis*-1,2-bis(trifluoromethyl)-ethylene-1,2-dithiolato)Co(II), $[(CF_3)_2C_2S_2CoS_2C_2(CF_3)_2]$ [70], di-μ-di-phenylphosphinito-bis[bis(acetylacetono)chromium(III)],

$$[(acac)_2Cr(OP(C_6H_5)_2O)_2Cr(acac)_2] \text{ [223].}$$

Trinuclear Basic Acetate [171, 75]: $Cr_3(CH_3COO)_6OCl \cdot 5H_2O$.

Tetranuclear Basic Acetate: $Be_4O(CH_3COO)_6$ [40, 218], $Zn_4O(CH_3COO)_6$ [138].

Tetramer: Diethyldithiocarbamatocopper(I) [112].

Polymeric Chains: Bis(thiourea)cadmium(II) formate [164], uranium acetate [127], $[M^{II}(N_2H_4)_2]Cl_2$ (M = Cd, Zn, Mn, [72, 73]) and manganese(II) croconate [94].

Polymeric Sheets: $CuCN \cdot NH_3$ [62].

Three-Dimensional Polymer: $Cu(HCO_2)_2$ [29].

G. Less Usual Coordination Numbers

***1. A Five-Coordinate Zinc Complex, Terpyridyl Zinc Chloride,* Zn(terpy)Cl_2** [52]. The molecular structure of this complex is shown in Fig. 1-36. The zinc atom is linked in a distorted trigonal bipyramid to three nitrogen atoms (at 2.2 A) and two chlorine atoms (at 2.29 A). The ligand molecule is flat within experimental error and lies in a plane approximately at right angles to the plane of the Zn—Cl bonds; thus the complex has approximately the symmetry C_{2v}–mm. The intersection of these two planes of symmetry bisects the Cl-Zn-Cl angle and the central pyridine ring. Distortion is imposed in part by the terpyridyl ligand which causes the apical nitrogen-zinc-equatorial nitrogen angles to decrease to 79° and 72°. The corresponding cadmium and copper complexes are isomorphous with the zinc complex. Five-coordination of Zn is found in bis(acetylacetono)zinc(II) hydrate, $Zn(acac)_2 \cdot H_2O$ [146]. The structure is perhaps best described as a distorted tetragonal pyramid, though the geometry cannot be precisely defined.

2. A Five-Coordinate Zn Complex, Bis(salicylaldehyde)ethylenediamine Zinc Monohydrate [101]. This complex (XXI) has been shown by x-ray analysis to adopt an unusual molecular shape in which the zinc atom has a coordination number of five. The nitrogen and oxygen atoms of the chelating ligand are coplanar and form bonds of normal length with the zinc atom. The zinc atom lies, not in this plane, but 0.34 A above it on a line perpendicular to the center of the coordination square. The water molecule lies on

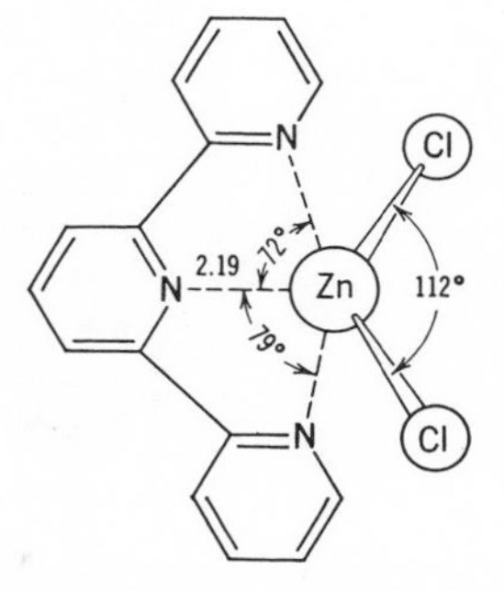

Fig. 1-36. A molecule of [Zn(terpy)Cl_2].

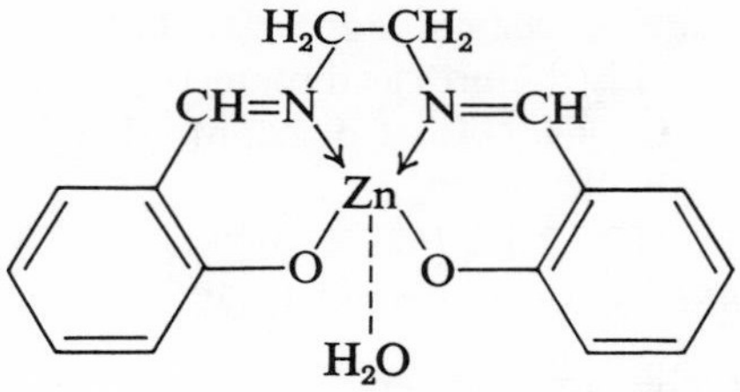

XXI

the same line, 2.13 A below the zinc atom. The water molecule is not only bonded to the zinc atom of one complex molecule, but it also forms strong hydrogen bonds (2.49 A) with the two oxygen atoms of another molecule. Closely linked chains of molecules run parallel to one of the crystal axes. Such an efficient packing arrangement may well stabilize the unusual configuration.

***3. A Five-Coordinate Copper(II) Complex, Bis(2,2′-bipyridyl)copper(II) iodide,* $Cu(bipy)_2I_2$ [32].** The crystal is built of positively charged iodo-bis-(2,2′-bipyridyl)copper(II) ions and negatively charged iodide ions (structure XXII). The copper atom is surrounded by four nitrogen atoms (at a distance of 2.02 A) and an iodine atom (at 2.71 A) at the corners of a distorted trigonal bipyramid. The iodine atom and two nitrogen atoms from different bidentate ligands are in the same plane as the copper atom. The other two nitrogen atoms lie on a line passing through the copper atom which makes an angle of about 9° with the normal to this plane. All five donor atoms are at normal covalent bond distances from the copper atom. In this respect the trigonal

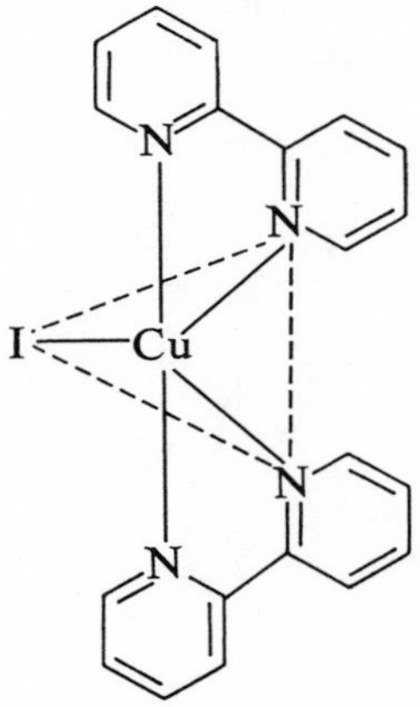

XXII

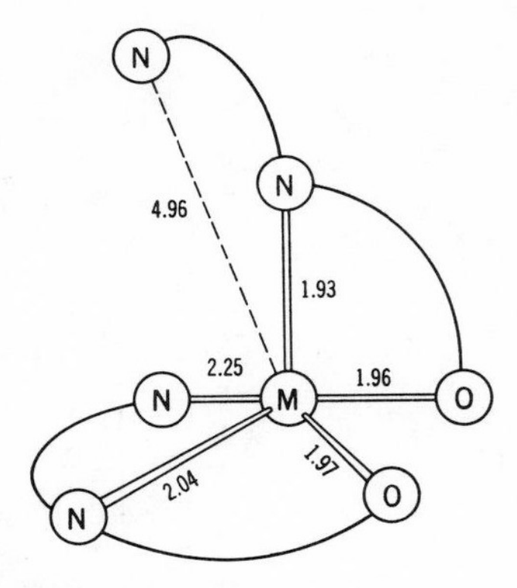

Fig. 1-37. A molecule of $M[ClC_6H_3(O)CH{=}N{-}C_2H_4{-}N(C_2H_5)_2]_2$. Salicylaldimine residues are schematically represented by solid lines.

bipyramidal configuration of the copper atom differs from the usual stereochemistry of bivalent copper, which involves four bonds of normal covalent length with a fifth, and sometimes a sixth donor atom, at a greater distance.

4. Five-Coordinate Nickel and Cobalt Complexes: Bis(N-β-diethylamino-ethyl-5-chlorosalicylaldimino)- nickel(II) and -cobalt(II) [189]. 5-chloro-salicylaldehyde forms with N,N-diethylethylenediamine a Schiff base of the formula $ClC_6H_3(OH)CH{=}N{-}C_2H_4{-}N(C_2H_5)_2$. In the Ni(II) and Co(II) complexes of this ligand the metal atom is five-coordinate. The structure consists of discrete molecules, and the coordination polyhedron is a distorted square pyramid. A sketch of the molecular structure is shown in Fig. 1-37, in which salicylaldimine residues are schematically represented by solid lines.

5. Ethylenediaminetetraacetate(EDTA) Complexes. A hexadentate EDTA complex in standard octahedral coordination is feasible only for small metal ions [222, 208, 115] and is quite impracticable for larger ions, such as rare earth cations. The constraints attending multiple as well as fused ring formation, together with the effective size of the central atom, take primary responsibility for the selection of the coordination number and the shape of the coordination polyhedron.

Stereochemical considerations based on infrared absorption spectra, acid-base titration curves, etc., have suggested that the complex ions of EDTA with Fe(III) and Mn(II) are seven-coordinate [114]. This suggestion was confirmed by x-ray structure analyses of $Rb[Fe(H_2O)(EDTA)]\cdot H_2O$, $Li[Fe(H_2O)(EDTA)]\cdot 2H_2O$ [113], and $Mn_3(H{-}EDTA)_2\cdot 10H_2O$ [181]. In Fig. 1-38 is shown an idealized model of the hexadentate seven-coordinate

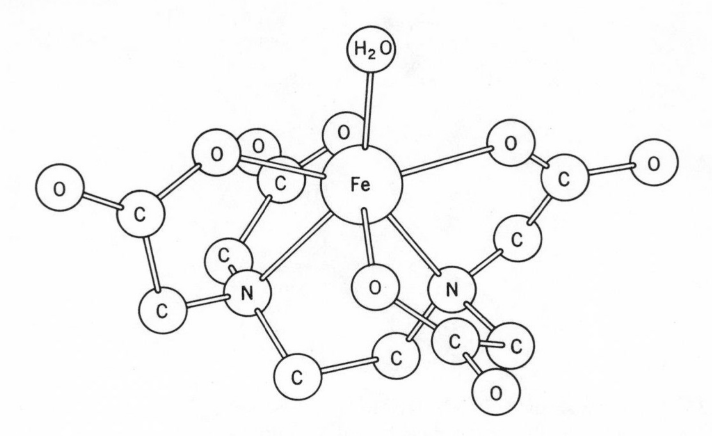

Fig. 1-38. $[Fe(H_2O)(EDTA)]^-$ [113].

$[Fe(H_2O)(EDTA)]^-$ ion. A water molecule, two nitrogen atoms, and two carbonyl oxygen atoms define the five-sided and roughly planar girdle in the crudely pentagonal bipyramidal coordination group.

In $La(H_2O)_4$(H–EDTA) [144] (Fig. 1-39), the four carboxylate oxygen atoms which are complexed to La^{3+} form a trapezoidal array that is planar; the lanthanum ion, however, lies slightly out of plane on the side away from the nitrogen atoms, thus relegating the entire chelating ligand to one hemisphere while leaving ample space for four water molecules in the other. The lanthanum and nitrogen atoms define a plane which serves rather accurately

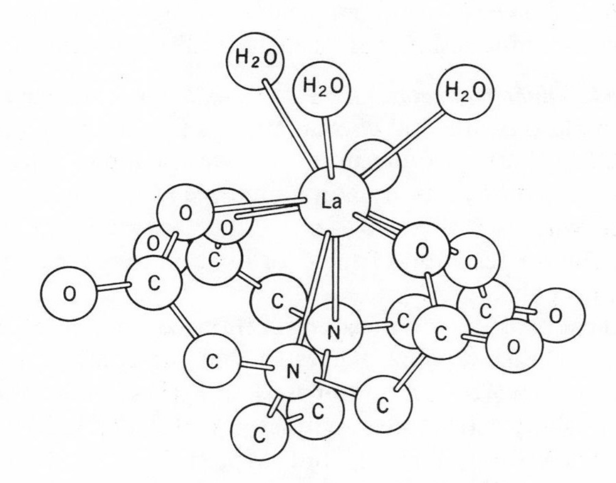

Fig. 1-39. $[La(H_2O)_4$(H-EDTA)] [144].

as a quasi-mirror for the coordination group, but not for the ethylene radical of the strongly puckered ethylenediamine ring. The nine-coordination group of D_{3h}–$\bar{6}2m$ symmetry, the uniquely satisfactory choice for La^{3+} in unconstrained circumstances, is quite incapable of meeting the stereochemical requirements of this chelating agent.

The chelation pattern in $[La(H_2O)_3(EDTA)]^-$ [118] found in crystals of $KLa(H_2O)_3(EDTA)\cdot 5H_2O$ is basically the same as that described for $La(H_2O)_4$(H–EDTA), but the loss of a proton from the ten-coordinate molecule is accompanied by a significant tightening of all chelating linkages and a rejection of one water molecule.

6. Trigonal Prismatic Coordination. Eisenberg and Ibers [67] recently proved that $Re[S_2C_2(C_6H_5)_2]_3$ has a trigonal prismatic structure. The refined data of their x-ray analysis has also been reported [68]. The six sulfur atoms are equidistant from the rhenium; all the S-S distances, both intra- and interligand, are approximately identical. The ReS_6 trigonal prism is accordingly almost perfect. The coordination symmetry is D_{3h}–$\bar{6}2m$ and the overall molecular symmetry, C_3–3. The phenyl groups are not coplanar with the five-membered rings, and so do not seem to be conjugated with them. Similar structures determined by x-ray analysis have been reported for $Mo[S_2C_2H_2]_3$ [206] and for $V[S_2C_2(C_6H_5)_2]_3$ [69]. The interligand S-S distances for these three chelates are very nearly alike (3.05–3.11 A) and the metal sulfur distances are all 2.32–2.34 A. Eisenberg et al. [69] suggest that the relatively short interligand S-S distance indicates stronger inter donor-atom bonding forces than in octahedral systems. They feel that S-S bonding may be an important factor in stabilizing trigonal prismatic coordination.

7. Other Examples of Less Usual Coordination Numbers.

Coordination Number 5: Bis(acetylacetono)oxovanadium(IV) [64], bis-(N,N-dipropyldithiocarbamato)copper(II) [177], and bis(diethyldithiocarbamato)-copper(II) and -zinc(II) [39].

Coordination Number 7: Sodium uranyl acetate [227] and bis(acetylacetono)oxouranium(VI) monohydrate [64].

Coordination Number 8: Uranyl nitrate hexahydrate [216, 104], tetrakis(dibenzoylmethanido)cerium(IV) [225], rubidium uranyl nitrate [34], and potassium bis(nitrilotriacetato)zirconate(IV) [117].

H. Metal-Metal Interaction

Metal-to-metal bonding in coordination compounds has been recognized during recent years in an ever-increasing number of examples of widely varying type. The existence of such bonding can be verified directly by crystal structure analysis, though other physical methods, like magnetic measurements, can predict such interactions. Unusually short distances between metal atoms are the direct evidence of metal-to-metal bonding. Some examples are discussed below.

1. Copper(II) Acetate Dihydrate and Related Compounds. The first example of a chelate shown to have metal-to-metal bonding was copper(II) acetate dihydrate. In crystals of copper(II) acetate dihydrate,

$$Cu_2(CH_3COO)_4 \cdot 2H_2O \text{ [220]},$$

a binuclear molecule such as that shown in Fig. 1-40 was found. The six nearest neighbors of a copper atom are four oxygen atoms belonging to four different acetate groups, a copper atom, and a water molecule. These six donor atoms form a distorted octahedral configuration about each of the two copper atoms. Each acetate group is planar within experimental error. The short Cu-Cu distance (2.64 A) indicates that a metal-metal bond may well exist in this structure. Detailed magnetic measurements of this substance [38] predicted independently the existence of isolated pairs of copper ions coupled by exchange forces, with each copper ion bonded to four oxygen atoms in a plane. This is exactly the arrangement found in this structure. Chromium(II)

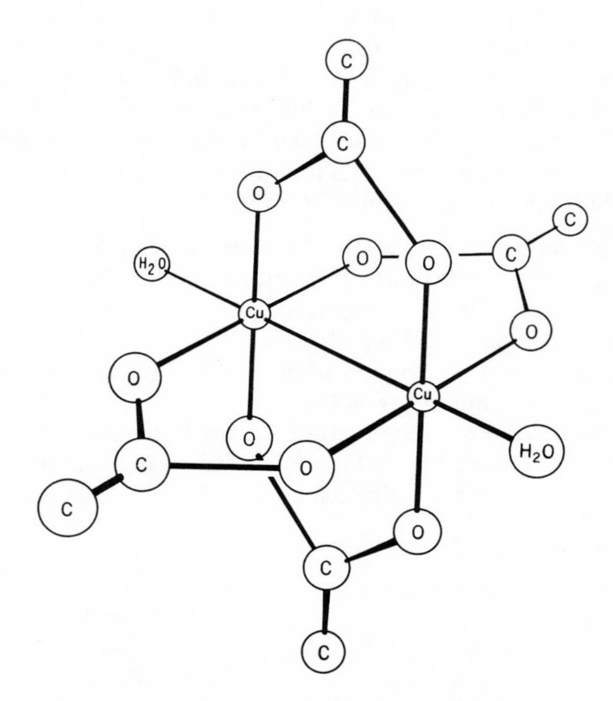

Fig. 1-40. A molecule of $Cu_2(CH_3COO)_4 \cdot 2H_2O$.

acetate, $Cr_2(CH_3COO)_4 \cdot 2H_2O$ [221], is isomorphous with the copper compound, and again the close approach (2.64 A) of the two chromium atoms in a molecule suggests direct interaction between such atoms.

In the molecule of cupric acetate pyridinate, $Cu_2(CH_3COO)_4 \cdot (C_5H_5N)_2$ [109, 30], the copper atoms are bridged in pairs by four acetate groups to form binuclear molecules similar to those described above. Instead of water molecules, two molecules of pyridine are coordinated with the copper atoms. Similar bridging acetate groups were found in molybdenum acetate, $[Mo(CH_3COO)_2]_2$ [143]. This molecule is centrosymmetric with a very short Mo···Mo distance of 2.11 A. In the crystals of Cu(II) succinate dihydrate [167] there exist infinite chains of covalently bonded binuclear units, each of which resembles copper(II) acetate dihydrate (structure XXIII). The Cu-Cu distance is 2.610 A.

XXIII

2. ***Bis(dimethylglyoximato)Ni(II) and Related Compounds.*** Weak interactions are expected between Ni atoms in crystals of $[Ni(dmg)_2]$ [95, 224]. The molecule is planar and contains four five-membered rings. The molecule is required by the space group to have the symmetry C_{2v}–$2/m$. The bond distances (Fig. 1-41) suggest that the principal electronic form contributing to the ground state is one in which all the bonds are essentially single, except

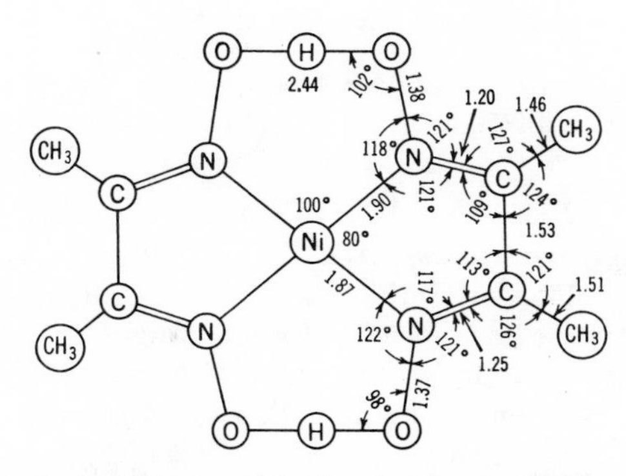

Fig. 1-41. A molecule of bis(dimethylglyoximato)nickel(II).

the C—N bonds, which are essentially double. All bonds are, however, somewhat short for this interpretation; this indicates resonance with other structures which give some double-bond character to the single bonds. Of special interest is the very short O—H—O bond, 2.44 A. It is noteworthy that the O—H stretching frequency for this hydrogen bond is only 1800 cm^{-1}, about half of the normal 3600 cm^{-1} [186].

In the crystalline state these molecules are arranged exactly parallel to (001), and stacked with the nickel atoms directly above each other and separated by 3.245 A. The methyl groups form the thickest portion of the flat molecule, so that alternate molecules are rotated 90° about Ni. The packing along the Ni-Ni axis resembles interlocking blocks. The bonding to nickel is square-planar; this should mean that nickel uses dsp^2 hybrid orbitals and possesses a vacant $4p$ orbital. Consequently Rundle [224] suggested that the first excited state consists essentially of the electronic configuration derived from promoting a $3d$ electron to the $4p$ level and forming two Ni—Ni bonds. This configuration would be octahedral, using d^2sp^3 hybrid orbitals. If some of this configuration contributes to the ground state, with some of the ground configuration mixed into the first excited state, there will be a small but important contribution of the Ni-Ni interaction to the ground state. This would stabilize the particular packing noted in the crystal, and might also account for the very slight solubility of this compound in various solvents. Recently Ingraham [123] made an MO calculation on bis(dimethylglyoximato)nickel(II). However, his result suggests that there is no Ni-Ni interaction.

Platinum [84] and palladium [175] complexes with dimethylglyoxime are isomorphous to the nickel complex, and their metal-to-metal distances are 3.23 A and 3.26 A, respectively. A weak interaction can also be expected between Au(I) and Au(III) (3.26 A) in $[Au(dmg)_2][AuCl_2]$ [187]. In bis-(methylethylglyoximato)Ni(II) [85], which has a *trans* configuration, a Ni···Ni bond does not, however, seem to exist.

3. Other Examples of Metal-Metal Interaction. bis(N-methylsalicylaldimino)copper(II) [153, 145] and $BaMoO_4(C_2O_4)_2 \cdot 5H_2O$ [60].

I. Intermolecular Interaction in the Crystal

1. Bis(dimethylglyoximato)copper(II) [83, 82, 44]. In a molecule of bis-(dimethylglyoximato)copper(II) the two ligands coordinated with the copper atom do not lie in the same plane but form an angle of 158°16′. The copper atom is surrounded by four N atoms at 1.94 A, and is displaced from the plane of the four N atoms in the direction of an oxygen of a neighboring molecule. Therefore Cu is five-coordinate, the Cu-O distance being 2.43 A. The resulting structure is a dimer, as shown in Fig. 1-42. In contrast to the

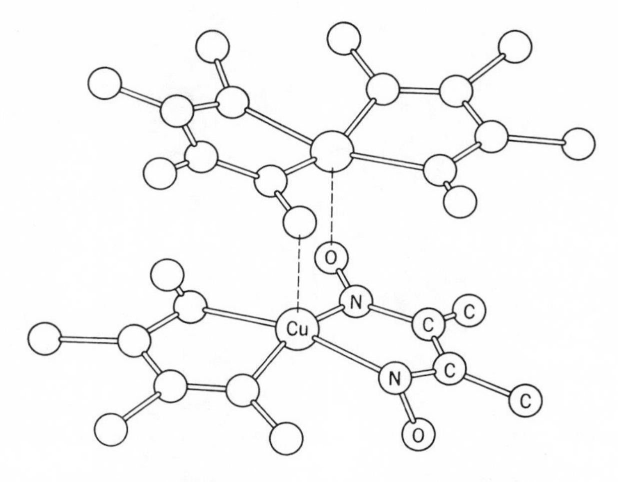

Fig. 1-42. Bis(dimethylglyoximato)copper(II).

nickel complex, the copper(II) complex is soluble in various solvents. The extra electron of Cu would prevent the type of interaction described in Sec. 1-2H, 2, and would, in part, explain the choice of a different packing arrangement for this compound.

2. Bis(acetylacetone)ethylenediiminocopper(II) [102, 103]. In this complex pairs of centrically related molecules appear to be bound together by polarization-type bonds. Molecules are stacked in pairs in such a way that two six-membered chelate rings overlay one another with a normal separation of 3.36 A. And the closest approach to the copper atom, normal to the coordination square, is by the γ-carbon atom of an acetylacetone group in the adjacent molecule; this Cu-C distance is 3.38 A.

3. An Authenticated Perchlorate Complex, **$[Co(CH_3S(CH_2)_2SCH_3)_2](ClO_4)_2$** [59]. As an example of the so-called weakly coordinating ligands, the structure of bis(2,5-dithiahexane)cobalt(II) perchlorate is now described. Figure 1-43 shows the structure of one formula unit and gives its principal dimensions. The cobalt atom lies at a crystallographic center of symmetry. The Co-O distance of 2.34 A indicates a well-defined bonding interaction, but it is perhaps 0.1 ~ 0.3 A longer than normal Co-O bonds in other Co(II) complexes. The infrared spectrum of the compounds is in excellent agreement with the unidentate coordination of the ClO_4^- groups.

4. Other Examples. Bis(salicylaldehyde)ethylenediiminocopper(II) [172], nitroso(dimethyldithiocarbonato)Co(II) [26], bis(salicylaldehydato)Cu(II) [152], and anhydrous bis(8-hydroxyquinolinato)copper(II) [174].

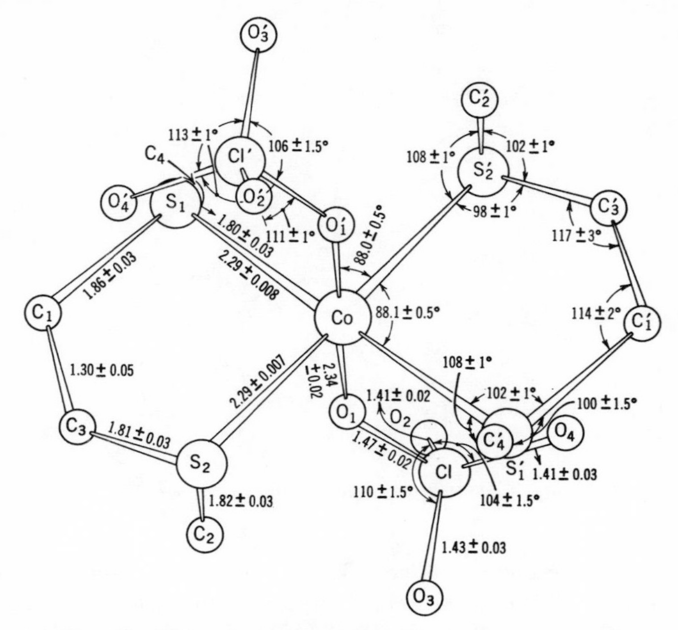

Fig. 1-43. Bis(2,5-dithiahexane)cobalt(II) perchlorate [59].

J. Distortion of the Chelate Ring

1. Bis(2,2′-bipyridylimino)palladium(II), **$Pd(C_{10}H_8N_3)_2$** [91]. Square planar coordination of palladium(II) is well established. In bis-(2,2′-bipyridylimino)palladium(II) (Fig. 1-44), a planar arrangement should result

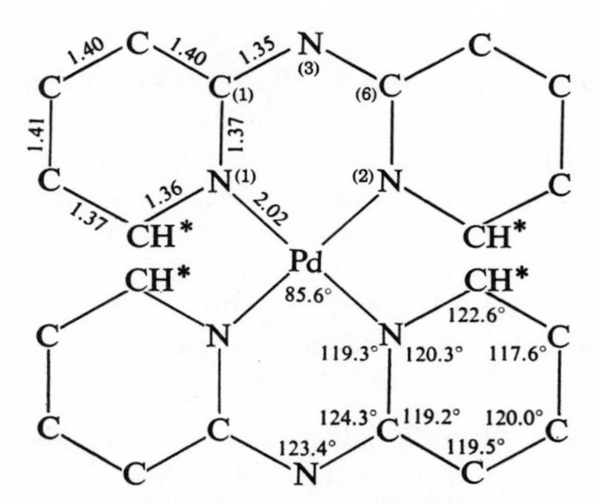

Fig. 1-44. Bis(2,2′-bipyridylimino)palladium(II) [91].

in considerable interligand repulsions between the hydrogen atoms indicated by asterisks. The structure has been solved and refined by three-dimensional least-squares methods. The molecule is centrosymmetric. The upper ligand in the figure is folded upwards (say) about the line Pd···N(3) until the angle between the normals to the two pyridyl rings is 38.2°. The chelate ring PdN(1)C(1)N(3)C(6)N(2) takes a boat form, and the atoms Pd and N(3) lie on the same side of the plane defined by N(1)N(2)C(6) and C(1). The distortions in the bottom ligand are in the opposite sense to that described. The pyridyl rings themselves are not strictly planar. The maximum deviations occur for the four atoms which are common to the pyridyl and chelate rings. The palladium-nitrogen bond length in this complex is 2.02 A, which is in agreement with the sum of the single-bond radii of Pd(II) and N, 1.31 and 0.70 A, respectively. Such a structure, in which a resonating system is deformed by interligand repulsions, is novel among chelate compounds, but is characteristic of overcrowded aromatic molecules like bianthronyl [111].

*2. **Bis(N-t-butylsalicylaldimino)copper(II)*** [49]. It has been established that copper(II) will adopt an approximately tetrahedral environment under certain circumstances. A recent investigation of bis-(N-*t*-butylsalicylaldimino)copper(II) (Fig. 1-45) has shown that in this molecule the metal-ligand

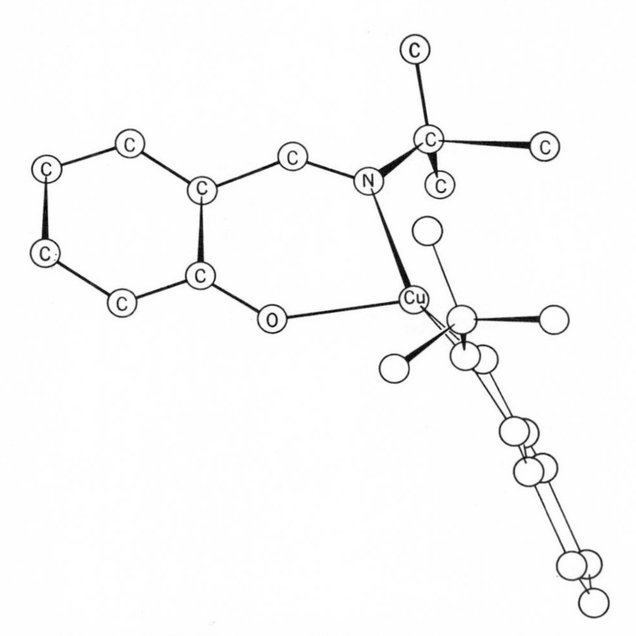

Fig. 1-45. Bis(*N*-*t*-butylsalicylaldimino)copper(II) [49].

framework is indeed tetrahedral. The angle between the mean planes containing the salicylaldimine residues is 80°, which indicates a basically tetrahedral environment of the copper atom. The most striking feature of the complex is that one half of the molecule contains an almost planar salicylaldimino copper group, while in the other half there are considerable deviations from planarity in the group. There is no obvious explanation for this quite unusual distortion. Similar tetrahedral coordination of copper(II) is also found in bis(N-isopropylsalicylaldimino)copper(II) [81].

3. Other Examples of Chelate Ring Distortion. Ni-Phthalocyanine [209] and biacetylbis(mercaptomethylimino)nickel(II) [71].

K. Biologically Interesting Compounds

1. Metal-Peptide Complexes. A series of x-ray analyses have been made by Freeman et al. and by others in order to obtain a detailed understanding of the steric relationships involved in metal-protein interactions. Figure 1-46 illustrates the shape and size of the complex found in glycylglycinocopper(II) trihydrate [210]. The copper atom is five-coordinate. Somewhat distorted square-planar coordination occurs as a result of fused chelate ring formation. The molecule contains two coordinated water molecules; one lies in the square-planar plane, and the other lies perpendicular to it. The square coordination around each copper atom is actually somewhat distorted from

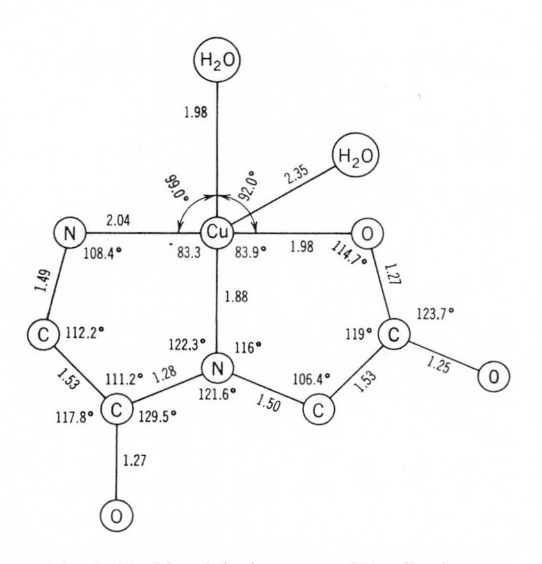

Fig. 1-46. Glycylglycinocopper(II) trihydrate.

perfect planarity in such a fashion that the copper atoms are moved toward the loosely bound water molecules. The atoms of the peptide group are almost planar. The third water molecules are not coordinated directly with the copper atoms but link the different parts of the structure by means of hydrogen bonds.

In molecules of glycylglycylglycinocopper(II) chloride sesquihydrate [88, 51], the metal atom is bonded to the terminal nitrogen and the oxygen atom of the first peptide residue (structure XXIV). The terminal carboxyl group is

$$
\begin{array}{l}
\quad\ \ \ \mathrm{Cl}\ \ \mathrm{H_2O} \qquad\qquad\qquad\qquad\qquad\qquad\qquad\qquad\qquad\qquad\ \ \mathrm{Cl}\ \ \mathrm{H_2O}\\
-\underset{\underset{\mathrm{O}}{\|}}{\mathrm{C}}-\mathrm{O}-\mathrm{Cu}-\mathrm{NH_2}-\mathrm{CH_2}-\underset{\underset{\mathrm{O}}{\|}}{\mathrm{C}}-\mathrm{NH}-\mathrm{CH_2}-\underset{\underset{\mathrm{O}}{\|}}{\mathrm{C}}-\mathrm{NH}-\mathrm{CH_2}-\underset{\underset{\mathrm{O}}{\|}}{\mathrm{C}}-\mathrm{O}-\mathrm{Cu}-
\end{array}
$$

XXIV

coordinated with a second copper atom so that the structure consists of infinite -Cu-peptide-Cu-peptide chains. These chains are crosslinked by a hydrogen-bond network through efficient use of the water molecules and chloride ions. In this compound the copper atom exhibits fivefold coordination. The oxygen atom of the coordinated water lies at the apex of a pyramid, of which the other four ligand atoms form an approximately planar base. The unusual coordination number five may in this instance be due to steric hindrance in the sixth octahedral coordination position. In the violet crystals of sodium glycylglycylglycinocuprate(II) monohydrate,

$$\mathrm{Na[CuNH_2(CH_2CON)_2CH_2COO]\cdot H_2O}\ [90],$$

the complex is a dimer consisting of two copper atoms and two polypeptide molecules. It has a twofold symmetry axis (structure XXV). Again

$$
\begin{array}{c}
\mathrm{O}-\overset{\overset{\mathrm{O}}{\|}}{\mathrm{C}}-\mathrm{CH_2}-\mathrm{N}=\overset{\overset{\mathrm{O}}{\|}}{\mathrm{C}}-\mathrm{CH_2}-\mathrm{N}-\overset{\overset{\mathrm{O}}{\|}}{\mathrm{C}}-\mathrm{CH_2}-\mathrm{NH_2}\\
\mathrm{Cu} \qquad\qquad\qquad\qquad\qquad \mathrm{Cu}\\
\mathrm{H_2N}-\mathrm{CH_2}-\underset{\underset{\mathrm{O}}{\|}}{\mathrm{C}}-\mathrm{N}-\mathrm{CH_2}-\underset{\underset{\mathrm{O}}{\|}}{\mathrm{C}}=\mathrm{N}-\mathrm{CH_2}-\underset{\underset{\mathrm{O}}{\|}}{\mathrm{C}}-\mathrm{O}
\end{array}
$$

XXV

the copper atoms are five-coordinate; each forms bonds with the three nitrogen atoms of one peptide and with a carboxyl oxygen atom of the other peptide molecule of the dimer. There is also a weaker bond between the copper and nitrogen of the second peptide molecule indicated by broken lines

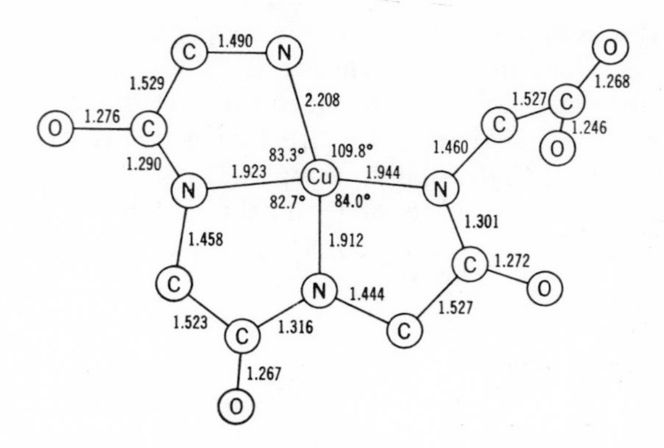

Fig. 1-47. A glycylglycylglycylglycinocuprate(II) ion.

in XXV. This weakly bound atom lies at the apex of a pyramid, while the corners of its approximately square base are occupied by the strongly bound ligand atoms. The complex ion as a whole carries a double negative charge, which must be distributed mainly over the peripheral oxygen atoms. It is these atoms which lie closest to the sodium ions.

The structure of the pink complex, disodium glycylglycylglycylglycinocuprate(II) decahydrate, $Na_2[CuNH_2(CH_2CON)_3CH_2COO]\cdot 10H_2O$, has also been determined [92, 89]. The environment of the copper atom is approximately square-planar. The discrete glycylglycylglycylglycinocuprate-(II) ions are extensively hydrogen-bonded to water molecules. The shape of the complex ion is shown in Fig. 1-47. The environment of the copper atom is similar to that of the copper in potassium bis(biureto)cuprate(II) tetrahydrate (structure XV) [87].

A significant structural feature which these compounds have in common is the planarity of their peptide groups. The fact that this characteristic feature of the uncomplexed peptides is retained implies that the complexes tend to form without sacrificing the resonance energy of the peptide groups. On the other hand, none of the chelated peptide residues is quite coplanar with the copper atom to which it is joined. A second property shared by these compounds is the five-coordination of the copper atoms. It remains to be seen whether this is characteristic of copper-peptide complexes or not.

2. Histidine-Zinc Complexes. Some evidence [215] suggests that, in the zinc-insulin complex, the metal is bonded to a histidine residue. The exceptional ability of histidine to bind metal ions is said to be due to the formation of six-membered chelate rings in which the metal is attached to an imidazole nitrogen atom and the α-amino group. In order to clarify this point

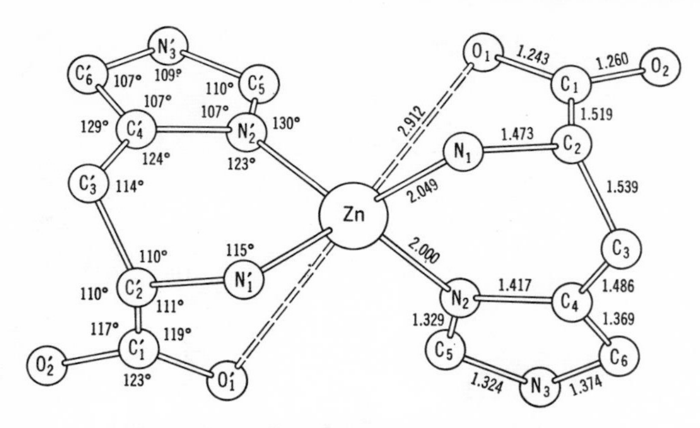

Fig. 1-48. Bis(*d,l*-histidino)zinc(II) [110].

the structures of bis(*d,l*-histidino)zinc(II) pentahydrate [110] and bis(L-histidino)zinc(II) dihydrate [140, 139] were investigated. Figure 1-48 shows a molecule of bis(*d,l*-histidino)zinc. The primary coordination around the zinc is approximately tetrahedral, and a carboxyl oxygen atom of each histidine approaches to within 3 A of the zinc to form a secondary coordination grouping.

*3. **Porphyrins.*** The tetrapyrrole nucleus has unique chelating properties. Through the effects of substituents on the porphyrin nucleus and of extra ligands which may be added perpendicularly to the porphyrin plane, the physicochemical properties of metal porphyrins may vary widely. In the natural compounds these properties lead to various specific biological activities of the complexes of iron (the hemoproteins) and of magnesium (the chlorophylls). Since metal porphyrins are essential to the life of bacteria, fungi, plants, and animals, extensive studies have been made of their organic chemistry, biochemistry, and biosynthesis.

That the porphyrin nucleus is likely to be planar had long been evident from the x-ray analysis of the related tetrazaporphyrin and phthalocyanin [183, 184]. Recent results obtained from three-dimensional x-ray diffraction data indicate that the porphyrin nucleus in free porphine is planar [221a]. However, the porphine skeleton is rather flexible and possesses an adaptability to a considerable range of stereochemical requirements. Thus the porphyrin nucleus in Ni-etioporphyrin-I [76] is not planar.

A particularly marked ruffling of the porphine skeleton, which is consistent with a fourfold axis of rotatory inversion, characterizes molecules of the copper(II) [77, 78], palladium(II) [78], and metal-free [116, 105] tetraphenylporphines as they exist in an isomorphous series of tetragonal crystals.

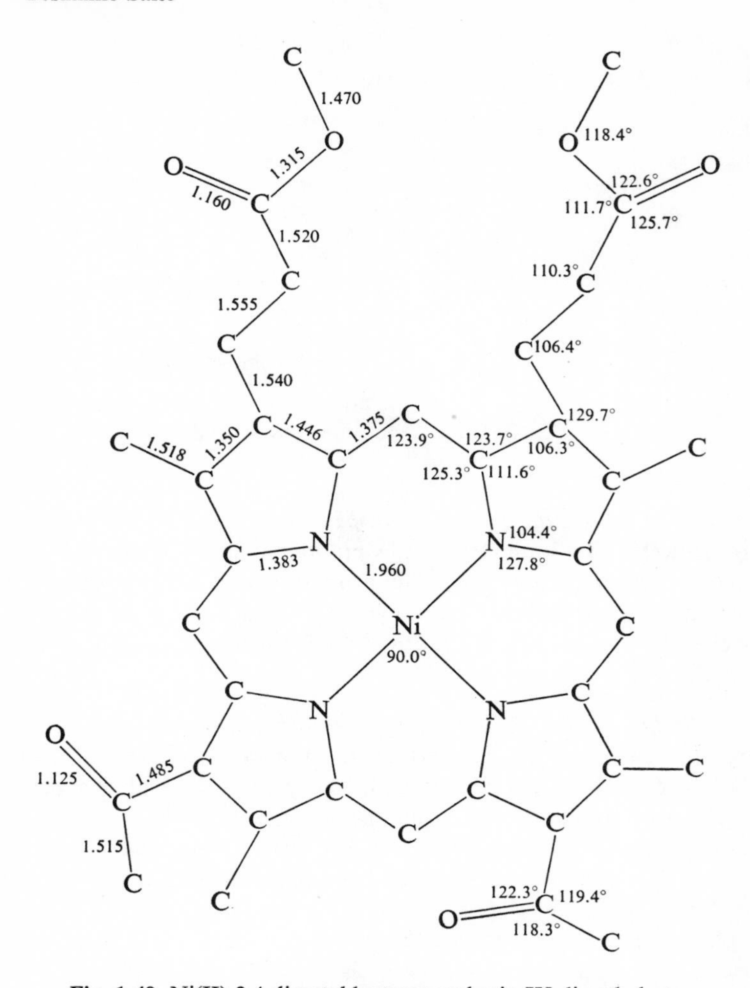

Fig. 1-49. Ni(II) 2,4-diacetyldeuteroporphyrin-IX dimethylester.

Significantly nonplanar porphine skeletons are observed also in crystals of a triclinic tetraphenylporphine [203] and of chlorohemin [134]. Figure 1-49 shows a skeletal diagram of the Ni(II) 2,4-diacetyldeuteroporphyrin-IX dimethyl ester molecule [106]. In Fig. 1-49 are shown the averaged bond distances in this molecule. This averaging was performed in agreement with $\boldsymbol{C}_{4v}$–$4mm$ or effectively $\boldsymbol{D}_{4h}$–$4/mm$ symmetry. In crystals of methoxyiron(III) mesoporphyrin-IX dimethyl ester [119], the porphine skeleton is slightly convex and the iron(III) atom, as in chlorohemin, is quite definitely five-coordinate. In the structure of myoglobin [129] the iron atom lies more than 0.25 A from the mean plane of the heme group.

4. Vitamin B_{12} [120, 121]. The crystal structure analysis of vitamin B_{12} by Hodgkin and her collaborators represents a most advanced x-ray analysis of an extremely complex structure. Vitamin B_{12}, cyanocobalamine, is noted for its effectiveness against pernicious anemia. This substance, a cobalt chelate, was first isolated in 1948 by Folkers et al. [182] of Merck Laboratories and a little later by Smith [207]. Specimens of several related crystals of similar structures have been studied.

Vitamin B_{12} is known, from chemical degradation, to contain cobalt, a cyanide group, and a nucleotide-like group containing phosphate, *d*-ribose, and dimethylbenzimidazole. Positions of the cobalt atoms were established

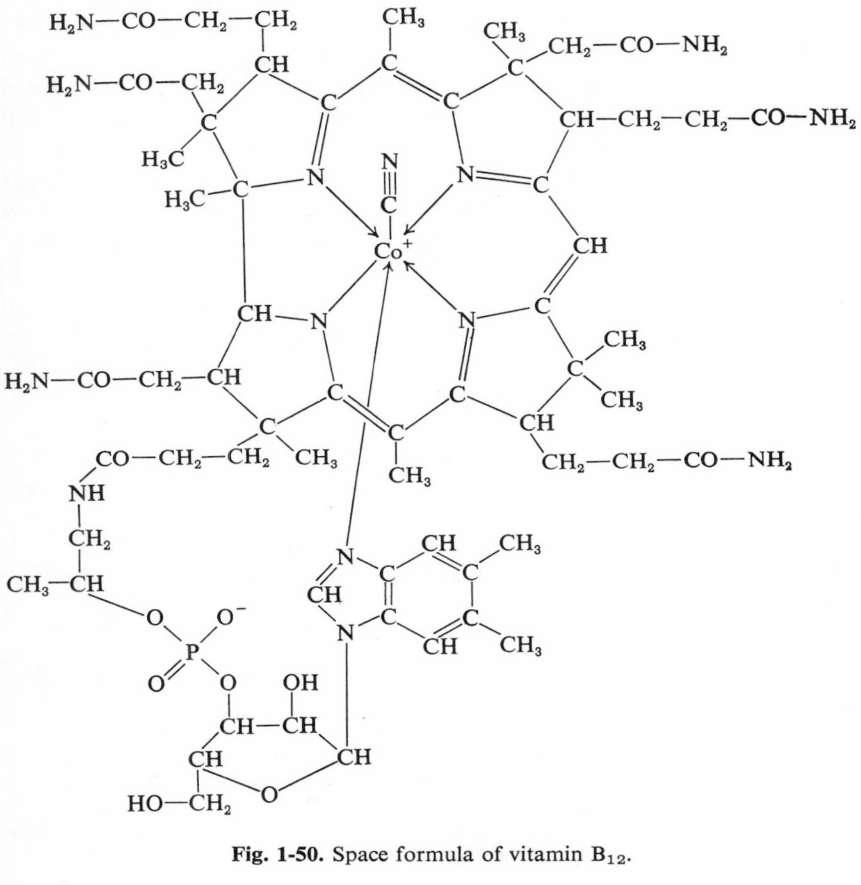

Fig. 1-50. Space formula of vitamin B_{12}.

by three-dimensional Patterson synthesis. Then phases were calculated on the basis of the contribution of these atoms—a very poor approximation to the truth since cobalt only contributes about 12% of the total x-ray scattering effect in these crystals—and three-dimensional Fourier maps were computed. From these the positions of the known part of the molecule could be recognized. In subsequent calculations on these and other crystals it became clear that the cobalt atom is surrounded by a porphyrin-like group which has the unusual feature that two of the five-membered rings are linked directly without an intermediate atom. This new type of nucleus is called the corrin nucleus. As each new atom or group was recognized on the maps as stereochemically reasonable, it was put into the next calculation of phases. If a peak was of low electron density or in an unlikely position, it was omitted. In this way, step by step, the atoms revealed their true positions. Figure 1-50 shows the space formula; in it the benzimidazole ring is perpendicular to the corrin nucleus. The most direct and accurate evidence for the existence of the corrin nucleus in natural vitamin B_{12} and related compounds was obtained from the compound known as cobyric acid [65, 122]. It contains the nucleus of vitamin B_{12} without the nucleotide and propanolamine. In order to find one carboxyl group among six amide groups, neutron diffraction measurements were performed. Oxygen and nitrogen differ in x-ray scattering power by one electron only, but nitrogen is almost twice as effective in neutron scattering as oxygen. Accordingly, these atoms can be more easily distinguished by neutron scattering experiments. About 1300 neutron diffraction spectra have been measured for this acid, and the three-dimensional map derived from them has been calculated. Thus the structure of cobyric acid has been established.

REFERENCES

Books, Catalogs of Data

[1] Bacon, G. E., *Neutron Diffraction*. Oxford: University Press, 1962, 426 pp.
[2] Bacon, G. E., *Applications of Neutron Diffraction in Chemistry*. Oxford: Pergamon, 1963, 141 pp.
[3] Bailar, J. C., Jr., ed., *The Chemistry of the Coordination Compounds*. New York: Reinhold, 1956, 834 pp.
[4] Bragg, W. L., *The Crystalline State*. Vol. 1: *A General Survey*. London: Bell, 1949, 352 pp.
[5] Buerger, M. J., *X-ray Crystallography*. New York: Wiley, 1958, 531 pp.
[6] Buerger, M. J., *Elementary Crystallography*. New York: Wiley, 1958, 528 pp.
[7] Buerger, M. J., *Crystal Structure Analysis*. New York: Wiley, 1960, 668 pp.
[8] Bunn, C. W., *Chemical Crystallography*. Oxford: University Press, 1961, 509 pp.
[9] Dwyer, F. P., and D. P. Mellor, eds., *Chelating Agents and Metal Chelates*. New York: Academic Press, 1964, 530 pp.

[10] Hückel, W., *Structural Chemistry of Inorganic Compounds*. Amsterdam: Elsevier, 1952. Vol. 1, 483 pp.; Vol. 2, 656 pp.
[11] *International Tables for X-ray Crystallography*. Birmingham: Kynoch Press, 1952, Vol. 1, 558 pp.; 1959, Vol. 2, 444 pp.; 1962, Vol. 3, 362 pp.
[12] James, R. W., *The Optical Principles of the Diffraction of X-rays* (*The Crystalline State*, Vol. 2). London: Bell, 1963, 640 pp.
[13] Lipson, H., and W. Cochran, *The Determination of Crystal Structures* (*The Crystalline State*, Vol. 3). London: Bell, 1966, 414 pp.
[14] Lonsdale, K., *Crystals and X-rays*. London: Bell, 1948, 199 pp.
[15] McLachlan, D., *X-ray Crystal Structure*. New York: McGraw-Hill, 1957, 416 pp.
[16] Pauling, L., *The Nature of the Chemical Bond*. Ithaca, N.Y.: Cornell University Press, 1960, 644 pp.
[17] Wells, A. F., *Structural Inorganic Chemistry*. Oxford: University Press, 3rd Ed., 1962, 1055 pp.
[18] Wheatley, P., *The Determination of Molecular Structure*. Oxford: University Press, 1959, 263 pp.

Catalogs of Data

[19] Wyckoff, R. W. G., *Crystal Structures*. New York: Interscience, 1963, 2nd Ed., Vol. 1, 467 pp.; 1964, Vol. 2, 588 pp.; 1965, Vol. 3, 981 pp.
[20] *Strukturbericht*. Leibzig: Akademische Verlagsgesellschaft. 1931–43, Vols. 1–7. A detailed list of all crystal structure determinations published in the period 1913–1939.
[21] *Structure Reports*. Utrecht: Oosthoek. 1951–. Vols. 8–23. A continuation of *Strukturbericht* in which the reports are even more comprehensive and detailed and in which they are supplemented by editorial comment. At the time of going to press, Vols. 8–23, covering the years 1940–1959 except 1957 (Vol. 21), are available. Publication continues.
[22] Sutton, L. E., ed., *Tables of Interatomic Distances and Configuration in Molecules and Ions*. London: Chemical Society (Special Publication No. 11), 1958, 259 pp. Supplement 1956–59. (Special Publication No. 18), 1965, 208 pp.

Review Articles

[23] Holm, R. H., G. W. Everett, Jr., and A. Chakravorty, "Metal Complexes of Schiff Bases and β-Ketoamines," in *Progress in Inorganic Chemistry*, F. A. Cotton, ed. New York: Interscience, 1965, Vol. VII, pp. 83–214.
[24] Livingstone, S. E., "Metal Complexes of Ligands Containing Sulphur, Selenium, or Tellurium as Donor Atoms," *Quart. Rev.* (*London*), **19**, 386–404 (1965).
[25] Muetterties, E. L., and R. A. Schunn, "*Pentacoordination*," *Quart. Rev.* (*London*), **20**, 245–299 (1966).

Other References

[26] Alderman, P. R. H., P. G. Owston, and J. M. Rowe, *J. Chem. Soc.*, 668 (1962).
[27] Amirthalingam, V., V. P. Padmanabhan, and J. Shankar, *Acta Cryst.*, **13**, 201 (1960).
[28] Antsyskina, A. S., and M. A. Porai-Koshits, *Dokl. Akad. Nauk SSSR*, **143**, 105 (1962).
[29] Barclay, G. A., and C. H. L. Kennard, *J. Chem. Soc.*, 3289 (1961).
[30] Barclay, G. A., and C. H. L. Kennard, *J. Chem. Soc.*, 5244 (1961).

[31] Barclay, G. A., and B. F. Hoskins, *J. Chem. Soc.*, 586 (1962).
[32] Barclay, G. A., B. F. Hoskins, and C. H. L. Kennard, *Nature*, **192**, 425 (1691); *J. Chem. Soc.*, 5691 (1963).
[33] Barclay, G. A., and B. F. Hoskins, *J. Chem. Soc.*, 1979 (1965).
[34] Barclay, G. A., and T. Sabine, *Acta Cryst.*, **19**, 205 (1965).
[35] Becker, K. A., G. Grosse, and K. Plieth, *Z. Krist.*, **112**, 375 (1959).
[36] Bevan, J. A., D. P. Graddon, and J. F. McConnel, *Nature*, **199**, 373 (1963).
[37] Bijvoet, J. M., A. F. Peerdeman, and A. J. van Bommel, *Nature*, **168**, 271 (1951).
[38] Bleaney, B., and B. Bowers, *Proc. Roy. Soc.* (*London*), **A214**, 451 (1952).
[39] Bonamico, M., G. Dessy, and L. Zambonelli, *Atti Accad. Nazl. Lincei, Rend.*, **35**, 338 (1963).
[40] Bragg, W. H., and G. T. Morgan, *Proc. Roy. Soc.* (*London*), **A104**, 437 (1923).
[41] Brintzinger, H., H. Thiele, and U. Müller, *Z. Anorg. Allgem. Chem.*, **251**, 285 (1943).
[42] Brown, B. W., and E. C. Lingafelter, *Acta Cryst.*, **16**, 753 (1963).
[43] Brown, B. W., and E. C. Lingafelter, *Acta Cryst.*, **17**, 254 (1964).
[44] Bua, E., and G. Schiavinanto, *Gazz. Chim. Ital.*, **81**, 847, 856 (1951).
[45] Bullen, G. J., *Acta Cryst.*, **12**, 703 (1959).
[46] Bullen, G. J., *Nature*, **177**, 537 (1961).
[47] Busch, D. H., and J. C. Bailar, Jr., *J. Am. Chem. Soc.*, **75**, 4574 (1953).
[48] Cavalca, L., M. Nardelli, and G. Fava, *Acta Cryst.*, **13**, 594 (1960).
[49] Cheeseman, T. P., D. Hall, and T. N. Waters, *Nature*, **205**, 494 (1965).
[50] Comyns, A. E., *Acta Cryst.*, **13**, 278 (1960).
[51] Cooper, T., H. C. Freeman, G. Robinson, and J. C. Schoone, *Nature*, **194**, 1237 (1962).
[52] Corbridge, D. E., and E. G. Cox, *J. Chem. Soc.*, 594 (1956).
[53] Corey, E. J., and J. C. Bailar, Jr., *J. Am. Chem. Soc.*, **81**, 2620 (1959).
[54] Coster, D., K. S. Knol, and J. A. Prins, *Z. Physik*, **63**, 345 (1930).
[55] Coster, D., and K. S. Knol, *Proc. Roy. Soc.* (*London*), **A139**, 459 (1933); *Z. Physik*, **75**, 340 (1934).
[56] Cotton, F. A., and R. C. Elder, *Inorg. Chem.*, **3**, 397 (1964).
[57] Cotton, F. A., and J. S. Wood, *Inorg. Chem.*, **3**, 245 (1964).
[58] Cotton, F. A., and R. C. Elder, *J. Am. Chem. Soc.*, **86**, 2294 (1964); *Inorg. Chem.*, **4**, 1145 (1965).
[59] Cotton, F. A., and D. L. Weaver, *J. Am. Chem. Soc.*, **87**, 4189 (1965).
[60] Cotton, F. A., and S. M. Morehouse, *Inorg. Chem.*, **4**, 1377 (1965).
[61] Cox, E. G., W. Wardlaw, and K. C. Webster, *J. Chem. Soc.*, 1475 (1935).
[62] Cromer, D. T., A. C. Larson, and R. B. Roof, Jr., *Acta Cryst.*, **19**, 192 (1965).
[63] Cruickshank, D. W. J., *Acta Cryst.*, **2**, 65 (1949).
[64] Dodge, R. P., D. H. Templeton, and A. Zalkin, *J. Chem. Phys.*, **35**, 55 (1961).
[65] Dunitz, J. D., and E. Meyer, Jr., *Proc. Roy. Soc.* (*London*), **A286**, 324 (1965).
[66] Dwyer, F. P., E. C. Gyrafas, and D. P. Mellor, *J. Phys. Chem.*, **59**, 296 (1955).
[67] Eisenberg, R., and J. A. Ibers, *J. Am. Chem. Soc.*, **87**, 3776 (1965).
[68] Eisenberg, R., and J. A. Ibers, *Inorg. Chem.*, **5**, 411 (1966).
[69] Eisenberg, R., E. I. Steifel, R. C. Rosenberg, and H. B. Gray, *J. Am. Chem. Soc.*, **88**, 2874 (1966).
[70] Enemark, J. H., and W. N. Lipscomb, *Inorg. Chem.*, **4**, 1729 (1965).
[71] Fernando, Q., and P. J. Wheatley, *Inorg. Chem.*, **4**, 1726 (1965).
[72] Ferrari, A., A. Braibanti, and G. Bigliardi, *Z. Krist.*, **117**, 241 (1962).
[73] Ferrari, A., A. Braibanti, and G. Bigliardi, *Acta Cryst.*, **16**, 498 (1963).

[74] Figgis, B. N., J. Lewis, R. F. Long, R. Mason, R. S. Nyholm, P. J. Pauling, and G. B. Robertson, *Nature*, **195**, 1278 (1962).
[75] Figgis, B. N., and G. B. Robertson, *Nature*, **205**, 694 (1965).
[76] Fleischer, E. B., *J. Am. Chem. Soc.*, **85**, 146 (1963).
[77] Fleischer, E. B., *J. Am. Chem. Soc.*, **85**, 1353 (1963).
[78] Fleischer, E. B., C. K. Miller, and L. E. Webb, *J. Am. Chem. Soc.*, **86**, 2342 (1964).
[79] Fleming, J. E., and H. Lynton, *Chem. Ind. (London)*, **79**, 1416 (1960).
[80] Foss, O., and K. Marøy, *Acta Chem. Scand.*, **13**, 201 (1959).
[81] Fox, M. R., E. C. Lingafelter, P. Orioli, and L. Sacconi, *Nature*, **197**, 1104 (1963).
[82] Frasson, E., R. Zannetti, R. Bardi, S. Bezzi, and G. Giacometti, *J. Inorg. Nucl. Chem.*, **8**, 452 (1958).
[83] Frasson, E., R. Bardi, and S. Bezzi, *Acta Cryst.*, **12**, 201 (1959).
[84] Frasson, E., C. Panattoni, and R. Zannetti, *Acta Cryst.*, **12**, 1027 (1959).
[85] Frasson, E., and C. Panattoni, *Acta Cryst.*, **13**, 893 (1960).
[86] Frasson, E., G. Bombieri, and C. Panattoni, *Nature*, **202**, 1325 (1964).
[87] Freeman, H. C., J. Smith, and J. Taylor, *Nature*, **184**, 707 (1959); *Acta Cryst.*, **14**, 407 (1961).
[88] Freeman, H. C., G. Robinson, and J. C. Schoone, *Acta Cryst.*, **17**, 719 (1964).
[89] Freeman, H. C., and M. R. Taylor, *Proc. Chem. Soc.*, 88 (1964).
[90] Freeman, H. C., J. C. Schoone, and J. G. Sime, *Acta Cryst.*, **18**, 381 (1965).
[91] Freeman, H. C., and M. R. Snow, *Acta Cryst.*, **18**, 845 (1965).
[92] Freeman, H. C., and M. R. Taylor, *Acta Cryst.*, **18**, 939 (1965).
[93] Freeman, H. C., and J. Smith, *Acta Cryst.*, **20**, 153 (1966).
[94] Glick, M. D., and L. F. Dahl, *Inorg. Chem.*, **5**, 289 (1966).
[95] Godycki, L. E., and R. E. Rundle, *Acta Cryst.*, **6**, 487 (1953).
[96] Graddon, D. P., and E. C. Watton, *Nature*, **190**, 907 (1961).
[97] Grdenic, D., and B. Matkovic, *Nature*, **182**, 465 (1958).
[98] Grdenic, D., and B. Matkovic, *Acta Cryst.*, **12**, 817 (1959).
[99] Haagensen, C. O., and S. E. Rasmussen, *Acta Chem. Scand.*, **17**, 1630 (1963).
[100] Hall, D., and M. D. Woufle, *Proc. Chem. Soc.*, 346 (1958).
[101] Hall, D., and F. H. Moore, *Proc. Chem. Soc.*, 256 (1960).
[102] Hall, D., A. D. Rae, and T. N. Waters, *Proc. Chem. Soc.*, 143 (1962).
[103] Hall, D., A. D. Rae, and T. N. Waters, *J. Chem. Soc.*, 5897 (1963).
[104] Hall, D., A. D. Rae, and T. N. Waters, *Acta Cryst.*, **19**, 389 (1965).
[105] Hamor, M. J., T. A. Hamor, and J. L. Hoard, *J. Am. Chem. Soc.*, **86**, 1938 (1964).
[106] Hamor, T. A., W. S. Caughey, and J. L. Hoard, *J. Am. Chem. Soc.*, **87**, 2305 (1965).
[107] Hanic, F., *Acta Chim. Acad. Sci. Hung.*, **32**, 305 (1962).
[108] Hanic, F., *Chem. Zvesti*, **17** (6), 365 (1963).
[109] Hanic, F., D. Stemelova, and K. Hanicova, *Acta Cryst.*, **17**, 633 (1964).
[110] Harding, M. M., and S. J. Cole, *Proc. Chem. Soc.*, 178 (1962); *Acta Cryst.*, **16**, 643 (1963).
[111] Harnic, E., and G. M. J. Schmidt, *J. Chem. Soc.*, 3295 (1954).
[112] Hesse, R., *Arkiv Kemi*, **20**, 481 (1963).
[113] Hoard, J. L., M. Lind, and J. V. Silverton, *J. Am. Chem. Soc.*, **83**, 2770 (1961).
[114] Hoard, J. L., G. S. Smith, and M. Lind, *Advances in the Chemistry of the Coordination Compounds*. New York: Macmillan, 1961, p. 296.
[115] Hoard, J. L., C. H. L. Kennard, and G. S. Smith, *Inorg. Chem.*, **2**, 1316 (1963).
[116] Hoard, J. L., M. J. Hamor, and T. A. Hamor, *J. Am. Chem. Soc.*, **85**, 2334 (1963).

[117] Hoard, J. L., E. Willstadter, and J. V. Silverton, *J. Am. Chem. Soc.*, **87**, 1610 (1965).
[118] Hoard, J. L., B. Lee, and M. D. Lind, *J. Am. Chem. Soc.*, **87**, 1612 (1965).
[119] Hoard, J. L., M. J. Hamor, T. A. Hamor, and W. S. Caughey, *J. Am. Chem. Soc.*, **87**, 2312 (1965).
[120] Hodgkin, D. C., J. Kamper, J. Lindsey, M. Mackay, J. Pickworth, J. H. Robertson, C. B. Shoemaker, J. G. White, R. J. Prosen, and K. N. Trueblood, *Proc. Roy. Soc. (London)*, **A242**, 228 (1957).
[121] Hodgkin, D. C., *Science*, **150**, 979 (1965); *Angew. Chem.*, **77**, 954 (1965).
[122] Hodgkin, D. C., *Proc. Roy. Soc. (London)*, **A288**, 294 (1965).
[123] Ingraham, L. L., *Acta Chem. Scand.*, **20**, 283 (1966).
[124] Iwasaki, H., and Y. Saito, *Bull. Chem. Soc. Japan*, **39**, 92 (1966).
[125] Jaeger, F. M., and L. Bijkerk, *Z. Anorg. Allgem. Chem.*, **233**, 97 (1937), and earlier articles by Jaeger.
[126] James, R. W., *The Optical Principles of the Diffraction of X-rays*. London: Bell, 1962, p. 608.
[127] Jelenic, I., D. Grdenic, and A. Bezjak, *Acta Cryst.*, **17**, 758 (1964).
[128] Jones, L. L., and H. Eyring, *J. Chem. Educ.*, **38**, 601 (1961).
[129] Kendrew, J. C., *Science*, **139**, 1259 (1963).
[130] Kinoshita, Y., I. Matsubara, and Y. Saito, *Bull. Chem. Soc. Japan*, **32**, 741 (1959).
[131] Kinoshita, Y., I. Matsubara, and Y. Saito, *Bull. Chem. Soc. Japan*, **32**, 1216 (1959).
[132] Kinoshita, Y., I. Matsubara, T. Higuchi, and Y. Saito, *Bull. Chem. Soc. Japan*, **32**, 1221 (1959).
[133] Kobayashi, M., *J. Chem. Soc. Japan*, **64**, 648 (1939).
[134] Koenig, D. F., *Acta Cryst.*, **18**, 663 (1965).
[135] Komiyama, Y., and E. C. Lingafelter, *Acta Cryst.*, **17**, 1145 (1964).
[136] Komiyama, Y., to be published.
[137] Koyama, H., Y. Saito, and H. Kuroya, *J. Inst. Polytech. Osaka City Univ., Ser. C.*, **4**, 43 (1953).
[138] Koyama, H., and Y. Saito, *Bull. Chem. Soc. Japan*, **27**, 112 (1954).
[139] Krestinger, R. H., R. F. Bryan, and F. A. Cotton, *Proc. Chem. Soc.*, 177 (1962).
[140] Krestinger, R. H., F. A. Cotton, and R. F. Bryan, *Acta Cryst.*, **16**, 651 (1963).
[141] Kuhn, W., *Z. Elektrochem.*, **56**, 506 (1952).
[142] La Villa, R. E., and S. H. Bauer, *J. Am. Chem. Soc.*, **85**, 3597 (1963).
[143] Lawton, D., and R. Mason, *J. Am. Chem. Soc.*, **87**, 921 (1965).
[144] Lind, M. D., B. Lee, and J. L. Hoard, *J. Am. Chem. Soc.*, **87**, 1611 (1965).
[145] Lingafelter, E. C., G. I. Simmons, and B. Morosin, *Acta Cryst.*, **14**, 1222 (1961).
[146] Lippert, E., and M. R. Truter, *J. Chem. Soc.*, 4996 (1960).
[147] Lydon, J. E., M. R. Truter, and R. C. Walting, *Proc. Chem. Soc.*, 193 (1964).
[148] Marsh, R. E., *Acta Cryst.*, **11**, 654 (1958).
[149] Masuko, A., T. Nomura, and Y. Saito, *Bull. Chem. Soc. Japan*, **40**, 511 (1967).
[150] Matkovic, B., *Acta Cryst.*, **16**, 456 (1963).
[151] Mazzi, F., *Rend. Soc. Mineralog. Ital.*, **9**, 148 (1953).
[152] McKinnon, A. J., T. N. Waters, and D. Hall, *J. Chem. Soc.*, 3290 (1964).
[153] Meuthen, B., and M. V. Stackelberg, *Z. Anorg. Allgem. Chem.*, **305**, 279 (1960).
[154] Moffitt, W., *J. Chem. Phys.*, **25**, 1189 (1956).
[155] Montgomery, H., and E. C. Lingafelter, *Acta Cryst.*, **16**, 740 (1963).
[156] Montgomery, H., and E. C. Lingafelter, *Acta Cryst.*, **16**, 748 (1963).
[157] Montgomery, H., and E. C. Lingafelter, *Acta Cryst.*, **17**, 1481 (1964).

[158] Morosin, B., and J. R. Brathovde, *Acta Cryst.*, **17**, 705 (1964).
[159] Morosin, B., *Acta Cryst.*, **19**, 131 (1965).
[160] Nakahara, A., Y. Saito, and H. Kuroya, *Bull. Chem. Soc. Japan*, **25**, 331 (1952).
[161] Nakatsu, K., Y. Saito, and H. Kuroya, *Bull. Chem. Soc. Japan*, **29**, 428 (1956).
[162] Nakatsu, K., M. Shiro, Y. Saito, and H. Kuroya, *Bull. Chem. Soc. Japan*, **30**, 158 (1957).
[163] Nakatsu, K., *Bull. Chem. Soc. Japan*, **35**, 832 (1962).
[164] Nardelli, M., G. Fava, and P. Boldrini, *Gazz. Chim. Ital.*, **92**, 1392 (1962); *Acta Cryst.*, **18**, 618 (1965).
[165] Nawata, Y., H. Iwasaki, and Y. Saito, *Bull. Chem. Soc. Japan*, **40**, 515 (1967).
[166] Nishikawa, S., and K. Matsukawa, *Proc. Imp. Acad. (Tokyo)*, **4**, 96 (1928).
[167] O'Conner, B. H., and E. N. Maslen, *Acta Cryst.*, **20**, 824 (1966).
[168] Ooi, S., Y. Komiyama, Y. Saito, and H. Kuroya, *Bull. Chem. Soc. Japan*, **32**, 263 (1959).
[169] Ooi, S., Y. Komiyama, and H. Kuroya, *Bull. Chem. Soc. Japan*, **33**, 354 (1960).
[170] Ooi, S., and H. Kuroya, *Bull. Chem. Soc. Japan*, **36**, 1083 (1963).
[171] Orgel, L. E., *Nature*, **187**, 504 (1960).
[172] Pachler, K., and M. V. Stackelberg, *Z. Anorg. Allgem. Chem.*, **305**, 286 (1960).
[173] Padmanabhan, V. M., *Proc. Indian Acad. Sci.*, **47A**, 329 (1958).
[174] Palenik, G. J., *Acta Cryst.*, **17**, 687, 696 (1964).
[175] Panattoni, C., E. Frasson, and R. Zannetti, *Gazz. Chim. Ital.*, **89**, 2132 (1959).
[176] Pauling, L., and R. G. Dickinson, *J. Am. Chem. Soc.*, **46**, 1615 (1924).
[177] Pignedoli, A., and G. Peyronel, *Gazz. Chim. Ital.*, **92**, 745 (1962).
[178] Piper, T. S., *J. Am. Chem. Soc.*, **83**, 3908 (1961).
[179] Porai-Koshits, M. A., and L. K. Minacheva, *Zh. Strukt. Khim.*, **5**, 642 (1962).
[180] Rasmussen, S. E., *J. Inorg. Nucl. Chem.*, **8**, 441 (1958); *Acta Chem. Scand.*, **13**, 2009 (1959).
[181] Richards, S., B. Pedersen, J. V. Silverton, and J. L. Hoard, *Inorg. Chem.*, **3**, 27 (1964).
[182] Rickes, E. L., N. G. Brink, F. R. Koniuszy, T. R. Wood, and K. Folkers, *Science*, **107**, 396 (1948).
[183] Robertson, J. M., *J. Chem. Soc.*, 615 (1935); 1195 (1936).
[184] Robertson, J. M., and I. Woodward, *J. Chem. Soc.*, 219 (1937); 236 (1940).
[185] Roof, R. B., Jr., *Acta Cryst.*, **9**, 781 (1956).
[186] Rundle, R. E., and M. Parasol, *J. Chem. Phys.*, **20**, 1487 (1952).
[187] Rundle, R. E., *J. Am. Chem. Soc.*, **76**, 3101 (1954).
[188] Ryan, T. D., and R. E. Rundle, *J. Am. Chem. Soc.*, **83**, 2814 (1961).
[189] Sacconi, L., P. L. Orioli, and M. DiVaira, *J. Am. Chem. Soc.*, **87**, 2059 (1965).
[190] Saito, Y., K. Nakatsu, M. Shiro, and H. Kuroya, *Acta Cryst.*, **8**, 729 (1955); *Bull. Chem. Soc. Japan*, **30**, 795 (1957).
[191] Saito, Y., and H. Iwasaki, *Bull. Chem. Soc. Japan*, **35**, 1131 (1962).
[192] Saito, Y., H. Iwasaki, and H. Ota, *Bull. Chem. Soc. Japan*, **36**, 1543 (1963).
[193] Saito, Y., to be published.
[194] Sargeson, A. M., "Optical Phenomena in Metal Chelates," in *Chelating Agents and Metal Chelates*, F. P. Dwyer and D. P. Mellor, eds. New York: Academic Press, 1964, Chapter 5, p. 216.
[195] Sasvari, K., *Acta Geol. Acad. Sci. Hung.*, **4**, 467 (1957).
[196] Sato, T., K. Nagata, M. Shiro, and H. Koyama, *Chem. Commun.*, 192 (1966).
[197] Scouloudi, H., and C. H. Carlisle, *Acta Cryst.*, **6**, 651 (1953); *Nature*, **166**, 357 (1950).

[198] Shibata, S., and S. Sone, *Bull. Chem. Soc. Japan*, **29**, 852 (1956).
[199] Shibata, S., *Bull. Chem. Soc. Japan*, **30**, 753 (1957).
[200] Shugam, E. A., *Dokl. Akad. Nauk SSSR*, **81**, 853 (1951).
[201] Shugam, E. A., *Kristallografiya*, **1**, 478 (1956).
[202] Shvelashvili, A. E., M. A. Porai-Koshits, and A. S. Antsyshkina, *Zh. Strukt. Khim.*, **5**, 147 (1964).
[203] Silvers, S., and A. Tulinsky, *J. Am. Chem. Soc.*, **86**, 927 (1964).
[204] Silverton, J. V., and J. L. Hoard, *Inorg. Chem.*, **2**, 243 (1963).
[205] Smirnoff, A. P., *Helv. Chim. Acta*, **3**, 177 (1930).
[206] Smith, A. E., G. N. Schrauzer, V. P. Mayweg, and W. Heinrich, *J. Am. Chem. Soc.*, **87**, 5798 (1965).
[207] Smith, E. L., *Nature*, **161**, 638 (1948).
[208] Smith, G. S., and J. L. Hoard, *J. Am. Chem. Soc.*, **81**, 556 (1959).
[209] Speakman, J. C., *Acta Cryst.*, **6**, 784 (1953).
[210] Strandberg, B., I. Lindqvist, and R. Rosenstein, *Z. Krist.*, **116**, 266 (1961).
[211] Sugano, S., *J. Chem. Phys.*, **33**, 1883 (1960).
[212] Swallow, A. G., and M. R. Truter, *Proc. Chem. Soc.*, 166 (1961); *Proc. Roy. Soc.* (*London*), **A266**, 527 (1962).
[213] Swink, L. N., and M. Atoji, *Acta Cryst.*, **13**, 639 (1960).
[214] Takano, T., Y. Sasada, and M. Kakudo, *Acta Cryst.*, **21**, 514 (1966).
[215] Tanford, C., and J. Epstein, *J. Am. Chem. Soc.*, **76**, 2170 (1954).
[216] Taylor, J. C., and M. H. Mueller, *Acta Cryst.*, **19**, 536 (1965).
[217] Tomita, K., *Bull. Chem. Soc. Japan*, **34**, 280, 286 (1961).
[218] Tulinsky, A., and C. R. Worthington, *Acta Cryst.*, **12**, 626 (1959).
[219] van Niekerk, J. N., and F. R. L. Schoening, *Acta Cryst.*, **5**, 196, 475, 499 (1952).
[220] van Niekerk, J. N., and F. R. L. Schoening, *Acta Cryst.*, **6**, 227 (1953).
[221] van Niekerk, and F. R. L. Scheoning, *Acta Cryst.*, **6**, 501 (1953).
[221a] Webb, L. E., and E. B. Fleischer, *J. Chem. Phys.*, **43**, 3100 (1965).
[222] Wiekliem, H. A., and J. L. Hoard, *J. Am. Chem. Soc.*, **81**, 549 (1959).
[223] Wilkes, C. E., and R. A. Jacobson, *Inorg. Chem.*, **4**, 99 (1965).
[224] Williams, D. E., G. Wohlauer, and R. E. Rundle, *J. Am. Chem. Soc.*, **81**, 755 (1959).
[225] Wolf, L., and H. Baringhausen, *Acta Cryst.*, **13**, 778 (1960).
[226] Woodward, P., L. F. Dahl, E. W. Akel, and B. C. Crosse, *J. Am. Chem. Soc.*, **87**, 5251 (1965).
[227] Zachariasen, W. H., and H. A. Pettinger, *Acta Cryst.*, **12**, 526 (1965).

2

VISIBLE AND ULTRAVIOLET SPECTROSCOPY

CURTIS R. HARE
State University of New York at Buffalo

The use of visible and ultraviolet spectroscopy to explain the structural aspects of chelates is a rather simple but powerful tool. Only since about 1955 have all the potentialities of the methods been realized, and this has largely come about through the collection of a vast amount of data. A complete survey of all these results is quite impossible, since over a thousand papers reporting spectral results are published each year. The purpose here is rather to survey the basic principles, discuss only a few specific references, and supply additional references.

The application of visible spectroscopy is limited to chelates of the transition metal ions, the lanthanides, and the actinides. Ultraviolet spectroscopy is more universal and can be useful in structural determinations of all chelates since they all absorb in this region. In a typical transition metal chelate the observed spectrum, in general, consists of a series of crystal field bands which are in the visible region and depend largely on the donor atom of the ligand and on the metal ion. The crystal field transitions are of two types: the more intense spin-allowed transitions and the lower intensity spin-forbidden transitions, which usually appear as shoulders on the spin-allowed transitions. The ultraviolet spectrum is complicated and consists of electronic transitions between the ligand and the metal (charge transfer), and also transitions within the ligand itself which are usually $\pi \rightarrow \pi^*$ or $\sigma \rightarrow \sigma^*$ transitions. The spectra of the non-transition metal ion chelates usually consist only of the charge transfer and ligand transitions. The ligand transitions in all cases are characteristic of the coordinated ligand and not of the free ligand; however, the

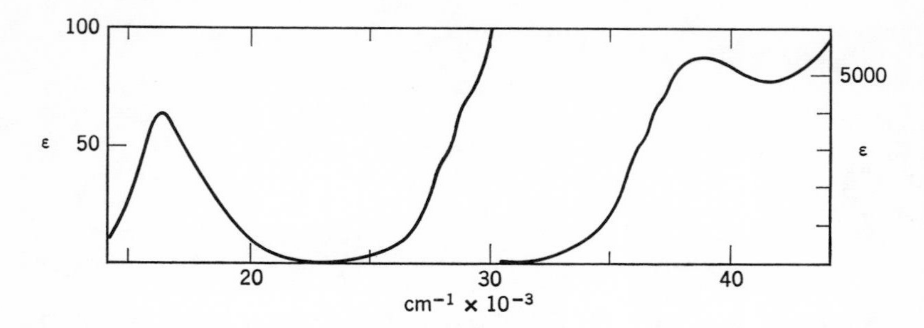

Fig. 2-1. The spectrum of bis(phenylglycino)copper(II).

spectrum of the free ligand aids in classifying the transitions of the coordinated ligand. The intensities of the crystal field transitions never exceed a molar extinction coefficient of 500 ($\varepsilon \leqslant 500$), whereas the charge transfer and ligand transitions in the ultraviolet usually exceed an extinction coefficient of 500 ($\varepsilon > 500$). The spectrum of bis(phenylglycino)copper(II) [250] (Fig. 2-1) is indicative of the types of transitions one may observe in a metal chelate. The transition at 16,000 cm^{-1} ($\varepsilon = 65$) is a crystal field transition centered largely on the metal ion (components of $^2E_g \rightarrow {}^2T_{2g}$), and the transitions at 36,000 cm^{-1} ($\varepsilon = 3000$) are characteristic of a bound phenyl group. The transition at 39,000 cm^{-1} ($\varepsilon = 6000$) has been assigned to a charge transfer apparently from the carboxyl group to the metal; the amino group-metal charge transfer and the carboxyl $\pi \rightarrow \pi^*$ transitions are at about 52,000 cm^{-1}. Very weak shoulders ($\varepsilon \sim 50$) in the 28,000 cm^{-1} region have been assigned to the spin-forbidden phenyl group transitions.

Interpretation of the results of such spectral determinations would require a complete molecular orbital treatment. Such treatments are rare, and the methods used for such computations are only approximate in nature. The present situation is that the spectral results are used to test the theories, and the correlation of the spectrum with the theory gives a greater understanding of the bonding and interactions in chelates. The visible spectra of transition metal ion chelates, however, can be well understood and described quantitatively by crystal field theory or its extension, ligand field theory. Thus this chapter is largely restricted to chelates of the transition metal ions because of the greater understanding which may be gained from them.

2-1 THEORY

The relationship between the spectrum of a transition metal ion complex (chelate or non-chelate) and its color has been known for a long time. The series of complexes $[Ni(H_2O)_6]^{2+}$ (green), $[Ni(gly)_3]^-$ (blue), $[Ni(en)_3]^{2+}$

(violet), and $[Ni(bipy)_3]^{2+}$ (red) exemplifies the change in color (spectrum) with the change in the coordinated ligand. The colors arise from the minima of their absorption spectra (i.e., the transmitted light) which are observed in the visible spectrum (Fig. 2-2). In the series mentioned, the features of the

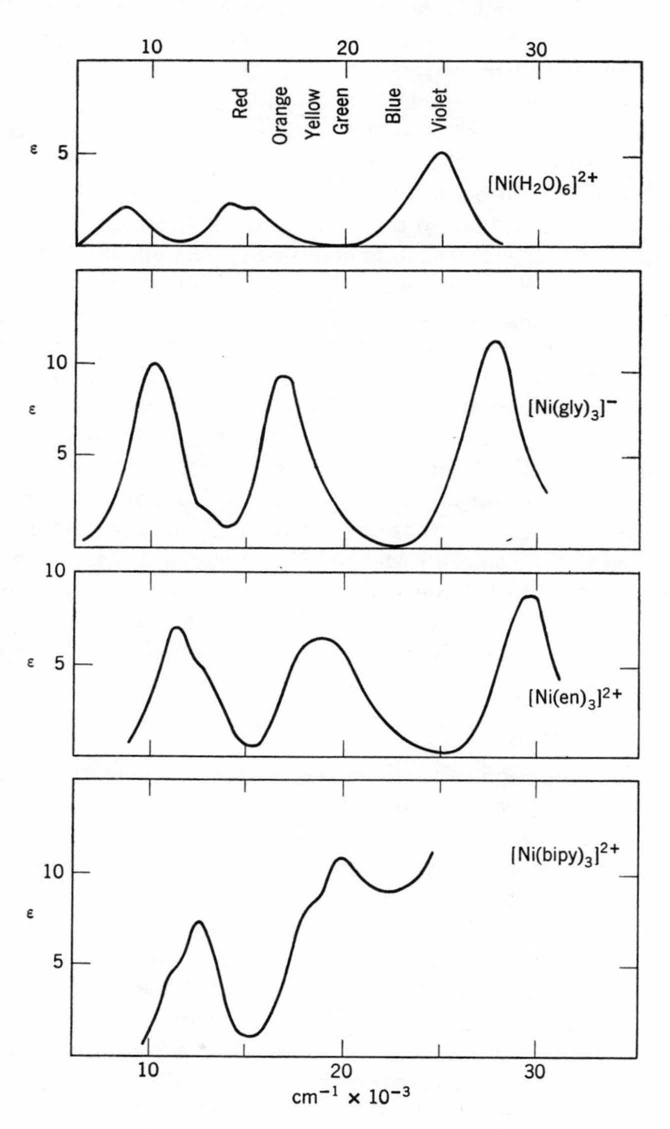

Fig. 2-2. The color and spectra of several nickel(II) chelates.

absorption bands are similar, and all the bands are progressively shifted to higher energies (as are the minima) in order of the complexes. Tsuchida [243] and Fajans [110] were the first to propose that the ligands could be listed in order of increasing energy of their absorption bands; this is the basis of the well-known spectrochemical series. The origins of the bands were first explained by Finkelstein and VanVleck [123] but not fully accepted until Ilse and Hartmann [150] demonstrated that the spectrum of $[Ti(H_2O)_6]^{3+}$ could be explained on the basis of Bethe's [51] splitting of atomic terms (states) by crystalline fields. Other workers—Orgel, Bjerrum, Jørgensen, and Ballhausen —then quickly showed that the concept was common to the explanation of the spectra of the transition metal ion complexes. The origins of the colors of these complexes were explained on the basis of an electrostatic (crystal field) model in which the charge donation could be correlated with position of the ligand in the spectrochemical series. The explanation of the spectrochemical series, that is, why a ligand occupies the position it does, has evolved from a more extensive theory (molecular orbital theory) which treats the complex as a whole. Few attempts have been made to correlate the ultraviolet spectra of complexes with available theories. "Charge transfer" spectra and "ligand" spectra are not as well understood as are crystal field transitions. Several comprehensive books treat crystal field, ligand field, and molecular orbital theory as applied to transition metal ions. In an approximate order of increasing difficulty they are those by Orgel [13], Dunn [4], Ballhausen and Gray [2], Dunn, Pearson, and McClure [5], Figgis [8], Jørgensen [10, 11], Ballhausen [1], Griffith [9], Stevenson [15], and Watanabe [16]. The review book by Jørgensen [12] is also highly recommended for those interested in the chemistry of inorganic complexes and the application of spectra to structural problems.

A. Crystal Field Theory

A complete description of the origins of crystal field theory is beyond the scope of this review; however, the basic concepts are really simple and are pertinent to understanding the origins of the spectra of transition metal chelates. Before a quantitative description of the theory is presented, it will be described qualitatively.

Consider first any of the transition metal ions in field-free space. It is known from atomic spectra that the five d orbitals are degenerate; that is, they have the same energy. These five orbitals are designated in Cartesian coordinates as the d_{z^2}, $d_{x^2-y^2}$, d_{xy}, d_{xz}, and d_{yz} orbitals. If the ion is exposed to a negative electrostatic field, the energy of the system (ion and field) is increased. If the field is spherical in shape, the only effect is an increase in the energy of the system, but, if the field is non-spherical, then the interaction has directional properties. The imposition of a field of point charges which has the shape of a regular octahedron increases the energy and removes the degeneracy of the set of *d* orbitals. If the octahedral field is imposed along the positive and negative directions of the original Cartesian coordinate system, the interaction is at a maximum in these directions and at a minimum in the other directions. The d_{z^2} and $d_{x^2-y^2}$ orbitals are directed along the Cartesian axes, and therefore their interaction with the field is maximal and their energy is increased. The d_{xy}, d_{xz}, and d_{yz} orbitals lie between the point charges; hence their interaction is minimal. Since the choice of *x*, *y*, and *z* axes is arbitrary, the orbitals d_{z^2} and $d_{x^2-y^2}$ are degenerate, and the orbitals, d_{xy}, d_{xz}, and d_{yz}, are also degenerate. The results depicted in Fig. 2-3 are independent of the coordinate system chosen: a set of two orbitals of higher energy and three of lower energy. The separation between them, Δ, is derived from the basis set

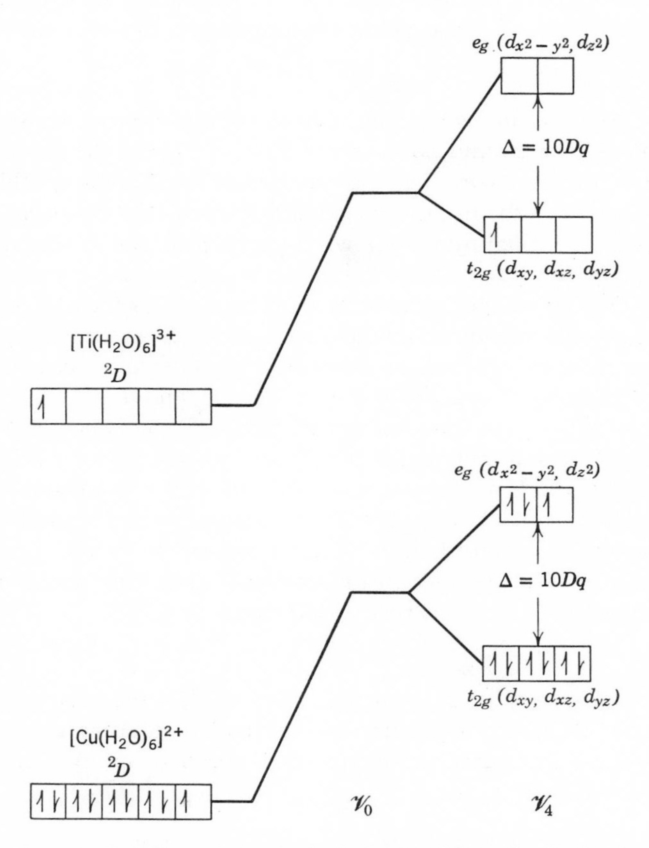

Fig. 2-3. The effect of the octahedral potential on the *d* orbitals.

of five orbitals. The same result is predicted from an application of group theoretic considerations. The sets of orbitals are conveniently labeled by their group theoretical representations; the doubly degenerate set is e_g, the threefold degenerate set is t_{2g}.

This is a qualitative description of Bethe's [51] splitting of atomic terms and forms the basis for the interpretation of the spectrum of $[Ti(H_2O)_6]^{3+}$ by Ilse and Hartmann [150]. The complex ion $[Ti(H_2O)_6]^{3+}$ is a system of one d electron interacting with an octahedral field of water dipoles. Population of one electron in the five degenerate d orbitals gives rise to a ^2D ground state term for Ti^{3+}. The octahedral field splits the orbitals as above, and the electron populates the orbitals of lowest energy, that is, the t_{2g} orbitals. The accepted notation is to label orbitals with small letters and the states derived by population of the orbitals with capital letters; thus the ground state of $[Ti(H_2O)_6]^{3+}$ is $^2T_{2g}(D)$. The transition $^2T_{2g}(D) \rightarrow {}^2E_g(D)$ is the absorption band at 20,000 cm^{-1}. The same description applies to the complex

$$[Cu(H_2O)_6]^{2+}.$$

This is a system of nine d electrons and it has one electron void, or hole, in the d orbitals. The ground state term of Cu^{2+} is ^2D and the orbitals are split as before by the octahedral field. Population of the t_{2g} and e_g orbitals leaves the hole in the e_g orbital, and the ground state of $[Cu(H_2O)_6]^{2+}$ is $^2E_g(D)$. If an electron is excited from the lower t_{2g} orbitals to the e_g orbitals, then the hole is left in the t_{2g} orbitals. This process corresponds to the transition $^2E_g(D) \rightarrow {}^2T_{2g}(D)$ which occurs at approximately 12,500 cm^{-1}. It illustrates an important principle in crystal field theory: the d^1 system is equivalent in description to d^9 except that the states (not orbitals) are inverted. The same principle applies to d^2 and d^8, d^3 and d^7, and d^4 and d^6.

The quantitative description of crystal field splitting is based on quantum mechanics. The Hamiltonian of the complex is considered to be $\mathscr{H} = \mathscr{H}_0 + \mathscr{V}$ in which $\mathscr{H}_0$ is the Hamiltonian of the free ion and $\mathscr{V}$ is the potential of the interacting field. The latter term is small compared with the total energy and can be treated as a perturbation. The generalized potential is expanded in spherical harmonics in which only the zero and fourth order terms are retained for an octahedral field interacting with d electrons:

$$\mathscr{V} = \mathscr{V}_0 + \mathscr{V}_4. \tag{2-1}$$

The term $\mathscr{V}_0$ is the spherical potential term which is the dominant term and contributes to the heat of formation of the complex (total energy) but, because of its spherical symmetry, is unimportant spectroscopically. In Cartesian coordinates the term $\mathscr{V}_4$ is

$$\mathscr{V}_4 = \frac{35eZ}{4R^5}\left(x^4 + y^4 + z^4 - \tfrac{3}{5}r^4\right), \tag{2-2}$$

in which e is the charge of each point, R is the distance from the ion to the charge, and x, y, z, and r are the coordinates of the electron. There are two solutions for $\langle \Psi | \mathscr{V}_4 | \Psi \rangle$, in which the Ψ's are d electron wave functions:

$$E_+ = 6Dq \qquad \text{and} \qquad E_- = -4Dq. \tag{2-3}$$

Here Dq is the crystal field parameter and is equal to $Ze^2\overline{r^4}/6R^5$ ($\overline{r^4}$ is the average of the fourth power of a d orbital radius). The eigenfunctions of E_+ are d_{z^2} and $d_{x^2-y^2}$, and d_{xy}, d_{xz}, and d_{yz} are eigenfunctions of E_-. This gives the separation Δ as $10Dq$, which may then be determined as an experimental parameter. Dq is then a measure of the point charge or dipole of the ligand and also depends on the central metal ion ($\overline{r^4}$). A knowledge of $\overline{r^4}$ (from Slater or Hartree-Fock orbitals), of the charge or dipole moment of the ligand, and of the metal-ligand distance allows a calculation of Dq. The values of Dq calculated by this method usually do not agree with the experimental values by an order of magnitude. Also, as the charge on the metal ion increases, $\overline{r^4}$ decreases more rapidly than R^5, and this model predicts lower Dq for the higher charged metal ions, whereas experimentally the opposite is true. The failure of this model has led to the adoption of ligand field theory which takes Dq as an experimentally determined parameter.

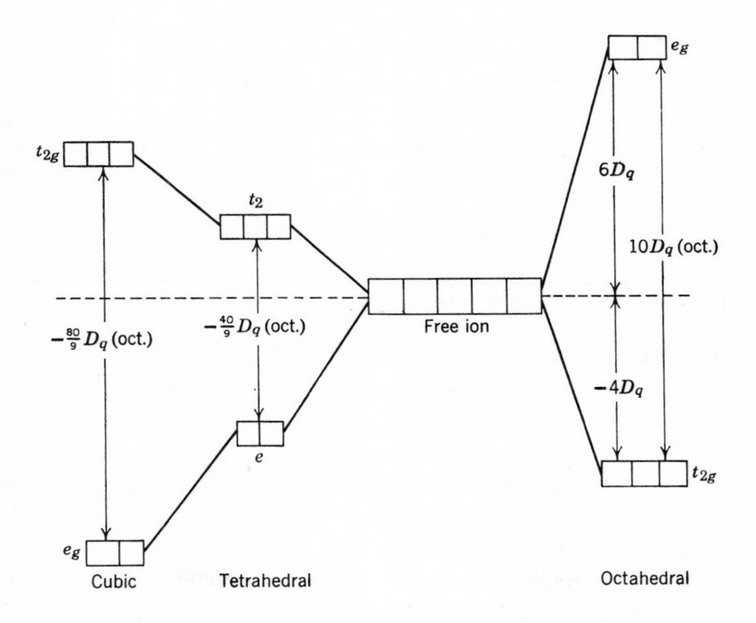

Fig. 2-4. Splittings of d orbitals in octahedral, tetrahedral, and cubic fields.

These arguments are specific for octahedral fields. Use of the same coordinate system for a cubic environment of eight points will lead to the result: $\mathscr{V}_{\text{cubic}} = -\frac{8}{9}\mathscr{V}_{\text{oct.}}$, which means that $Dq(\text{oct.}) = -\frac{9}{8}Dq(\text{cubic})$. Also, since a tetrahedron can be considered to be half a cube, $Dq(\text{oct.}) = -\frac{9}{4}Dq(\text{tet.})$. This expression means that for the same metal ion and ligand a cubic or tetrahedral environment gives a reversal of the order of the orbitals (and states), and that the splittings of the orbitals are $\frac{8}{9}$ or $\frac{4}{9}$ of the octahedral values if the metal-ligand distances remain the same (Fig. 2-4). The octahedral values of Dq for the more common donor and metal ions are given in Table 2-1.

Table 2-1 Values of Dq (cm^{-1}) for Octahedral Symmetry of Selected Transition Metal Ions and Ligands[a]

		Ligands			
Configuration	Ion	Cl^-	H_2O	NH_3	en
d^1	Ti^{3+}	1550	2030		
d^2	V^{3+}	1300	1840		
d^3	V^{2+}		1235		
	Cr^{3+}	1318	1750	2150	2188
d^4	Cr^{2+}	(1300)	1410		1800
	Mn^{3+}	(2000)	2100		
d^5	Mn^{2+}	750	850		1010
	Fe^{3+}	(1100)	1430		
d^6	Fe^{2+}		(1040)		
	Co^{3+}		(1820)	2290	2320
	Rh^{3+}	2032	2700	3410	3460
	Ir^{3+}	2500			4140
d^7	Co^{2+}		950	1050	(1100)
d^8	Ni^{2+}	750	860	1080	1120
d^9	Cu^{2+}		1250	(1500)	(1640)

[a] Parentheses indicate less certain values.

B. Ligand Field Theory

The differences between ligand field theory and crystal field theory are subtle. Crystal field theory considers the ligand only as a point dipole and neglects any other interaction such as covalent bonding. Ligand field theory considers the extent of metal-ligand interaction by the parameter Dq, which is an experimentally determined parameter. Theoretically Dq can be evaluated only by taking all metal-ligand interactions into consideration.

The splitting of the metal ion terms of all atomic configurations by the field of the ligands is treated in a more sophisticated manner in ligand field theory. The theoretical treatment which is most general and useful for the interpretation of the visible spectra of chelates is that of Tanabe and Sugano [241]. This highly successful theory depends on only three parameters for the interpretation: *Dq*, the crystal field or ligand field parameter, which has the same significance here as in crystal field theory, and the Racah [200] electron repulsion parameters, *B* and *C*. The Racah parameters for the free metal ion are determined from the term separations as observed in the atomic spectrum of the ion. Theoretically these parameters are the numerical values of the coulombic repulsion and exchange integrals between electrons in various states. They are a measure of the natural separation of the terms of the metal ion. The values of *B* and the ratio *C*/*B* for free, uncoordinated ions are given in Table 2-2. The ratio *C*/*B* is essentially constant at a value of 4.4 for the first row transition metal ions.

Table 2-2 The Racah Electron Repulsion Parameters (*B*) (cm^{-1}) and Ratios *C*/*B* for Gaseous (Free) Transition Metal Ions[a]

Ion	2+		3+		4+	
Metal	*B*	*C*/*B*	*B*	*C*/*B*	*B*	*C*/*B*
Ti	695	4.19				
Zr	539	3.3				
Hf						
V	755	4.31	861	4.43		
Nb	532	2.54	602	4.4		
Ta						
Cr	810	4.40	918	4.50	1039	4.07
Mo			(610)		(682)	
W						
Mn	860	4.48	965	4.61	(1064)	
Tc						
Re					(650)	
Fe	917	4.41	1015	4.73	1144	3.90
Ru	(620)					
Os					(700)	
Co	970	4.40	1065	4.81		
Rh	(620)		(720)			
Ir			(660)			
Ni	1030	4.71	1115	4.89		
Pd	(683)		(720)			
Pt			(660)		(720)	
Cu	1238	3.78				
Ag						
Au						

[a] Values obtained by extrapolation and uncertain values are given in parentheses.

Tanabe and Sugano [241] have computed the matrix elements for the state energies (diagonal elements) and the repulsion terms (off-diagonal elements) for each state that is possible in cubic symmetry for the d^2, d^3, d^4, and d^5 electronic configurations. Because these matrices are given in the original paper [241] and in several books [9, 23, 15], only the matrices for d^2 are given here (Table 2-3). The matrices for d^2, d^3, and d^4 are identical with

Table 2-3 Tanabe-Sugano Energy Matrices for the States of d^2 in Cubic Symmetry

State	Cubic Field Configuration	Matrix	
$^3A_2$	e^2	$-8B + 12Dq$	
$^3T_2$	t_2e	$-8B + 2Dq$	
$^3T_1$	$t_2{}^2$	$-5B - 8Dq$	$6B$
	t_2e	$6B$	$4B + 2Dq$
^1E	$t_2{}^2$	$B + 2C - 8Dq$	$-2\sqrt{3}\,B$
	e^2	$-2\sqrt{3}\,B$	$2C + 12Dq$
$^1T_2$	$t_2{}^2$	$B + 2C - 8Dq$	$2\sqrt{3}\,B$
	t_2e	$2\sqrt{3}\,B$	$2C + 2Dq$
$^1T_1$	t_2e	$4B + 2C + 2Dq$	
$^1A_1$	$t_2{}^2$	$10B + 5C - 8Dq$	$\sqrt{6}\,(2B + C)$
	e^2	$\sqrt{6}\,(2B + C)$	$8B + 4C + 12Dq$

those for d^8, d^7, and d^6, respectively, and may be used for the latter configurations, but it must be recalled that, in the latter configurations, the values of Dq are negative for octahedral symmetry. These matrices may also be used for tetrahedral or cubic systems, but again the appropriate sign convention for the parameter Dq must be observed. The matrices for the states of d^2, shown in Table 2-3, are simple and serve as good examples for the application of the theory. These equations are quadratic, and the roots (eigenvalues) may be obtained by use of the binomial theorem. The matrices for the higher d electron configurations are of higher order and are best solved by using computers. It can be seen from the diagonal elements of the d^2 matrices that one of the states of $^3T_1$ has the lowest energy (most negative) and is the ground state when Dq is positive (e.g., V^{3+}). However, when Dq is negative (e.g., Ni^{2+}), the ground state is $^3A_2$. Solution of the determinantal equations is achieved by diagonal multiplication and equating to zero. For

Ni^{2+} complexes (d^8) the ground state $^3A_2$ is chosen to have zero energy, and with the simplification $-10\,Dq = \Delta$ the energies of the important states become

$$
\begin{aligned}
^3A_2 &= 0, \\
^3T_2 &= \Delta, \\
{}^3_aT_1 &= \frac{15B}{2} + \frac{3\Delta}{2} - \frac{1}{2}[(9B - \Delta)^2 + 144B^2]^{1/2}, \\
{}^3_bT_1 &= \frac{15B}{2} + \frac{3\Delta}{2} + \frac{1}{2}[(9B - \Delta)^2 + 144B^2]^{1/2}, \\
{}^1_aE &= \frac{17B}{2} + 2C + \Delta - \frac{1}{2}[(B + 2\Delta)^2 + 48B^2]^{1/2}, \\
{}^1_aT_2 &= \frac{17B}{2} + 2C + \frac{3\Delta}{2} - \frac{1}{2}[(B + \Delta)^2 + 48B^2]^{1/2}, \\
{}^1_aA_1 &= 17B + \frac{9C}{2} + \Delta - \frac{1}{2}[(10B + 5C + 2\Delta)^2 - \Delta(32B + 16C)]^{1/2}.
\end{aligned}
\tag{2-4}
$$

These equations predict three transitions to the spin-allowed triplet states $^3T_2$, ${}^3_aT_1$, and ${}^3_bT_1$ in order of increasing energy, and the energy of the first transition is Δ (or $-10\,Dq$). Only the singlet states contain the parameter C, and separation of the two $^3T_1$ states is $[(9B - \Delta)^2 + 144B^2]^{1/2}$. Equations (2-4) may be applied to the d^2 case (e.g., V^{3+}), where the lowest $^3T_1$ state is the ground state:

$$
\begin{aligned}
{}^3_aT_1 &= 0, \\
^3T_2 &= -\frac{15B}{2} + \frac{\Delta}{2} + \frac{1}{2}[(9B + \Delta)^2 + 144B^2]^{1/2}, \\
^3A_2 &= -\frac{15B}{2} + \frac{3\Delta}{2} + \frac{1}{2}[(9B + \Delta)^2 + 144B^2]^{1/2}, \\
{}^3_bT_1 &= [(9B + \Delta)^2 + 144B^2]^{1/2}, \\
{}^1_aE &= B + 2C + \frac{\Delta}{2} + \frac{1}{2}[(9B + \Delta)^2 + 144B^2]^{1/2} \\
&\quad - \frac{1}{2}[(B - 2\Delta)^2 + 48B^2]^{1/2}, \\
{}^1_aT_2 &= B + 2C + \frac{1}{2}[(9B + \Delta)^2 + 144B^2]^{1/2} \\
&\quad - \frac{1}{2}[(B - \Delta)^2 + 48B^2]^{1/2}, \\
{}^1_aA_1 &= \frac{19B}{2} + \frac{9C}{2} + \frac{\Delta}{2} + \frac{1}{2}[(9B + \Delta)^2 + 144B^2]^{1/2} \\
&\quad - \frac{1}{2}[(10B + 5C - 2\Delta)^2 + \Delta(32B + 16C)]^{1/2}.
\end{aligned}
\tag{2-5}
$$

The energy difference between $^3T_2$ and $^3A_2$ is, of course, Δ (or $10Dq$), but it is the difference in energy between the first and second spin-allowed bands.

It is well known that Tanabe-Sugano matrices do not fit the observed spectrum of a complex with high accuracy if the free ion values are used for the parameters B and C. The values of these parameters used to describe the energy levels of a complex are always less than those for a free ion; however,

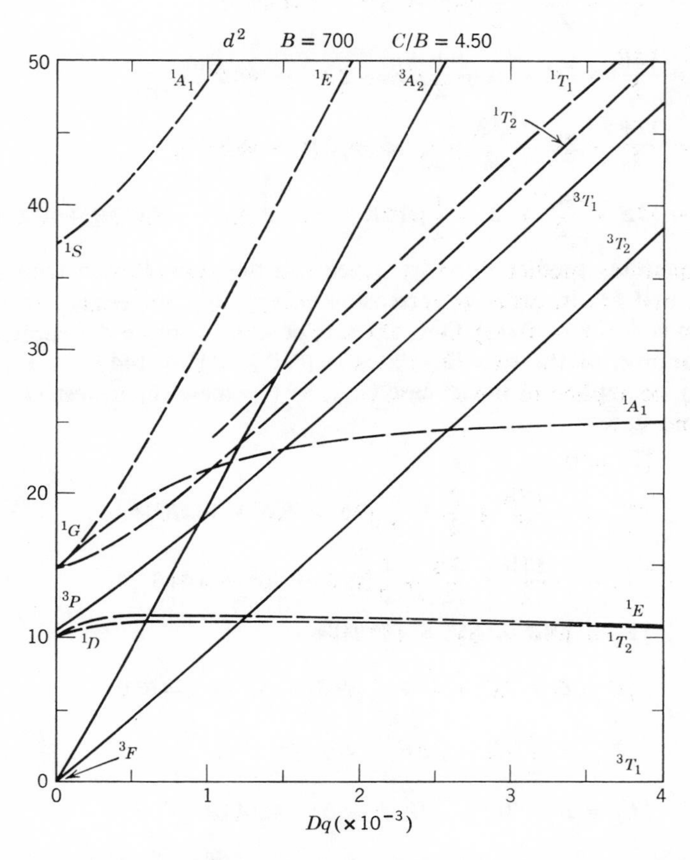

Fig. 2-5. Energy levels of the d^2 configuration as a function of Dq for octahedral fields. The ground state is $^3T_{1g}(F)$, and three spin-allowed transitions to $^3T_{2g}(F)$, $^3A_{2g}(F)$, and $^3T_{1g}(P)$, respectively, are possible. In addition, weak shoulders may appear in the spectrum; they would correspond to the spin-forbidden transitions. $^1T_{2g}(D)$, $^1E_g(D)$, and $^1A_{1g}(G)$ may appear in the visible spectrum. The energy separation between $^3T_{2g}(F)$ and $^3A_{2g}(F)$ is 10 Dq. This diagram may be used for tetrahedral complexes of the d^8 configuration.

the ratio C/B is usually very close to the free ion value. The reduction of these parameters from the free ion values has been attributed to covalent effects by Jørgensen [21]. He defines the ratio $\beta = B'/B$, in which B' is the parameter determined from the spectrum of the complex and B is the free ion value, as a measure of covalency in the complex. This ratio is, then, a measure of the extent of reduction of the separation of atomic terms on forming the complex and also is the basis of the nephelauxetic series, which is discussed later.

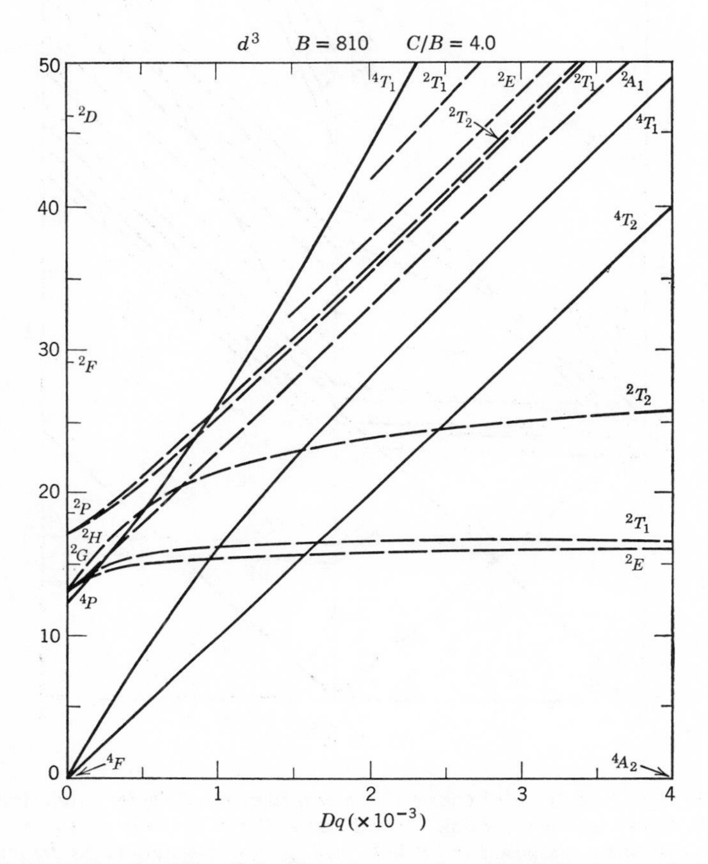

Fig. 2-6. Energy levels of the d^3 configuration as a function of Dq for octahedral fields. The ground state is $^4A_{2g}(F)$, and three spin-allowed transitions to $^4T_{2g}(F)$, $^4T_{1g}(F)$, and $^4T_{1g}(P)$ are possible (usually only the lowest two are observed). Weak shoulders may appear in the visible spectrum; they would correspond to the spin-forbidden transitions $^2E_g(G)$, $^2T_{1g}(G)$, and $^2T_{2g}(G)$. The first spin-allowed transition $^4A_{2g}(F) \rightarrow {}^4T_{2g}(F)$ is $10Dq$. This diagram may also be used to correlate the spectra of tetrahedral complexes of the d^7 configuration.

The values of β are always less than unity and usually range between 0.60 and 0.80.

The Tanabe-Sugano matrices have been solved for the d^2 through d^8 configurations and are presented in graphical form in Figs. 2-5 to 2-11. These

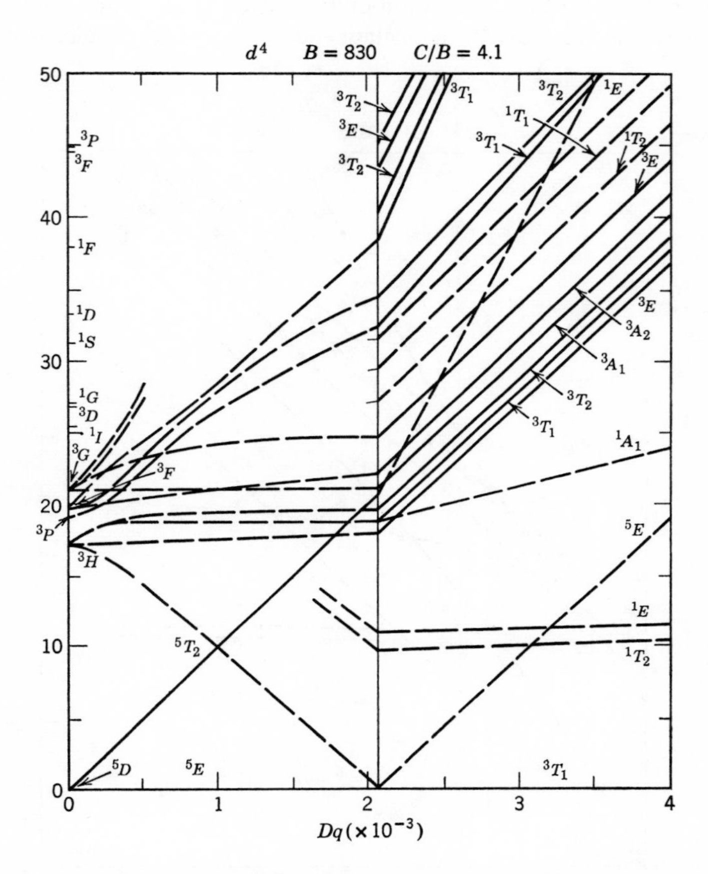

Fig. 2-7. Energy levels of the d^4 configuration as a function of Dq for octahedral fields. This configuration has two possible ground states. For low values of Dq the ground state is $^5E_g(D)$ and corresponds to the high spin $t_{2g}{}^3e_g$ configuration. As Dq increases, the $^3T_{1g}(H)$ state is stabilized and becomes the ground state; this corresponds to the low spin (spin-paired) $t_{2g}{}^4$ configuration. Neglecting off-diagonal elements, the crossover point is at $-10Dq + 6B + 5C = 0$ (indicated by a vertical line). For high spin one transition $^5E_g(D) \rightarrow {}^5T_{2g}(D)$ is expected, and this corresponds to $10Dq$; any others would be weak, spin-forbidden transitions. For low spin the high energy transitions to $^3E_g(D)$, $^5T_{1g}(P)$, $^3T_{2g}(D)$, $^3A_{1g}(G)$, $^3A_{2g}(F)$ would be observed with spin-forbidden transitions in the visible region. This diagram may be used for tetrahedral d^6 complexes.

graphs are plots of the energy of a given state in units of a thousand wave numbers ($cm^{-1} \times 10^3$) as a function of the parameter Dq, which is also given in units of a thousand wave numbers. The states which have the same multiplicity as the ground state are represented by solid lines, and states which differ from the ground state in multiplicity, by dashed lines to indicate that

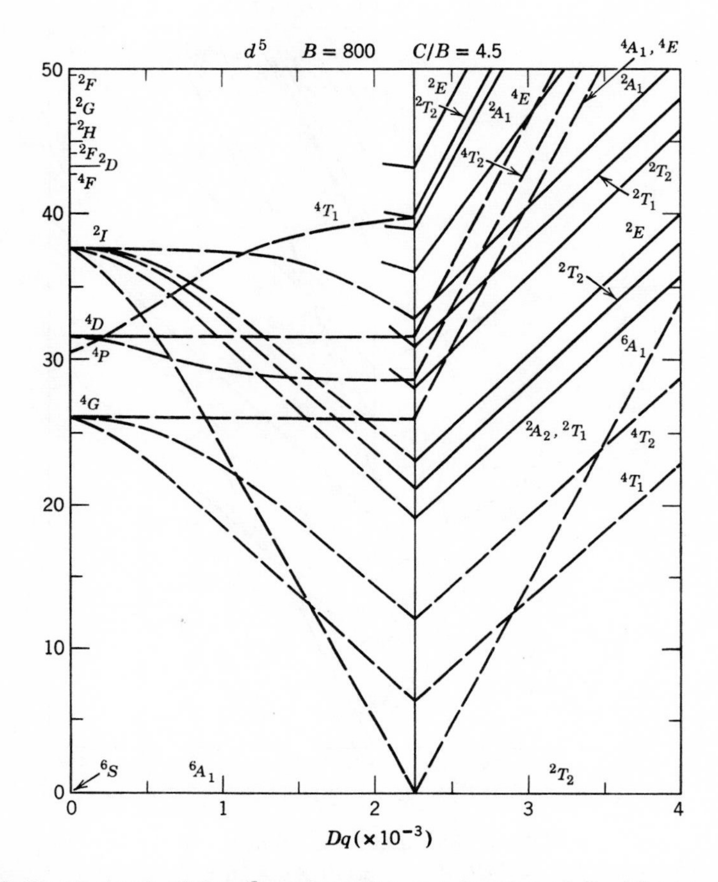

Fig. 2-8. Energy levels of the d^5 configuration as a function of Dq. The two possible ground states for this configuration correspond to the high spin, $^6A_{1g}(S)$ $(t_{2g}{}^3e_g{}^2)$, and low spin, $^2T_{2g}(I)$ $(t_{2g}{}^5)$, states. The crossover point is at $-20Dq + 15B + 10C = 0$, neglecting off-diagonal matrix elements. All excited states are spin-forbidden for the high spin configuration, and the spectrum will consist of weak transitions. For the low spin configuration the intense transitions will be at high energy and will correspond to transitions to $^2A_{2g}$, $^2T_{1g}$, ^2E_g, $^2T_{2g}$. Spin-forbidden transitions may be observed in the visible region. This diagram applies to both octahedral and tetrahedral complexes of the d^5 configuration.

transitions to these states are spin-forbidden.† The values of B which were used are reduced from the free ion values, while C/B has been given the free ion value. These values were selected to be useful to chelate chemists. It should be emphasized that these solutions are sensitive to the values of B as

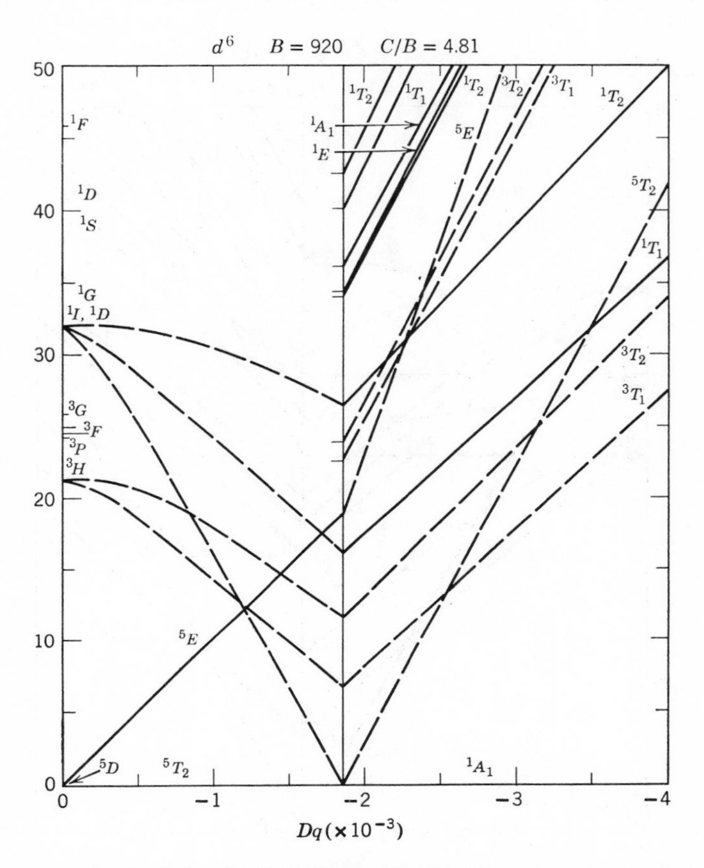

Fig. 2-9. Energy levels of the d^6 configuration as a function of Dq for octahedral fields. The crossover point for the high spin, $^5T_{2g}(D)$ $(t_{2g}{}^4e_g{}^2)$, and low spin, $^1A_{1g}(I)$ $(t_{2g}{}^6)$, ground states occurs at $20Dq + 5B + 8C = 0$. For high spin only one spin-allowed transition $^5T_{2g}(D) \rightarrow {}^5E_g(D)$ is observed and it corresponds to $10Dq$. Two high-energy spin-allowed transitions from $^1A_{1g}$ to $^1T_{1g}$ and $^1T_{2g}$ will be observed for low spin. The spin-forbidden transitions to $^3T_{1g}$ and $^3T_{2g}$ may appear as shoulders on the $^1T_{1g}$ transition. This diagram may be used for d^4 tetrahedral complexes.

† The aid of Thomas P. Sleight in the solution of the matrices and preparation of the graphs is acknowledged.

are the more usual graphs of E/B versus Dq/B. The free ion atomic terms arising from the configuration of d electrons are given on the ordinate. For the sake of simplicity several of the spin-forbidden states have been deleted from the more complicated diagrams, as have been the g subscripts for octahedral states.

Liehr and Ballhausen [165] have extended the Tanabe-Sugano theory by the inclusion of the spin-orbit interaction (called complete matrices) in the

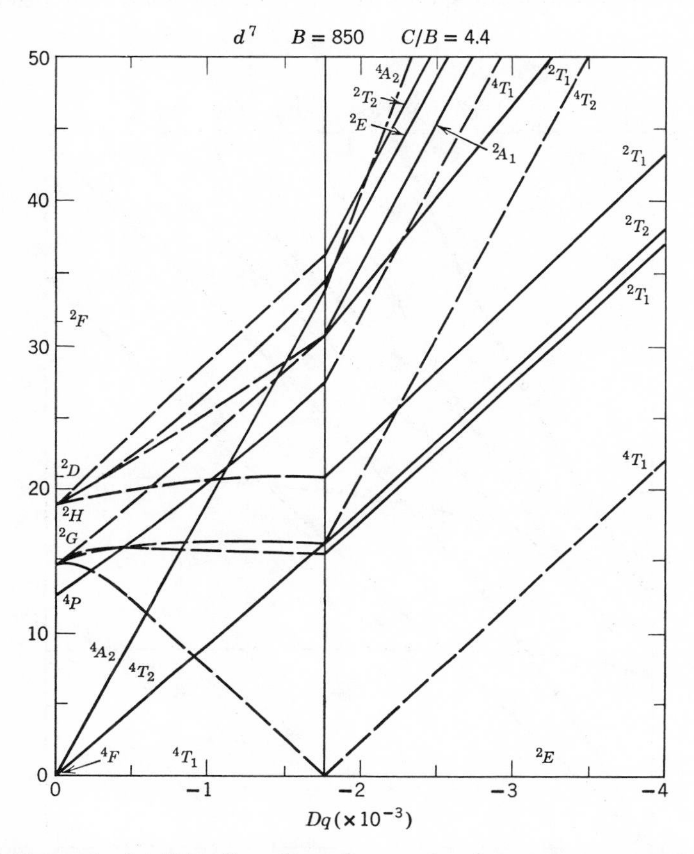

Fig. 2-10. Energy levels of the d^7 configuration as a function of Dq for octahedral fields. The crossover point for the high spin, ${}^4T_{1g}(F)$ $(t_{2g}{}^5e_g{}^2)$, and low spin, ${}^2E_g(H)$ $(t_{2g}{}^6e_g)$, ground states occurs at approximately $10\ Dq + 4B + 4C = 0$. For high spin the spin-allowed transitions to ${}^4T_{2g}(F)$, ${}^4A_{2g}(F)$, and ${}^4T_{1g}(P)$ will be observed. The energy difference between ${}^4T_{2g}(F)$ and ${}^4A_{2g}(F)$ is $10\ Dq$. Three high-energy spin-allowed transitions from ${}^2E_g$ to ${}^2T_{1g}$, ${}^2T_{2g}$, and ${}^2T_{1g}$ will be observed for low spin. One spin-forbidden transition ${}^4T_{1g}$ may be observed. This diagram may be used for d^3 tetrahedral complexes.

d^2 and d^8 systems. The spin-orbit matrix elements for the d^3 and d^7 configurations have been given by Runciman and Schroeder [206], Eisenstein [103], and Weaklein [249]. Schroeder has computed matrix elements for d^4 and d^6 configurations [230] and also the d^5 configuration [229]. Liehr [167] has given graphical solutions of the spin-orbit matrices of d^3 and d^7. Inclusion of the spin-orbit interactions complicates the graphical presentation, but they are important for the explanation of the low temperature spectra and shoulders observed in many spectra [165]. Also, the values of B' determined

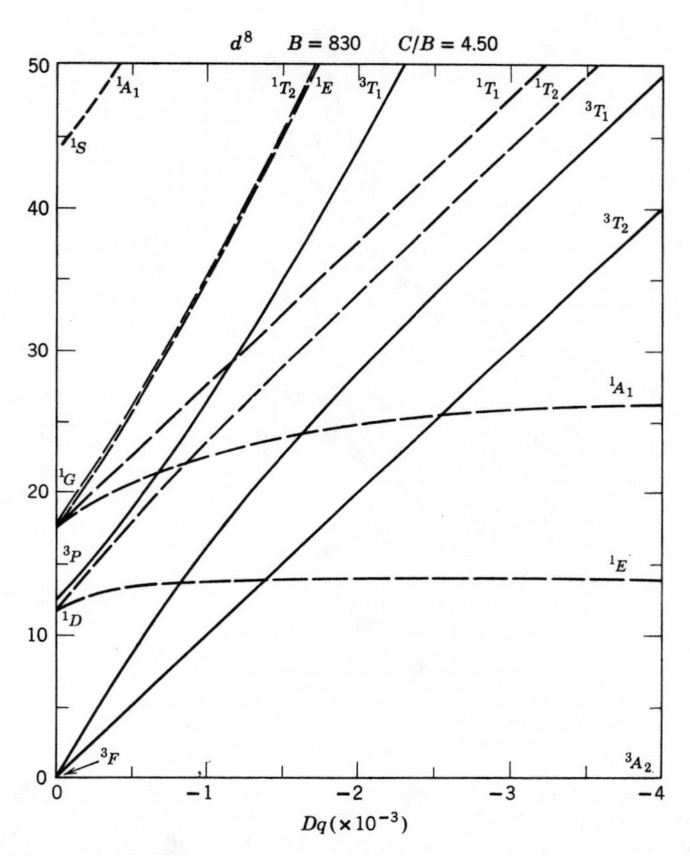

Fig. 2-11. Energy levels for the d^8 configuration as a function of Dq for octahedral fields. The ground state is $^3A_{2g}(F)$, and three spin-allowed transitions to $^3T_{2g}(F)$, $^3T_{1g}(F)$, and $^3T_{1g}(P)$ may be observed. In addition the spin-forbidden states, ^1E_g, $^1A_{1g}$, and $^1T_{2g}$, may appear as shoulders on the spin-allowed transitions. This diagram may be used for d^2 tetrahedral complexes.

by use of the spin-orbit matrices are usually smaller than those determined from the Tanabe-Sugano theory and are therefore less than the free ion values. The Tanabe-Sugano theory is hardly applicable to d^1 or d^9, since no repulsion of terms exists and Dq is the only parameter required. Liehr [166] has given the spin-orbit matrices for d^1 and d^9.

These theoretical treatments of Liehr and Ballhausen or Tanabe and Sugano are very good for the prediction of spectra of the first row transition metal ion chelates of approximate cubic symmetry up to about 30,000 cm^{-1}. At higher energy, more intense charge transfer and ligand transitions take place, and they are not predictable by these theories.

C. Molecular Orbital Theory

A complete molecular orbital theory for complexes should account for all interactions of the metal ion and the ligand. This would require a large basis set of metal and ligand orbitals and the solution of the Schrödinger equation for a very large number of electrons. Such a rigorous computation is beyond the means of present-day computational methods. The most rigorous molecular orbital computation for a transition metal ion complex is that of Watson and Freeman [248] on $KNiF_3$. Their treatment was specifically to determine $10Dq$ and was based on a Hartree-Fock Hamiltonian:

$$\mathscr{H} = -\tfrac{1}{2}\nabla^2 + V_m + V_l + V_s + V_\gamma + V_{sl}, \tag{2-6}$$

where ∇^2 is the kinetic energy and V_m and V_l are the metal and ligand coulombic and exchange interaction potentials. The potential V_s arises from the non-orthogonality of the metal and ligand orbitals, and the term V_{sl} treats the non-orthogonality of the ligand orbitals. V_γ is the covalent term; it considers contributions from bonding orbitals which have no antibonding partners, and accounts for the transfer of charge from the ligand to the metal by covalent mixing. The molecular orbitals (Ψ) were constructed from linear combination of SCF metal ion orbitals ψ_m and ligand functions ψ_l:

$$\Psi = N^{-1/2}(\psi_m + \lambda\psi_l). \tag{2-7}$$

The coefficients N and λ were determined by self-consistent methods, and all two- and three-center integrals were evaluated. Their results show that the important contributions to $10Dq$ are the off-diagonal (covalency or exchange) terms, and that even in a fluoride complex covalency is large. The σ-bonding dominates the energetics of the system, but the π-admixture is important in accounting for the reduction in the spin-orbit coupling constant, the value of the Racah parameter B, and the intensity of the spectral transitions. The results of Watson and Freeman [248] differ from those of similar computations of Sugano and Shulman [240]. In the Sugano and Shulman approach the antibonding electrons dominate the contributions to the physical observables, but Watson and Freeman contend that the bonding electrons, which have no antibonding partners, account for the covalent effects in the observables. Table 2-4 compares the results of Watson and Freeman with those of Sugano and Shulman for the various contributions to $10Dq$ in $KNiF_3$.

Table 2-4 Overlap and Covalent Contributions to $10Dq$ (cm^{-1}) for $KNiF_3$

	Watson-Freeman [248]	Sugano-Shulman [240]
s Contributions		
Overlap	2870	2070
Covalent	240	790
Total	3110	2860
$p\sigma$ Contributions		
Overlap	4680	2400
Covalent	5035	6170
Total	9715	8570
$p\pi$ Contributions		
Overlap	−1535	−730
Covalent	−1440	−1680
Total	−2975	−2410
Total overlap	6015	3740
Total covalent	3835	5380
Total	9850	9020

Observed $10Dq$ for $KNiF_3$ = 7250 cm^{-1}

The Wolfsberg-Helmholz [255] method for the computation of the molecular energy levels of a complex has been most widely used. This approximate, semiempirical method is described in detail by Ballhausen [1] and by Ballhausen and Gray [2]. In this method linear combinations of appropriate metal and ligand valence orbitals are taken (2-7) and adapted to the symmetry of the complex. The metal orbitals belong to specific representations of the symmetry group, and the ligand orbitals also conform to this representation. The coefficient λ is determined by a consistent Mulliken [181] charge population analysis and is related to N by

$$N^{-2} = 1 + 2\lambda S_{ml} + \lambda^2, \tag{2-8}$$

where S_{ml} is the overlap integral

$$S_{ml} = \int \psi_m^* \psi_l \, d\tau. \tag{2-9}$$

A secular determinant is then constructed for each symmetry representation derived from the original basis set:

$$|\mathscr{H}_{ij} - S_{ij}E| = 0. \tag{2-10}$$

The $\mathscr{H}_{ii}$ are the coulomb integrals:

$$\mathscr{H}_{ii} = \int \psi_i^* \psi_i \, d\tau, \tag{2-11}$$

and are estimated from the valence state ionization energies [13] of the metal and the ligand.

The exchange integrals are then determined by the relation

$$\mathscr{H}_{ij} = \mathscr{H}_{ji} = FS_{ij} \frac{\mathscr{H}_{ii} + \mathscr{H}_{jj}}{2}. \tag{2-12}$$

The coefficient F is normally given a value of 2. Several alternative methods have been suggested for the estimation of the exchange (or resonance) integrals. Ballhausen and Gray [39] have suggested the use of the geometric mean:

$$\mathscr{H}_{ij} = 2S_{ij}(\mathscr{H}_{ii} \cdot \mathscr{H}_{jj})^{1/2}, \tag{2-13}$$

and Cusachs [78] has suggested the theoretically more justifiable

$$\mathscr{H}_{ij} = \tfrac{1}{2}S_{ij}(\mathscr{H}_{ii} + \mathscr{H}_{jj})(F - |S_{ij}|). \tag{2-14}$$

Again, F is usually set equal to 2. Iteration of (2-8) and (2-10) to consistent charge, which is of the order of +0.5 for the trivalent complexes, gives the λ's, which are a measure of the ligand character of the state, and also gives the energies of the states (see Fig. 2-12). The results of these calculations usually correlate well with the spectrum, and this is usually a criterion of the validity of the computations. However, the method has been criticized [113, 10] for being too empirical and not correlating with other experimental observables except the spectrum. Cotton and Haas [71] have also discussed the sensitivity of the solutions to the choice of parameters $\mathscr{H}_{ii}$ and F and to the method of calculating $\mathscr{H}_{ij}$. The wavefunctions used for molecular orbital calculations have been the more exact Watson [247] Hartree-Fock orbitals, the double exponential orbitals of Richardson et al. [204, 205], the "best" Slater-type orbitals of Clementi and Raimondi [68], or Slater [232] orbitals.

Fenske and his co-workers [114] have recently extended the Sugano-Shulman method to include excited states, but analytical approximation techniques are used for the evaluation of some of the off-diagonal integrals. They have kept the Mulliken self-consistent charge analysis of the Wolfsberg-Helmholz technique but with a more limited basis. Reasonable charges (+2 for trivalent ions) have been calculated by this method, and it gives better

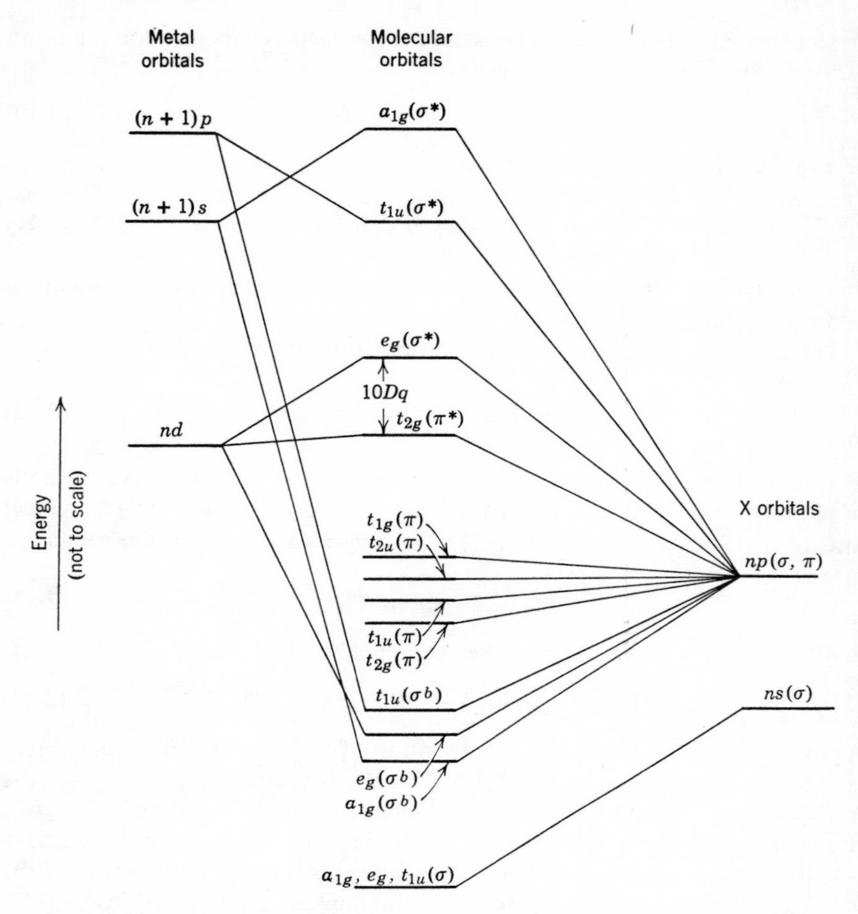

Fig. 2-12. Molecular orbital scheme for MX_6^{n-} complex (X = halide ion).

agreement with more observables than does the Wolfsberg-Helmholz method. It will, perhaps, be extended to more systems.

Another approximate scheme which has had some success in the explanation of the ultraviolet spectrum of metal β-diketone chelates [43, 107] is the Hückel LCAO-MO method. This method considers only the π electrons of the complex. In Barnum's [43] treatment of the trivalent metal acetylacetonates a secular determinant similar to (2-10) is set up for the metal t_{2g} and ligand π electrons. This determinant is reduced by symmetry considerations, and overlap between even adjacent orbitals is neglected. Then only the coulombic ($\mathscr{H}_{ii}$) and exchange ($\mathscr{H}_{ij}$) integrals remain to be evaluated in the secular determinant. The exchange integrals are estimated from bond energies. For example, the exchange integral between two carbon atoms in the chelate is the difference between the bond energy of a carbon-carbon double bond and single bond. The metal-oxygen exchange integral is left as a parameter to be varied to give best fit with the spectrum. The coulombic integrals are estimated from the electronegativities of the atoms, and only one chelate ring is considered.

Figure 2-13, taken from Barnum's results on Co(acac)$_3$, shows the effect on the energy levels of an increase in the metal-oxygen exchange (covalency) interaction. Correlation of the spectrum with this diagram at $\mathscr{H}_{ij} = -14{,}000$ is satisfactory (Table 2-5). Again metal-ligand covalency is predicted to be large.

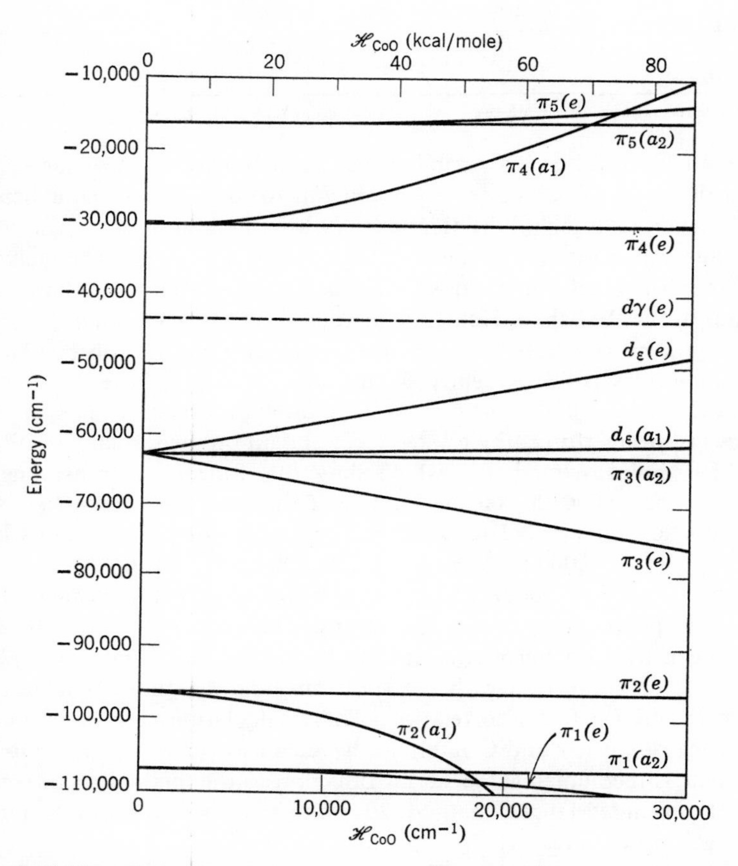

Fig. 2-13. The effect of metal-oxygen π-bonding on the energy levels of Co(acac)$_3$.

Table 2-5 Comparison of Calculated and Observed Transition Energies in some Tris(acetylacetono) Complexes[a] [43]

Transition	Complex $\mathscr{H}_{MO}$ (cm^{-1})[b]					
	Ti(acac)$_3$ $\mathscr{H}_{MO} > -2000$	V(acac)$_3$ $\mathscr{H}_{MO} = -6000$	Cr(acac)$_3$ $\mathscr{H}_{MO} = -11,000$	Mn(acac)$_3$ $\mathscr{H}_{MO} = -12,000$	Fe(acac)$_3$ $\mathscr{H}_{MO} = -12,500$	Co(acac)$_3$ $\mathscr{H}_{MO} = -14,00$
$\pi_3(e) \rightarrow \pi_4(a_1)$	32,400 (32,400)	~35,900 (35,600)	39,300 (40,600)	39,500 (42,400)	42,500 (41,800)	43,900 (44,000)
$\pi_3(e) \rightarrow \pi_4(e)$ $\pi_3(a_2) \rightarrow \pi_4(a_1)$		~34,500 (34,500) (33,500)	37,100 (37,100) (37,000)	~36,800 (38,300) (36,400)	36,800 (37,200) (37,400)	39,000 (38,800) (37,500)
$\pi_3(a_2) \rightarrow \pi_4(e)$		— (32,500)	33,900 (32,600)	— (32,300)	— (32,500)	~33,900 (32,300)
$d\varepsilon(a_1) \rightarrow \pi_4(e)$ $d\varepsilon(e) \rightarrow \pi_4(a_1)$	27,500 (30,000)	29,200 (30,400) (29,500)	30,200 (30,800) (29,800)	31,400 (32,000) (30,700)	28,500 (30,000) (29,500)	31,000 (31,700) (31,000)
$d\varepsilon(e) \rightarrow \pi_4(e)$		— (28,400)	26,300 (26,800)	~25,000? (26,700)	32,300? (25,000)	~25,000? (25,600)

[a] Calculated transition energies are in parentheses. [b] $\mathscr{H}_{MO}$ is the metal-ligand exchange integral.

This model is based on π overlap between the metal and the acetylacetonate ion, and the behavior is not exactly analogous to that of an organic heterocycle. These crude calculations are, however, very helpful in making spectral assignments and in estimating the bonding energies in chelates. The inclusion of overlap in the Hückel method is called the extended Hückel method and is analogous to the Wolfsberg-Helmholz method. The extended Hückel method has been quite successful in the calculation of several physical observables in organic molecules [147]. Recently Cotton and Wise [72] have applied the extended Hückel method to Cu(acac)$_2$ in an attempt to clarify the spectrum. Their results are reinforced by crystal spectral data on the similar Cu(DPM)$_2$ (H–DPM = dipivaloylmethane), which show that four $d \rightarrow d$ transitions are observed in the 15,000–20,000 cm^{-1} range. This result shows that the tetragonal distortion of the chelate is large. The higher energy transitions have been assigned: 26,000 cm^{-1} $\sigma \rightarrow d_{xy}$, 33,500 cm^{-1} (doublet) $\pi \rightarrow \pi^*$, 40,000 cm^{-1} $\sigma \rightarrow d_{xy}$, 48,600 cm^{-1} $d_{yz} \rightarrow \pi^*$ † on the basis of the molecular orbital energies and polarization of the transitions. The extended Hückel method has also been used on porphyrin chelates by Zerner and Gouterman [262]. Their results correlate well with the spectrum, and assignments have been made on the basis of this correlation. These calculations are of particular interest because Zerner and Gouterman have considered the chelate moiety. Their article is recommended as an introduction to the method. An iterative form of the extended Hückel method [203] which has been introduced should be applicable to chelates.

† This assignment differs from that previously proposed [43, 107].

Schläfer and König [226] have used a very simple method to assign the charge transfer transitions in complexes of pyridine, bipyridine, and *o*-phenanthroline. Their method treats the central metal ion as a point charge of $+Z$ which perturbs the π electrons of the ligands. Their results are in agreement with previous assignments by Jørgensen [156], namely, that the transitions are $d \rightarrow \pi$. An extension of their method has been made by Mataga and Mataga [177]. Other simple methods which may have potential usefulness to chelates are those of Day and Jørgensen [82] and Schäffer and Jørgensen [224].

The complete assignment of the spectrum on the basis of a theoretically derived energy level scheme is generally considered to validate the calculations and the bonding scheme. The most critical states in the correlation are the charge transfer states, that is, the transitions from bonding orbitals, usually of high ligand character, to the metal t_{2g} and e_g orbitals. From the proper assignment of the bonding levels, one can determine the nature of the bond (electrostatic, σ covalent and/or π covalent) and the stability of the chelate. The preceding discussion leads to the conclusion that the best results indicate that metal complexes contain largely σ covalent bonds with additional stability from π covalent bonds. The π bonds are energetically less important than the σ bonds, but the π-bonding is an important consideration in many experiments.

D. Spectrochemical and Nephelauxetic Series

The spectrochemical series [110, 243] provides a convenient arrangement of the various ligands in order of increasing Dq for a fixed metal ion. A partial listing is

$$\begin{aligned} &I^- < Br^- < Cl^- < \text{dsep}^- \sim S^{2-} < \text{dtp}^- < F^- < \text{dtc}^- \\ &< OH^- < \text{ox}^{2-} \sim \text{mal}^{2-} \sim H_2O < NCS^- < \text{gly}^- < \text{py} \\ &\sim NH_3 < \text{en} \sim \text{den} < NO_2^- \sim \text{bipy} \sim o\text{-phen} \ll CN^-. \end{aligned}$$

In order of donor atoms of the ligands,

$$I < Br < Cl < F < O < N \ll C.$$

This order of listing is not consistent with point-dipole moments, as it would have to be to meet the requirement of a purely electrostatic model. The series may, however, be divided into groups according to the ligand's ability to act as a π donor or acceptor. A ligand-to-metal π bond, in which the ligand π orbitals are lower in energy and overlap the metal t_{2g} orbitals, destabilizes the t_{2g} metal orbitals and gives rise to a small value of Dq. Examples of these ligands are at the beginning of the series (e.g., Br^-). If the ligand π orbitals are higher in energy than the metal t_{2g} orbitals and they overlap, then the t_{2g} orbitals are stabilized and Dq is larger. Examples of these ligands are found at the end of the series (e.g., bipy). The intermediate region consists of the non-π-bonding ligands (e.g., ox^{2-} and en), in which the ligands are roughly in the order of their dipole moments. The intermediate region is flanked by the weak π donors (e.g., OH^-) and the weak acceptors (e.g., *o*-phen).

The detailed computation of Sugano and Shulman [240] correlates with this rationalization. Their results show that the exchange or covalent terms make the largest contribution to Dq and that the molecular orbitals have considerable π admixture.

The reduction of the Racah parameter B in the complex from the free ion values prompted Jørgensen [21] to develop the nephelauxetic series (cloud-expanding series). This is a listing of the ligands in order of the extent to which the B value of the complex diverges from the free ion B value for a fixed metal ion (decreasing β):

$$\text{No ligands} > F^- > H_2O > NH_3 > \text{en} \sim ox^{2-} > NCS^- > Cl^- \sim CN^- > Br^- > S^{2-} \sim \text{dtp}^- > I^- > \text{dsep}^-,$$

which roughly corresponds to the electronegativity order of donor atoms:

$$F > O > N > Cl > Br > S \sim I > Se.$$

This series is largely attributed to two effects: (1) the reduction of the effective charge on the metal by covalent bond formation, and (2) the mixing of metal with ligand orbitals. Both effects tend to decrease the repulsion of the electrons in metal states. Sugano and Shulman [240] and Fenske and co-workers [114] have shown that the ratio $\beta = B'/B$ may be calculated from molecular orbitals.

E. Intensities

The intensities of spectroscopic bands are usually measured in terms of the maximum molar extinction coefficient (ε_{max}) as defined by the Beer-Lambert Law. Qualitatively, the extinction coefficient is a measure of the probability of the electronic transition, and it may be also used to classify the nature of the transition, since it is related to the event probability called the oscillator strength (f):

$$f = 4.60 \times 10^{-9} \varepsilon_{max}\, \delta, \qquad (2\text{-}15)$$

where δ is the width of the band (in cm^{-1}) measured at half-height. Equation (2-15) is an approximation for the total area of the absorption band, which is in turn a measure of the probability. Almost all the transitions in metal complexes have been shown to be electric dipole in nature [1] (as opposed to the weaker magnetic dipole and electric quadrapole transitions), and therefore f is related to the electric dipole moment (μ) by

$$f = 1.085 \times 10^{-5} \bar{\nu} |\mu|^2, \qquad (2\text{-}16)$$

where

$$\mu = \int \Psi_1^*(er)\Psi_2 \, d\tau \tag{2-17}$$

and $\bar{\nu}$ is the energy of the transition (in cm^{-1}). If the functions Ψ_1 and Ψ_2 are orthogonal, as for the metal states of the same configuration, then the transition moment is zero. This means that the ligand field transitions are forbidden and their extinction coefficients should be zero. In practice they are exemplified by low extinction coefficients (ε = 1–500). The non-zero values are due to interactions which remove the orthogonality. These interactions in order of importance are: (1) odd vibrations which remove the symmetry restrictions and couple the transitions with allowed states (vibronic coupling); (2) odd, non-centric ligand fields (as for many chelates), which also couple allowed states with the forbidden; and (3) mixing of vibronic metal-ligand states, which couples charge transfer states with the forbidden by a vibronic mechanism. Covalency [240] appears also to be a major factor in the intensity.

The charge transfer bands have extinction coefficients which range from 1000 to 10,000, and the ligand transitions can even be more intense. These intensity ranges are approximately valid only if the spin state (multiplicity) does not change during the transition. A change in spin orientation is not favored spectroscopically, and its probability is low. The spin-forbidden transitions are usually considerably less intense than their spin-allowed analogs. The spin selection rule may be broken down by spin-orbit coupling, and many spin-forbidden transitions do appear (usually as shoulders) in spectra of transition metal ion chelates.

For the odd ligand fields of chelates, an estimation of the intensity can be made on the basis of direct products of the group representations, since the transition moment integral (2-17) has a numerical value only if the integral is invariant under the group operations. The transitions of non-centric tetrahedral complexes are usually 10 to 100 times more intense than those of their octahedral analogs. The odd fields of pseudo-octahedral complexes may enhance the intensity by as much as 2–3 times that of the octahedral analog, but usually their intensity is smaller. Discussions of intensity may be found in the literature [1, 17, 37, 38, 104, 115, 164].

F. Lower Symmetry Fields

The most important geometry of transition metal ion chelates would seem to be octahedral, but octahedral symmetry is impossible for chelates; the seemingly octahedral chelates are really octahedral-like or pseudo-octahedral. Inclusion of the spanning groups in the chelate ring reduces the symmetry to a group lower than octahedral even when the six donor atoms are identical. For example, $[Co(ox)_3]^{3-}$ would be trigonal ($\mathbf{D}_3$) and $[1,2,6\text{-}Co(den)_2]^{3+}$, rhombic ($\mathbf{D}_2$). The low resolution spectra of these complexes would be those predicted for octahedral Co(III) in $Co(O)_6$ or $Co(N)_6$. The high resolution crystal spectrum (at low temperatures) would reflect the trigonal or rhombic fields. Mixed complexes, that is, complexes in which the binding sites are not the same, can never be octahedral. However, they may have typical octahedral spectra under conditions of low resolution. It is possible to observe the spectroscopic effects of the difference in the binding sites in such cases even under conditions of moderate resolution.

A commonly encountered lower field is the tetragonal field. In crystal field theory the tetragonal field is treated as a perturbation of the octahedral levels, and the perturbation (tetragonal potential) is added to the Hamiltonian (2-1). Two parameters, *Ds* and *Dt*, are then derived for *d* electron systems by expansion techniques (see [1], p. 100, for the details of this derivation). The splitting of the *d* orbitals in terms of these tetragonal parameters is given in Fig. 2-14. The signs of *Ds* and *Dt* are arbitrary. Ballhausen [1] uses the convention of a negative sign for axial stabilization or compression (greater d_{z^2} interaction), whereas Liehr and Perumareddi [168] have suggested the opposite convention.

Tetragonal matrix elements have been derived for low spin Co(III) by Ballhausen and Moffitt [35], and for high spin Cr(III) and Ni(II) by Hare and Ballhausen [140]. The parameters derived neglect configurational and spin-orbit interactions. For these systems the tetragonal splitting of the first band is $\frac{35}{4}Dt$, that of the second band is $-6Ds + \frac{5}{4}Dt$, and the splitting of the third is $3Ds - 5Dt$. Thus the splitting of the first band defines *Dt*, and those of the second and third bands define *Ds*. If *Ds* and *Dt* have the same

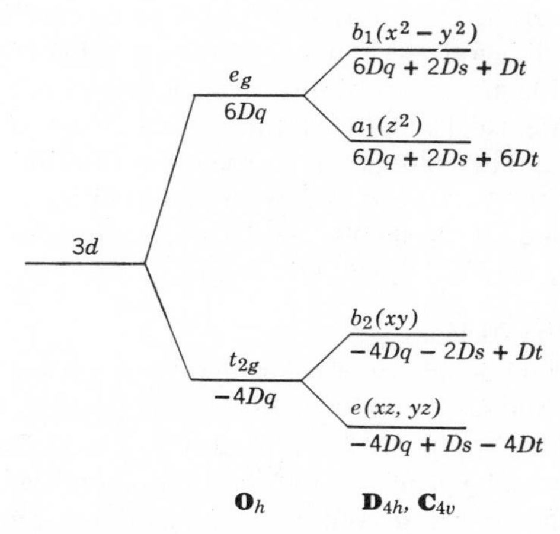

Fig. 2-14. The splittings of the *d* orbitals in terms of the tetragonal parameters.

sign, as they usually do, the splitting of the first band is large and those of the others are smaller.

Fenske, Martin, and Ruedenberg [112] have pointed out that, if Dt is $\frac{4}{7}Dq$, the complex is square-planar. For non-square-planar tetragonal fields Wentworth and Piper [251] have shown that the following equation is approximately valid:

$$Dt = -\tfrac{4}{7}(Dq_{xy} - Dq_z), \tag{2-18}$$

where Dq_{xy} is the planar interaction and Dq_z is the axial interaction. Thus a knowledge of Dq for the planar and axial substituents gives an approximate value of the splitting parameter. Equation (2-18) has further utility. Consider the coordinate system in Fig. 2-15, in which substituents are at each of the octahedral positions. If, for example, the interaction along the x axes can be separated and we can write

$$Dqx\bar{x} = \tfrac{1}{2}(Dqx + Dq\bar{x}), \tag{2-19}$$

then

$$Dq(\text{complex}) = \tfrac{1}{3}(Dqx\bar{x} + Dqy\bar{y} + Dqz\bar{z}). \tag{2-20}$$

If all the ligands are the same, then Dq(complex) is the octahedral Dq, but, if $x = y = z \neq \bar{x} = \bar{y} = \bar{z}$ (i.e., a 1,2,3-trisubstituted complex), their Dq(complex) is the average of the Dq of the substituents. The complex is essentially "octahedral" because

$$Dt = -\tfrac{4}{7}\left[\frac{Dqx + Dqy + Dq\bar{x} + Dq\bar{y}}{4} - \frac{Dqz + Dq\bar{z}}{2}\right] = 0 \tag{2-21}$$

in this case and no splitting of the first band occurs.

The same argument may be applied to *cis*-MA_4B_2 complexes in which $x = y = z = \bar{z} \neq \bar{x} = \bar{y}$. Application of (2-21) gives

$$Dt = -\tfrac{4}{7}(\tfrac{1}{2})(Dq\bar{x}\bar{y} - Dqz\bar{z}). \tag{2-22}$$

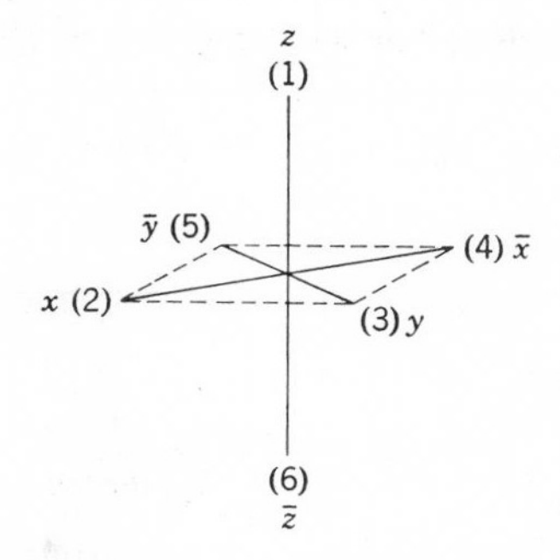

Fig. 2-15. The coordinate system for octahedral complexes.

Comparison of this with (2-18) shows that Dt for a *cis* complex is half that for a *trans* complex and also, since the substituents have changed coordinate positions, that a reversal of sign takes place. This is in accord with the observation that the splitting of the bands for the *cis* isomer is half that for the corresponding *trans* isomer. A similar result was predicted by Ballhausen and Jørgensen [34] on the basis of a point charge model. It should be noted that the splittings of the *trans*-tetragonal isomers are the largest. Equations (2-18) to (2-21) may be used to estimate the magnitude of the splittings or, if the splittings are observed, to estimate the Dq of the substituents.

Wentworth and Piper [251] have derived and used (2-18) to relate Dt and Dq for a number of monoacido and diacido tetragonal complexes of Co(III) and Cr(III). They have shown that the tetragonal parameter Dt for the monoacido complexes (approximate C_{4v} symmetry) is

$$Dt(C_{4v}) = -\tfrac{2}{7}(Dq_{xy} - Dq_z), \tag{2-23}$$

which is half the value of the diacido tetragonal parameter. The values of Dq for the substituents are derived from the observed spectral splittings of the first band (approximately $\tfrac{35}{4}Dt$) and application of (2-18) and (2-20). A partial listing of these results is given in Table 2-6; the derived values of Dq agree quite well with known values. Baker and Phillips [33] have carried out similar correlations for the *trans*-diacidobis(ethylenediamine)chromium(III) complexes.

Liehr [166] has given the tetragonal parameters for the d^1 and d^9 systems with and without inclusion of the spin-orbit interaction. Liehr and Perumareddi [168] have given a comprehensive treatment of the d^2,d^8 tetragonal system including the spin-orbit interaction, and Perumareddi [193] has extended this work to d^3 and d^7. Goode [132] has reported tetragonal matrix elements for only the low-lying states of d^5.

Table 2-6 Absorption Spectra, Splitting Parameters, and *Dq* for *trans*-Diacido Bisethylenediamine Complexes of Co(III)

Complex	Band Maxima ($cm^{-1} \times 10^{-3}$)			Dt (cm^{-1})	Calculated *Dq* (cm^{-1})
	IA	IB	II		
$[Co(en)_3]^{3+}$	21.47		29.50		2530 (en)
$[Co(en)_2F_2]^+$	17.20	22.62	27.60	488	1676 (F)
$[Co(en)_2Cl_2]^+$	16.12	22.49	25.92	612	1459 (Cl)
$[Co(en)_2Br_2]^+$	15.21	21.68	masked	716	1277 (Br)
$[Co(en)_2(H_2O)_2]^{3+}$	18.2	22.5	29.0	374	1875 (H_2O)
$[Co(en)_2(NCS)Cl]^+$	17.8	22.4	masked	419	2050 (NCS)

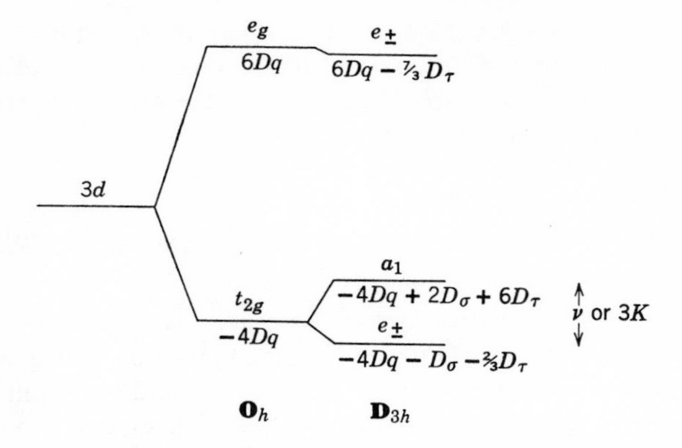

Fig. 2-16. The splittings of the *d* orbitals in terms of trigonal parameters (diagonal approximation).

For trigonal fields a set of parameters d_σ and d_τ may be derived by expansion of the trigonal potential. These parameters are given in Fig. 2-16. Pryce and Runciman [199] have pointed out that, if off-diagonal elements are neglected, one parameter is required for the trigonal system (ν or K). The parameter $\nu = -3K = \frac{1}{3}(9D_\sigma + 20D_\tau)$. This parameter is most widely used because of its simplicity, for the splitting of the first band is usually $\pm\nu$ or $\mp 3K$ or simple multiples of these parameters. The derivation of the trigonal parameters including and not including the spin-orbit interaction is given in several papers. The importance of the off-diagonal elements has been stressed by MacFarlane [173]; for accurate results they should not be neglected.

2-2 APPLICATIONS

A. Pseudo-octahedral Fields

Most chelates that are formed with transition metal ions have a pseudo-octahedral geometry; that is, they are six-coordinate and, if the donor atoms were the same and if the spanning chelate moiety were neglected, they would be geometrically regular octahedra. Thus the fictitious complex

$AsMe_2$
PH OH
M
S NH
NH

which has no symmetry except the trivial C_1 axis can be treated as if the metal ion M "felt" a field that is the average of the contributions of the donors: $AsMe_2$, PH, S, two NH, and OH. The crystal field parameter for the complex, *Dq*(complex), is, to a first approximation,

$$Dq(\text{complex}) = \frac{Dq(\text{AsMe}_2) + Dq(\text{PH}) + Dq(\text{S}) + 2\,Dq(\text{NH}) + Dq(\text{OH})}{6}. \quad (2\text{-}24)$$

This is the so-called *rule of average environment*; it is a very powerful tool for the determination of binding sites and the geometry of transition metal ion complexes. The energy levels of the "average" complex should correspond fairly well to the Tanabe-Sugano energy level schemes at the average *Dq* (complex). The rule of average environment holds well for solution spectra at room temperature whenever the detailed structure of the absorption bands would be smeared out by thermal and solvent effects. In the reflectance spectra obtained in the solid state more detail is usually observed in the absorption bands, as in polarized crystal spectra also (see Sec. 2-2G). In these spectra the average of the bands roughly obeys the rule of average environment. The relative positions of the donors in the spectrochemical series also dictate whether the rule is strictly obeyed. If the donors differ greatly in their positions in the series, then the deviations from the rule are greatest, since the position in the spectrochemical series is not a linear function; that is, the stronger ligands dominate the weaker ligands. If the donors do not differ greatly in their positions in the series (e.g., O and N donors), adherence to the rule of average environment is greater. Also, on the basis of the discussion in Sec. 2-1F on the parameters of low fields, for similar donors the lower the symmetry of the complex, the smaller are the splittings and thus the more strictly is the rule of average environment obeyed.

1. Like Donor Fields. The determination of the geometry of a chelate by use of visible absorption spectra is easily achieved by comparing the chelate spectrum with that of its unidentate analog. The best example of such a determination is that of $Ni(acac)_2$. Before the advent of ligand field theory $Ni(acac)_2$ was believed to be a typical example of tetrahedral nickel(II) [26]. The spectrum of $Ni(acac)_2$, however, was directly comparable with that of $[Ni(H_2O)_6]^{2+}$, and on this basis alone it was shown that $Ni(acac)_2$ must be six-coordinate and octahedral [36, 154]. The complete x-ray structure determination of this compound [59] showed that it is a trimer and that the coordination is indeed octahedral. This direct comparison is not expected to be exact because it has definitely been shown that a chelating ligand and its unidentate analog are not necessarily in the same position in the spectrochemical series. A comparison of the spectra of $[Ni(H_2O)_6]^{2+}$ and $Ni(acac)_2$

shows that the acetylacetonate ligand position is higher than water in the spectrochemical series. Similar arguments have been used to show that nitrato complexes of Ni(II) must also be bidentate [238].

	Bands in Wave Numbers			
$[Ni(H_2O)_6]^{2+}$	8500	13,500	15,400	25,300
$Ni(acac)_2$	8800	12,900	15,250	—

A comprehensive study of multidentate amine complexes with Ni(II) by Jørgensen [155] has shown that the chelate effect (i.e., the stability of the chelate is greater than that of its unidentate analog) is manifest in the spectrum. In Table 2-7 a partial listing of Jørgensen's observations are given on the

Table 2-7 Nickel Amine Complexes[a]

Complex	D	R	$^3T_{2g}(F)$	$^3T_{1g}(F)$	$^3T_{1g}(P)$
$[Ni(NH_3)_6]^{2+}$	1	0	10,750(4.0)	17,500(4.8)	28,200(6.3)
$[Ni_2(tren)_3]^{4+}$	4	5	10,700(9.4)	18,300(7.2)	28,700(9.3)
$[Ni(tn)_3]^{2+}$	2	6	10,900(5.9)	17,800(7.7)	28,200(10.8)
$[Ni(bdn)_3]^{2+}$	2	6	10,950(6.5)	18,000(7.0)	28,600(10.0)
$[Ni(en)_3]^{2+}$	2	5	11,200(7.3)	18,350(6.7)	29,000(8.6)
$[Ni(ptn)_2]^{2+}$ [b]	3	5	11,200(4.7)	19,000(4.5)	29,700(4.6)
$[Ni(den)_2]^{2+}$	3	5	11,500(12.9)	18,700(7.7)	29,100(10.8)
$[Ni(bipy)_3]^{2+}$	2	5	12,650(7.1)	19,200(11.6)	—
$[Ni(phen)_3]^{2+}$	2	5	12,700(6.8)	19,200(11.9)	—

[a] The energies of the states are given in wave numbers and the extinction coefficients are given in parentheses. D is the number of donors per ligand and R is the ring size.
[b] ptn = 1,2,3-triaminopropane

basis of the octahedral assignments. Ethylenediamine is higher in the spectrochemical series than ammonia, as it should be, in accordance with the chelate effect. These results also show that trimethylenediamine is between en and NH_3 in the spectrochemical series. Jørgensen points out that the tn complex is less stable than that of en even though the donor strengths should be similar. He attributes this difference in part to enthalpy contributions to the chelate effect as well as to the more familiar entropy contributions. The comparison of tn with diethylenetriamine (den) shows that the secondary amine donor is higher in the spectrochemical series than the primary. The ligands containing nitrogen as part of an aromatic system are similar to the amines but are higher in the spectrochemical series. The higher intensities of the chelate spectra are in part accounted for by the lower symmetry of those complexes, but Jørgensen [153] also points out that the larger the organic substituents, the greater is the observed intensity. Pavkovic and Meek [192] have studied a series of N-alkylated ethylenediamines which shows that the more bulky groups have lower Dq than the smaller groups. These authors claim that steric factors may outweigh inductive factors for the contributions to Dq.

In a separate study of Co(III) chelates with oxygen donors, Jørgensen [154] has found a marked difference between carbonate, oxalate, and acetylacetonate ions as ligands. His results are summarized in Table 2-8. The ordering in

Table 2-8 The Spectra of Cobalt(III) Chelates with Oxygen Donors[a]

Compound	Ring	$^1T_{1g}$	$^1T_{2g}$	$(^1T_{2g} - {}^1T_{1g})$
$[Co(CO_3)_3]^{3-}$	4	15,700	22,800	7100
$[Co(ox)_3]^{3-}$	5	16,500	23,800	7300
$Co(acac)_3$	6	16,900	masked	—
$[Co(H_2O)_6]^{3+}$	0	16,600	24,900	8300

[a] Energies in cm^{-1}.

the spectrochemical series is $CO_3^{2-} < ox^{2-} < H_2O < acac^-$, and the low position of CO_3^{2-} is probably due to two effects: (1) the lower stability of the four-membered ring, and (2) the greater electron delocalization in the carbonate ion. Both factors should make carbonate a weaker ligand. The energy difference between $^1T_{2g}$ and $^1T_{1g}$ is nearly the same for the oxalate and

Table 2-9 Absorption of Cupric Dicarboxylate Complexes

	Dicarboxylate Ion	λ (mμ)	ε
1	Oxalate	700	33.5
2	Malonate	700	36.6
3	Methylmalonate	695	48.3
4	Ethylmalonate	700	51.1
5	*n*-Propylmalonate	697	47.5
6	*n*-Butylmalonate	695	55.4
7	Diethylmalonate	680	58.4
8	Ethyl-*n*-butylmalonate	680	69
9	Phenylmalonate	690	61
10	Phenyl methylmalonate	710	69
11	Chloromalonate	720	40.9
12	Bromomalonate	710	41.8
13	Dibromomalonate	705	40.3

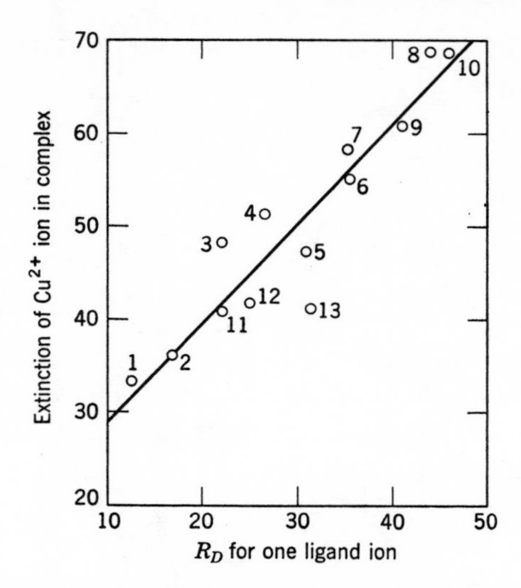

Fig. 2-17. Relationship of extinction of Cu^{2+} band to the refraction equivalent of the ligand. The ligands are numbered as in Table 2-9.

carbonate chelates and is about 1000 cm^{-1} less than for the aquo complex. This result shows that the nephelauxetic effect for the chelates is greater than for the unidentate analogs. [The generality of this statement has not been verified. It is also true for the nickel amines (see Table 2-7), but the effect is very small.] Jørgensen [158] in a study of a few sulfur donor chelates has pointed out that there is a direct correlation between Dq and the number of lone pairs on the donor. The anion R—S^- with three lone pairs has a lower Dq than the sulfide R_2S with only two lone pairs, whereas the unidentate SO_3^{2-} with one lone pair has an even higher Dq. The nephelauxetic effect for these sulfur-containing ligands is quite marked, and β may be as low as 0.25.

Graddon [136] has made a very interesting study of the correlation of band intensity with chelate substituents for a number of copper malonate complexes. His results are given in Table 2-9, and a correlation of the molar refractivity (R_D) of the ligand with the intensity (ε) is given in Fig. 2-17. This correlation suggests that the increase in intensity is related to the size (molar volume) of the coordinated anion, and that electronic delocalization and inductive effects are of lesser importance to the intensity (dibromo malonate is a possible exception). This correlation apparently does not hold for chelates with seven-membered rings such as the phthalate, maleate, and succinate complexes, and Graddon concludes that the formation of the unstable seven-membered rings changes the environment sufficiently to preclude a correlation of the refractivity with the molar volume for these chelates.

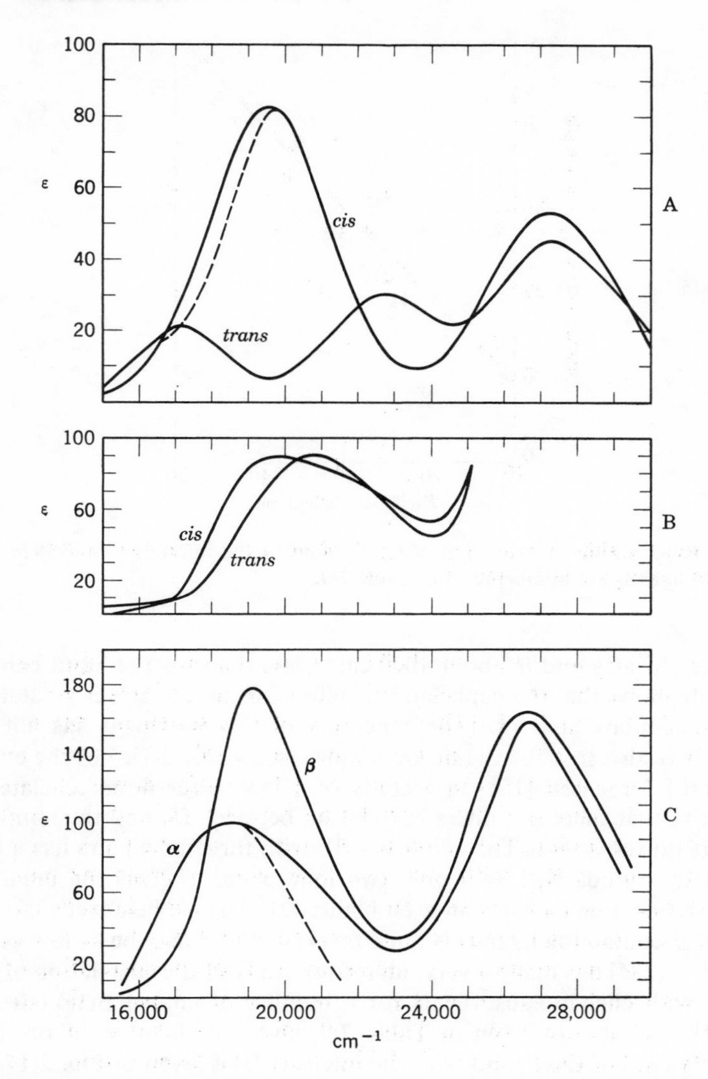

Fig. 2-18. Molar extinction coefficient versus wavenumber in cm^{-1}. (A) *cis*-[$Co(en)_2F_2$]-NO_3, *trans*-[$Co(en)_2F_2$]NO_3. (B) *cis*-[$Co(en)_2(NO_2)Cl$]Cl, *trans*-[$Co(en)_2(NO_2)Cl$]-NO_3. (C) α-[$Co(NH_2CH_2COO)_3$]$\cdot 2H_2O$, β-[$Co(NH_2CH_2COO)_3$]$\cdot H_2O$.

2. Mixed Donor Fields (Geometrical Isomers). If the donor atoms of a pseudo-octahedral chelate are not the same, it is possible that geometrical isomers may exist. The application of absorption spectroscopy to the differentiation of geometric isomers was first carried out by Basolo, Ballhausen, and Bjerrum [44]. The α and β isomers of tris(glycino)cobalt(III) were assigned

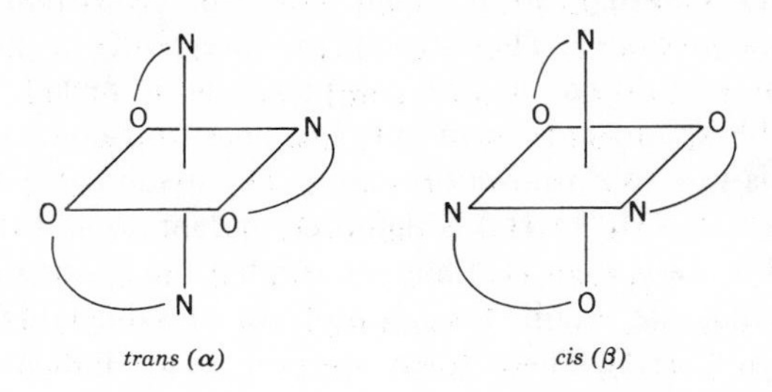

trans (α) *cis* (β)

the *trans* and *cis* configurations on the basis of similarity of spectra with those of known *cis* and *trans* isomers, (MA_4B_2) and (MA_4BC), given in Fig. 2-18. The rationale was based on the argument [44] that the *cis* isomer is more like an octahedral complex and should have a more symmetric absorption band, while the *trans* isomer should have a broad absorption band at lower energy and a lower molar extinction coefficient. The total area under both bands is essentially the same. The rule of average environment was also shown to be valid for the $[Co(en)_2F_2]^+$ and $[Co(en)_2(NO_2)Cl]^+$ isomers, which are included in Fig. 2-18. Dunn, Nyholm, and Yamada [93] have more recently used the same method to determine the geometric isomers of the cobalt(III) chelates of *o*-phenylenebisdimethylarsine (diarsine). The spectra of $[Co(diarsine)_3]^{3+}$, *trans*-$[Co(diarsine)_2X_2]^+$, and *cis*-$[Co(diarsine)_2X_2]^+$ (X = halide ion) are remarkably similar to the ethylenediamine analogs. Diarsine lies higher on the spectrochemical series than en and also has a larger nephelauxetic effect.

Jørgensen [158] has also used the rule of average environment to assign the structure of $[Ni(daes)_2](ClO_4)_2$ [daes = 2,2′-di(aminoethyl)sulfide] in the solid state and in aqueous solution. The reflectance spectrum of the anhydrous solid and that of the aqueous solution are similar, and the splitting and intensity of the bands are quite similar to those of the *trans*-NiO_2N_4 complexes previously investigated [56, 155]. On this basis it was assumed that the complex has the *trans* configuration. In the same study, Jørgensen [158] points out

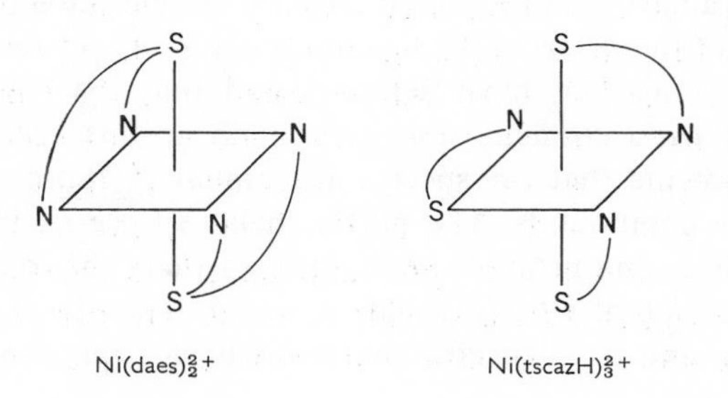

$Ni(daes)_2^{2+}$ $Ni(tscazH)_3^{2+}$

that the nickel chelates of thiosemicarbazide (tscazH), $Ni(tscazH)_3^{2+}$, have spectra similar to that of $Ni(daes)_2^{2+}$, and he concludes that $Ni(tscazH)_3^{2+}$ has a "*trans*-like" configuration which would be consistent with the 1,2,6 configuration shown above. These results are interesting in that these ligands form high spin pseudo-octahedral complexes with nickel, whereas other thioacids and thioaminoacids form only low spin four-coordinate complexes [$Ni(tscazH)_2^{2+}$ is low spin four-coordinate]. The results also show that thiosemicarbazide ($NH_2NHCSNH_2$) is definitely bidentate and that the binding sites are N and S. Jørgensen [12] has interpreted the spectra of Dwyer et al. [94] of the sexadentate, Schiff base complexes of salicylaldehyde and thioetheramines with Co(III). These green compounds are indicative of a Co(III) $N_2O_2S_2$ chromophore.

The complex $Ni(NTA)_2^{4-}$ (NTA = nitrogen triacetate) has been shown by Jørgensen [155] to have a spectrum characteristic of a complex with a center of symmetry and therefore the nitrogen atoms must be *trans*. The complex $Ni(tren)(NCS)_2$ has been shown by x-ray analysis to have a *cis* configuration [201]. The aqueous solution spectrum is identical with the solid state reflectance spectrum, and therefore the complex $[Ni(tren)(H_2O)_2]^{2+}$ has a *cis* configuration [155]; similarly $[Ni(en)_2(H_2O)_2]^{2+}$ also has the *cis* configuration in solution. A more recent interpretation [111] has shown that the solids containing the chromophore $[Ni(en)_2(H_2O)_2]^{2+}$ with large anions such as BPh_4^- and ClO_4^- display a tetragonal splitting of the first band and have been assigned a *trans* configuration. In solution these complexes give spectra characteristic of the *cis* configuration. Melson and Wilkins [180] have shown that the nickel chelates of the hexadentate ligands, penten [N,N,N′,N′-tetra-(2-aminoethyl)ethylenediamine] and tren, may be successively protonated:

$$[Ni(penten)]^{2+} \xrightarrow{H^+} [Ni(H\ penten)(H_2O)]^{3+}$$
$$\xrightarrow{H^+} [Ni(H_2\ penten)(H_2O)_2]^{4+}$$

The spectra are indicative of water replacing the protonated amino group in the coordination sphere, the second water apparently being added *cis* to the first.

Visible spectra have also been used recently to determine the configuration of the chelates of the macrocyclic ligand cyclam (1,4,8,11-tetraazacyclotetradecane). Tobe et al. [55] have demonstrated that the Co(III) and Ni(II) chelates have a *trans* configuration with acido groups occupying the axial positions by showing that the spectra are similar to those of known *trans*-ethylenediamine compounds. The Ni(II) chelates have an interesting anion effect. The chloride and bromide are paramagnetic in the solid state and give spectra characteristic of a *trans*-diacido complex. The perchlorate and iodide are diamagnetic and have spectra characteristic of square-planar Ni(II). In

```
        H2C——————CH2
         |        |
   Me2C—NH       N=CMe
   /      ↘2+↙      \
 H2C        Ni       CH2
   \      ↗   ↖      /
   MeC=N       HN—CMe2
       |        |
      H2C——————CH2
```

I

```
        H2C——————CH2
         |        |   H
   Me2C—NH       HN—CMe
   /      ↘2+↙      \
 H2C        Ni       CH2
   \      ↗   ↖      /
   MeC—NH       HN—CMe2
   H    |        |
       H2C——————CH2
```

IIa and IIb

```
        H2C——————CH2
         |        |
   Me2C—NH       NH2
   /      ↘2+↙
 H2C        Ni
   \      ↗   ↖
   MeC=N       NH2
       |        |
      H2C——————CH2
```

III

solution all these Ni(II) chelates give spectra characteristic of square-planar geometry. A similar anion effect has been observed by Goodgame and Venanzi [133] in N,N′-diethylethylenediamine chelates, but the anion coordinated in the solid state remains coordinated on solution. This anion effect has been attributed to steric considerations, and it is argued that the bulky anions or the solvent is unable to approach the metal closely enough to exert sufficient axial interaction to be considered coordinated. Curtis [73] has also noted this anion effect in other macrocyclic, tetradentate chelates. It is also possible for macrocyclic tetradentate chelates to exist in a *cis* configuration. Curtis [76] has prepared such complexes by reduction of the cyclic Schiff base-amine complex (I) with sodium borohydride to produce the isomers in II and the non-cyclic III. Complexes containing the borohydride anion and other anions have been separated, and the spectra (Table 2-10) of

$$\text{(IIb)}\ BH_4\cdot ClO_4\cdot \tfrac{1}{2}H_2O, \quad \text{(IIb)}\ (BH_4)_2, \quad \text{(III)}\ BH_4\cdot ClO_4,$$

$$\text{(IIb)}\ NO_3\cdot ClO_4\cdot \tfrac{1}{2}H_2O, \quad \text{and} \quad \text{(IIb)}\ CH_3COO\cdot ClO_4\cdot \tfrac{1}{2}H_2O$$

have been shown to be similar to that of *cis* complex $[Ni_2(en)_4Cl_2]Cl_2$ [30]. On the basis of this information and other physical measurements, Curtis [76] concludes that the borohydrides of (IIb) and (III) have the structure (IVa), which contains the bidentate borohydride anion. The complexes (IIa)X_2 ($X = Cl^-$, CH_3COO^-, NO_3^-, BH_4^-) all have similar spectra which are indicative of the *trans* configuration of the chelate, and in conjunction

IVa IVb

with other physical measurements the borohydride has been assigned the structure (IVb), which contains unidentate borohydride. Curtis and Curtis [74] have also shown that complexes (IIb)NO_3X ($X = NO_3^-$, ClO_4^-) contain bidentate nitrate and are therefore in the *cis* configuration, as is structure IVa, while the nitrates of (IIa) are unidentate and similar to structure IVb. They have also shown that the complexes

$$[Ni(en)_2NO_3]X \quad \text{and} \quad [Ni(tn)_2NO_3]X$$

($X = I^-$, BF_4^-, and ClO_4^-) also exist as *cis* isomers, the nitrate anion acting as a bidentate ligand. The compound $Ni(en)_2(NO_3)_2$, however, is *trans* and contains unidentate nitrate. The configuration of the ligand plays an important role in the formation of the geometrical isomers of the monocyclic ligands. Discussions of ligand isomerization may be found in [32, 55, 76, 77].

Table 2-10 Reflectance Spectra ($cm^{-1} \times 10^{-3}$)[a]

Compound	ν_1	ν_2	ν_3
(IIa)$(BH_4)_2$	9.0, 14.0sh	19.8	29.0
(IIa)$(NO_3)_2$	8.1, 15.7	19.0	28.7
(IIa)Cl_2	8.1, 14.3	18.4	28.0
(IIb)$BH_4 \cdot ClO_4 \cdot \frac{1}{2}H_2O$	7.6sh, 10.3	18.1	27.7
(IIb)$(BH_4)_2$	8.0sh, 10.4	17.8	27.4
(III)$BH_4 \cdot ClO_4$	8.0sh, 10.6	18.6	28.3
(IIb)$NO_3 \cdot ClO_4 \cdot \frac{1}{2}H_2O$	10.7	17.5	27.6
$[Ni_2(en)_4Cl_2]Cl_2$	10.2	16.9	26.9

[a] $\nu_1 = {}^3A_{2g} \rightarrow {}^3T_{2g}$, $\nu_2 = {}^3A_{2g} \rightarrow {}^3T_{1g}(F)$, $\nu_3 = {}^3A_{2g} \rightarrow {}^3T_{1g}(P)$ for O_h symmetry.

In their determination of the crystal field parameters, Busch and his co-workers [58] have applied the equations of Wentworth and Piper [251] to a series of *trans* diacido (tetragonal) Ni(II) complexes of several of the macrocyclic ligands. From the tetragonal splittings of the first band they obtain the spectrochemical series

$$Br^- < Cl^- \sim NO_3^- < N_3^- < F^- < ONO^- < NCS^- < CN^-,$$

which is in accord with the normal spectrochemical series. The larger tetragonal splittings occur at both ends of this series, since the macrocyclic ligands are in the middle of the spectrochemical series. References [178, 183, 186, 260] are also of interest as regards application of spectra to the determination of isomers.

B. Pseudo-tetrahedral Fields

Tetrahedral geometry is not common in chelates, and it has only been in past few years that well-characterized examples have been reported. The divalent metal ions of Mn, Fe, Co, Ni, Cu, and Zn and trivalent Fe all have been shown to form tetrahedral complexes, and chelate examples exist for most of these ions. Tetrahedral geometry is very rare for the second and third row transition metal ions. In general the Tanabe-Sugano diagrams (Figs. 2-5 to 2-11) can be used for the tetrahedral chelates even though they have been derived for octahedral geometry. The crystal field parameter Dq (tet.) for tetrahedral geometry is $-\frac{4}{9}Dq$ (oct.); this again means an inversion of the states of a given configuration or, more simply, that the hole equivalence may be used. Thus for a d^7 tetrahedral example one may use the d^3 octahedral Tanabe-Sugano diagram. The fit of the spectrum occurs at the appropriate value of Dq. The value of Dq which best fits the spectrum and energy level diagram is not, in all likelihood, $\frac{4}{9}$ of the Dq of the octahedral analog. In general a value of $\frac{1}{2}$ or greater is observed. The reason is that the metal-ligand interaction is greater for tetrahedral than for octahedral complexes, as is manifest in the known shortening of bond lengths in the case of tetrahedral complexes. This metal-ligand interaction, largely by virtue of enhanced π-bonding, also lowers the Racah parameter B more than for the octahedral analog. As an example, if $\beta = 0.80$ for an octahedral chelate, its tetrahedral analog may have $\beta = 0.60$, and as a result strict adherence to the Tanabe-Sugano diagrams is not observed; the predicted energy difference between the observed bands is too high, even though the level ordering is correct.

A few examples of tetrahedral chelates with like donor atoms are known. Cotton and co-workers [69, 70] have recently described the tetrahedral chelate $Co(DPM)_2$ (H–DPM = dipivaloylmethane, a *t*-butyl β-diketone). The spectrum of this complex is given in Fig. 2-19, and, although it is not exactly

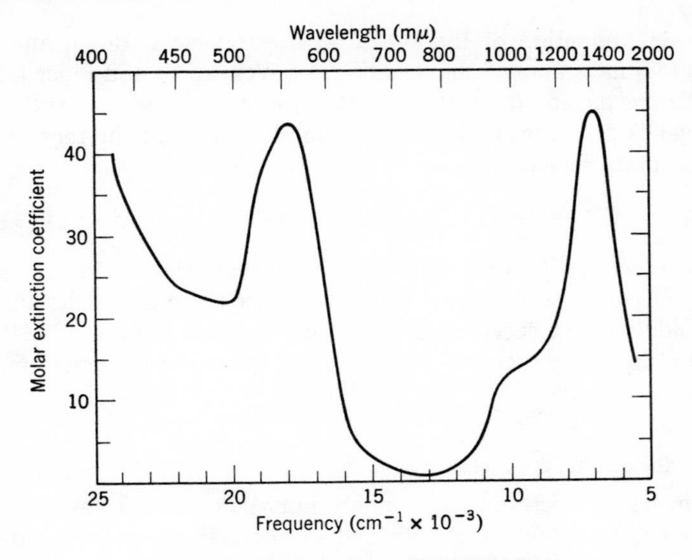

Fig. 2-19. The spectrum of Co(DPM)$_2$.

typical of tetrahedral symmetry, the gross features are nevertheless tetrahedral. Cotton and Wood [70] have shown that Co(DPM)$_2$ is isostructural with Zn(DPM)$_2$, and that the symmetry of the chelate is not $\boldsymbol{T_d}$ but $\boldsymbol{D_{2d}}$. This distortion from regular tetrahedral symmetry causes reduction of the band intensities; the observed ratio $f(\nu_3)/f(\nu_2)$ is 1 and in normal tetrahedral complexes $f(\nu_3)/f(\nu_2)$ is usually 5–10 (ν_1 is not observed in Fig. 2-19). Fackler et al. [109] have recently prepared Fe(DPM)$_2$, which is the first example of tetrahedral Fe(II) with oxygen donors. The compound is isomorphous with Zn(DPM)$_2$ [70] and the *d-d* spectra are masked by ligand overtones and low energy, charge transfer bands. Cotton and Soderberg [69] have also given spectral evidence that Co(acac)$_2$ may exist as a tetrahedral monomer at high

EtO$_2$C Me Me CO$_2$Et
Me NH N Me

V

EtO$_2$C
EtO$_2$C Me Me
Me NH N Br

VI

temperatures. According to Eaton, Phillips, and Caldwell [99], the Ni(II) aminotroponeimineates and the N,N′-diethyl chelate are tetrahedral because they exhibit absorption bands at 8800 cm^{-1} ($\varepsilon = 150$) and 7550 cm^{-1} ($\varepsilon = 85$) which are characteristic of NiN_4 tetrahedral spectra. Tetrahedral Fe(II) chelates of aminotroponeimines have also been reported [99].

Fergusson and Ramsay [121] have observed that the spectra of the Co(II), Ni(II), Cu(II), and Zn(II) complexes of the dipyrromethenes V and VI are characteristic of tetrahedral geometry. Their results and assignments are given in Table 2-11. In addition to the *d-d* spectra, very intense charge transfer bands are observed in these complexes.

Table 2-11 Spectra and Assignments of Chelates of the Dipyrromethenes[a]

Complex	ν (cm^{-1})	Molar extinction coeff.	Assignment
$Co(MMPM)_2$	9,170sh		$^4A_2 \rightarrow {}^4T_1(F)$
	10,260	54	
	13,830	187	$^4A_2 \rightarrow {}^4T_1(P)$
	15,000	176	
$Co(MBrPM)_2$	9,300sh		$^4A_2 \rightarrow {}^4T_1(F)$
	10,400	55	
	13,010	498	$^4A_2 \rightarrow {}^4T_1(P)$
	14,090	383	
$Ni(MMPM)_2$	6,050	24	$^3T_1(F) \rightarrow {}^3A_2$
	11,100	very weak	
	13,500	345	$^3T_1(F) \rightarrow {}^3T_1(P)$
	14,600	337	
	15,750sh	shoulder	
$Ni(MBrPM)_2$	6,450	26.9	$^3T_1(F) \rightarrow {}^3A_2$
	12,310	645	$^3T_1(F) \rightarrow {}^3T_1(P)$
	13,590	496	
	14,800sh	weak	
$Cu(MMPM)_2$	11,700	210	$^2T_2 \rightarrow {}^2E$
$Cu(MBrPM)_2$	10,300	327	$^2T_2 \rightarrow {}^2E$

[a] H–MMPM = V, H–MBrPM = VI.

Sacconi and co-workers [209] were first to show the existence of tetrahedral Ni(II) chelates. Since 1962 a large variety of N-alkylsalicylaldimine chelates (structure VII) of Co(II), Ni(II), Cu(II), and Zn(II) have been prepared, and they constitute the best examples of pseudo-tetrahedral chelates with mixed donor atoms. A characteristic of these complexes is that a complex may exist in associated polymeric (octahedral), square-planar, and tetrahedral forms,

VII

and the tetrahedral form is favored at elevated temperatures. The equilibria associated with these interesting compounds are discussed later. The reflectance spectra of a series of Ni(II) chelates [210] of VII with R = *n*-propyl, iso-propyl, and *t*-butyl are given in Fig. 2-20. The spectra *B* and *C* (R = iPr and tBu, respectively) are characteristic of tetrahedral Ni(II). The transition ${}^3T_1 \rightarrow {}^3A_2$ (ν_2) is at $\sim$6700 cm^{-1}, the bands at 14,100 to 16,900 cm^{-1} are assigned to ${}^3T_1 \rightarrow {}^3T_1$ (ν_3), the weak band at 10,900 cm^{-1} is assigned to ${}^3T_1 \rightarrow {}^1T_2$ or ${}^1E$. The spectrum of the *n*-propyl chelate is not characteristic of tetrahedral symmetry but of square-planar. The solution spectra of all the compounds are more indicative of mixtures of square-planar and tetrahedral forms. At elevated temperature the tetrahedral form is favored for R = nPr

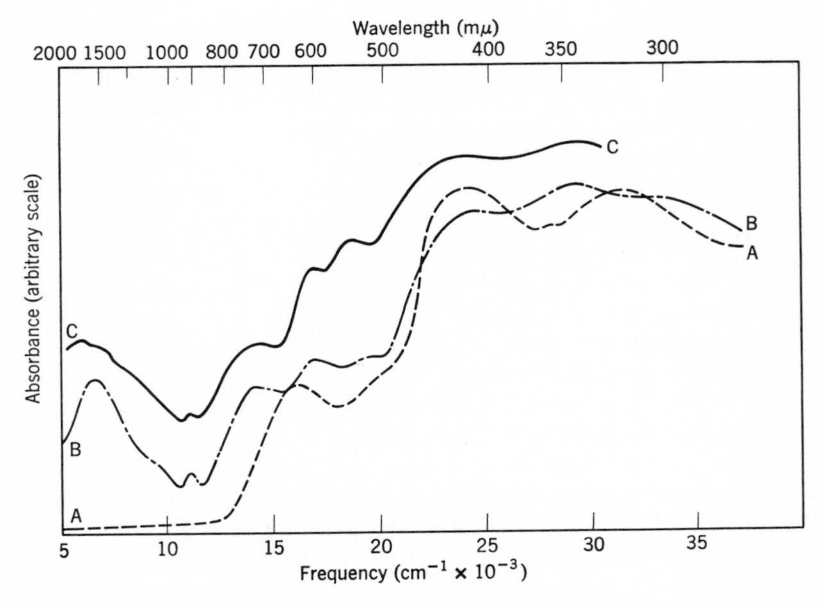

Fig. 2-20. Reflectance spectra of (A) bis-(N-*n*-propylsalicylaldimino)nickel(II), (B) bis-(N-*i*-propylsalicylaldimino)nickel(II), and (C) bis-(N-*i*-butylsalicylaldimino)nickel(II).

and iPr; the opposite is true for R = tBu. Fox, Orioli, Lingafelter, and Sacconi [126] have confirmed by x-ray diffraction the tetrahedral structure for the Ni(II) chelate of N-isopropylsalicylaldimine. The spectra of the Co(II) chelates of the N-alkylsalicylaldimines [208] are totally indicative of tetrahedral geometry in the solid state and in solution, but the spectra of the Cu(II) chelates [214] are not. The Cu(II) chelates are also not isomorphous with the analogous known tetrahedral chelates of these ligands. The x-ray structure of bis(N-isopropylsalicylaldimino)copper(II) [190] shows the chelate to be a flattened tetrahedron, and the spectral deviations may be attributed to this distortion. The bands of this chelate are at 8900, 14,000, and 20,500 cm^{-1}, respectively. A band in the distorted tetrahedral nonchelate $[Cu(TMG)_4](ClO_4)_2$ [171] is observed at 6600 cm^{-1}, corresponding to the tetrahedral predictions of Liehr [166]. The variation of the energy levels of Cu(II) complexes from square-planar to tetrahedral has been treated by Lohr and Lipscomb [169]. Sacconi and Bertini [221] have recently prepared Ni(N,N,N′,N′-tetramethylpropylenediamine)X_2 ($X = Br^-, I^-$) whose spectra indicate tetrahedral geometry for the molecules. Goodgame and Goodgame [134] have recently given a compilation of near-infrared data of many nonchelate tetrahedral complexes of Co(II) and Ni(II), and Sacconi [28] has also recently reviewed the tetrahedral coordination of Schiff base chelates.

C. Five-Coordination

Pentacoordination has been reviewed recently by Muetterties and Schunn [25]. They note the danger of assuming from the stoichiometric formula that a given compound is five-coordinate. A compound may actually be five-coordinate in the gaseous state, but six-coordinate in the solid state or in solution because of association or solvation. Or an apparently four-coordinate species may by the same means attain five-coordination in the solid state or in solution. They point out also that, although a trigonal bipyramidal structure is usually favored for a five-coordinate species (at least for d^0, d^8, d^{10} configurations), the energy difference between this and the square-pyramidal structure is in general rather small. Therefore the latter may actually be favored under certain circumstances, especially when the ligands are large and bulky. They also note that of the 26 five-coordinate chelates for which accurate solid state, structural data are had, 15 are known to have a square-pyramidal structure. They discuss the known compounds and point out possible reasons for the apparent preference of chelates for this structure. For an extensive discussion of the stereochemistry of five-coordination from the point of view of the theory of valency-shell electron pair repulsions, see the articles of Gillespie [131].

Ciampolini [65] has calculated the term splittings for Ni(II) in fields arising from five point dipoles arranged in trigonal-bipyramidal and square-pyramidal configurations. He employed the weak field scheme with configuration interaction. His results (Fig. 2-21 and Fig. 2-22) indicate that diamagnetism is expected only for large field strengths or extensive nephelauxetic effects.

Electronic spectroscopy has been used in conjunction with other physical methods in the study of five-coordinate molecules. In some cases a five-coordinate structure has been assigned to a given species on the basis of similarity of its electronic absorption spectrum to that of a compound known from x-ray analysis to possess such a structure. For example, Mair, Powell, and Venanzi [174] have shown by x-ray analysis that the cation in

$$[Pt(QAS)I][B(C_6H_5)_4], \; QAS = \left(o\text{-}C_6H_4(As(C_6H_5)_2)- \right)_3 As$$

has basically a trigonal bipyramidal structure. The spectra of this and similar complexes differ significantly from those of square-planar complexes of Pt(II). The former compounds have a lower energy band at 20,000–28,000 cm^{-1} with a molar extinction coefficient of the order of 10^3; this band is about 10^4 cm^{-1} lower than the corresponding band in square-planar Pt(II) complexes [57]. Analogous compounds of Pt(II) with the tridentate ligand

$$TAS = \left(o\text{-}C_6H_4(As(C_6H_5)_2)- \right)_2 As(C_6H_5)$$

which are expected to be square-planar, show spectra very like that of $[Pt(NH_3)_3Cl)]^+$ and quite different from those of the QAS complexes [57]. The five-coordinate structures mentioned here were found to persist in the solid state as well as in solution. A similar series of Pd(II)-QAS complexes showed spectra which are consistent with a trigonal bipyramidal structure [223]. The low energy band typical of this structure appears in this series between 16,000 and 20,000 cm^{-1}.

This line of investigation has been extended to the Pd(II) and Pt(II) complexes of the phosphorus ligands, QP and TP, which are analogous to QAS and TAS [144], and to the low spin Ni(II) complexes of QAS and QP [96]. "The differences in visible and ultraviolet spectra of corresponding compounds of nickel, palladium, and platinum are in accord with the increase in orbital splitting in the series Ni < Pd < Pt, thus establishing that the nickel(II) complexes have trigonal bipyramidal structure like their palladium(II) and platinum(II) analogs" [96]. The Ni-(QAS) complexes show bands at

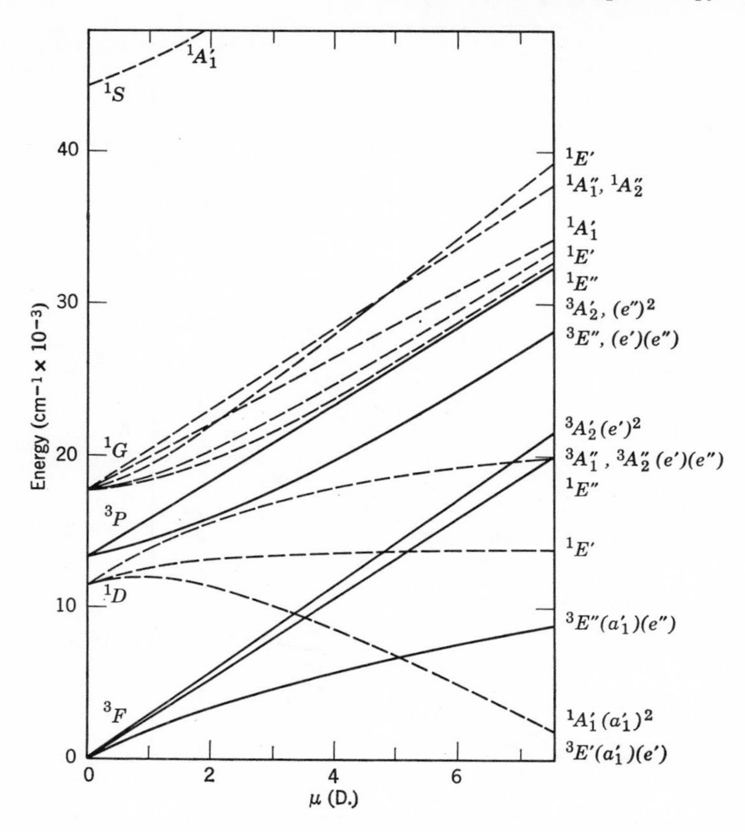

Fig. 2-21. Energy level diagram for the nickel(II) ion in fields of five equivalent dipoles of strength μ arranged in a trigonal bipyramidal configuration. The $^3E'$ state is taken as the zero of energy.

~16,000 cm^{-1} (the more intense) and ~21,000 cm^{-1}. In the QP complexes the bands appear at ~17,500 and ~27,000 cm^{-1}, respectively.

The splitting of the d orbitals in a trigonal bipyramidal field is shown in Fig. 2-23. The two observed transitions in the low spin Ni complexes can accordingly be ascribed to the two $^1A_1 \rightarrow {}^1E$ transitions, which correspond to electronic transitions from the two e levels to the a_1 level. This assignment is made despite the high intensity of the bands; ε = 3500–10,000 for the lower energy band at about 16,000 cm^{-1} and 200–1900 for the band at about 25,000 cm^{-1}. However, Dyer and co-workers [96] note that, because in C_{3v} the symmetry of both the d_{z^2} and the p_z orbitals on the metal is a_1, they may mix; this would cause the transition to be more allowed and would

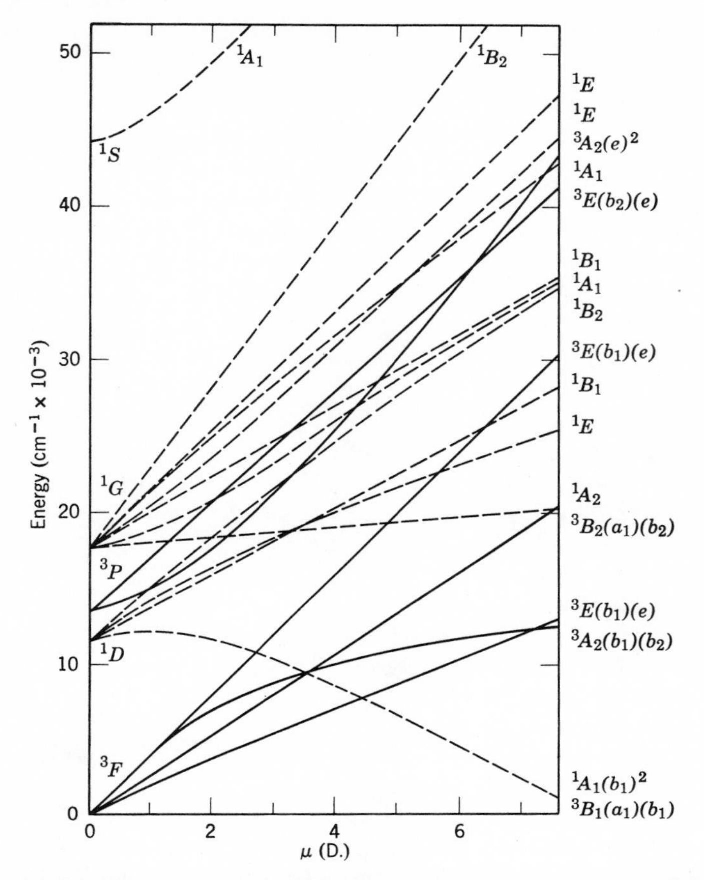

Fig. 2-22. Energy level diagram for Ni(II) in fields of five equivalent dipoles of strength μ arranged in a tetragonal pyramidal configuration. The $^3B_1$ state is taken as the zero of energy.

explain in part the band intensity. Band intensity could also be gained by mixing of metal e' with ligand orbitals.

Dyer and Meek [95] have also studied analogous Ni(II) complexes with

$$\text{TSP} = \left(\text{C}_6\text{H}_4(\text{SCH}_3) \right)_3 \text{P}$$

$[Ni(TSP)X]^+$ ($X = I^-, Br^-, Cl^-, NCS^-$) and $[Ni(TSP)L]^{2+}$ [L = thiourea, $P(C_6H_5)_3$, $CH_3P(C_6H_5)_2$]. The spectra are similar to those of the QAS compounds, but of lower intensity; ε_{max} = 915–2470 for the lower frequency band and 250–1015 for the higher frequency band. They noted that the lower energy band is Gaussian in shape, an indication that the Ni atom is in a symmetrical trigonal field. Distortions of this field would lift the degeneracy of the $d_{x^2-y^2}$ and d_{xy} orbitals and cause this band to appear as an unresolved doublet. The intensity of this lower energy band was found to vary only slightly with a change in X or L. If the band's intensity reflects the extent of π back-donation of electrons from metal to the sulfur's empty *d* orbitals, then it is clear that metal-ligand π-bonding varies little in this series of compounds and is much less extensive than in the arsenic series of complexes. Dyer and Meek [98] have also prepared the selenium analog TSeP, and have found that the chelates $[Ni(TSeP)X]^+$ have spectra similar to the corresponding TSP compounds except at slightly lower energy. These diamagnetic compounds correspond to the Ciampolini [65] energy level diagram except that β is low.

Inspection of the spectra of $[M(QAS)Br](ClO_4)$, where M = Ni, Pd, and Pt, shows that the lower energy band in the Ni chelate is Gaussian in shape, but quite irregular in shape for the Pd and Pt chelates. Dyer and Venanzi [97] explain this by assuming a trigonal distortion in the latter Pd and Pt chelates. The primary effect of such a distortion would be the removal of the degeneracy of the e' orbitals; a lesser splitting of the e'' would also occur (see Fig. 2-23). The difference between the Ni and the Pd or Pt chelates may be due to steric effects; Pd and Pt have the same covalent radius, which is larger than that of Ni. "The observation that the splitting of the e' level increases slightly with the ligand field indicates that electronic effects may also be responsible for the observed distortions" [97]. Benner, Hatfield, and Meek [50] have also shown, in part from spectral studies, that Ni(II) forms trigonal bipyramidal complexes with the ligand $P[CH_2CH_2CH_2As(CH_3)_2]_3$. Also, several complexes of *o*-phenylenebis(dimethylarsine) with Au(I) and Au(III) [142] and divalent Ni, Pd, and Pt [143] have been shown by Harris, Nyholm, and

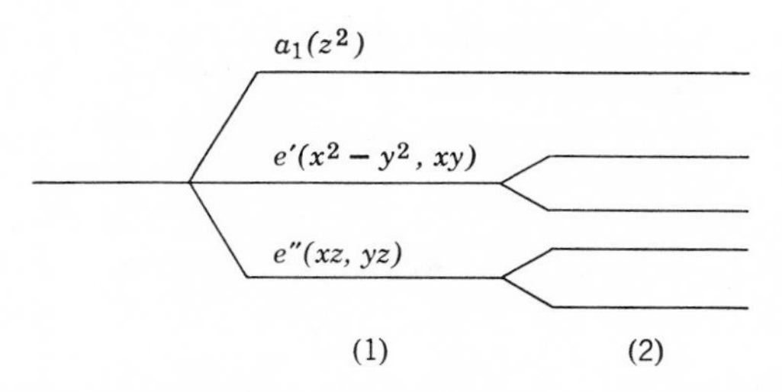

Fig. 2-23. Splittings of the orbitals in trigonal bipyramidal ligand fields. (1) Regular trigonal bipyramid. (2) Distorted trigonal bipyramid.

Phillips to contain five-coordinate metal atoms. Barclay, Nyholm, and Parish [42] have suggested that the tendency of a metal to form five-coordinate species should increase with a decrease of the formal charge and effective nuclear charge on the metal atom, and with an increase in the polarizability of the anion. Dyer and Meek [95] have noted also that the five-coordinate complexes of d^8 transition metals are usually formed from ligands capable of extensive π-bonding.

Various salicylaldimine Schiff bases have been shown by Sacconi and co-workers to form five-coordinate chelates with many metals. As for the QAS derivatives, the structures of a few of the compounds [188, 189] have been determined by x-ray analysis, and then from isomorphism and spectral similarities similar structures have been assigned to analogous compounds. For example, Orioli, DiVaira, and Sacconi [188] report that the high spin bis-chelate of VIII (X = 5-Cl, R = R′ = C_2H_5) with Ni(II) has a distorted square-pyramidal structure with one $N(C_2H_5)_2$ group not coordinated with the nickel. The isomorphous Co(II) chelate is assigned the same structure [217]. And, from similarities in the spectra, Sacconi and co-workers deduce a similar structure for the bis-chelates of Ni(II) and VIII (X = 3-Cl or

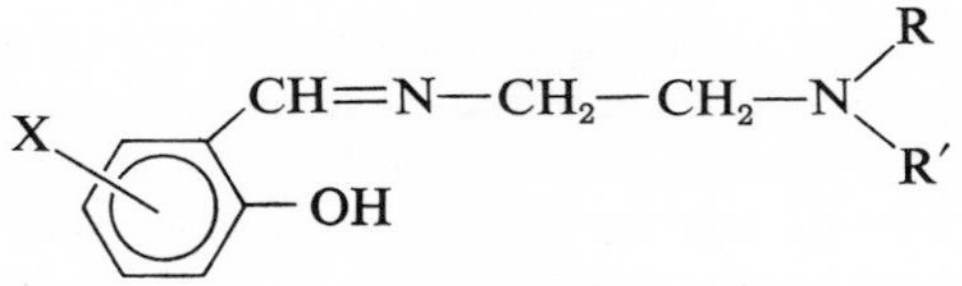

X–C_6H_3(OH)–CH=N—CH_2—CH_2—N(R)(R′)

VIII

3,4-benzo, R = R′ = C_2H_5). To similar nickel chelates with other X substituents they assign square-planar or octahedral configurations [216]. Typical spectra of the five-coordinate compounds may be found in Sec. 2-2C.

Several other studies of salicylaldimine complexes have indicated five-coordinate structures. The metals involved were Cu(II) [139, 221], vanadyl-(IV) [220], Zn(II) [189, 219], Co(II) [219], Mn(II) [219], and Ni(II) [219].

Ciampolini and Nardi [66] have reported the spectra of the high spin complexes of the divalent metals from Mn to Zn with the quadrivalent ligand, tris(2-dimethylaminoethyl)amine, and Ciampolini [64] discusses a Cr(II) compound of this ligand. He notes that the latter is the first high spin, five-coordinate Cr(II) compound to be reported. A trigonal bipyramidal structure is proposed for all these compounds. A study of the chelates of Mn to Zn with bis(2-dimethylaminoethyl)methylamine (trenMe) has been made by Ciampolini and Speroni [67]. The spectra of [Co(trenMe)Cl]Cl and [Ni(trenMe)Cl]Cl are given in Fig. 2-24 and Fig. 2-25, respectively. The assignments of these spectra are listed in Table 2-12 and are based on the crystal field calculations of Ciampolini [65], and of Ciampolini, Nardi, and

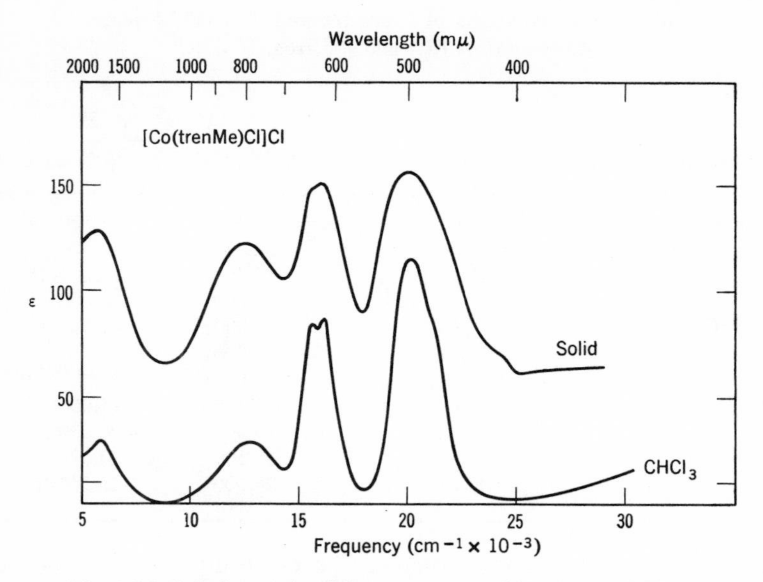

Fig. 2-24. Solution and solid state spectra of [Co(trenMe)Cl]Cl.

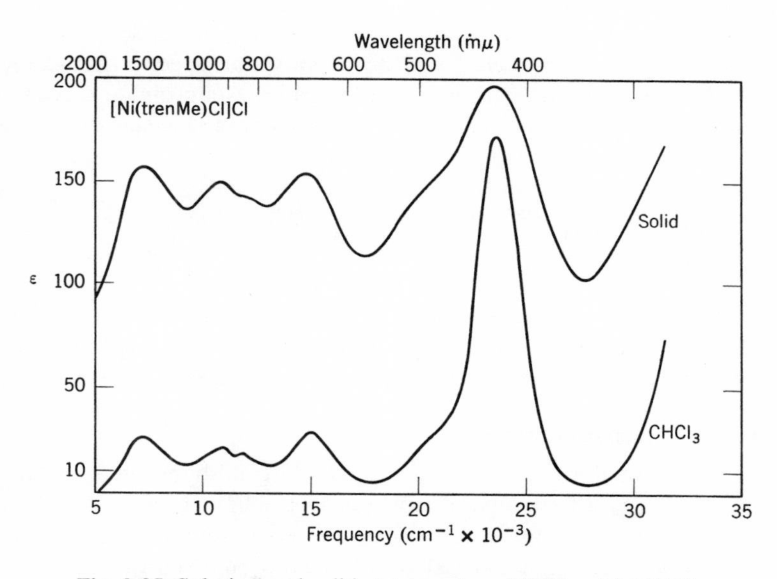

Fig. 2-25. Solution and solid state spectra of [Ni(trenMe)Cl]Cl.

Table 2-12 Comparison of Predicted and Observed Spectra of [Co(trenMe)Cl]Cl and [Ni(trenMe)Cl]Cl

Compound	Transition	Frequencies (cm^{-1})	
		Calculated	Observed
[Co(trenMe)Cl]Cl	$^4A_2'(F) \rightarrow {}^4E''(F)$	4,000	5,500
(μ = 1.2 a.u.)	$^4E''(F)$	12,000	12,500
	$^4A_2'(P)$	15,800	15,600–16,100
	$^4E''(P)$	20,000	20,000
[Ni(trenMe)Cl]Cl	$^3E'(F) \rightarrow {}^3E''(F)$	7,000	7,200
(μ = 2.2 a.u.)	$^1E'(F)$	12,900	11,000–12,000w
	$^3A_1''(F)$; $^3A_2''(F)$	14,000	
	$^3A_2'(F)$	15,300	15,000
	$^1E''(D)$	19,000	≈20,000sh
	$^3E''(P)$	23,500	23,500
	$^3A_2'(P)$	26,000	≈25,500sh

Speroni [18]. The spectra and structure of five-coordinate Cu(II) chelates of 6-methyl-2-picolylamine have also been reported recently [244]. Dori and Gray [92] indicate that their study of complexes of 1,1,7,7-tetraethyldiethylenetriamine, (Et_4den), suggests that "Ni(Et_4den)Cl_2 exists as a high spin, five-coordinate species in certain solutions and that Co(Et_4den)Cl_2 is five-coordinate and high spin both in the solid state and in solution."

Fackler [108] has also offered evidence from visible spectra that $Co(acac)_2$ reacts in benzene solution with pyridine to yield five-coordinate species. And Langford and co-workers [162] have shown that some bis-chelates of cobalt with

$$\left(\mathrm{N{\equiv}C{-}C(S){=}C(S){-}C{\equiv}N}\right)^{2-} \quad \text{and} \quad \left(\mathrm{CH_3{-}C_6H_3S_2}\right)^{2-}$$

form five- and six-coordinate adducts in solutions to which pyridine, *o*-phen, $As(C_6H_5)_3$, and other bases are added. The ultraviolet and visible spectra aided in the determination of the stoichiometry but not of the structure.

D. Square-Planar Coordination

The elongation along the *z* axis of two metal-ligand bonds in an octahedral complex results in a tetragonal distortion. In the limit of a very large tetragonal distortion the complex can best be treated as square-planar, with only four ligands surrounding the metal atom. The *d* orbital splitting in tetragonal fields has been discussed in detail by Ballhausen [1, pp. 99–103] and is shown in Fig. 2-26. The degeneracy of the d orbitals is completely lifted except for the lowest lying e_g state. Gray [20] notes that the exact order of the orbitals in a square-planar field is not certain, but the one shown in Fig. 2-26 is most consistent with results obtained in studies of square-planar cyanides and halides [138]. The observation that there are one unstable ($d_{x^2-y^2}$) and three stable orbitals is well-established.

Square-planar configurations are most important in the coordination of Cu(II), Ni(II), Pd(II), and Pt(II). Some square-planar complexes are also known for Co(II), Rh(I), Rh(II), Ir(I), Au(III), and Fe(I). All pertinent electronic spectral studies to about 1961 are discussed by Ballhausen [1]. Most of the more recent studies have been reviewed by Gray [20] (see also references cited in [254]).

An example of the study of the spectral properties of square-planar complexes is that of the Ni, Pd, and Pt complexes of maleonitriledithiolate anion, (N≡C—C(S)=C(S)—C≡N)$^{2-}$, [(MNT)$^{2-}$] [231]. The anion in the diamagnetic complex $[(CH_3)_4N]_2[Ni(MNT)_2]$ was shown by x-ray analysis [100] to have a square-planar configuration. Because the spectral and magnetic properties of the compound are essentially the same in the solid state and in various solvents, the ion was concluded to be square-planar in solution also. The analogous Pd(II) and Pt(II) complexes are diamagnetic and isomorphous with

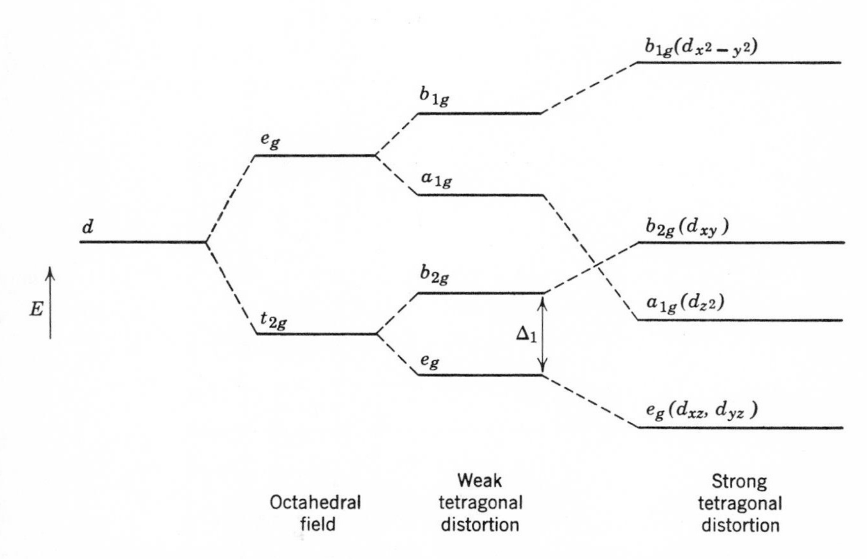

Fig. 2-26. The splittings of the d orbitals in weak and strong tetragonal fields.

various other square-planar complexes. Therefore they were also assigned a square-planar configuration. A complete MO calculation was carried out for the $[Ni(MNT)_2]^{2-}$ ion ($\boldsymbol{D}_{2h}$ symmetry) by a method reported earlier [246]. The calculation yielded all the orbital energies; the energy levels of greatest interest are shown in Fig. 2-27. According to this calculation the ground state of the complex is $(4b_{2g})^2(4a_g)^2 = {}^1A_g$. The actual spectrum of the ion in acetonitrile solution is shown in Fig. 2-28, and the assignments are listed in Table 2-13. From its intensity, the lowest energy band was assigned to a

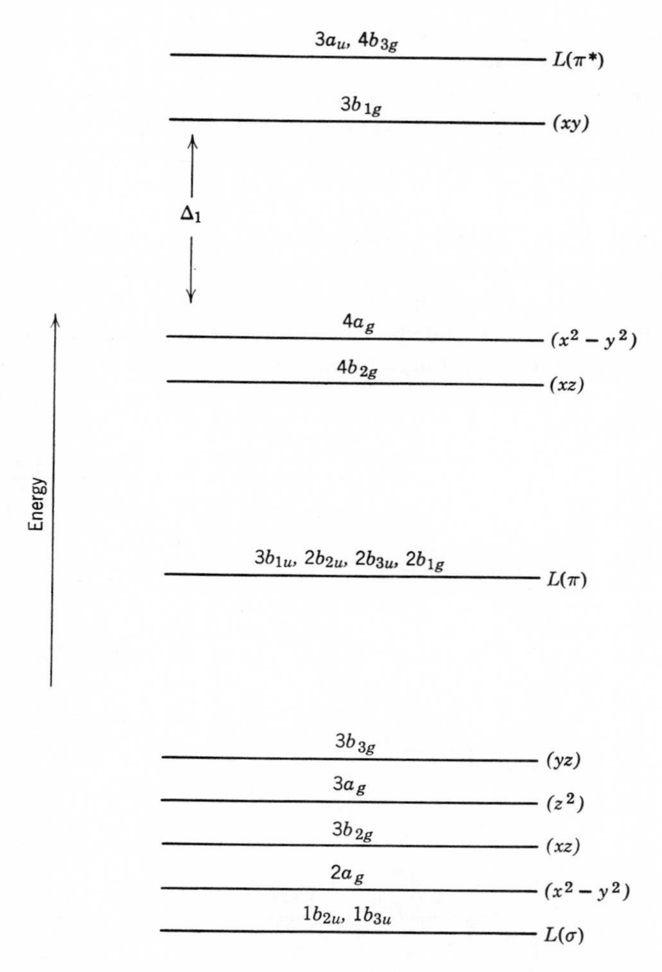

Fig. 2-27. The most important energy levels in $[Ni(MNT)_2]^{n-}$ complexes.

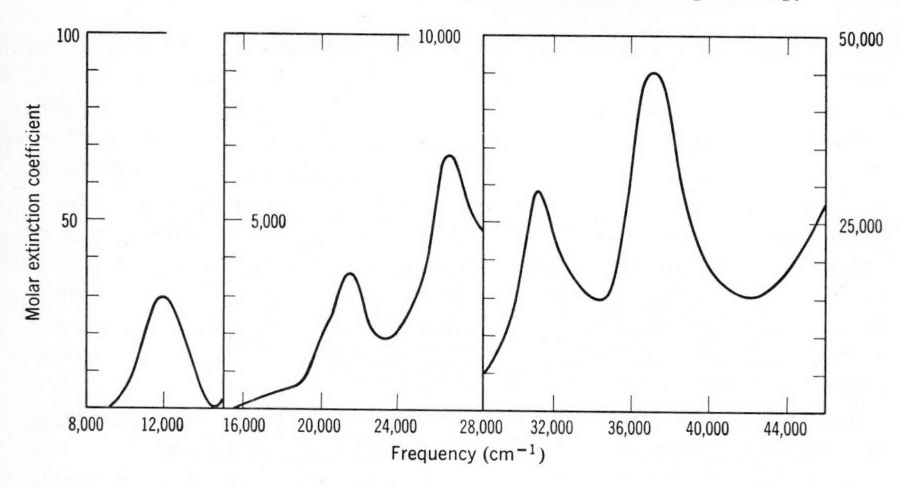

Fig. 2-28. The spectrum of $[Ni(MNT)_2]^{2-}$ in acetonitrile solution.

Laporte forbidden $d \rightarrow d$ transition. The other types of bands found in the spectrum are the allowed transitions on the ligand ($L \rightarrow L^*$), and the charge transfer transitions, ($M \rightarrow L$) and ($L \rightarrow M$). Because the first band in the free ligand occurs at 30,000 cm^{-1}, the band at 31,300 in the nickel chelate was assigned to an $L \rightarrow L^*$ transition. Work on square-planar halide complexes [138] has shown that they usually show two $L \rightarrow M$ charge transfer bands separated by about 10,000 cm^{-1}, the higher energy band being much

Table 2-13 The Electronic Spectrum of $[Ni(MNT)_2]^{2-}$

Band Maxima,[a] (cm^{-1})	ε	Assignments	Calculated Energies (cm^{-1})
11,690	30	${}^1A_g \rightarrow {}^1B_{1g}(4a_g \rightarrow 3b_{1g})$	17,000[b]
17,500sh	570	${}^1A_g \rightarrow {}^1B_{3g}(4b_{2g} \rightarrow 3b_{1g})$	17,600[b]
19,250sh	1,250	${}^1A_g \rightarrow {}^1A_u(4a_g \rightarrow 3a_u)$	17,600[c]
21,000	3,800	${}^1A_g \rightarrow {}^1B_{2u}(4b_{2g} \rightarrow 3a_u)$	18,400[b]
26,400	6,600	${}^1A_g \rightarrow {}^1B_{2u}, {}^1B_{3u}(2b_{3u}, 2b_{2u} \rightarrow 3b_{1g})$	26,800
31,300	30,000	${}^1A_g \rightarrow {}^1B_{2u}, {}^1B_{3u}(3b_{1u} \rightarrow 4b_{3g})$	30,000
37,000	50,000	${}^1A_g \rightarrow {}^1B_{2u}, {}^1B_{3u}(1b_{3u}, 1b_{2u} \rightarrow 3b_{1g})$	36,800

[a] Measured in acetonitrile solutions of $[(n\text{-}C_4H_9)_4N]_2[Ni(MNT)_2]$.
[b] Corrected for interelectronic repulsion energy, using $F_2 = 10F_4 = 800$ cm^{-1}.
[c] Correct relative energy of the $4a_g$ orbital is about $-88,000$ cm^{-1} from the position of the ${}^1A_g \rightarrow {}^1B_{1g}$ band.

more intense. Thus the bands at 26,400 and 37,000 cm^{-1} could be assigned to $L \rightarrow M$ charge transfer. The specific assignments were based on the MO calculation; for example, 26,400 cm^{-1} is a transition from an in-plane "ligand" π orbital to a σ^* "metal" MO, and 37,000 cm^{-1} is a transition from a "ligand" σ orbital to a σ^*MO. The band at 21,000 cm^{-1} was assigned to the first $M \rightarrow L$ charge transfer band, again in analogy with earlier work [138]. The other assignments are less certain and were made on the basis of MO calculation; similar assignments were made for the Pd and Pt complexes (Table 2-13). The authors present evidence for assigning the highest filled orbital in $\boldsymbol{D}_{2h}$ as the $d_{x^2-y^2}(4a_g)$ in these complexes. The value of Δ_1 was found to increase in the expected order, Ni < Pd < Pt, in the complex ions $[M(MNT)_2]^{2-}$.

The same molecular orbital calculation was later used as an aid in assigning the electronic structures of several other square-planar dithiooxalate complexes of Ni(II), Pt(II), and Au(III) [163], and of some toluenedithiolate $(TDT)^{2-}$ chelates [254]. The authors of the latter paper note that in the course of studies on planar chelates of MNT^{2-}, TDT^{2-}, and of *cis*-stilbenedithiolate, $[(C_6H_5)_2C_2S_2]^{2-}$, and *cis*-1,2-bis(trifluoromethyl)ethylenedithiolate, $[(CF_3)_2C_2S_2]^{2-}$, many unusual electronic states have been found, as, for example, a singlet state for copper, a doublet state for Ni [80], Pd [80], Pt [80], and Rh [53], a singlet and triplet state for Co [52], and a quartet state for Fe.

A study of complexes of 1,1-dicyanoethylene-2,2-dithiolate,

$$\left[\begin{array}{c} N{\equiv}C{-}\underset{\displaystyle CN}{\underset{|}{C}}{-\!-\!-}\underset{\displaystyle S}{\underset{|}{C}}{-}S \end{array}\right]^{2-} \equiv (i\text{-MNT})^{2-},$$

has also appeared [253]. The spectral and magnetic properties of the complexes are compared with those of the isomeric maleonitriledithiolate complexes. The square-planar Co, Ni, and Cu chelates of benzene-1,2-dithiolate and related ligands have also been the subject of a recent similar study [40]. Schrauzer and Mayweg [227] have presented an extensive study of bis(dithioglyoxal)nickel(II), $NiS_4C_4H_4$, a planar, diamagnetic complex, and of some related compounds. The electronic spectra of $[NiS_4C_4H_4]^{0,1-,2-}$ are given and interpreted by means of MO calculations. These compounds are of interest because they represent molecules in which there is a high degree of ground state π electron delocalization. Their data are also compared with those on Ni-MNT complexes [231].

For a review of square-planar complexes containing S donors, see [20]. A recent review by Livingstone [22] of metal complexes of ligands containing sulfur, selenium, and tellurium as donor atoms is also interesting.

Baddley et al. [31] have studied the square-planar complexes $[Au(den)X]^{2+}$ and $[Au(denH)X]^{+}$ and have discussed the assignment of the spectra based on a molecular orbital scheme for $[M(den)X]^{n+}$ complexes. The high energy band in all the generalized complexes of the latter type is at about 30,000 cm^{-1} ($\varepsilon = \sim 500$) and is assigned to $\pi^* \rightarrow d_{x^2-y^2}$. This transition is at about 36,000 cm^{-1} ($\varepsilon = \sim 2500$) in the Au(I) complex; an additional band at 27,000 cm^{-1} ($\varepsilon = \sim 2000$) has been assigned to $p\pi$(den) $\rightarrow d_{x^2-y^2}$.

Tentative assignments have been made of the spectra of *cis*- and *trans*-$[Pt(gly)_2]$ by Balzani et al. [41] .The bands and their extinction coefficients are given in Table 2-14. The assignments of these bands have followed the

Table 2-14 Spectra of *cis*- and *trans*-Bis(glycino)platinum(II)

cis		*trans*	
cm^{-1}	ε	cm^{-1}	ε
30,800	28	~31,200	36
40,000sh	100	37,000sh	100

energy level scheme of Martin and Lenhardt [175]. The local symmetry of the *cis* isomer is C_{2v}, and the two bands are assigned to ${}^1A_1 \rightarrow {}^1A_1$, and ${}^1A_1 \rightarrow {}^1A_2$. For the *trans* isomer the bands have been assigned for D_{2h} symmetry as ${}^1A_g \rightarrow {}^1A_g$ and ${}^1A_g \rightarrow {}^1B_{1g}$, respectively. Other spectra of square-planar chelates may be found in Sec. 2-21; for square-planar Cu(II) see the earlier discussion [72] (p. 96).

E. Trigonal Prismatic Coordination

X-ray analysis of $Re[S_2C_2(C_6H_5)_2]_3$ [101] and of several similar molecules has shown that they possess a trigonal prismatic structure (p. 53); a complete listing of the x-ray analyses is given in [234]. Trigonal prismatic complexes have only been observed with bidentate, unsaturated sulfur donor ligands such as *cis*-stilbenedithiolate, benzene-1,2-dithiolate, and toluene-3,4-dithiolate. The ligand undoubtedly is very important in confirming this unusual geometry, and it might be anticipated that similar low electronegativity donor ligands (P, As, Sb, Se, Te) may also form complexes of this type. These complexes display a variability of oxidation states, but it is believed [129] that oxidation-reduction involves the entire complex rather than the central metal ion. The electronic spectra of these complexes are quite similar and are characterized by two intense bands at about 15,000 cm^{-1} ($\varepsilon \sim 25{,}000$) and 23,000 cm^{-1} ($\varepsilon \sim 15{,}000$). Two conflicting [228, 234] molecular orbital schemes have been developed to make assignments of the spectra and correlate the properties of these complexes. Extensive discussion of these assignments is not warranted at this time. Other discussions of trigonal prismatic coordination may be found in [129, 187].

F. Equilibria

Equilibria between the octahedral, tetrahedral, square-planar and the five-coordinate configurations have been a subject of considerable interest in the past few years. Absorption spectroscopy has played a very important role in the elucidation of these equilibria largely because of the simplicity of the experiments and usually straightforward interpretations of the results. The study of configurational equilibria has not been restricted to the N-substituted salicylaldimine chelates (VII, p. 116), but these chelates constitute the most extensively investigated examples. The equilibria are dependent on the substituents R and X and also on the solvent. Sacconi [212] clearly demonstrated that the temperature dependence of the spectrum of bis(N-decylsalicylaldimino)nickel(II) (Fig. 2-29) indicates planar $\rightleftarrows$ tetrahedral equilibrium and that increasing temperature favors the tetrahedral configuration. The bands at 7200 cm^{-1} and 11,200 cm^{-1} are characteristically tetrahedral. The intensity of these bands increases with temperature, whereas the 16,000 cm^{-1} (square-planar) band disappears. Sacconi, Ciampolini, and Nardi [213] have studied the dependence of the equilibria on the nature of R

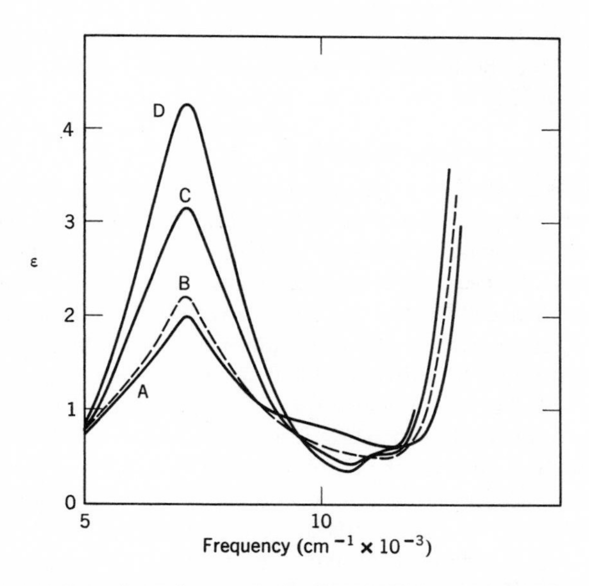

Fig. 2-29. Temperature dependence of the ligand-field spectrum of molten bis(N-*n*-decylsalicylaldimino)nickel(II). A, 100°; B, 130°; C, 170°; D, 200°.

and X (VII). The results for the solid Ni(II) chelates are summarized in Table 2-15. The diamagnetic planar chelates are olive green and the paramagnetic tetrahedral chelates are brown. The configuration for the crystalline

Table 2-15 Structure of the X-substituted Bis-(R-N-salicylaldimino)nickel(II) Complexes in the Solid State from the Reflectance Spectra[a]

R	X									
	H	3-Me	5-Me	3-Cl	5-Cl	3-Br	$3\text{-}NO_2$	$5\text{-}NO_2$	5,6-Benzo	3,4-Benzo
n-Propyl	*P*	*P*	*P*	*P*	*P*			*P*	*P*	*P*
i-Propyl	*T*	*P*	*P*	*P*	*P*	*P*	*P*	*P*	*P*	*T*
t-Butyl	*T*			*T*	*T*	*T*	*T*	*T*	*T*	*T*

[a] *P*, planar form; *T*, tetrahedral form. $\bar{\nu}_{max}$ of the planar form, ca. 16,000 cm^{-1}. $\bar{\nu}_{max}$ of the tetrahedral form, ca. 6700, 9500, 10,900, 14,000, 16,900, 19,600 cm^{-1}.

chelates was determined by magnetic and reflectance spectral measurements. The configuration in the solid state is dependent on the bulk of the group R. The more bulky groups favor the tetrahedral configuration, but for R = propyl or butyl the electronic, geometric, and crystal packing factors associated with the substituent X are also important. The temperature dependence of the spectra of these chelates in non-coordinating solvents, such as *m*-xylene and dibenzyl, has also been investigated by these authors, and except for R = tBu the tetrahedral configuration is favored at increased temperatures. In the range of temperatures between 80° and 170°, log *k* for the equilibria varies inversely with temperature. This variation facilitated a determination of the thermodynamic parameters for these equilibria from the spectra. These results are summarized in Table 2-16. The free energy difference between the configurations is small, and the enthalpy of formation favors the planar configuration, whereas the entropy favors the tetrahedral configuration for both series of R. The favorable entropy has been attributed to the change in spin multiplicity and the greater freedom of R group rotation in the tetrahedral configuration. When R = methyl, neither the planar nor the tetrahedral configuration is favored [207], but the magnetic and spectral properties are more consistent with an octahedral configuration achieved by polymerization. Holm and Swaminathan [148] have observed that the Ni(II) N-arylsalicylaldimine chelates exhibit the equilibrium planar $\rightleftarrows$ octahedral (polymeric) for R = phenyl and para-substituted phenyl, and that the ortho-substituted

Table 2-16 Thermodynamic Functions for the Planar ⇄ Tetrahedral Equilibrium of Substituted Bis-(R-N-salicylaldimino)nickel(II) Complexes at 120° in Dibenzyl Solutions

R	X	ΔF (kcal/mole)	ΔH (kcal/mole)	ΔS (e.u.)
n-Propyl	H	2.9	4.6	4
	3-Cl[a]	—	—	—
	5-Me	2.8	4.6	4
	5-Cl	2.7	5.2	6
i-Propyl	H	0.52	3.2	10
	3-Cl	0.35	3.0	7
	5Me	0.20	2.5	6
	5-Cl	0.16	2.0	4

[a] No spectroscopic evidence for the presence of the tetrahedral form.

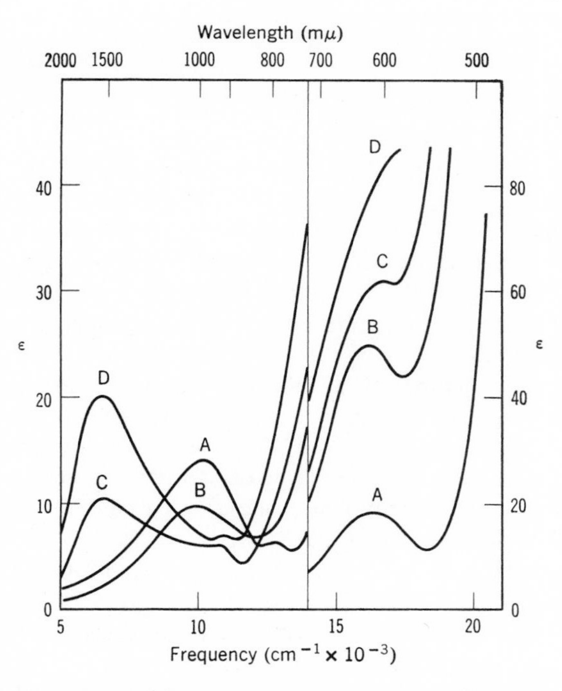

Fig. 2-30. Temperature dependence of the spectrum of bis(N-3-tolylsalicylaldimino)-nickel(II). A, in xylene at 25°; B, in xylene at 80°; C, in dibenzyl at 120°; D, in dibenzyl at 180°.

phenyl chelates are planar. Sacconi and Ciampolini [211] have extended this study, and from the temperature dependence of the spectra and magnetic susceptibilities they have demonstrated that the equilibria octahedral (polymeric) $\rightleftarrows$ planar $\rightleftarrows$ tetrahedral prevail except for the ortho-substituted aryl chelates, which are planar at all temperatures. The spectral results of Sacconi and Ciampolini [211] are given in Fig. 2-30 and Table 2-17.

Table 2-17 Spectrophotometric Data for Bis(N-arylsalicylaldimino)-nickel(II) Complexes in Dibenzyl at 180°

Aryl	Concentration (mmolal)	ν_{max} (cm^{-1})	ε_{max}	Tetrahedral (%)[a]
C_6H_5	21.5	6600	17	40
$3\text{-}CH_3C_6H_4$	19.8	6900	20	48
$4\text{-}CH_3C_6H_4$	21.6	6900	14	33
$3\text{-}ClC_6H_4$	19.8	6900	10	24
$4\text{-}ClC_6H_4$	19.1	6700	11	26
$\beta\text{-}C_{10}H_7$	19.5	6500	16	38

[a] Calculated from the formula % tetrahedral = $100 \times \varepsilon_{max}/42$.

At room temperature (spectrum A, Fig. 2-30) the chelate exists in the octahedral associated form (more than 80%) and the planar form (less than 20%) while the tetrahedral form is negligible. Increased temperature (spectra C and D) produces an increase in the characteristic tetrahedral absorption at 6700 cm^{-1}, and the percentage of the tetrahedral form is estimated (Table 2-17) from the anticipated maximum extinction coefficient [212] of 42 for 100% tetrahedral. The magnetic susceptibilities indicate that all three forms coexist even at the higher temperatures. Sacconi's group [215, 216, 218] has extended these studies to the N-substituted and N,N-substituted ethylene-diamine Schiff bases of salicylaldehyde; structure VII (p. 116). The N-substituted chelates with Ni(II) [215] display the equilibrium

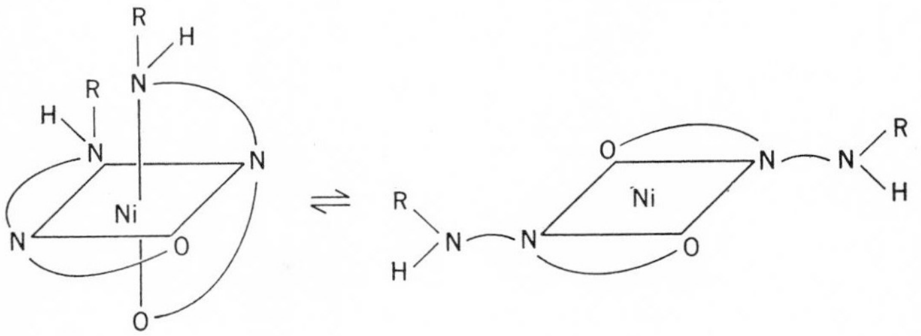

$\rightleftharpoons$

in non-coordinating solvents, the planar form being favored at higher temperatures. The N,N-substituted chelates of Ni(II) have more complex equilibria, in which six-, five-, and four-coordinate species coexist.

The existence of the tetragonal pyramidal complex has been confirmed by x-ray [218] and spectral evidence. The spectra of the Ni(II) chelate with R = R′ = ethyl and X = 5Cl are given in Fig. 2-31. Spectrum A in $CHCl_3$ is indicative of a mixture of all three forms. The bands at 7000, 9800, 12,700, and 16,200 cm^{-1} are indicative of the five-coordinate species, but these bands overlap those of the six-coordinate species (bis-tridentate), which predominates in pyridine solution (spectrum C); the latter species is characterized by absorptions at 10,000 and 16,800 cm^{-1}. In benzene (spectrum B) the four-coordinate species predominates, but five- and six-coordinate forms are also present. Increased temperature

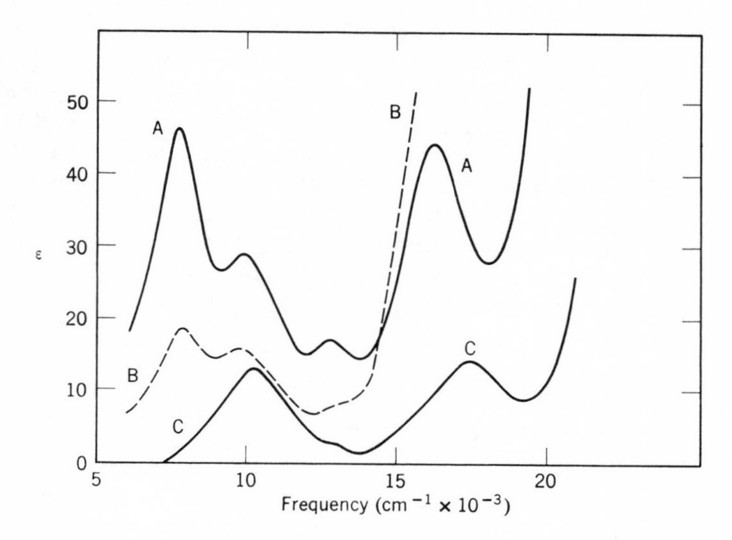

Fig. 2-31. Absorption spectra of the [5-Cl-SALen-NN$(C_2H_5)_2]_2$Ni complex at room temperature. A, in chloroform; B, in benzene; C, in pyridine.

favors formation of the planar form, and for a given X substituent the planar species is favored in the series

$$—N(CH_3)_2 < \overline{\text{cyclic}}N(CH_2)_4 < —N(C_2H_5)_2 < \overline{\text{cyclic}}N(CH_2)_5,$$

which is the order of the donor power of the groups except for the piperidine group, displaced for steric reasons. Similar results with these ligands have been obtained with Co(II), except that in this case equilibrium mixtures of octahedral, pyramidal, and tetrahedral species have been observed in the spectra. Other spectral studies of configurational equilibria in salicylaldimines and similar ligands may be found in [61, 62, 99, and 149]. A comparison of the thermodynamic parameters derived from spectral and NMR data [99] for the aminotroponeimineates shows that the two methods give essentially the same results. The pressure dependence of the spectra of the salicylaldimine and aminotroponeimine complexes has recently been studied by Ewald and Sinn [105]; their results show that the planar form has a smaller volume in the aminotroponeimineates and the octahedral associated form is favored by increased pressure in the salicylaldimineates.

Jørgensen [157] has shown that solutions of *cis*-$[Ni(trien)(H_2O)_2]^{2+}$ change color from blue to yellow on addition of inert salts. He showed from the spectra that the equilibrium

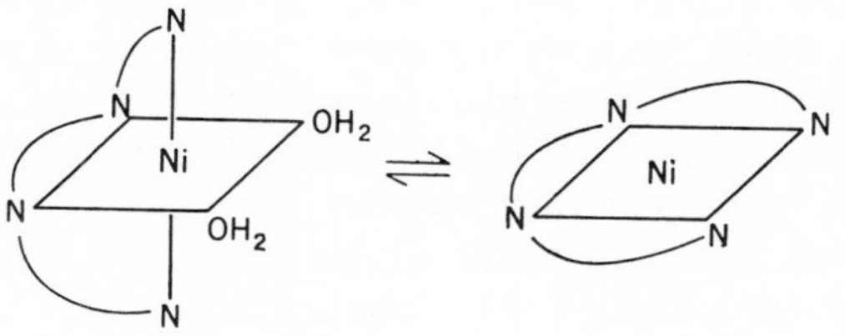

gives rise to these color changes. Sone and Kato [233] have observed similar equilibria for $[Ni(en)_2(H_2O)_2]^{2+}$ and $[Ni(pn)_2(H_2O)_2]^{2+}$ in which the planar form is favored at higher temperatures, and at a given temperature the propylenediamine chelate contains more of the planar form. Curtis and House [75] have shown that $[Ni(trien)(py)_2]^{2+}$ also exhibits the same equilibrium except for the apparent loss of the aromatic amine instead of water.

Higginson, Nyburg, and Wood [146] have studied the octahedral $\rightleftharpoons$ planar equilibrium in bis(meso-stilbenediamine)nickel(II) dichloroacetate. In ethanol solution at room temperature one-third of the complex exists as the planar form, and the dichloroacetate is coordinated axially in the remainder. Crystals formed from these solutions may be either blue or yellow. The blue form is composed of *trans*-bis(dichloroacetato)bis(meso-stilbenediamine)-nickel(II) exclusively. The yellow crystals [184], however, are extremely interesting in that they contain one-third planar (cationic) and two-thirds octahedral (anionic) complexes.

The thermochromism of bis(N,N-diethylethylenediamine)copper(II) perchlorate has been discussed by Hatfield, Piper, and Klabunde [145]. The change from the red (room temperature) form to the blue (44°) form in this chelate has been ascribed to a conformational change of the chelate ring with concomitant alterations in the ligand field. This interesting proposal has not been pursued further. Another type of thermochromism is due to thermal equilibrium between two possible electronic ground states (electronic isomerism?). König and Madeja [161] claim that changes in the spectrum of $[Fe(phen)_2(NCS)_2]$ arise from equilibrium between the possible $^5T_2$ and $^1A_1$ ground states in the vicinity of the crossing point in the d^6 Tanabe-Sugano diagram (cf. p. 88). Ewald et al. [106] had made similar observations on Fe(III) chelates of N,N-dialkyldithiocarbamates which are apparently in the vicinity of the $^6A_2$–$^2T_2$ crossover point for d^5. Stoufer et al. [236, 237] have discussed the thermal equilibrium at the crossing point of d^7 Co(II) complexes.

Equilibria associated with the stepwise formation of a labile chelate may also be studied spectrophotometrically. The determination of stability constants by spectrophotometry has been comprehensively reviewed by Rossotti and Rossotti [14]. A simple method for the determination of stability constants has been described by Vareille [245]. This method utilizes the isosbestic points (points at which two species have the same extinction coefficient) in the spectra determined at different values of pH. A plot of the extinction coefficient (or optical density) at the isosbestic wavelength versus pH may be used to determine the equilibrium constant between the two species. The absorption spectrum of the last member of an isosbestic family may, under the proper conditions, be indicative of an intermediate species. Examples of this method may be found in Sec. 2-2H. DeWitt and Watters [83] have used a different method to deduce the stability constants for the mixed chelate [Cu(ox)(en)] and have used the constants to construct the spectrum.

G. Crystal Spectra

The spectroscopy of crystalline chelates can give considerable information about the energy levels of the chromophore, particularly if the chromophore is well oriented in one of the crystallographic directions and the spectra are determined with polarized light. The selection rules of the transitions observed with polarized light usually allow the splittings of the lower symmetry ligand fields to be determined with moderate accuracy. These observations are enhanced when the experiments are carried out at low temperatures. It has also been shown by Ballhausen and Moffitt [35] that the observed selection rules for the polarized crystal spectra of centric complexes (e.g., $[Co(en)_2Cl_2]ClO_4$)

may best be interpreted using the vibronic mechanism. Reviews by McClure [23] and Ferguson [19] should be consulted for more details.

An example of the study of polarized crystal spectra of metal chelates is afforded by the work of Piper and Carlin [196]. The compounds investigated were $NaMg[Al(ox)_3]\cdot 9H_2O$ in which part of the aluminum atoms were replaced by trivalent Ti, V, Cr, Mn, Fe, and Co. The resulting six-coordinate ions, $[M(ox)_3]^{3-}$, have $\boldsymbol{D}_3$ symmetry. The irreducible representations of this point group are A_1, A_2, and E. Since there is no center of inversion, the metal 3d orbitals mix with odd atomic and molecular orbitals, and the electronic transitions are no longer Laporte-forbidden. The selection rules for the spectra with respect to the major axis of $\boldsymbol{D}_3$ are

$$A_{1,2} \overset{\perp}{\longleftrightarrow} E; \qquad A_1 \overset{\parallel}{\longleftrightarrow} A_2;$$
$$A_{1,2} \not\longleftrightarrow A_{1,2}; \qquad E \overset{\parallel,\perp}{\longleftrightarrow} E; \tag{2-25}$$

where $\parallel$ and $\perp$ refer to polarization of the electric vector of the incident light parallel and perpendicular to the trigonal axis (C_3). The spectra obtained with the electric vector $\parallel$ and $\perp$ to the trigonal axis are called π and σ spectra, respectively. An example of the results of this study is the spectrum of the Cr^{3+} complex at 77°K (Fig. 2-32 and Table 2-18). In a pure octahedral field three spin-allowed transitions are expected: ${}^4A_2 \rightarrow {}^4T_2$, ${}^4T_1(F)$, ${}^4T_1(P)$. Each

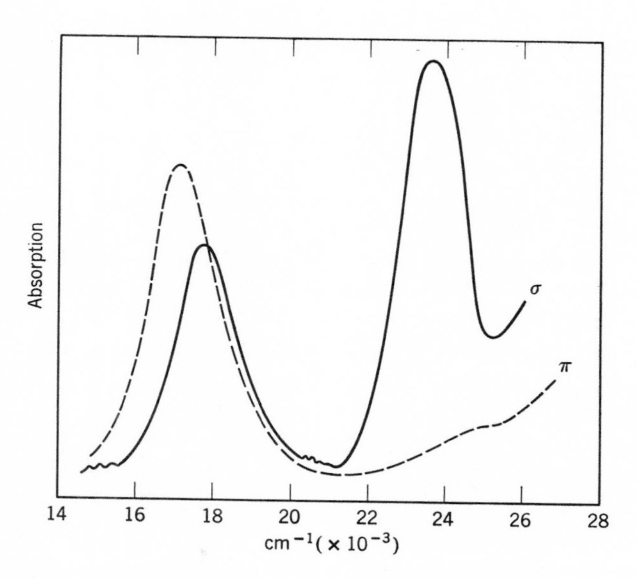

Fig. 2-32. Polarized spectra of Cr(III) in $NaMgAl(C_2O_4)_3\cdot 9H_2O$. $T = 77$°K.

Table 2-18 Spectra of Cr(III) in $NaMgAl(ox)_3 \cdot 9H_2O$

Ion and Ground State	Band Max. (cm^{-1})	$\Gamma(D_3)$	$\Gamma(O_h)$	Remarks
Cr(III)	14,455 }		^2E	$f = 1.2 \times 10^{-7}$(axial)
($^4A_2$)	14,476 }			$f = 1.3 \times 10^{-7}$(axial)
	15,216 }			
	15,284 }		$^2T_1$	all doublets at 77°K
	15,323 }			
	17,316 π	$^4A_1$	$^4T_2$	$\varepsilon_{\parallel}/\varepsilon_{\perp} = 1.3$ at 25°C
	17,620 σ	^4E_a		
	20,555 }			
	20,704 }		$^2T_2$	axial
	21,119 }			
	23,670 σ	^4E_b		$\varepsilon_{\parallel}/\varepsilon_{\perp} = 0.06$ at 25°
	—	$^4A_{2b}$	$^4T_1$	

of these will be further split in D_3 as follows: T_2 becomes $A_1 + E$, and T_1 becomes $A_2 + E$. The third transition is not observed in the spectra, because it appears in the ultraviolet and is obscured by more intense ultraviolet transitions. The two components of the first transition are observed and are assigned on the basis of the selection rules; only one of the two components of the second band is observed and is assigned on the same basis. From the spectra the authors have calculated Dq for the $[Cr(ox)_3]^{3-}$ ion to be 1770 cm^{-1} and the trigonal field splitting parameter K to be +270 cm^{-1}. A comparison of the σ and π spectra on the $(10\bar{1}1)$ face with the axial spectrum on the (0001) face shows that the transitions are electric dipolar and not magnetic dipolar in character. The very weak bands are the spin-forbidden chromium absorptions which have been extensively studied in ruby [172] and emerald [256]. Piper and Carlin [196] also found weaker and more diffuse lines at 14,771 and 14,792 cm^{-1}. These lines are 316 cm^{-1} above the ^2E doublet (14,455 and 14,476 cm^{-1}) and are ascribed to a vibrational transition superimposed on the electronic transition. The authors suggest that it is one of the Cr-O stretching vibrations. Normal coordinate analysis of $K_3[Cr(ox)_3] \cdot 3H_2O$ shows that it is not one of the Cr-O modes, but may possibly be an out-of-plane vibration. Tris(acetylacetono)chromium(III) has been the subject of several crystal studies. Chakravorty and Basu [60] noted that the σ spectrum has a band at 17,100 cm^{-1}, whereas in the π spectrum it is at 18,000 cm^{-1}. The two components of different polarization are the result of the trigonal splitting of the octahedral field. Forster and DeArmond [124] have studied the luminescence of this compound, and Piper and Carlin [194] have made a further study of its polarized crystal spectrum. The latter have calculated Dq and K for this compound to be 1810 and 530 cm^{-1}, respectively. They noted considerable fine structure in the weak bands around 13,000–15,000 cm^{-1}. The analysis of this region (12,000–15,000 cm^{-1}) carried out by Forster and Armendarez [125] at 4°K led to the conclusion that the fine structure of the bands are vibronic components of the $^4A_2 \rightarrow {}^2E$ transition. For example, they noted that in the 12,882–12,931 cm^{-1} region the spacing and polarization of the lines are repeated at 455 and 677 cm^{-1} intervals. Nakamoto and co-workers [182] observed IR bands at 459 and 677 cm^{-1} and assigned them by normal coordinate analysis to the Cr-O stretch, and ring def. + Cr-O stretch, respectively. Piper and Carlin [197] have noted that, although the crystal field ionic model is satisfactory in accounting for the sign of K and the intensity ratios for $[Cr(ox)_3]^{3-}$, it is not helpful for $Cr(acac)_3$. They attribute this to covalency in the latter compound, and especially to π covalency in the chelate ring.

Ferguson [116] has also reported the polarized crystal spectra of bis-(salicylaldimino)nickel(II) and the N-methyl analog. Analysis of the spectra made with two different crystal faces, (001) and (100), indicates that the crystal field is rhombic ($\boldsymbol{D}_{2h}$), and that the in-plane symmetry axes lie between the metal-ligand bonds and not along the bonds. Ferguson [118] has also reported the crystal spectra of five-coordinate salicylaldehydeethylenediamine copper(II) chelates. The interpretation of Ferguson's results on the Schiff base chelates has been criticized by Belford and Piper [49]. Differences in the interpretation of the spectrum of $Cu(acac)_2$ [29, 85, 117, 119, 120, 198] may be due to the lack of ideal alignment of the molecules in this crystal and to the equivocal selection rules. The results of Cotton and Wise [72] on the analogous $Cu(DPM)_2$, which does have ideal molecular alignment, have apparently resolved the discrepancy. Basu et al. [45] have also reported observing four $d \rightarrow d$ transitions in the polarized crystal spectrum of bis(3-phenylacetylacetono)copper(II). Their results are similar to those of Cotton and Wise [72]. Piper [195] has reported a large trigonal splitting in the first band of $Co(acac)_3$. The band in the $\perp$ spectrum is at 17,000 cm^{-1} (^1E), and in the $\parallel$ spectrum at 16,200 cm^{-1} ($^1A_2$), and K was found to be 600 cm^{-1}. Vanadyl acetylacetonate has been studied (unpolarized) by Basu, Yeranos, and Belford [46] and at 90°K a 1000 cm^{-1} vibrational progression is observed in the higher energy bands (24,000 cm^{-1}). No detailed assignments have been made. Dingle [86] has assigned the transition at 17,500 cm^{-1} in $Mn(acac)_3$ to the low field ($\boldsymbol{D}_2$ symmetry) components of $^5E_g \rightarrow {}^5T_{2g}(\boldsymbol{O}_h)$. The low energy band at 9000 cm^{-1} may be either a spin-forbidden transition ($^5E_g \rightarrow {}^3T_{2g}$) or a charge transfer band [89]. The near-infrared crystal spectrum of $Os(acac)_3$ has a highly structured band at about 4000 cm^{-1}, and Dingle [88] has assigned this as a transition to a spin-orbit component of the ground state configuration. The structure in this band has been assigned to vibrational progressions of Os—O (300 cm^{-1}), and C—O (1400 cm^{-1}) stretching modes. Trigonal effects are also discussed by Dingle.

The crystal spectrum of bis(dimethylglyoximato)nickel(II) is characterized by a strong band in the 20,000 cm^{-1} region which is not present in the solution spectrum. This band had previously been assigned [257] to an interaction of the metal ions in the solid (metal-metal bond); however, more recent investigations by Basu et al. [47] on the pure and diluted solid have precluded this interpretation. The assignment of the sharp band at 18,750 cm^{-1} in $Ni(dmg)_2$ to a $d \rightarrow p$ transition [261] has also been refuted by Basu et al. [47] on the basis of intensity.

Basu and Belford [48] have also reported polarized crystal spectra of a series of square-planar nickel salicylaldimines. The bands at about 19,000 cm^{-1} are temperature dependent and are apparently coupled to a 320 cm^{-1} vibration. Uninterpreted fine structure at about 21,000 cm^{-1} has been observed in one direction ($\| c$) in the spectrum of bis(N-methylsalicylaldimino)nickel(II). This structure disappears on addition of small amounts of Cu(II) impurities.

Dingle [87] has investigated the crystal spectrum of

$$dl\text{- and } d\text{-}Na[Co(en)_3]Cl_3 \cdot 6H_2O$$

in the possible orientations at temperatures down to that of liquid helium. His results show that the observed transitions are almost entirely electric dipolar in nature and that the trigonal splitting is negligible. The slight differences which exist between the crystal spectrum of the racemic and optically active crystals have been attributed to the differences in the lattice site symmetry of the two chromophores. Progressions in the t_{1u} and t_{2u} non-totally symmetric vibrations (185, 345, and $\sim$400 cm^{-1}) have been observed in addition to the 255 cm^{-1} totally symmetric vibration. The static contribution to the total intensity has been estimated from the vibrational pattern; it contributes approximately 10–15% of the total intensity in the racemic crystal and considerably less in the optically active crystal. This chromophore had previously been examined by Yamada and Tsuchida [259]. Dingle [90] has recently studied the crystal spectra of

$$trans\text{-}[Co(en)_2Cl_2]ClO_4, \qquad trans\text{-}[Co(en)_2Cl_2]Cl \cdot HCl \cdot 2H_2O,$$

and also

$$trans\text{-}[Co(l\text{-}pn)_2Cl_2]Cl \cdot HCl \cdot 2H_2O.$$

The spectra of these three crystals are quite similar even down to 25°K, and no vibrational progressions have been observed. At 25°K the bands and intensities are 16,350(32), 27,100(15), 28,500(2100) cm^{-1} parallel to the Cl-Co-Cl axis and 16,600(10), 22,500($\sim$23), 24,250(30), 27,100(25), and 28,500($>$100) cm^{-1} perpendicular. Dq is -2175 cm^{-1} with $Ds = 600$ cm^{-1} and $Dt = 700$ cm^{-1}. The intensities are very temperature-dependent and the results have been interpreted using a vibronic model under $\boldsymbol{D}_{4h}$ symmetry.

Conformational effects apparently play no role in the spectra. Wentworth [252] and Yamada et al. [258] had previously studied these *trans* systems.

Palmer and Piper [191] have carried out a comprehensive investigation of divalent Cu, Ni, Co, Fe, and Ru complexes of the types $[M(bipy)_3]Br_2 \cdot 6H_2O$ and $[M(bipy)_3]SO_4 \cdot 7H_2O$. Their results are particularly interesting for Cu(II), because they observe two bands at 6400 cm^{-1} and 14,400 cm^{-1}, and attribute them to a large splitting of $^2T_{2g}(O_h)$. The trigonal splittings in Ni(II) and Co(II) are small; on the basis of octahedral symmetry $Dq = -1280$ and -1270 cm^{-1}, respectively, and $B = 710$ and 790 cm^{-1}, respectively. The $d \rightarrow d$ transitions in the Fe(II) crystals are apparently obscured by the charge transfer bands. Crystal spectra of Co(II) chelates of tris(1-pyrazolyl)methane have been reported by Jesson [152] along with comprehensive ESR studies. The spectra have been interpreted on the basis of an octahedral model.

The crystal spectrum of $[Cu(ac)_2 \cdot H_2O]_2$ in which the acetate anion is bridging bidentate has been studied by Tonnet et al. [242] and by Reimann et al. [202]. Three bands are observed at 11,000, 14,400, and 27,000 cm^{-1}. The latter band may not be a $d \rightarrow d$ transition; the assignment of the first band is either $d_{x^2-y^2} \rightarrow d_{z^2}$ or d_{xy}, and that of the 14,400 cm^{-1} band is $d_{x^2-y^2} \rightarrow d_{xz}, d_{yz}$. Dijkgraaf [84] has studied the crystal spectra of copper alaninate and copper α-aminobutyrate, and the crystal spectrum of copper phthalocyanine has been discussed by Fielding and McKay [122] and by Day et al. [81]. The results for Cu phthalocyanine are very interesting as they indicate a Davydov [3] resonance interaction between the translationally non-equivalent molecules in the unit cell.

H. Biological Applications

Metalloporphyrins play an important role in photosynthesis, nitrogen fixation in plants, and oxygen transport in the blood, and they have received considerable attention from biochemists. Electronic spectroscopy has proved to be a useful tool in elucidating structure and bonding in these compounds [6, 7].

The square-planar chelates of porphyrins with divalent metals (Fig. 2-33) exhibit the three bands listed in Table 2-19. In general, the α and β bands are of moderate intensity and the Soret band is much stronger than the others. The stable chelates such as those of Ni(II), Co(II), and Cu(II) have the order of intensities $\alpha > \beta$, while the ratio, α/β, is decreased or the order is reversed in the less stable chelates, such as those of Mg(II), Zn(II), and Cd(II). It is known that the thermodynamic stabilities of metalloporphyrins are

$$\mathrm{Ni(II)} > \mathrm{Co(II)} > \mathrm{Ag(II)} > \mathrm{Cu(II)} > \mathrm{Zn(II)} > \mathrm{Cd(II)}.$$

Table 2-19 indicates that the bands are shifted to the red as the stability decreases with concomitant decrease in the ratio of intensities, α/β.

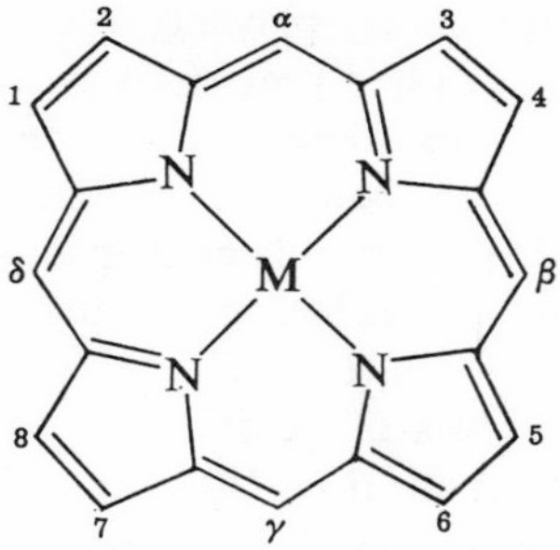

Fig. 2-33. Structure of a metal porphyrin complex.

A number of theoretical analyses have been carried out to explain the origin of these bands [135, 170, 27]. None of them is conclusive as yet. According to Gouterman [135] the important MO orbitals which account for these transitions are the a_{1u} and a_{2u} (top filled orbitals) and the e_g (lowest empty orbitals), which are shown in Fig. 2-34. In metal chelates the p_π orbitals of the metal interact with the π electrons of the ring. Because of the nodal properties shown in Fig. 2-34, only the a_{2u} orbital can interact with the p_π orbitals. This raises the energy of the a_{2u} orbital. Thus the visible bands (α and β) are assigned to $a_{2u} \rightarrow e_g$, and the Soret band is assigned to the $a_{1u} \rightarrow e_g$ transition. It can be shown that, as the metal becomes more electropositive, the a_{2u} orbital rises in energy and the visible bands shift to the red and become more intense.

The origin of splitting of the $a_{2u} \rightarrow e_g$ transition into two bands (α and β) is not clear. According to Platt [27] the α band is a forbidden $0 \rightarrow 0$ transition,

Table 2-19 Spectra of Metal Chelates of Protoporphyrin[a] in CCl_4 Solution (mμ)[b]

	α	β	Soret
Co(II)	561.5	528.5	403
Ni(II)	561	525.5	403
Cu(II)	573	534	409
Ag(II)	570	534	417.5
Zn(II)	579	541	411.5
Cd(II)	587	559	414

[a] In Fig. 2-33, 1, 3, 5, 8 = methyl; 2, 4 = vinyl; 6, 7 = CH_2CH_2COOH.
[b] From [6].

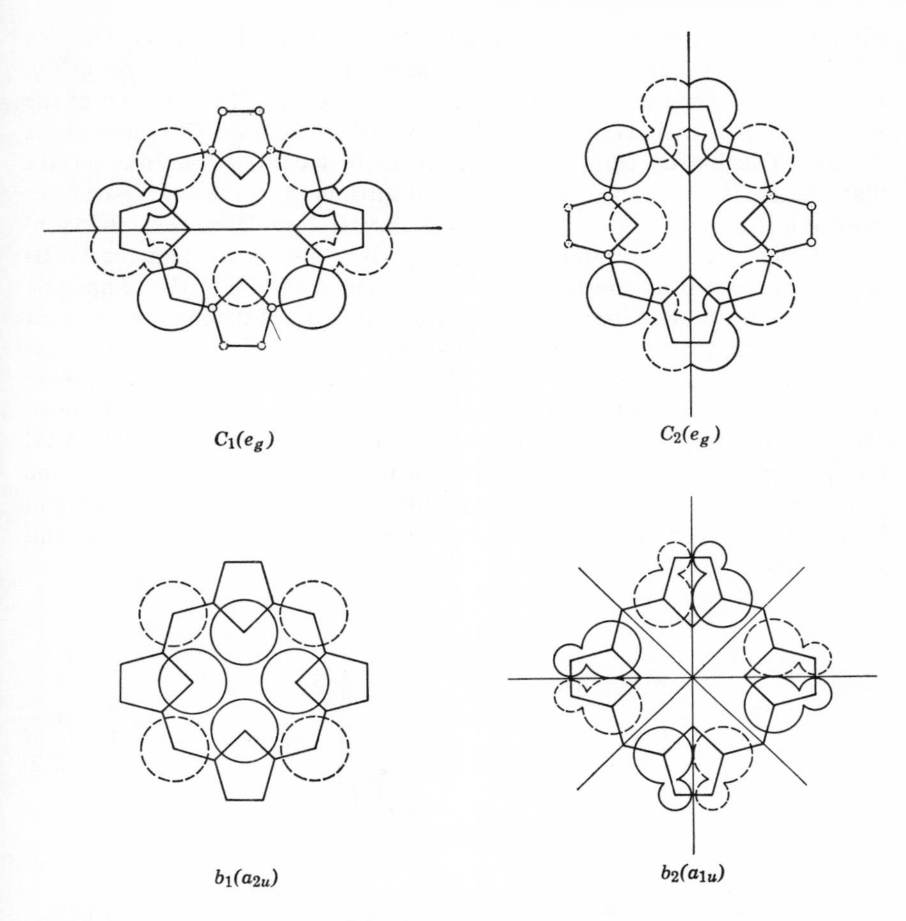

Fig. 2-34. Porphin MO's. The atomic orbital coefficients are proportional to the size of the circles; solid or dashed circles indicate sign. Symmetry nodes are drawn in heavy lines [135].

whereas the β band is an allowed $0 \rightarrow 1$ transition of the same electronic transition. Thus the α band is sensitive to substitution on the porphyrin nucleus which lowers the symmetry, whereas the β band is not so sensitive. This is exemplified in the pyridine adducts of the hemochromes, in which the β band is very weak or absent.

The Fe(II) porphyrins react readily with extra ligands such as pyridine, H_2O, CO, O_2, CN^-, OH^-, and halide ions to form octahedral complexes. Figure 2-35 illustrates the effect of this "perpendicular coordination" on the spectra [7]. The α band is stronger than the β band in the $Fe(II)(py)_2$ complex (Fig. 2-35c). However, this trend is reversed in the $Fe(II)(CN^-)(CO)$, $Fe(II)(CN^-)_2$, $Fe(II)(OH^-)_2$ complexes (Fig. 2-35,f–h). The spectrum of the Fe(II) complex itself (Fig. 2-35d) is very diffuse because the molecule is dimeric. The Fe(III) complexes (Fig. 2-35,a,b) have more diffuse spectra than the Fe(II) complexes. This has been attributed to the charge transfer from the ligand to Fe(III). In fact, the Fe(III) complexes (Fig. 2-35,a,b) exhibit charge transfer bands near 630 mμ. It is also noted that the Fe(II) complexes absorb at shorter wavelengths than do the Fe(III) complexes. This is due to the stabilization of the filled t_{2g} orbitals of the metal as a result of π interaction with the ligand. In a series of porphyrin complexes of the same metal, the bands usually shift to the red as the electron-attracting power of side chains increases (as the basicity of porphyrin decreases). For example, Table 2-20 compares the spectra of Co(II) complexes of mesoporphyrin IX and protoporphyrin IX. The red shifts of the bands from the meso to the proto may be attributed to the substitution of the ethyl groups of the meso by the vinyl groups in the proto. For more details of the discussion and spectral data, see [7, 128, and 185].

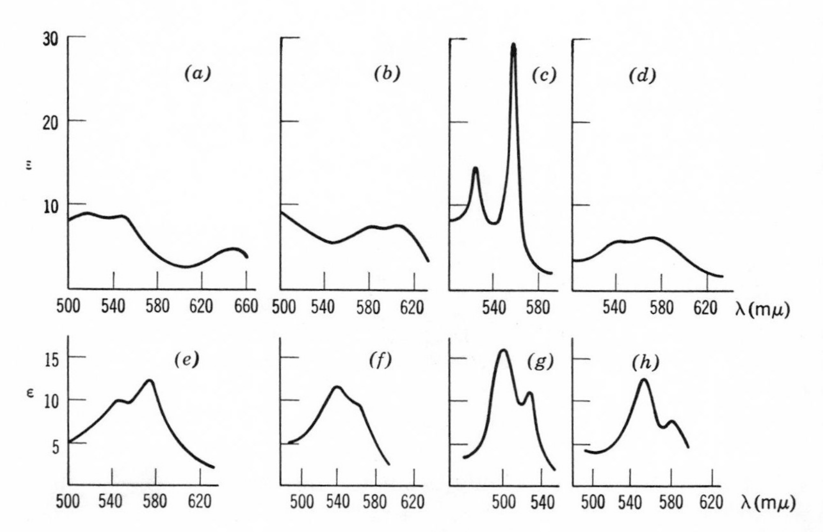

Fig. 2-35. Visible spectra of iron-protoporphyrin chelates. (a) hemin (Fe(III)-porphyrin chloride, measured in 2.5% CTAB); (b) hematin (Fe(III)-porphyrin hydroxide, measured in 2.5% CTAB); (c) pyridine hemochrome (bispyridine-Fe(II)-porphyrin); (d) dimeric heme (Fe(II)-porphyrin in 0.1 *N* NaOH); (e) bisaquoheme (Fe(II)-porphyrin in 2.5% CTAB, 0.1 *N* in respect to NaOH); (f) CO—CN^--Fe(II)-porphyrin; (g) bis-CN^--Fe(II)-porphyrin; (h) bishydroxyl heme (Fe(II)-porphyrin in 50% ethanol, 1.3 *N* in respect to NaOH). CTAB, cationic detergent solution [7].

Table 2-20 Effect of Side Chains on the Spectra of Metalloporphyrins (mμ)

		α	β	Soret	Solvent
Meso IX dimethyl ester Co(II)	λ	552	518	—	Benzene
	ε	(24.19)	(10.88)		
Proto IX dimethyl ester Co(II)	λ	563	529	404	Benzene
	ε	(23.1)	(11.86)	(165)	

The metal chelates of the amino acids and peptides have also been studied by spectroscopic means. Many of the amino acid and peptide complexes are optically active, and their properties are discussed in Chapter 3. The chelates of Cu(II) and Ni(II) with these ligands have been most extensively studied, and spectroscopic data are lacking for other metal ions. The addition of the amino acid anion (A^-) to Cu(II) follows the formation sequence

$$Cu^{2+} \xrightarrow{A^-} CuA^+ \xrightarrow{A^-} CuA_2,$$

in which water of hydration is replaced by the bidentate amino acids. The spectra of this sequence have well-defined isosbestic points which are indicative of the isolated equilibria; the extremum of an isosbestic family can be taken as a measure of the property of an individual species. The visible spectra of a number of simple α-amino acid complexes of Cu(II) have been studied by Hare et al. [141] as a function of pH. The spectrum of Cu(l-ala)$_2$ is typical and is given in Fig. 2-36. The species CuA^+ (HA = amino acid) all have maximum extinction coefficients of about 30 at 725 mμ. The extremum of the isosbestic family is achieved at about pH = 3, and the isosbestic point is at 940 mμ. The species CuA_2 have extinction coefficients between 60 and 75 at 620 mμ, the wavelength of maximum absorbance. The pH of the maximum extinction coefficient is variable, but lies generally between 9 and 10; and the isosbestic points for the formation of the species are at about 710 mμ. These generalizations apply to the Cu(II) chelates of the following amino acids: glycine, alanine, α-aminobutyric acid, proline, glutamic acid, aspartic acid, valine, leucine, phenylglycine, phenylalanine, norvaline, isoleucine, norleucine, tyrosine, and tryptophane.

A comparison of the spectra of Cu(II) chelates of the optically active and racemic amino acids has also been carried out [84, 130, 141]. A small stereochemical preference has been observed spectroscopically [84, 141] for the addition of the second ligand to the CuA^+ species in the case of proline. This effect is best observed by different pH sequences, isosbestic points, and extinction coefficients. The data can be interpreted to indicate a preferential stability of the meso (d-A plus l-A) CuA_2 species as opposed to the racemic

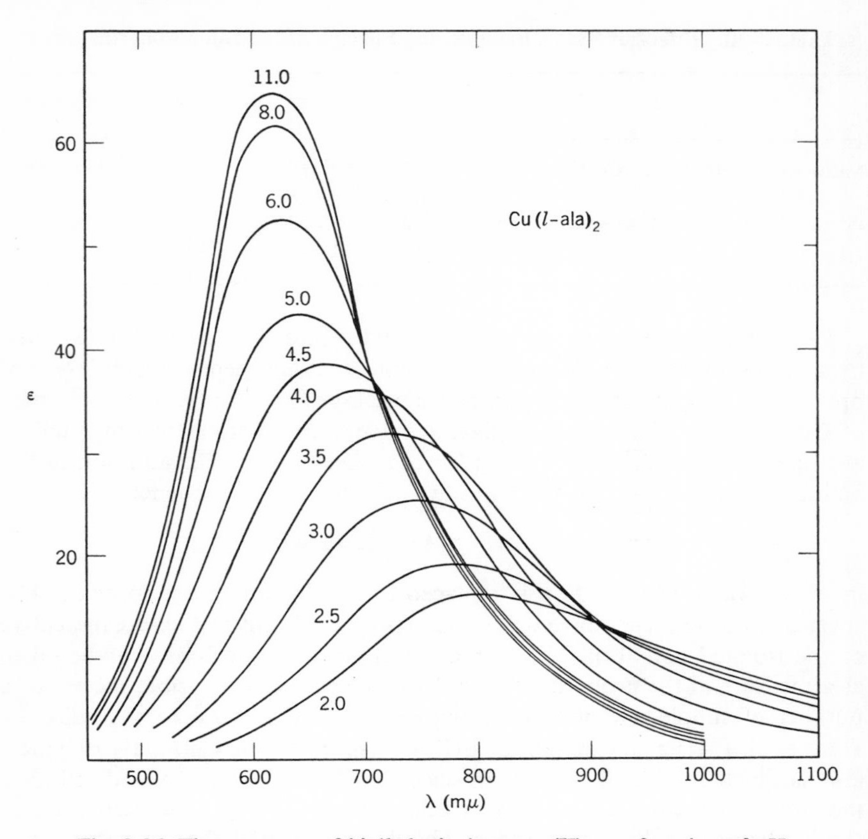

Fig. 2-36. The spectrum of bis(*l*-alanino) copper(II) as a function of pH.

(equimolar mixture of *d*-CuA_2 and *l*-CuA_2) CuA_2 species. These observations are not in agreement with those of Gillard et al. [130], who find no differences between the spectra of the optically active and racemic chelates at pH 8.86. The formation constants for these systems have been determined by Gillard and co-workers [130], and they are the same within experimental error. The spectral differences can be accounted for only by entropy differences, but the discrepancy remains to be resolved. Graddon and Munday [137] have also studied the spectra of the Cu(II) α-amino acid chelates and have observed that the results are best interpreted on the basis of a *trans* configuration for the two chelate rings and axial coordination of water molecules. This is in agreement with the predictions of Bjerrum and co-workers [54] and has been tested by Hare et al. [141]. The latter point out that a model *cis* chelate $Cu(EDDA)_2$(H_2EDDA, ethylenediaminediacetic acid) has an absorption band at 675 mμ ($\varepsilon = 97$), and the *trans* model, bis(N,N-dimethylglycino)copper(II), has an absorption band at 620 mμ (ε = 120).

The Cu(II) complexes of glycylglycine have been studied comprehensively by Kim and Martell [159]. From potentiometric and spectrophotometric data they have deduced the spectra for the various species of the chelate given in Table 2-21. Similar results have been reported by Dobbie and Kermack [91]. Koltun and co-workers [160] have studied some additional Cu(II) peptides; they report slightly different values for the species with glycylglycine as given in Table 2-22.

Table 2-21 Absorption Characteristics of Cu(II)-Glycylglycine[a] Complexes [159]

Species	λ_{max} (mμ)	ε
CuL^+	~780	~46
CuA	645	90
$CuAOH^-$	640	78
$Cu_2A_2OH^-$	630	175
$CuAL^-$	615	84

[a] Glycylglycine = HL = H_2A.

Table 2-22 Absorption Spectra of Cu(II) Peptide Species [160]

	CuL^+		CuA		$CuAL^-$	
Peptide	λ_{max}	ε	λ_{max}	ε	λ_{max}	ε
Glycylglycine	735	65	635	84	625	82
Prolylglycine	735	34	635	100	645	105
Glycylvaline	760	27	640	95	625	87
Valylglycine	715	100	635	92	620	91

Jennings [151] has tried with only partial success to use spectral means to determine the binding sites of Cu(II) with ovalbumin. He has also reported the spectra of Cu(II) with tri- and tetraglycine. Martin and co-workers [176] have studied the formation constants of Ni(II) with glycylglycine, glycylsarcosine, glycyl-*l*-proline, glycyl-*l*-valine, *l*-valylglycine, glycylglycine ethyl ester, glycinamide, triglycine, and tetraglycine. The first six peptides form blue octahedral species in the range of the titrations. Solutions of the latter three are blue up to pH 9, but turn yellow as the base content is increased.

The spectra of the yellow species are characteristic of square-planar molecules with maxima at 438, 430, and 412 mμ, and molar extinction coefficients of 61, 240, and 215 for glycinamide, triglycine, and tetraglycine chelates, respectively. Chamberlain and Wilkins [63] have extended this work and have observed that glycylglycinamide, *l*-alaninamide, *l*-α-aminobutyrylamide, *l*-prolinamide, picolinamide, glycylglycylalanine, and *l*-leucinamide also give rise to the yellow planar species. Apparently, removal of the amido or peptide proton is required for formation of these species. Figure 2-37 shows the

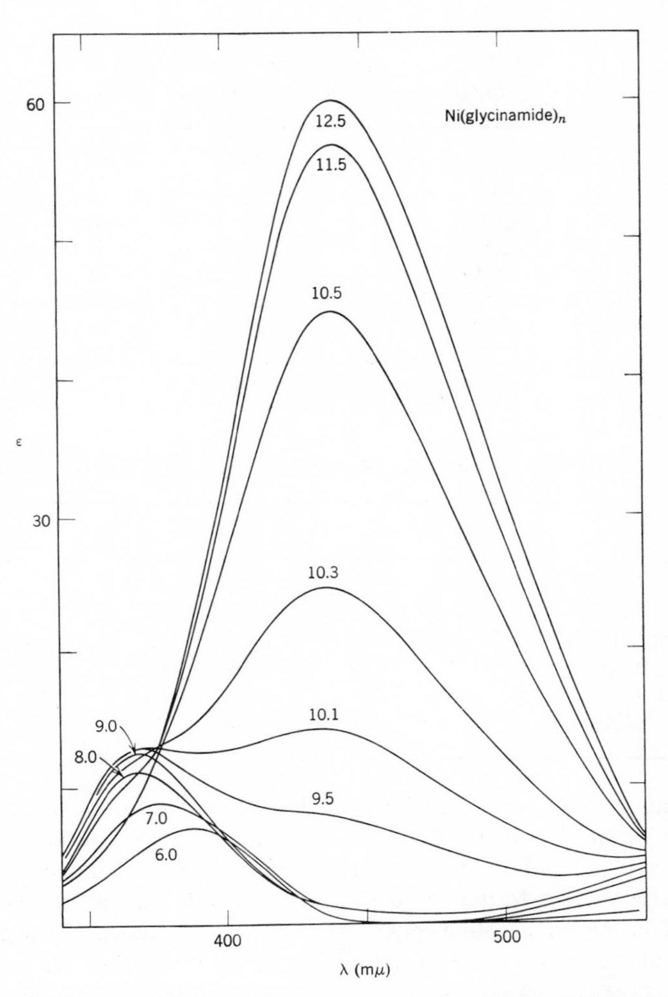

Fig. 2-37. pH dependence of the spectrum of a solution of nickel(II) and glycinamide.

formation of the blue octahedral species of Ni(II) with glycinamide (λ_{max} = 366 mμ, $\varepsilon = 13$, $pH_{max} = 9.0$) and the yellow species ($\lambda_{max} = 438$, $\varepsilon = 61$, $pH_{max} = 12.5$). The isosbestic points for the equilibria are at 373 mμ and 558 mμ.

REFERENCES

Books

[1] Ballhausen, C. J., *Introduction to Ligand Field Theory*. New York: McGraw-Hill, 1962.
[2] Ballhausen, C. J., and H. B. Gray, *Molecular Orbital Theory*. New York: Benjamin, 1964.
[3] Davydov, A. S. (transl. by M. Kasha and M. Oppenheimer), *Theory of Molecular Excitons*. New York: McGraw-Hill, 1962.
[4] Dunn, T. M., in *Modern Coordination Chemistry*, J. Lewis and R. G. Wilkins, eds. New York: Interscience, 1960.
[5] Dunn, T. M., R. G. Pearson, and D. S. McClure, *Crystal Field Theory*. New York: Harper and Row, 1965.
[6] Falk, J. E., and D. D. Perrin, in *Haematin Enzymes*, J. E. Falk, R. Lemberg, and R. K. Morton, eds. London: Pergamon, 1961, p. 56.
[7] Falk, J. E., *Porphyrins and Metalloporphyrins*. New York: Elsevier, 1964.
[8] Figgis, B. N., *Introduction to Ligand Fields*. New York: Interscience, 1966.
[9] Griffith, J. S., *The Theory of Transition-Metal Ions*. London: Cambridge University Press, 1961.
[10] Jørgensen, C. K., *Absorption Spectra and Chemical Bonding in Complexes*. Reading, Mass.: Addison-Wesley, 1962.
[11] Jørgensen, C. K., *Orbitals in Atoms and Molecules*. New York: Academic, 1962.
[12] Jørgensen, C. K., *Inorganic Complexes*. New York: Academic, 1963.
[13] Orgel, L. E., *An Introduction to Transition-Metal Chemistry*. New York: Wiley, 1960.
[14] Rossotti, F. J. C., and H. Rossotti, *The Determination of Stability Constants*. New York: McGraw-Hill, 1961.
[15] Stevenson, R., *Multiplet Structure of Atoms and Molecules*. Philadelphia, Pa.: Saunders, 1965.
[16] Watanabe, H., *Operator Methods in Ligand Field Theory*. Englewood Cliffs, N.J.: Prentice-Hall, 1966.

Reviews

[17] Ballhausen, C. J., *Progr. Inorg. Chem.*, **2**, 251 (1960).
[18] Ciampolini, M., N. Nardi, and G. P. Speroni, *Coord. Chem. Rev.*, **1**, 222 (1966).
[19] Ferguson, J., *Rev. Pure Appl. Chem.*, **14**, 1 (1964).
[20] Gray, H. B., *Progr. Transition Metal Chem.*, **1**, 240 (1965).
[21] Jørgensen, C. K., *Progr. Inorg. Chem.*, **4**, 73 (1963).
[22] Livingstone, S. E., *Quart. Rev. (London)*, **19**, 386 (1965).
[23] McClure, D. S., *Solid State Phys.*, **9**, 399 (1959).
[24] Moffitt, W., and C. J. Ballhausen, *Ann. Rev. Phys. Chem.*, **7**, 107 (1956).
[25] Muetterties, E. L., and R. A. Schunn, *Quart. Rev. (London)*, **20**, 245 (1966).

[26] Nyholm, R. S., *Quart. Rev. (London)*, **7**, 377 (1953).
[27] Platt, J. R., in *Radiation Biology*, A. Hollaender, ed., Vol. 3, Chapter 2. New York: McGraw-Hill, 1956.
[28] Sacconi, L., *Coord. Chem. Rev.*, **1**, 126 (1966); *Essays on Coordination Chemistry Exp. Suppl.*, **14**, 148 (1966).

Other References

[29] Allen, H. C., *J. Chem. Phys.*, **45**, 553 (1966).
[30] Antsyshkina, A. S., and M. A. Porai-Koshits, *Dokl. Akad. Nauk SSSR*, **143**, 105 (1962).
[31] Baddley, W. H., F. Basolo, H. B. Gray, C. Nölting, and A. J. Poe, *Inorg. Chem.*, **2**, 921 (1963).
[32] Bailey, M. F., and I. E. Maxwell, *Chem. Commun.*, 908 (1966); R. R. Ryan, B. T. Kilbourn, and J. D. Dunitz, *Chem. Commun.*, 910 (1966).
[33] Baker, W. A., and M. G. Phillips, *Inorg. Chem.*, **5**, 1042 (1966).
[34] Ballhausen, C. J., and C. K. Jørgensen, *Kgl. Danske Videnskab. Selskab, Mat.-Fys. Medd.*, **29**, No. 14, (1955).
[35] Ballhausen, C. J., and W. Moffitt, *J. Inorg. Nucl. Chem.*, **3**, 178 (1956).
[36] Ballhausen, C. J., *Rec. Trav. Chim.*, **75**, 665 (1956).
[37] Ballhausen, C. J., and A. D. Liehr, *J. Mol. Spectry.*, **2**, 342 (1958).
[38] Ballhausen, C. J., and A. D. Liehr, *Mol. Phys.*, **2**, 123 (1959).
[39] Ballhausen, C. J., and H. B. Gray, *Inorg. Chem.*, **1**, 111 (1962).
[40] Baker-Hawkes, M. J., E. Billig, and H. B. Gray, *J. Am. Chem. Soc.*, **88**, 4870 (1966).
[41] Balzani, V., V. Carassiti, L. Moggi, and F. Scandola, *Inorg. Chem.*, **4**, 1243 (1965).
[42] Barclay, G. A., R. S. Nyholm, and R. V. Parish, *J. Chem. Soc.*, 4433 (1961).
[43] Barnum, D. W., *J. Inorg. Nucl. Chem.*, **21**, 221 (1961); **22**, 183 (1961).
[44] Basolo, F., C. J. Ballhausen, and J. Bjerrum, *Acta Chem. Scand.*, **9**, 810 (1955).
[45] Basu, G., R. L. Belford, and R. E. Dickerson, *Inorg. Chem.*, **1**, 439 (1962).
[46] Basu, G., W. Yeranos, and R. L. Belford, *Inorg. Chem.*, **3**, 929 (1964).
[47] Basu, G., G. H. Cook, and R. L. Belford, *Inorg. Chem.*, **3**, 1361 (1964).
[48] Basu, G., and R. L. Belford, *J. Mol. Spectry.*, **17**, 167 (1965).
[49] Belford, R. L., and T. S. Piper, *Mol. Phys.*, **5**, 251 (1962).
[50] Benner, G. S., W. E. Hatfield, and D. W. Meek, *Inorg. Chem.*, **3**, 1544 (1964).
[51] Bethe, H. A., *Ann. Physik*, **3**, 133 (1929).
[52] Billig, E., H. B. Gray, S. I. Shupack, J. H. Waters, and R. Williams, *Proc. Chem Soc.*, 110 (1964).
[53] Billig, E., S. I. Shupack, J. H. Waters, R. Williams, and H. B. Gray, *J. Am. Chem. Soc.*, **86**, 926 (1964).
[54] Bjerrum, J., C. J. Ballhausen, and C. K. Jørgensen, *Acta Chem. Scand.*, **8**, 1275 (1954).
[55] Bosnich, B., C. K. Poon, and M. L. Tobe, *Inorg. Chem.*, **4**, 1102 (1965); B. Bosnich, M. L. Tobe, and G. A. Webb, *Inorg. Chem.*, **4**, 1110 (1965).
[56] Bostrup, O., and C. K. Jørgensen, *Acta Chem. Scand.*, **11**, 1223 (1957).
[57] Brewster, J. A., C. A. Savage, and L. M. Venanzi, *J. Chem. Soc.*, 3699 (1961).
[58] Brubaker, G. R., and D. H. Busch, *Inorg. Chem.*, **5**, 2114 (1966); J. L. Karn and D. H. Busch, 152nd Meeting American Chemical Society, Pittsburgh, March, 1966.
[59] Bullen, G. J., R. Mason, and P. Pauling, *Nature*, **189**, 291 (1961).

[60] Chakravorty, A., and S. Basu, *J. Chem. Phys.*, **33**, 1266 (1960).
[61] Chakravorty, A., and R. H. Holm, *Inorg. Chem.*, **3**, 1010 (1964).
[62] Chakravorty, A., J. P. Fennessey, and R. H. Holm, *Inorg. Chem.*, **4**, 26 (1965).
[63] Chamberlain, P., and R. G. Wilkins, unpublished results.
[64] Ciampolini, M., *Chem. Commun.*, 47 (1966).
[65] Ciampolini, M., *Inorg. Chem.*, **5**, 35 (1966).
[66] Ciampolini, M., and N. Nardi, *Inorg. Chem.*, **5**, 41, 1150 (1966).
[67] Ciampolini, M., and G. P. Speroni, *Inorg. Chem.*, **5**, 45 (1966).
[68] Clementi, E., and D. L. Raimondi, *J. Chem. Phys.*, **35**, 2686 (1963).
[69] Cotton, F. A., and R. H. Soderberg, *Inorg. Chem.*, **3**, 1 (1964).
[70] Cotton, F. A., and J. S. Wood, *Inorg. Chem.*, **3**, 245 (1964).
[71] Cotton, F. A., and T. E. Haas, *Inorg. Chem.*, **3**, 1004 (1964).
[72] Cotton, F. A., and J. J. Wise, *J. Am. Chem. Soc.*, **88**, 3451 (1966).
[73] Curtis, N. F., *J. Chem. Soc.*, 2644 (1964).
[74] Curtis, N. F., and Y. M. Curtis, *Inorg. Chem.*, **4**, 804 (1965).
[75] Curtis, N. F., and D. A. House, *J. Chem. Soc.*, 6194 (1965).
[76] Curtis, N. F., *J. Chem. Soc.*, 924 (1965).
[77] Curtis, N. F., Y. M. Curtis, and H. J. Powell, *J. Chem. Soc.*, 1015 (1966).
[78] Cusachs, L. C., *J. Chem. Phys.*, **43**, 5157 (1965).
[79] Cusachs, L. C., and J. W. Reynolds, *J. Chem. Phys.*, **43**, 5160 (1965).
[80] Davison, A., N. Edelstein, R. H. Holm, and A. H. Maki, *J. Am. Chem. Soc.*, **85**, 2029 (1963).
[81] Day, P., G. Scregg, and R. J. P. Williams, *J. Chem. Phys.*, **38**, 2778 (1963).
[82] Day, P., and C. K. Jørgensen, *J. Chem. Soc.*, 6226 (1964).
[83] DeWitt, R., and J. I. Watters, *J. Am. Chem. Soc.*, **76**, 3810 (1954).
[84] Dijkgraaf, C., *Spectrochim. Acta*, **20**, 1227 (1964).
[85] Dijkgraaf, C., *Theoret. Chim. Acta*, **3**, 38 (1965).
[86] Dingle, R., *J. Mol. Spectry.*, **9**, 426 (1962).
[87] Dingle, R., *Chem. Commun.*, 304 (1965).
[88] Dingle, R., *J. Mol. Spectry.*, **18**, 276 (1965).
[89] Dingle, R., *Acta Chem. Scand.*, **20**, 33 (1966).
[90] Dingle, R., *J. Chem. Phys.*, **46**, 1 (1967).
[91] Dobbie, H., and W. O. Kermack, *Biochem. J.*, **59**, 257 (1955).
[92] Dori, Z., and H. B. Gray, *J. Am. Chem. Soc.*, **88**, 1394 (1966).
[93] Dunn, T. M., R. S. Nyholm, and S. Yamada, *J. Chem. Soc.*, 1564 (1962).
[94] Dwyer, F. P., N. S. Gill, E. C. Gyarfas, and F. Lions, *J. Am. Chem. Soc.*, **74**, 4188 (1952); **75**, 2443 (1953).
[95] Dyer, G., and D. W. Meek, *Inorg. Chem.*, **4**, 1398 (1965).
[96] Dyer, G., J. G. Hartley, and L. M. Venanzi, *J. Chem. Soc.*, 1293 (1965).
[97] Dyer, G., and L. M. Venanzi, *J. Chem. Soc.*, 2771 (1965).
[98] Dyer, G., and D. W. Meek, *Inorg. Chem.*, **6**, 149 (1967).
[99] Eaton, D. R., W. D. Phillips, and D. J. Caldwell, *J. Am. Chem. Soc.*, **85**, 397 (1963).
[100] Eisenberg, R., J. A. Ibers, R. J. H. Clark, and H. B. Gray, *J. Am. Chem. Soc.*, **86**, 113 (1964).
[101] Eisenberg, R., and J. A. Ibers, *Inorg. Chem.*, **5**, 411 (1966).
[102] Eisenberg, R., E. I. Stiefel, R. C. Rosenberg, and H. B. Gray, *J. Am. Chem. Soc.*, **88**, 2874 (1966).
[103] Eisenstein, J. C., *J. Chem. Phys.*, **34**, 1628 (1961).
[104] Englman, R., *Mol. Phys.*, **3**, 48 (1960).

[105] Ewald, A. H., and E. Sinn, *Inorg. Chem.*, **6**, 40 (1967).
[106] Ewald, A. H., R. L. Martin, I. G. Ross, and A. H. White, *Proc. Roy. Soc.* (*London*), **A280**, 235 (1964).
[107] Fackler, J. P., F. A. Cotton, and D. W. Barnum, *Inorg. Chem.*, **2**, 97 (1963).
[108] Fackler, J. P., *Inorg. Chem.*, **2**, 266 (1963).
[109] Fackler, J. P., D. G. Holah, D. A. Buckingham, and J. T. Henry, *Inorg. Chem.*, **4**, 920 (1965).
[110] Fajans, K., *Naturwissenschaften*, **11**, 165 (1923).
[111] Farago, M. E., and J. M. James, *Chem. Commun.*, 470 (1965).
[112] Fenske, R. F., D. S. Martin, and K. Ruedenberg, *Inorg. Chem.*, **1**, 441 (1962).
[113] Fenske, R. F., *Inorg. Chem.*, **4**, 33 (1965).
[114] Fenske, R. F., K. G. Caulton, D. D. Radtke, and C. C. Sweeney, *Inorg. Chem.*, **5**, 951, 960 (1966).
[115] Fenske, R. F., *J. Am. Chem. Soc.*, **89**, 252 (1967).
[116] Ferguson, J., *J. Chem. Phys.*, **34**, 611 (1961).
[117] Ferguson, J., *J. Chem. Phys.*, **35**, 1612 (1961).
[118] Ferguson, J., *J. Chem. Phys.*, **34**, 2206 (1961).
[119] Ferguson, J., R. L. Belford, and T. S. Piper, *J. Chem. Phys.*, **37**, 1569 (1962).
[120] Ferguson, J., *Theoret. Chim. Acta*, **3**, 287 (1965).
[121] Fergusson, J. E., and C. A. Ramsay, *J. Chem. Soc.*, 5222 (1965).
[122] Fielding, P. E., and A. G. McKay, *J. Chem. Phys.*, **38**, 2777 (1963).
[123] Finkelstein, R., and J. H. VanVleck, *J. Chem. Phys.*, **8**, 790 (1940).
[124] Forster, L. S., and K. DeArmond, *J. Chem. Phys.*, **34**, 2193 (1961).
[125] Forster, L. S., and P. X. Armendarez, *J. Chem. Phys.*, **40**, 273 (1964).
[126] Fox, M. R., P. L. Orioli, E. C. Lingafelter, and L. Sacconi, *Acta Cryst.*, **17**, 1159 (1964).
[127] Fujita, J., A. E. Martell, and K. Nakamoto, *J. Chem. Phys.*, **36**, 324 (1962).
[128] George, P., J. Beetlestone, and J. S. Griffith, *Rev. Mod. Phys.*, **36**, 441 (1964).
[129] Gerloch, M., S. F. A. Kettle, J. Locke, and J. A. McCleverty, *Chem. Commun.*, 29 (1966).
[130] Gillard, R. D., H. M. Irving, R. M. Parkins, N. C. Payne, and L. D. Pettit, *J. Chem. Soc.*, *A*, 1159 (1966).
[131] Gillespie, R. J., *J. Chem. Soc.*, 4672, 4679 (1963).
[132] Goode, D. H., *J. Chem. Phys.*, **43**, 2830 (1965).
[133] Goodgame, D. M. L., and L. M. Venanzi, *J. Chem. Soc.*, 5909 (1963).
[134] Goodgame, D. M. L., and M. Goodgame, *Inorg. Chem.*, **4**, 139 (1965).
[135] Gouterman, M., *J. Mol. Spectry.*, **6**, 138 (1961).
[136] Graddon, D. P., *J. Inorg. Nucl. Chem.*, **7**, 93 (1958).
[137] Graddon, D. P., and L. Munday, *J. Inorg. Nucl. Chem.*, **23**, 231 (1961).
[138] Gray, H. B., and C. J. Ballhausen, *J. Am. Chem. Soc.*, **85**, 260 (1963).
[139] Hall, D., S. V. Sheat, and T. N. Waters, *Chem. Commun.*, 436 (1966).
[140] Hare, C. R., and C. J. Ballhausen, *J. Chem. Phys.*, **89**, 758 (1963).
[141] Hare, C. R., B. S. Manhas, T. G. Mecca, W. Mungall, and K. M. Wellman, *Proc. IX Intern. Conf. on Coord. Chem.*, St. Moritz, Switzerland, 1966, p. 199.
[142] Harris, C. M., and R. S. Nyholm, *J. Chem. Soc.*, 63 (1957).
[143] Harris, C. M., R. S. Nyholm, and D. J. Phillips, *J. Chem. Soc.*, 4379 (1960).
[144] Hartley, J. G., L. M. Venanzi, and D. C. Goodall, *J. Chem. Soc.*, 3930 (1963).
[145] Hatfield, W., T. S. Piper, and U. Klabunde, *Inorg. Chem.*, **2**, 629 (1963).
[146] Higginson, W. C. E., S. C. Nyburg, and J. S. Wood, *Inorg. Chem.*, **3**, 463 (1964).
[147] Hoffmann, R., *J. Chem. Phys.*, **39**, 1397 (1963).

[148] Holm, R. H., and K. Swaminathan, *Inorg. Chem.*, **1**, 599 (1962).
[149] Holm, R. H., A. Chakravorty, and L. J. Theroit, *Inorg. Chem.*, **5**, 625 (1966).
[150] Ilse, E., and H. Hartmann, *Z. Physik. Chem. (Leipzig)*, **197**, 239 (1951).
[151] Jennings, A. C., *Australian J. Chem.*, **16**, 1006 (1963).
[152] Jesson, J. P., *J. Chem. Phys.*, **45**, 1049 (1966).
[153] Jørgensen, C. K., *Acta Chem. Scand.*, **9**, 405 (1955).
[154] Jørgensen, C. K., *Acta Chem. Scand.*, **9**, 1362 (1955).
[155] Jørgensen, C. K., *Acta Chem. Scand.*, **10**, 887 (1956).
[156] Jørgensen, C. K., *Acta Chem. Scand.*, **11**, 151, 166 (1957).
[157] Jørgensen, C. K., *Acta Chem. Scand.*, **11**, 399 (1957).
[158] Jørgensen, C. K., *J. Inorg. Nucl. Chem.*, **24**, 1571 (1962).
[159] Kim, M. K., and A. E. Martell, *Biochemistry*, **3**, 1169 (1964).
[160] Koltun, W. L., M. Fried, and F. R. N. Gurd, *J. Am. Chem. Soc.*, **82**, 233 (1960).
[161] König, E., and K. Madeja, *Inorg. Chem.*, **6**, 48 (1967).
[162] Langford, C. H., E. Billig, S. I. Shupack, and H. B. Gray, *J. Am. Chem. Soc.*, **86**, 2958 (1964).
[163] Latham, A. R., V. C. Hascall, and H. B. Gray, *Inorg. Chem.*, **4**, 788 (1965).
[164] Liehr, A. D., and C. J. Ballhausen, *Phys. Rev.*, **106**, 1161 (1957).
[165] Liehr, A. D., and C. J. Ballhausen, *Ann. Phys. (N.Y.)*, **6**, 134 (1959).
[116] Liehr, A. D., *J. Phys. Chem.*, **64**, 43 (1960).
[167] Liehr, A. D., *J. Phys. Chem.*, **67**, 1314 (1963).
[168] Liehr, A. D., and J. R. Perumareddi, 150th Meeting, American Chemical Society, Atlantic City, September, 1965.
[169] Lohr, L. L., and W. N. Lipscomb, *Inorg. Chem.*, **2**, 911 (1963).
[170] Longuet-Higgins, H. C., C. W. Rector, and J. R. Platt, *J. Chem. Phys.*, **18**, 1174 (1950).
[171] Longhi, R., and R. S. Drago, *Inorg. Chem.*, **4**, 11 (1965).
[172] Low, W., *J. Chem. Phys.*, **33**, 1162 (1960).
[173] MacFarlane, R. M., *J. Chem. Phys.*, **39**, 3118 (1963); **40**, 373 (1964).
[174] Mair, G. A., H. M. Powell, and L. M. Venanzi, *Proc. Chem. Soc.*, 170 (1961).
[175] Martin, D. S., and C. A. Lenhardt, *Inorg. Chem.*, **3**, 1368 (1964).
[176] Martin, R. B., M. Chamberlin, and J. T. Edsall, *J. Am. Chem. Soc.*, **82**, 495 (1960).
[177] Mataga, S., and N. Mataga, *Z. Physik. Chem. (Frankfurt)*, **33**, 374 (1962).
[178] Matsuoka, N., J. Hidaka, and Y. Shimura, *Bull. Chem. Soc. Japan*, **39**, 1257 (1966).
[179] McClellan, W. R., and R. E. Benson, *J. Am. Chem. Soc.*, **88**, 5165 (1966).
[180] Melson, G. A., and R. G. Wilkins, *J. Chem. Soc.*, 2662 (1963).
[181] Mulliken, R. S., *J. Chem. Phys.*, **23**, 1833 (1955).
[182] Nakamoto, K., P. J. McCarthy, A. Ruby, and A. E. Martell, *J. Am. Chem. Soc.*, **83**, 1066 (1961).
[183] Nannelli, P., and L. Sacconi, *Inorg. Chem.*, **5**, 246 (1966).
[184] Nyburg, S. C., and J. S. Wood, *Inorg. Chem.*, **3**, 468 (1964).
[185] Offenhartz, P. O. D., *J. Chem. Phys.*, **42**, 3566 (1965).
[186] Ohkawa, K., J. Hidaka, and Y. Shimura, *Bull. Chem. Soc. Japan*, **39**, 1715 (1966).
[187] Olsen, D. C., V. P. Mayweg, and G. N. Schrauzer, *J. Am. Chem. Soc.*, **88**, 4876 (1966).
[188] Orioli, P. L., M. DiVaira, and L. Sacconi, *J. Am. Chem. Soc.*, **87**, 2059 (1965); **88**, 4383 (1966).
[189] Orioli, P. L., M. DiVaira, and L. Sacconi, *Chem. Commun.*, 103 (1965); 300 (1966).

[190] Orioli, P. L., and L. Sacconi, *J. Am. Chem. Soc.*, **88**, 277 (1966).
[191] Palmer, R. A., and T. S. Piper, *Inorg. Chem.*, **5**, 864 (1966).
[192] Pavkovic, S. F., and D. W. Meek, *Inorg. Chem.*, **4**, 20 (1965).
[193] Perumareddi, J. R., *J. Phys. Chem.*, **71**, 3144 (1967).
[194] Piper, T. S., and R. L. Carlin, *J. Chem. Phys.*, **33**, 1208 (1960).
[195] Piper, T. S., *J. Chem. Phys.*, **35**, 1240 (1961).
[196] Piper, T. S., and R. L. Carlin, *J. Chem. Phys.*, **35**, 1809 (1961).
[197] Piper, T. S., and R. L. Carlin, *J. Chem. Phys.*, **36**, 3330 (1962).
[198] Piper, T. S., and R. L. Belford, *Mol. Phys.*, **5**, 169 (1962).
[199] Pryce, M. H. L., and W. A. Runciman, *Discussions Faraday Soc.*, **26**, 34 (1958).
[200] Racah, G., *Phys. Rev.*, **62**, 438 (1942).
[201] Rasmussen, S. E., *Acta Chem. Scand.*, **13**, 2009 (1959).
[202] Reimann, C. W., G. F. Kokoszka, and G. Gordon, *Inorg. Chem.*, **4**, 1082 (1965).
[203] Rein, R., N. Fukuda, H. Win, G. A. Clarke, and F. E. Harris, *J. Chem. Phys.*, **45**, 4743 (1966).
[204] Richardson, J. W., W. C. Nieuwpoort, R. R. Powell, and W. F. Edgell, *J. Chem. Phys.*, **36**, 1057 (1962).
[205] Richardson, J. W., R. R. Powell, and W. C. Nieuwpoort, *J. Chem. Phys.*, **38**, 796 (1963).
[206] Runciman, W. A., and K. Schroeder, *Proc. Roy. Soc. (London)*, **A265**, 489 (1962).
[207] Sacconi, L., P. Paoletti, and R. Cini, *J. Am. Chem. Soc.*, **80**, 3583 (1958); *J. Inorg. Nucl. Chem.*, **8**, 492 (1958).
[208] Sacconi, L., M. Ciampolini, F. Maggio, and F. P. Cavasino, *J. Am. Chem. Soc.*, **84**, 3246 (1962).
[209] Sacconi, L., P. L. Orioli, P. Paoletti, and M. Ciampolini, *Proc. Chem. Soc.*, 255 (1962).
[210] Sacconi, L., P. Paolotti, and M. Ciampolini, *J. Am. Chem. Soc.*, **85**, 411 (1963).
[211] Sacconi, L., and M. Ciampolini, *J. Am. Chem. Soc.*, **85**, 1750 (1963).
[212] Sacconi, L., *J. Chem. Soc.*, 4608 (1963).
[213] Sacconi, L., M. Ciampolini, and N. Nardi, *J. Am. Chem. Soc.*, **86**, 819 (1964).
[214] Sacconi, L., and M. Ciampolini, *J. Chem. Soc.*, 267 (1964).
[215] Sacconi, L., P. Nannelli, and U. Campigli, *Inorg. Chem.*, **4**, 818 (1965).
[216] Sacconi, L., P. Nannelli, N. Nardi, and U. Campigli, *Inorg. Chem.*, **4**, 943 (1965).
[217] Sacconi, L., M. Ciampolini, and G. P. Speroni, *Inorg. Chem.*, **4**, 1116 (1965).
[218] Sacconi, L., P. L. Orioli, and M. DiVaira, *J. Am. Chem. Soc.*, **87**, 2059 (1965).
[219] Sacconi, L., M. Ciampolini, and G. P. Speroni, *J. Am. Chem. Soc.*, **87**, 3102 (1965).
[220] Sacconi, L., and U. Campigli, *Inorg. Chem.*, **5**, 606 (1966).
[221] Sacconi, L., and I. Bertini, *Inorg. Chem.*, **5**, 1520 (1966).
[222] Sacconi, L., and I. Bertini, *Inorg. Nucl. Chem. Letters*, **2**, 29 (1966).
[223] Savage, C. A., and L. M. Venanzi, *J. Chem. Soc.*, 1548 (1962).
[224] Schäffer, C. E., and C. K. Jørgensen, *Mol. Phys.*, **9**, 401 (1965).
[225] Schläfer, H. L., *Z. Physik. Chem., N.F.*, **8**, 373 (1956).
[226] Schläfer, H. L., and E. König, *Z. Physik. Chem., N.F.*, **19**, 265 (1959).
[227] Schrauzer, G. N., and V. P. Mayweg, *J. Am. Chem. Soc.*, **87**, 3585 (1965).
[228] Schrauzer, G. N., and V. P. Mayweg, *J. Am. Chem. Soc.*, **88**, 3235 (1966).
[229] Schroeder, K., *J. Chem. Phys.*, **37**, 1587 (1962).
[230] Schroeder, K., *J. Chem. Phys.*, **37**, 2553 (1962).
[231] Shupack, S. I., E. Billig, R. J. H. Clark, R. Williams, and H. B. Gray, *J. Am. Chem. Soc.*, **86**, 4594 (1964).

[232] Slater, J. C., *Phys. Rev.*, **36**, 57 (1930).
[233] Sone, K., and M. Kato, *Z. Anorg. Allgem. Chem.*, **301**, 277 (1959).
[234] Stiefel, E. I., R. Eisenberg, R. C. Rosenberg, and H. B. Gray, *J. Am. Chem. Soc.*, **88**, 2956 (1966).
[235] Stein, A., and M. Dezelic, *Z. Physik. Chem. (Leipzig)*, **180**, 131 (1937).
[236] Stoufer, R. C., D. W. Smith, E. A. Clevenger, and T. E. Norris, *Inorg. Chem.*, **5**, 1167 (1966).
[237] Stoufer, R. C., W. B. Hadley, and D. H. Busch, *J. Am. Chem. Soc.*, **83**, 3732 (1961).
[238] Straub, D. K., R. S. Drago, and J. T. Donoghue, *Inorg. Chem.*, **1**, 848 (1962).
[239] Sugano, S., and Y. Tanabe, *J. Phys. Soc. Japan*, **13**, 880 (1958).
[240] Sugano, S., and R. G. Shulman, *Phys. Rev.*, **130**, 517 (1963).
[241] Tanabe, Y., and S. Sugano, *J. Phys. Soc. Japan*, **9**, 753, 766 (1964).
[242] Tonnet, M. L., S. Yamada, and I. G. Ross, *Trans. Faraday Soc.*, **60**, 840 (1964).
[243] Tsuchida, R., *Bull. Chem. Soc. Japan*, **13**, 388, 436 (1938).
[244] Utsuno, S., and K. Sone, *J. Inorg. Nucl. Chem.*, **28**, 2647 (1966).
[245] Vareille, L., *Bull. Soc. Chim. France*, 870 (1955).
[246] Viste, A., and H. B. Gray, *Inorg. Chem.*, **3**, 1113 (1964); C. J. Ballhausen and H. B. Gray, *Inorg. Chem.*, **1**, 111 (1962).
[247] Watson, R. E., *Phys. Rev.*, **118**, 1036 (1960); **119**, 1934 (1960).
[248] Watson, R. E., and H. J. Freeman, *Phys. Rev.*, **134A**, 1526 (1964).
[249] Weakleim, H. A., *J. Chem. Phys.*, **36**, 2117 (1962).
[250] Wellman, K. M., C. R. Hare, and S. Bogdansky, unpublished results.
[251] Wentworth, R. A. D., and T. S. Piper, *Inorg. Chem.*, **4**, 709 (1965).
[252] Wentworth, R. A. D., *Inorg. Chem.*, **5**, 496 (1966).
[253] Werden, B. G., E. Billig, and H. B. Gray, *Inorg. Chem.*, **5**, 78 (1966).
[254] Williams, R., E. Billig, J. H. Waters, and H. B. Gray, *J. Am. Chem. Soc.*, **88**, 43 (1966).
[255] Wolfsberg, M., and L. Helmholz, *J. Chem. Phys.*, **20**, 837 (1952).
[256] Wood, D. L., *J. Chem. Phys.*, **42**, 3404 (1965).
[257] Yamada, S., and R. Tsuchida, *Bull. Chem. Soc. Japan*, **27**, 156 (1954).
[258] Yamada, S., A. Nakahara, Y. Shimura, and R. Tsuchida, *Bull. Chem. Soc. Japan*, **28**, 222 (1955).
[259] Yamada, S., and R. Tsuchida, *Bull. Chem. Soc. Japan*, **33**, 98 (1960).
[260] Yasui, T., and Y. Shimura, *Bull. Chem. Soc. Japan*, **39**, 604 (1966).
[261] Zahner, J. C., and H. G. Drickamer, *J. Chem. Phys.*, **33**, 1625 (1960).
[262] Zerner, M., and M. Gouterman, *Inorg. Chem.*, **5**, 1699, 1707 (1966).

3

OPTICAL ROTATORY DISPERSION AND CIRCULAR DICHROISM

JUNNOSUKE FUJITA AND YOICHI SHIMURA
Tohoku University ***Osaka University***

Some kinds of naturally occurring or synthetic substances rotate the plane of polarization of linearly polarized light which traverses them. Such substances are said to be optically active. The phenomenon of optical rotatory power was discovered in quartz by Biot in 1812. It was also discovered in solution and in vapor states successively. In the period 1848–1854 Pasteur predicted that optical activity would be attributed to a molecular or microscopic dissymmetry in the structure of a substance. His prediction was fulfilled after about 20 years in the light of the theores of van't Hoff and Le Bel. At that time the studies of optical activity contributed much to organic chemistry; organic stereochemistry was supported mainly by the concept of the asymmetric carbon atom. Inorganic stereochemistry, on the other hand, was founded by Werner [179] in 1893, and his theory was confirmed conclusively in 1911 by the successful resolution of the optically active octahedral complexes *cis*-$[CoX(NH_3)(en)_2]X_2$ ($X = Cl^-, Br^-$) [180]. A metal complex containing no carbon atoms, the optically active hexol complex

$$[Co\{(OH)_2Co(NH_3)_4\}_3]X_6,$$

was also resolved in 1914 [185]. This was the keystone in the arch of Werner's coordination theory.

The usefulness of optical activity in chemistry has increased with the years. Recently expanded regions of the technique include the optical rotatory dispersion (variation of optical rotatory power with the wavelength of light) and the Cotton effect (circular dichroism and the anomalous dispersion of rotation in the vicinity of an absorption band). From the theoretical point of view the fundamental theory of optical activity poses many difficult problems; it requires very accurate information about the electronic states and structures of individual molecules or ions. For complicated compounds such as metal chelates, it is rather difficult to obtain such information at present. With the development of the theory, however, the study of optical rotatory dispersion or circular dichroism will provide much important information about the electronic states or structures of metal chelates. Recently, instrumental advances have been very rapid [6, 7], and the data measured by automatic recording instruments are now rapidly accumulating [191–246]. Not much theoretical work has been carried out, however. Coordination chemists are earnestly seeking applications of this technique to the studies of absolute configuration, reaction mechanisms and kinetics, asymmetric or stereospecific syntheses, thermodynamic properties of dissolved species, and many other stereochemical problems.

Some excellent books [1–5] and general review articles [6–15] have been published on optical rotatory dispersion and circular dichroism; they complement this chapter, especially as regards studies made before about 1960.

3-1 THEORY

A. Definitions and Units

The direction of vibration of the electric vector of light emitted by a light source varies with time. Light whose vibration is confined in a plane is called linearly polarized, or plane polarized, light. The plane of polarization of linearly polarized light is determined by the direction of vibration of the magnetic (or electric) vector and the direction of propagation of the light. More generally, elliptically polarized light is produced by the composition of two linearly polarized beams of light of different vibrating planes (xz and yz planes) which propagate along the z axis; the electric or magnetic vector of the composed light vibrates elliptically in the xy plane. As shown in Fig. 3-1, the ellipticity of the ellipse is determined by the phase difference $\delta = \delta_1 - \delta_2$ of the two component beams; the elliptical vibration is crushed into a rectilinear one (linearly polarized light) when $\delta = k\pi$ ($k = 0, \pm 1, \ldots,$) and the ellipse expands into a circle (circularly polarized light) when

$$\delta = (2k + 1)(\pi/2) \quad (k = 0, \pm 1, \ldots),$$

provided the two component vibrations have equal amplitudes.

1. Optical Rotatory Power and Circular Birefringence. Given a linearly polarized beam of monochromatic light of frequency ν, which falls perpendicularly from a vacuum on the front surface of a sample containing an optically active substance, and passes through the sample back into a vacuum.

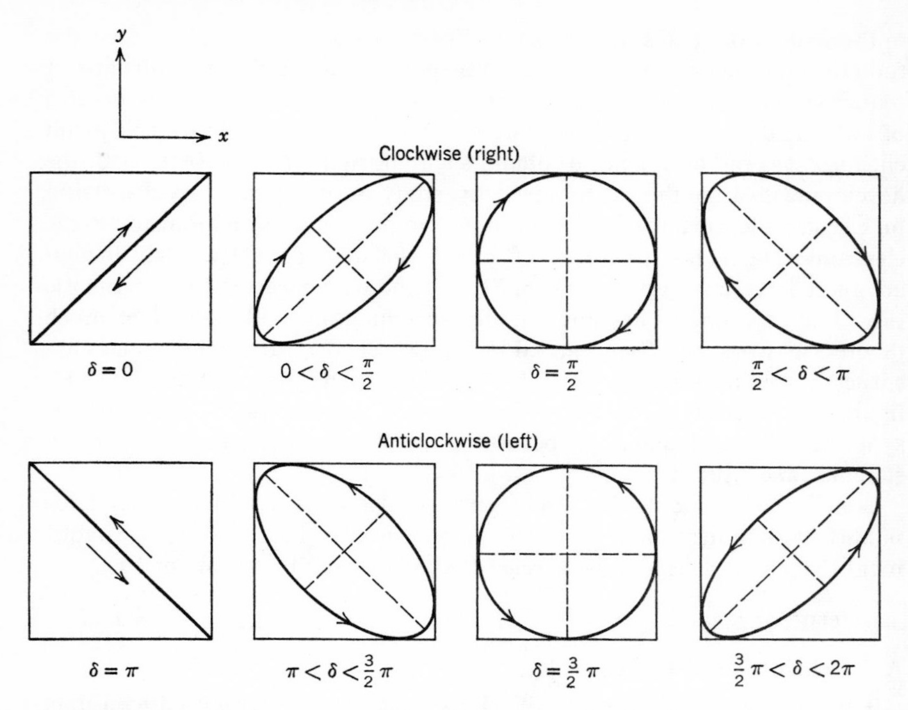

Fig. 3-1. Composition of two linearly polarized light beams of different vibrating planes (*xz* and *yz* planes) and of equal amplitudes.

If the back surface of the sample is parallel to and d cm from the front surface, and if the plane of polarization of the linearly polarized resultant light makes an angle α (degrees) or α' (radians) with that of the incident light, then the optical rotatory power of a 1-cm layer of this sample is defined as

$$\alpha^0 = \frac{\alpha'}{d} = \left(\frac{\pi}{\lambda}\right)(n_l - n_r) \quad \text{rad}\cdot\text{cm}^{-1}, \tag{3-1}$$

$$\alpha = 180\cdot\frac{\alpha'}{\pi} \quad \text{deg.} \tag{3-2}$$

Here $\lambda = C/\nu$, where C is the velocity of light in a vacuum, and n_r or n_l is the refractive index of the sample for the right or left circularly polarized light of which the incident light is composed. The quantity $n_l - n_r$ is termed circular birefringence. Circularly polarized light is defined as right when its electric or magnetic vector rotates clockwise as viewed by an observer facing the direction of the propagation of light. Also, the sign of the optical rotation

α is defined as (+) when the plane of polarization rotates clockwise as viewed by an observer facing the direction of the propagation of light. In this case the sample is said to be dextrorotatory; in the opposite case, levorotatory.

If, in addition, the sample exhibits circular dichroism, the light which has passed through the substance is not linearly, but elliptically, polarized. Then (3-1) and (3-2) hold for the angle α between the plane of the linearly polarized incident light and the plane of the major axis of the elliptically polarized resultant light.

If the concentration of an optically active substance is expressed by c' (g cm^{-3}) (density of a pure substance), the specific rotation $[\alpha]$ is defined as†

$$[\alpha] = \frac{\alpha}{c' \cdot d'} \quad \text{deg} \cdot \text{dm}^{-1} \cdot \text{cm}^3 \cdot \text{g}^{-1}, \tag{3-3}$$

where d' is the thickness of a layer of the sample (in dm); $d' = d/10$. The molar (or molecular) rotation $[M]$ is defined as

$$[M] = M[\alpha] \cdot 10^{-2} = \frac{M\alpha}{c' \cdot d'} \cdot 10^{-2} = \frac{\alpha}{c \cdot d} \cdot 10^{2} \quad \text{deg} \cdot \text{cm}^{-1} \cdot \text{l} \cdot \text{mole}^{-1} \cdot 10^{-2} \tag{3-4}$$

where M (g mole^{-1}) is the formula weight or molecular weight of the optically active compound, and c (mole l^{-1}) is its molar concentration; $c = 10^3 c'/M$.

2. Ellipticity and Circular Dichroism. The phenomenon that a substance absorbs right and left circularly polarized light differently is called circular dichroism. All optically active substances exhibit circular dichroism in the region of their appropriate electronic absorption bands. When linearly polarized monochromatic light travels through such a sample, elliptically polarized light emerges. The projection of the electric or magnetic vector of the resultant light on a plane perpendicular to the light beam makes an elliptical vibration (Fig. 3-2), and the following relations exist:

$$\begin{aligned} \varphi^0 &= \frac{\varphi'}{d} \simeq \tan\left(\frac{\varphi'}{d}\right) = \left(\frac{\pi}{\lambda}\right)(k'_l - k'_r) = \frac{1}{4}(k_l{}^0 - k_r{}^0) \\ &= \frac{2.3026}{4}(k_l - k_r) \quad \text{rad} \cdot \text{cm}^{-1}. \end{aligned} \tag{3-5}$$

Here the angle φ' is expressed in radians and the subscripts r and l refer to right and left circularly polarized light. The three absorption coefficients k', k^0, and k are defined respectively by $I = I_0 \cdot \exp(-4\pi/\lambda \cdot k'd)$, $I = I_0 \exp(-k^0 d)$, and $I = I_0 \cdot 10^{-kd}$, where I_0 and I are the intensity of the incident and the resultant light, respectively. Tan φ' is the ratio of the major to the minor axis

† The specific rotation of a crystal is also defined frequently as the rotation of a 1-mm layer of the sample.

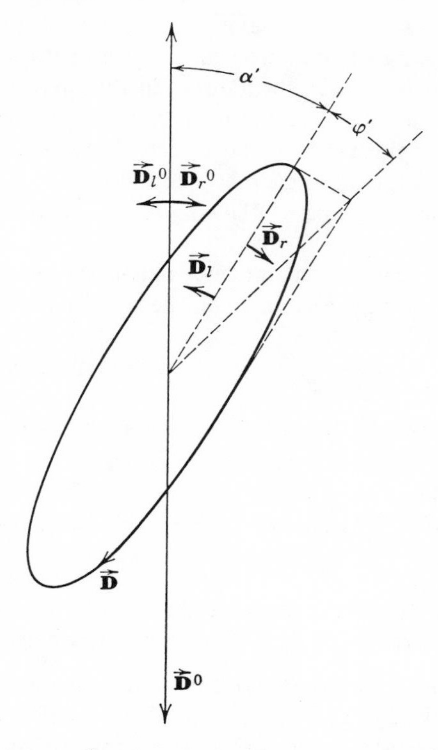

Fig. 3-2. Circular dichroism: $\vec{\mathbf{D}}^0$ is the vibration of the linearly polarized incident light and is equal to $(\vec{\mathbf{D}}_l^0 + \vec{\mathbf{D}}_r^0)$; $\vec{\mathbf{D}}$ is the vibration of the elliptically polarized resultant light and is equal to $(\vec{\mathbf{D}}_l + \vec{\mathbf{D}}_r)$; α' is the angle of optical rotation and φ' the angle of ellipticity.

of the ellipse and is termed its "ellipticity"; in practice it can be approximated by φ'. Ellipticity, like rotatory power, is measured in degrees; the angle φ (degrees) is defined

$$\varphi = 180 \cdot \frac{\varphi'}{\pi} \quad \text{deg.} \tag{3-6}$$

The specific ellipticity $[\varphi]$ and the molar ellipticity $[\theta]$ are defined as

$$[\varphi] = \frac{\varphi}{c' \cdot d'} \quad \text{deg} \cdot \text{dm}^{-1} \cdot \text{cm}^3 \cdot \text{g}^{-1}, \tag{3-7}$$

$$\begin{aligned} [\theta] &= M[\varphi] \cdot 10^{-2} = \frac{M\varphi}{c' \cdot d'} \cdot 10^{-2} \\ &= \frac{\varphi}{c \cdot d} \cdot 10^2 \quad \text{deg} \cdot \text{cm}^{-1} \cdot \text{l} \cdot \text{mole}^{-1} \cdot 10^{-2}. \end{aligned} \tag{3-8}$$

Then, using the molar extinction coefficient ε ($cm^{-1}\cdot l\cdot mole^{-1}$), which is determined by a decadic equation $I = I_0 \cdot 10^{-\varepsilon cd}$, the molar circular dichroism $\varepsilon_l - \varepsilon_r$ is defined as

$$\varepsilon_l - \varepsilon_r = \frac{k_l - k_r}{c} = \frac{4}{2.3026} \cdot \frac{\pi}{180} \cdot [\theta] \cdot 10^{-2}$$

$$= 0.3032 \cdot 10^{-3}[\theta] \quad cm^{-1}\cdot l \cdot mole^{-1}. \tag{3-9}$$

3. Optical Rotatory Dispersion and the Cotton Effect. The optical rotatory power and the circular dichroism both vary with the wavelength of light. By plotting the rotatory power $[\alpha]$ or $[M]$ against wavelength λ, wave number σ (or $\tilde{\nu}$), or frequency ν, the optical rotatory dispersion curve (or spectrum) is obtained; by similarly plotting $[\theta]$ or $\varepsilon_l - \varepsilon_r$, the circular dichroism curve (or spectrum) is obtained. In the remainder of this chapter, optical rotatory dispersion will be abbreviated ORD and circular dichroism, CD.

As between the dispersion of the refractive index and the absorption spectrum, the Kronig-Kramers relationship exists between the ORD and the CD [133]; this analogy is evident from (3-1) and (3-5). If only a Kth electronic transition is considered, a somewhat simplified form of the Kronig-Kramers relationship is written

$$[M_K(\lambda')] = \frac{2}{\pi} \int_0^\infty [\theta_K(\lambda)] \cdot \frac{\lambda}{(\lambda')^2 - \lambda^2} \, d\lambda. \tag{3-10}$$

As a first approximation let us consider the case in which the CD of the Kth transition can be approximated (on the wavelength scale) by a Gaussian curve of the form

$$[\theta_K(\lambda)] = [\theta_K^\circ] \cdot \exp\left\{-\frac{(\lambda - \lambda_K{}^{CD})^2}{(\Delta_K{}^{CD})^2}\right\}, \tag{3-11}$$

where $[\theta_K^\circ]$ is the extremum (maximum or minimum) value of the Kth CD curve, $\lambda_K{}^{CD}$ is the wavelength of the extremum, and $\Delta_K{}^{CD}$ is a $(1/e)$-halfwidth of the band (i.e., the wavelength interval in which the circular dichroism falls to $1/e = 0.3679$ of the extremum value). The so-called "halfwidth" $\Delta(\frac{1}{2})_K{}^{CD}$ is half the interval between the wavelengths at which the value falls to $\frac{1}{2}$ of the extremum value, and a relation $\Delta_K{}^{CD} = 1.201\ \Delta(\frac{1}{2})_K{}^{CD}$ exists. Substituting (3-11) into (3-10) and evaluating the integral on the assumption that $\lambda_K{}^{CD} \gg \Delta_K{}^{CD}$, it is found that [134]

$$[M_K(\lambda)] = \frac{2[\theta_K^\circ]}{\sqrt{\pi}} \left\{ \exp\left[-\frac{(\lambda - \lambda_K{}^{CD})^2}{(\Delta_K{}^{CD})^2}\right] \int_0^{(\lambda - \lambda_K{}^{CD})/\Delta_K{}^{CD}} e^{x^2}\, dx - \frac{\Delta_K{}^{CD}}{2(\lambda + \lambda_K{}^{CD})} \right\} \tag{3-12}$$

Figure 3-3 shows the curves of (3-11) and (3-12) calculated by using a set of actual figures; the ORD changes sign at the wavelength of the CD extremum. This type of dispersion was called inversive dispersion. Since the

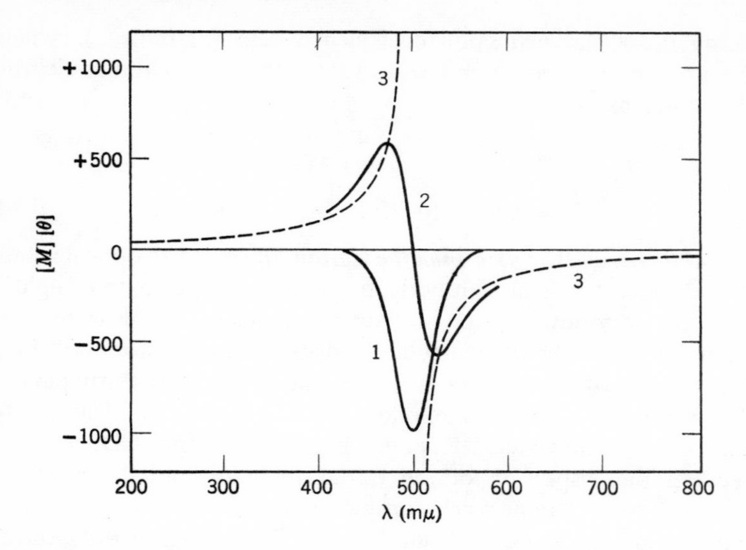

Fig. 3-3. Typical plots of Eqs. (3-11), (3-12), and (3-14): $\lambda_K{}^{CD} = 500$ mμ, $\Delta_K{}^{CD} = 30$ mμ, and $[\theta_K{}^0] = -1000°$. 1. CD: Eq. (3-11), 2. ORD: Eq. (3-12). 3. Drude formula: Eq. (3-14).

CD and the inversive dispersion of ORD in the vicinity of an absorption band were first discovered by Cotton [37], each of the two phenomena or the combined effect is called the "Cotton effect".

As (3-10) or Fig. 3-3 shows, a positive CD band produces a $(+) \rightarrow (-)$ inversive dispersion of ORD from longer to shorter wavelengths, whereas a negative CD band produces a $(-) \rightarrow (+)$ one; this constitutes the so-called Natanson's rule. Equation (3-12) takes two extremum values,

$$[M_K]_{\text{ext}} = \pm 0.6105[\theta_K^\circ],$$

at the wavelengths

$$(\lambda_{\text{ext}})_K = \lambda_K{}^{\text{CD}} \pm 0.9233\Delta_K{}^{\text{CD}} = \lambda_K{}^{\text{ORD}} \pm 0.9233\Delta_K{}^{\text{CD}}.$$

Only a Kth transition is considered above. If all the electronic transitions are considered, there is a relation

$$[M(\lambda)] = \sum_K [M_K(\lambda)]. \tag{3-13}$$

And, when λ goes far from a $\lambda_K{}^{\text{CD}}$, (3-12) becomes

$$[M_K(\lambda)] \simeq \frac{2}{\sqrt{\pi}}[\theta_K^\circ]\cdot\lambda_K{}^{\text{CD}}\cdot\frac{\Delta_K{}^{\text{CD}}}{\lambda^2 - (\lambda_K{}^{\text{CD}})^2} = \frac{K_K}{\lambda^2 - (\lambda_K{}^{\text{CD}})^2}. \tag{3-14}$$

When λ goes far from all the $\lambda_K{}^{CD}$, $[M(\lambda)]$ becomes

$$[M(\lambda)] \simeq \sum_K \frac{K_K}{\lambda^2 - (\lambda_K{}^{CD})^2}, \tag{3-15}$$

where the equation is identical with a multitermed dispersion formula of Drude. In a narrower range of wavelengths, (3-15) can be approximated well by a one-term Drude equation:

$$[M(\lambda)] = \frac{K}{\lambda^2 - \lambda_0^2}. \tag{3-16}$$

Although (3-11) to (3-16) were all derived on the assumption of a Gaussian CD curve, they have been found to agree well with the curves observed experimentally (see Sec. 3-1C).

4. Rotational Strength and the Dissymmetry Factor. The rotational strength R can be defined experimentally by

$$R = \frac{3hC}{8\pi^3 N'} \int_0^\infty \varphi^\circ(\lambda) \cdot \frac{1}{\lambda}\, d\lambda, \tag{3-17}$$

where h is Planck's constant, N' is the number of molecules or formula units of the optically active substance in a space of 1 cm^3. Equation (3-17) corresponds to "dipole strength," D in (3-18), which defines the strength of electronic

$$D = \frac{3hC}{8\pi^3 N'} \int_0^\infty k^0\,(\lambda) \cdot \frac{1}{\lambda}\, d\lambda \tag{3-18}$$

transitions. Substituting (3-5) and (3-9) in (3-17), we obtain (considering only a Kth transition)

$$\begin{aligned} R_K &= \frac{2.3026 \times 3 \times 10^3}{4} \cdot \frac{hC}{8\pi^3 N} \int_0^\infty [\varepsilon_l(\lambda) - \varepsilon_r(\lambda)]_K \cdot \frac{1}{\lambda}\, d\lambda \\ &= 22.9 \times 10^{-40} \int_0^\infty [\varepsilon_l(\lambda) - \varepsilon_r(\lambda)]_K \cdot \frac{1}{\lambda}\, d\lambda \quad \text{cgs}, \end{aligned} \tag{3-19}$$

where N is Avogadro's number; $N = 10^3 N'/c$. Similarly

$$\begin{aligned} D_K &= 2.3026 \times 3 \times 10^3 \cdot \frac{hC}{8\pi^3 N} \int_0^\infty \varepsilon_K(\lambda) \cdot \frac{1}{\lambda}\, d\lambda \\ &= 91.8 \times 10^{-40} \int_0^\infty \varepsilon_K(\lambda) \cdot \frac{1}{\lambda}\, d\lambda \quad \text{cgs} \end{aligned} \tag{3-20}$$

is also obtained.

For a Gaussian CD curve,

$$\int_0^\infty [\varepsilon_l(\lambda) - \varepsilon_r(\lambda)]_K \cdot \frac{1}{\lambda}\, d\lambda \simeq \sqrt{\pi}(\varepsilon_l - \varepsilon_r)_K{}^0 \cdot \frac{\Delta_K{}^{CD}}{\lambda_K{}^{CD}}. \tag{3-21}$$

Therefore

$$R_K \simeq 40.6 \times 10^{-40}(\varepsilon_l - \varepsilon_r)_K{}^0 \cdot \frac{\Delta_K{}^{CD}}{\lambda_K{}^{CD}}, \tag{3-22}$$

where $(\varepsilon_l - \varepsilon_r)_K{}^0$ is the extremum value of the Kth CD curve. For a Gaussian absorption band, $\varepsilon_K(\lambda)$ is expressed

$$\varepsilon_K(\lambda) = \varepsilon_K{}^0 \cdot \exp\left[-\frac{(\lambda - \lambda_K{}^{AB})^2}{(\Delta_K{}^{AB})^2}\right]. \tag{3-23}$$

Therefore it follows that

$$\int_0^\infty \varepsilon_K(\lambda) \cdot \frac{1}{\lambda}\, d\lambda \simeq \sqrt{\pi} \cdot \varepsilon_K{}^0 \cdot \frac{\Delta_K{}^{AB}}{\lambda_K{}^{AB}}. \tag{3-24}$$

Therefore

$$D_K = 163 \times 10^{-40} \varepsilon_K{}^0 \cdot \frac{\Delta_K{}^{AB}}{\lambda_K{}^{AB}}. \tag{3-25}$$

The dissymmetry factor G_K of the Kth electronic transition is defined as

$$G_K = \frac{4R_K}{D_K} = \frac{\int_0^\infty [\varepsilon_l(\lambda) - \varepsilon_r(\lambda)]_K \cdot \frac{1}{\lambda}\, d\lambda}{\int_0^\infty \varepsilon_K(\lambda) \cdot \frac{1}{\lambda}\, d\lambda}. \tag{3-26}$$

In other cases the dissymmetry factor $g(\lambda)$, which is a function of λ, is defined as

$$g(\lambda) = \frac{[\varepsilon_l(\lambda) - \varepsilon_r(\lambda)]}{\varepsilon(\lambda)}. \tag{3-27}$$

Since $\varepsilon(\lambda) = \frac{1}{2}[\varepsilon_l(\lambda) + \varepsilon_r(\lambda)]$, the absolute values of G_K and $g(\lambda)$ do not exceed 2 under their maximum conditions.

B. Symmetry Requirements for Optical Activity

Optical rotatory power originates in the three following situations: (1) the molecule of a substance is dissymmetrical in its structure; (2) there is a dissymmetrical arrangement of molecules or ions in a crystal; (3) the substance is affected by an outside field. Examples of (2) are found in SiO_2 and $NiSO_4 \cdot 6H_2O$ crystals, and an example of (3) is the Faraday effect produced by an external magnetic field. The term "molecular dissymmetry" means that a molecule or ion lacks the second kind of symmetry elements denoted by $\bar{n}$; they include a center of symmetry $\bar{1}$, a plane of symmetry $\bar{2}$, and a fourfold rotatory inversion axis, $\bar{4}$; etc. The bis(diamine)Cu(I) complex ion (diagram I) is not optically active, because it has a fourfold rotatory inversion axis (a planar structure of chelate rings was assumed). Whether or not a molecule possesses the first kind of symmetry elements (pure rotation axis),

I

n, has no influence on its optical activity. Frequently "asymmetry" and "dissymmetry" are differentiated in the following way. "Asymmetry" means that there is no symmetry element except a onefold rotation axis, as for an organic "asymmetric carbon atom." However, two or more asymmetric atoms can be found in a non-optically active molecule, such as a so-called meso isomer. Such a molecule taken as a whole has one of the second kind of symmetry elements and consequently possesses no optical activity. "Dissymmetry," on the other hand, is taken to mean that a molecule or ion has the first but not the second kind of symmetry elements; an example is $[Co(en)_3]^{3+}$ (II), which has one threefold and three twofold rotation axes but lacks $\bar{1}$, $\bar{2}$, and $\bar{4}$.

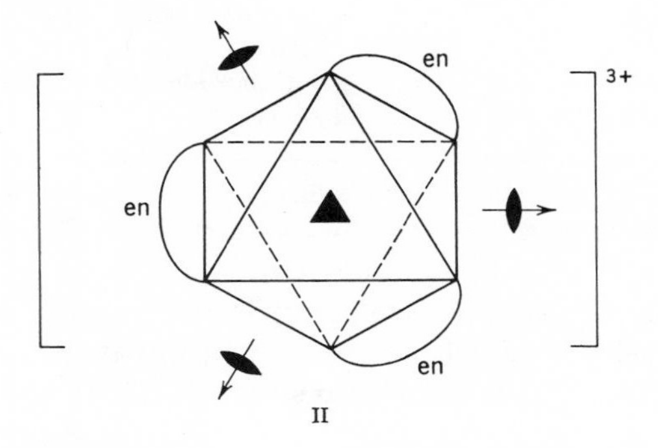

II

Most of the metal complexes resolved to date are six-coordinate octahedral ones, especially of the types $[M(AA)_3]$ or $[M(AA)_2ab]$ (Fig. 3-4), where AA represents a bidentate ligand. In Table 3-1 are presented the different types and typical examples of the octahedral complexes resolved thus far.

There are no examples of transition metal analogs of an asymmetric carbon atom, tetrahedral [M(a)(b)(c)(d)], whose optical activity has been

Table 3-1 Different Types of Resolved Octahedral Metal Complexes[a]

Type	Example	References
$[M(AA)_3]$	$[Co(en)_3]^{3+}$	181
$[M(AA)_2a_2]$	*cis*-$[Co(en)_2(NH_3)_2]^{3+}$	183
$[M(AA)_2ab]$	*cis*-$[Co(en)_2(NH_3)Cl]^{2+}$	180
$[M(AA)_2(BB)]$	$[Co(en)_2CO_3]^{+}$	182
$[M(AA)_2(BC)]$	$[Co(en)_2(gly)]^{2+}$	129
$[M(AA)a_2b_2]$	$[Co(en)(NH_3)_2Cl_2]^{+}$	17
$[M(AA)a_2bc]$	$[Pt(en)Cl_2(NH_3)(NO_2)]^{+}$	32, 34
$[M(AA)abcd]$	$[Pt(en)(Cl)(Br)(NH_3)(NO_2)]^{+}$	33
$[M(AA)(BB)a_2]$	*cis*-$[Co(en)CO_3(NH_3)_2]^{+}$	17
$[M(AA)(BC)a_2]$	*cis*-$[Co(en)(i\text{-bn})(NO_2)_2]^{+}$	35
$[M(AA)(BC)_2]$	*trans*-$[Co(ox)(gly)_2]^{-}$	77
$[M(AB)_3]$	$[Co(gly)_3]$	93
$[M(ABC)_2]$	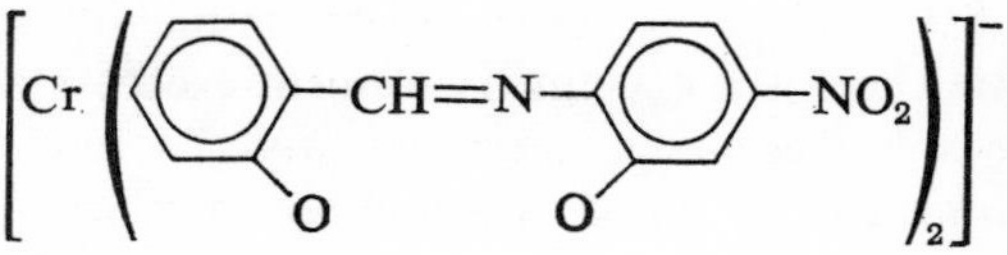 $[Cr(C_6H_4(O)\text{–CH=N–}C_6H_3(O)\text{–}NO_2)_2]^{-}$	145
$[M(ABBA)a_2]$	$[Co(trien)Cl_2]^{+}$	160
$[M(ABBA)ab]$	$[Co(trien)(OH_2)Cl]^{2+}$	159
$[M(ABBA)(CC)]$	$[Co(trien)CO_3]^{+}$	160
$[M(ABCBA)a]$	$[Co(tetraethylenepentamine)Cl]^{2+}$	57
$[M(A_2BCD)a]$	$[Co(H\text{–}EDTA)Br]^{-}$	29
$[M(A_2BBA_2)]$	$[Co(EDTA)]^{-}$	49
$[M(ABCCBA)]$	[Co{1,8-bis(salicylideneamino)-3,6-dithiaoctane}]$^{+}$	47
$[M(ABCDEF)]$	[Co{1,8-bis(salicylideneamino)-3-oxa-6-thiaoctane}]$^{+}$	48

[a] Complexes containing optically active ligands are not included. All the octahedral complexes resolved to date contain at least one chelate ligand.

proved. All the tetrahedral metal complexes previously resolved belong to the type $[M(AB)_2]$ (Fig. 3-5); an example of this type is bis(8-quinolinol-5-sulfonic acid)zinc(II) [108], shown in structure III.

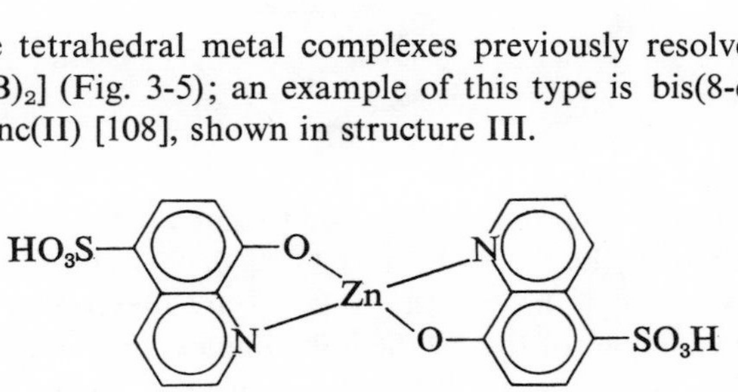

HO_3S–(quinolinolate)–O, N → Zn ← N, O–(quinolinolate)–SO_3H

III

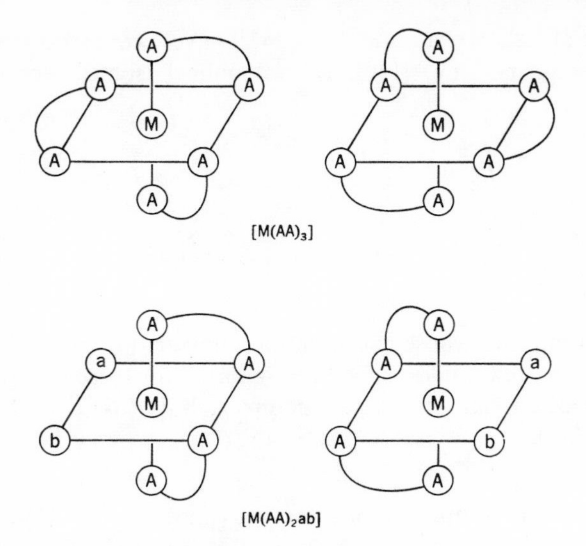

Fig. 3-4. Optical isomers of $[M(AA)_3]$- and $[M(AA)_2ab]$-type complexes.

Only a few planar complexes are known to be optically active; their optical activity originates in a more or less special arrangement of ligands. A classical example of this type is $[Pt(i\text{-bn})(\text{meso-stien})]^{2+}$ (IV) [131].

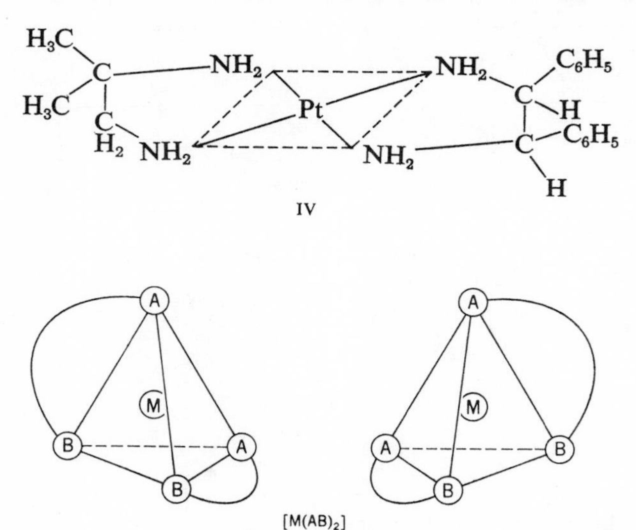

Fig. 3-5. Optical isomers of a tetrahedral complex of the type $[M(AB)_2]$.

Recently several other examples [139–141] have been reported, such as $[PtCl_3(\textit{trans}\text{-}2\text{-butene})]^-$ (V) [140]. Several optical isomers are also to be

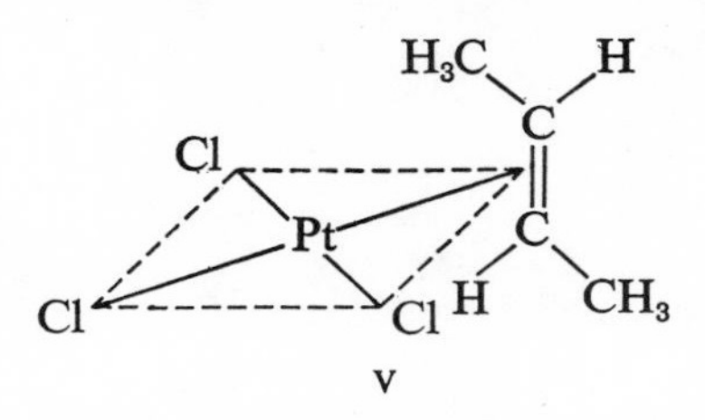

V

expected for complexes whose coordination numbers are other than six and four. Nevertheless, few studies have been reported for these groups, except a classical study of an eight-coordinate complex, $K_4[U(ox)_4]$, by Marchi and McReynolds [111].

Several binuclear complexes of the type $[(en)_2Co\langle{}^{a}_{b}\rangle Co(en)_2]^{n+}$ are known in optically active forms, and two tetranuclear hexol complexes, $[Co\{(OH)_2Co(NH_3)_4\}_3]^{6+}$ and $[Co\{(OH)_2Co(en)_2\}_3]^{6+}$. For the binuclear

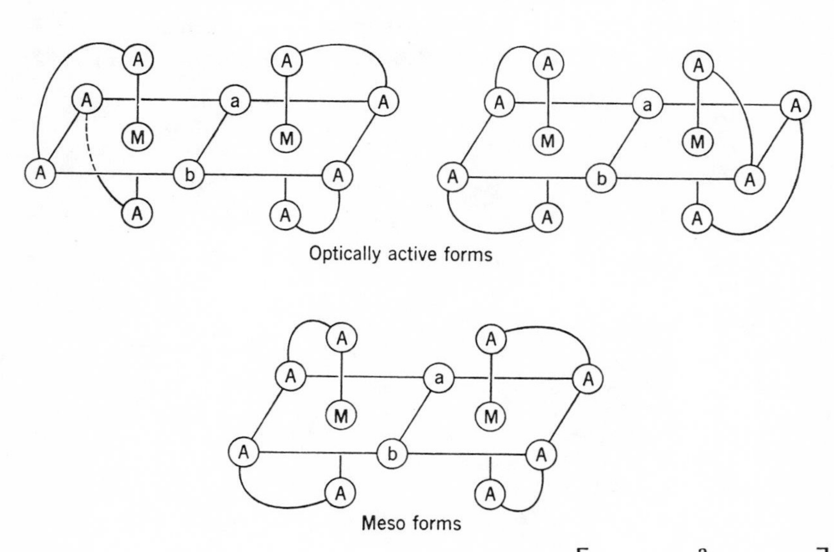

Fig. 3-6. Three isomers of a binuclear complex of the type $\left[(AA)_2M\langle{}^{a}_{b}\rangle M(AA)_2\right]$.

complexes, theory predicts three isomers, namely, two optically active and one meso form as in tartaric acid. The three isomers were actually separated

NH_2

by Werner [184] for $[(en)_2Co$ $Co(en)_2]^{4+}$ (Fig. 3-6).

NO_2

A few cases are known in which a new dissymmetric structure is produced by coordination of an optically inactive ligand. The dissymmetric structure is due either to an asymmetrically coordinated atom (VI) [114] or to an asymmetric atom in a chelate ring (VII) [113]. Another example is the fixation of an asymmetric nitrogen atom by coordination as in an N-ethyl-sarcosine complex of Pt(II) [94] (structure VIII).

$CH_2CH_2NH_2$
$\ddot{S}^*$—CH_2
Cl_4Pt
H_2N—CH_2

VI

H_2N—CH_2
Cl_4Pt Cl
H_2N—$\overset{*}{C}HCH_2NH_3$

VII

O—CO
K $(NO_2)_2Pt$
$\overset{*}{N}$—CH_2
H_3C C_2H_5

VIII

Finally, as is well known, ethylenediamine and other diamines are coordinated with a metal producing a non-planar puckered chelate ring [65]. The puckered chelate ring is dissymmetric and can be an origin of optical activity also. This type of dissymmetry is discussed in Sec. 3-2B.

C. Curve Analysis

Different types of curve analysis have been made by several authors in their studies of ORD and CD spectra. In principle they are divided into four groups: (1) curve fitting to a normal dispersion type of ORD; (2) curve correlation between ORD and CD; (3) the separation of observed ORD and CD curves into the partial curves which correspond to each of the individual electronic transitions; and (4) the separation of observed ORD and CD curves into the partial curves which originate in each of the asymmetric centers.

An excellent discussion of the curve analysis of the ordinary electronic absorption curves of metal complexes has been given by Jørgensen [87, 88].

1. Normal ORD Dispersion (Plain Curve). The ORD of an optically active substance outside the absorption band region is usually called a normal dispersion or a plain curve; the rotation increases gradually in a normal fashion to a rise in the ultraviolet region. Such dispersions can for the most part be well fitted to a one-term Drude equation (3-16). This is not true, however, when a very large range of wavelengths is concerned. To obtain the values of the parameters λ_0 and K in this (3-16) a "Lowry plot" is made [3, 109] by plotting $1/[M]$ against λ^2. If $1/[M]$ against λ^2 gives a straight line, the dispersion can be represented by one term of the Drude equation, with $1/K$ the slope and $\lambda_0{}^2$ the intercept on the λ^2 axis. Usually λ^2 is expressed in units of μ^2. An example of a Lowry plot is shown in Fig. 3-7. If the Lowry plot is not linear, two terms of the Drude equation or more complicated equations are required.

2. Curve Correlation between ORD and CD. As mentioned in Sec. 3-1A the Kronig-Kramers relationship (3-10) exists between the ORD and the

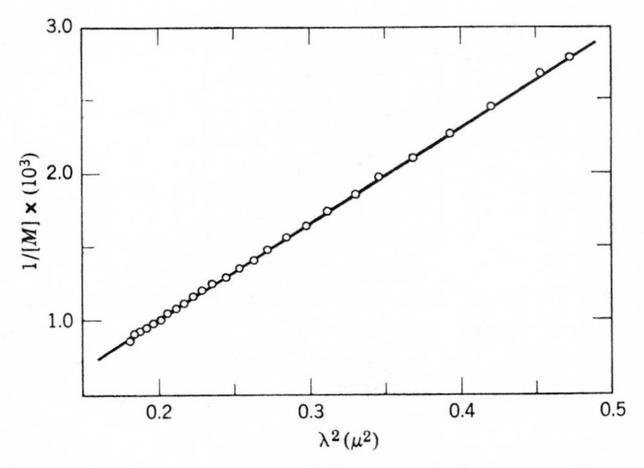

Fig. 3-7. An example of a Lowry plot: $[Pt^{II}\{(-)\text{-chxn}\}_2]Cl_2$ [84]. $\lambda_0{}^2 = 0.046$; $\lambda_0 = 0.214$; $K = 152$.

CD spectra. For the special case of a Gaussian CD band, the application of (3-10) yields [134] equation (3-12), the "Moscowitz equation." This equation is almost identical with the equation of Kuhn and Braun [95], which was obtained in 1930 from their pioneering work on a coupled oscillator model. It has been established that the curves calculated by means of (3-12) are in good agreement with the measured curves [24, 92, 190]. The so-called Kuhn integral,

$$F(u) = e^{-u^2} \int_0^u e^{x^2} \, dx, \tag{3-28}$$

which is needed in the calculation of (3-12), has been tabulated by Woldbye [15], who quotes from the calculations of Engberg. The maximum of the integral exists at $u_{\max} = 0.9233$, and the value of $F(u)_{\max}$ equals 0.54105.

A difficulty in practical applications of these equations is that the individual CD curves deviate more or less from the Gaussian type. It was recommended by Yasui et al. [190] that the method of Jørgensen be used, and curve analyses made separately on the longer and shorter wavelength sides of the individual inversive center, λ_K^{ORD}, using different widths, $\Delta_K^{\mathrm{CD}}(\text{red})$ and $\Delta_K^{\mathrm{CD}}(\text{blue})$, respectively (see Fig. 3-8). The rotational strength R_K^{CD} in this case is given by

$$R_K^{\mathrm{CD}} \simeq 20.3 \times 10^{-40} (\varepsilon_l - \varepsilon_r)_K{}^0 \left\{ \frac{\Delta_K^{\mathrm{CD}}(\text{red}) + \Delta_K^{\mathrm{CD}}(\text{blue})}{\lambda_K^{\mathrm{CD}}} \right\}. \tag{3-29}$$

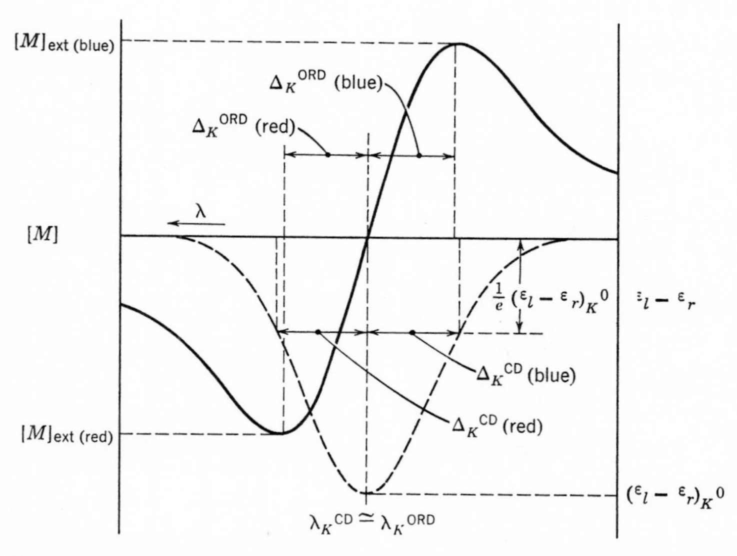

Fig. 3-8. Rotational parameters of a non-symmetrical CD or ORD band.

3. Partial Curves that Correspond to Each of the Individual Electronic Transitions. Since the symmetry of an optically active metal complex is necessarily rather low, the degeneracy of the $d_\varepsilon \rightarrow d_\gamma$ absorption bands of its O_h parentage is broken down to produce several component absorption bands, which are usually located rather close to each other. It is very difficult therefore to analyze the observed ORD and CD curves into the individual component curves. Only when the absorption band is widely split can the analysis be made more easily [39, 67, 190]. A good example is the study of *trans*-$[Co(en)_2(L\text{-}amH)_2]^{3+}$ by Yasui et al. [190] (L-amH = L-threonine and L-leucine). The curves for the threonine complex are shown in Fig. 3-9. The curve analyses were made as follows. First, the CD bands, which were assumed to be Gaussian on the wavelength scale, were separated from the observed CD curve, and the corresponding ORD curves were calculated. Then the sum of the individual curves was subtracted from the observed ORD curve (see Fig. 3-10). From the residual curve the constants in the Drude equation, λ_0 and K, were estimated by the Lowry plot. In the second step,

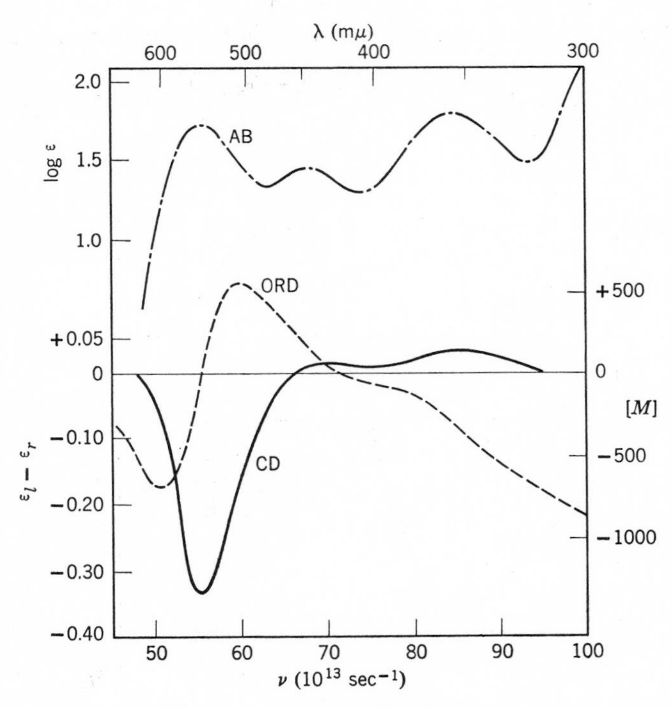

Fig. 3-9. Absorption (–·–·–), CD(———), and ORD(– – – – –) curves of the *trans*-$[Co(en)_2(L\text{-}thrH)_2]^{3+}$ ion (L-thrH = L-threonine).

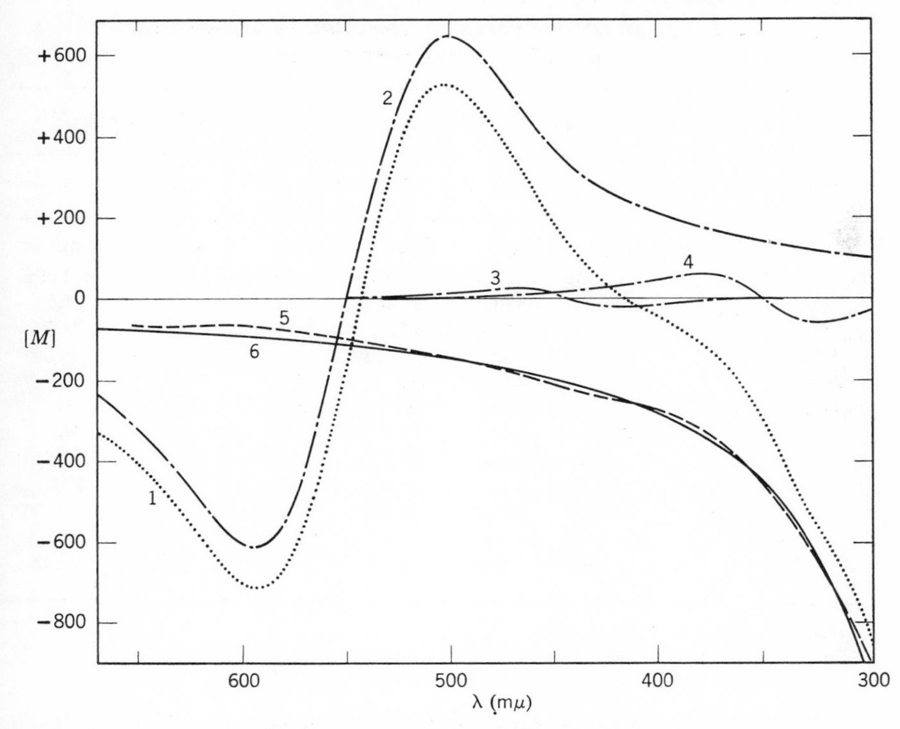

Fig. 3-10. The analyses of the ORD curves of *trans*-$[Co(en)_2(L\text{-thrH})_2]^{3+}$ ion. 1 (·····): observed ORD curve; 2, 3, and 4 (-·-·-): individual ORD curve of each absorption component; 5 (-----): observed curve minus the sum of the individual ORD curves; 6 (———): Drude curve.

the Drude curve was subtracted from the observed ORD curve, and the individual ORD curves of the absorption components were again determined by analyzing the residual curve. The results are shown in Fig. 3-10 and in Table 3-2, which also lists the results for the L-leucine complex. For these ions, therefore, it is possible to estimate the rotational strengths from the CD bands in the first step, and also from the ORD parameters in the second step. The rotational strength from the ORD data, R_K^{ORD}, is given by

$$R_K^{\text{ORD}} = 0.00546 \times 10^{-40}\{[M]_{\text{ext(red)}} - [M]_{\text{ext(blue)}}\} \times \left\{\frac{\Delta_K^{\text{ORD}}(\text{red}) + \Delta_K^{\text{ORD}}(\text{blue})}{\lambda_K^{\text{ORD}}}\right\}, \quad (3\text{-}30)$$

where the parameters are defined as in Fig. 3-8, and a relation

$$\Delta_K^{\text{ORD}} \simeq 0.9233\Delta_K{}^{\text{CD}}$$

Table 3-2 Rotational Parameters and Rotational Strengths of *trans*-$[Co(en)_2(\text{L-amH})_2]^{3+}$ Ions

	L-Leucine Complex[a]			L-Threonine Complex[b]		
Absorption Band	Ia	Ib	II	Ia	Ib	II
λ_K^{CD} (mμ)	552	445	349	550	440	350
$(\varepsilon_l - \varepsilon_r)_K^0$	−0.330	−0.042	+0.055	−0.340	+0.011	+0.030
Δ_K^{CD}(red) (mμ)	49.0	33.0	26.0	49.5	30.0	30.5
Δ_K^{CD}(blue) (mμ)	49.0	39.0	25.0	51.8	29.0	29.2
R_K^{CD} (10^{-40} cgs)	−1.19	−0.138	+0.163	−1.27	+0.0299	+0.104
λ_K^{ORD} (mμ)	550	443	351	551	448	353
$[M]_{ext(red)}$	−615	−89	+93	−609	+22	+66
$[M]_{ext(blue)}$	+635	+76	−91	+650	−20	−64
Δ^{ORD}(red) (mμ)	45.0	32.8	24.2	45.0	30.0	25.5
Δ^{ORD}(blue) (mμ)	46.2	33.3	25.0	49.0	26.0	22.0
R_K^{ORD} (10^{-40} cgs)	−1.13	−0.134	+0.141	+1.17	+0.0287	+0.0955
R_K^{ORD}/R_K^{CD}	0.95	0.97	0.87	0.92	0.96	0.92

[a] Drude term: $\lambda_0 = 0.246\ \mu$, $K = +7.5$.
[b] Drude term: $\lambda_0 = 0.245\ \mu$, $K = -29.0$.

exists for the ideal case in which the Moscowitz equation holds perfectly. As Table 3-2 shows, the experimental values of R_K^{CD} are approximately equal to those of R_K^{ORD}, and furthermore the positions of λ_K^{ORD} coincide well with those of λ_K^{CD} within the range of ± 5 mμ. Some arbitrary factors are inevitable in this kind of curve analysis, especially in the case of ORD [72–74, 166]. A characteristic feature of many CD analyses is a cancellation between the (+) and the (−) components; in such cases it is very difficult to carry out a curve analysis [26].

4. Additivity for the Asymmetric Centers. In complexes such as $[Co(en)_2(\text{L-am})]X_2$ (L-am $= NH_2 \cdot CHR \cdot COO^-$), there are contributions to the optical activity or to the circular dichroism from the spiral configuration of the complex as a whole and from the vicinal effect of the optically active ligand. If the two contributions are additive, it is possible to separate them by analyzing the observed curves of the Δ and Λ complexes (Fig. 3-12), provided both contain the same optically active ligand, (L-am$^-$) in our example. This problem has been thoroughly discussed by Douglas and his collaborators [40, 41, 107], who concluded that the configurational and the vicinal effects are separable on the CD curves. A few examples have also been reported [81, 107] for ORD curves.

D. Optical Rotatory Power of Metal Complexes

1. General Theory. Fresnel derived (3-1) starting from the hypothesis that linearly polarized light is composed of right and left circularly polarized light of equal amplitude. When such linearly polarized light passes through an optically active medium, the right and the left circularly polarized components propagate with different velocities. Resultant linearly polarized light is produced by recombination of the two circularly polarized lights. But between the two circularly polarized components a phase difference, equal to twice the angle of rotation of the plane of polarization, has been introduced. The classical theory of optical rotatory power was beset with the problems of explaining why the oppositely circular-polarized light beams propagate with different velocities or with different refractive indices, and of determining which physical quantities of the optically active substance govern the refractive indices, n_r and n_l and also the difference $n_l - n_r$.

To explain these things Kuhn [95], for example, postulated an interaction between oscillators which are located some distance apart in a dissymmetric molecule. These so-called "coupled oscillators" interact differently with right and left circularly polarized light. From this theory it was shown qualitatively that, when a molecule has non-parallel components of a vibrating moment corresponding to an absorption band, the molecule shows optical activity. Kuhn derived an equation identical with the Drude equation (3-16) and clarified the characteristics of the constant K in this equation; when $K = 0$, the absorption band (the mean wavelength exists at λ_0) does not contribute to the optical activity. Thus for optical activity it is necessary that at least one oscillator have an absorption band, and that in this absorption band right and left circularly polarized light be differently absorbed (circular dichroism).

Besides the coupled oscillator model, another theory based on a one-electron model [8] has been developed. In classical electromagnetic theory the electrons in a molecule are set in motion by interacting with the electromagnetic field of incident linearly polarized light; thereby both an electric dipole moment and a magnetic dipole moment are induced. The induced electric and magnetic moments vary proportionally with the variation of the magnetic and electric fields of the incident light, respectively. Thus it has been shown that optical activity can be explained through consideration of only one electron. If we adopt a helical model of a dissymmetric molecule, an equation similar to the Drude formula can be obtained.

In the quantum theory [4, 8, 154] the rotational strength R_{ab} is determined by

$$R_{ab} = \frac{e^2}{2mC} I_m\langle a\mathbf{Q}b\rangle\langle b\mathbf{L}a\rangle, \tag{3-31}$$

where a and b are the wavefunctions of the a and b states of the molecule or ion, and $e\langle a\mathbf{Q}b\rangle$ and $e/2mC\cdot\langle b\mathbf{L}a\rangle$ are the induced electric and magnetic moments, respectively, of an electronic transition $a \rightarrow b$; I_m is the imaginary part of the scalar product of these two moments. Thus the calculation of the optical activity can be reduced theoretically to the calculation of the matrix elements, $\langle a\mathbf{Q}b\rangle$ and $\langle b\mathbf{L}a\rangle$. At present, however, it is almost impossible to know accurately the eigenfunctions for molecules or ions as complex as metal chelates.

Equation (3-31), which is a general conclusion from quantum mechanical considerations, includes in a qualitative fashion all the necessary conditions for optical activity [56]. When a molecule or ion has a center of symmetry, the electric moment operator will have a non-vanishing matrix element only between odd and even states. On the other hand, the magnetic moment operator will have matrix elements only between two odd or two even states. Therefore there are no two states for which the scalar product of the electric and the magnetic matrix elements does not vanish. An analogous condition holds for a plane of symmetry and other symmetry elements of the second kind. It is also clear from (3-31) that an optically active isomer has a value of R_{ab} equal, but opposite in sign, to that of the corresponding enantiomer.

A more simplified and practical expression of the rotational strength is

$$R_{ab} = \rho\mu \cos\theta, \tag{3-32}$$

where ρ and μ are the real electric and magnetic dipole moments of the transition, respectively, and θ is the angle between the directions of the two moments. The necessary condition for optical activity is that the non-zero electric and magnetic transition moment vectors be not orthogonal. An example of such orthogonal electric and magnetic moments may be found in the case of some planar metal complexes.

***2. An Example:* $[Co(en)_3]^{3+}$.** A classical attempt to calculate the rotatory power of a metal complex had been made by Kuhn in 1934 for the $[Co(ox)_3]^{3-}$ ion [10]. The first quantum mechanical treatment of this problem was done by Moffitt [132] in 1956 for complexes of D_3 symmetry, such as $[Co(en)_3]^{3+}$ or $[Cr(ox)_3]^{3-}$, on the basis of the ligand field theory of the $d \rightarrow d$ absorption bands.

Most of the electronic absorption bands of metal complexes in the near-infrared, visible, and near-ultraviolet regions are due to transitions between the d levels. As is well known, the d orbitals of an octahedral six-coordinate complex, $[M(X)_6]$, of symmetry O_h are split into two groups, d_ε and d_γ; the separation in energy between these two groups is denoted by Δ or $10\,Dq$. The series of ligands [173] or metals [88] arranged in increasing order of the parameter Dq is known as the spectrochemical series. The excitation of d

electrons from the lower-lying d_ε orbitals to the higher-lying d_γ orbitals usually produces two or more absorption bands. The various reasons for this are repulsions among electrons, occurrence of different configurations of electrons in the two subshells, occurrence of states with different spin multiplicities, spin-orbit couplings, and so on [18, 88].

For example, a low-spin cobalt(III) complex having an electronic configuration $(d_\varepsilon)^6$ has in general four absorption bands in the region from the near infrared to the near ultraviolet [58]. Of these four, two shorter wavelength bands are known as the first (I) and the second (II) bands [173]. Two others are situated in the near-infrared or longer wavelength visible region and have very weak intensities (about 1/100 of the I or the II band). Linhard [106] named the longer wavelength one the A band, and the shorter wavelength one the B band. Theoretically these four bands are assigned [171], respectively, to

$$\begin{aligned} \mathrm{A}:{}^1A_{1g} &\longrightarrow {}^3T_{1g}, \\ \mathrm{B}:{}^1A_{1g} &\longrightarrow {}^3T_{2g}, \\ \mathrm{I}:{}^1A_{1g} &\longrightarrow {}^1T_{1g}, \\ \mathrm{II}:{}^1A_{1g} &\longrightarrow {}^1T_{2g}. \end{aligned}$$

All four excited states have the same electronic configuration, $d_\varepsilon{}^5 d_\gamma{}^1$, and they vary in energy with the variation of Dq. All these excited states are orbitally triply degenerate and are therefore split when the symmetry of the complex is lowered. Accordingly the $d \rightarrow d$ absorption bands of cobalt(III) complexes are shifted and split by ligand substitution [161, 176, 177, 187]. Substitution by chelate ligands frequently has a great influence on the ORD or CD spectra, but less on the absorption spectra. For example, the spectrum of $[Co(en)_3]^{3+}$, whose symmetry is lowered to D_3 as compared with O_h of $[Co(X)_6]$, shows a great resemblance to that of the corresponding O_h complex $[Co(NH_3)_6]^{3+}$, since the visible absorption spectrum is largely determined by the kinds of atoms directly coordinated with the metal.

On the contrary, the CD spectrum (Fig. 3-11) of this chelate complex clearly shows a splitting in the I band region (20,300 and 23,400 cm^{-1}). The selection rules for magnetic dipole transitions in O_h and D_3 symmetries are shown in Table 3-3. It is concluded that the CD spectrum of the complex is governed as regards its main features by the selection rules for O_h symmetry; the weakness of the CD in the II band is typical evidence for this conclusion. Nevertheless, the existence of two CD components (+ and −) in the I band region clearly shows the effect of a trigonal field. The magnetic dipole selection rules for D_3 symmetry predict the occurrence of two CD components, ${}^1A_1 \rightarrow {}^1A_2$ and ${}^1A_1 \rightarrow {}^1E_a$, in the I band, and only one component, ${}^1A_1 \rightarrow {}^1E_b$, in the II band; another component, ${}^1A_1 \rightarrow {}^1A_1$, is magnetic dipole forbidden. All these predictions are confirmed by the experimental curve of Fig. 3-11.

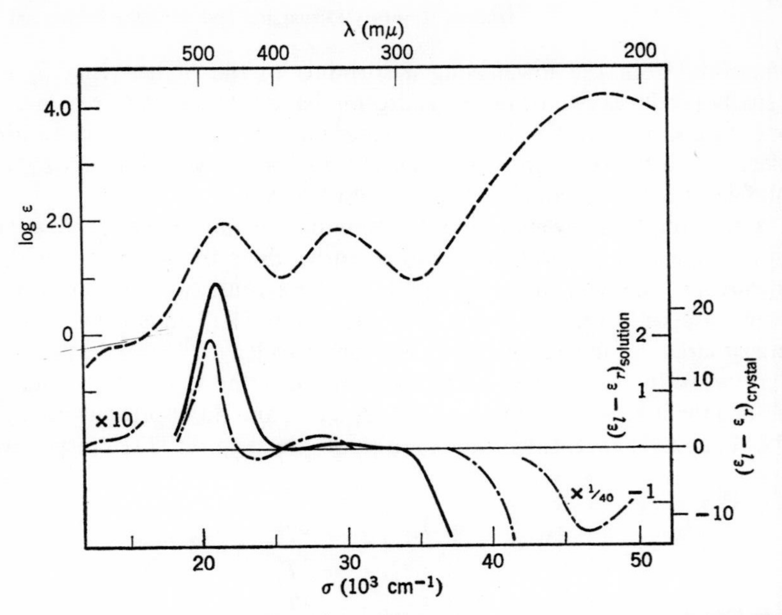

Fig. 3-11. The CD (–·–·–) and absorption spectra (-----) of Δ (+)-$[Co(en)_3]^{3+}$ in aqueous solution, and the CD (———) of the crystal Na(+)-$[Co(en)_3]_2Cl_7 \cdot 6H_2O$. After McCaffery and Mason [123].

Table 3-3 Magnetic Dipole Selection Rules in O_h and D_3 Point Groups[a]

	O_h				
	A_{1g}	A_{2g}	E_g	T_{1g}	T_{2g}
A_{1g}	×	×	×	○	×
A_{2g}	×	×	×	×	○
E_g	×	×	×	○	○
T_{1g}	○	×	○	○	○
T_{2g}	×	○	○	○	○

	D_3		
	A_1	A_2	E
A_1	×	○(‖)	○(⊥)
A_2	○(‖)	×	○(⊥)
E	○(⊥)	○(⊥)	○(‖, ⊥)

[a] ○ = allowed; × = forbidden; ‖ and ⊥ = allowed with the polarization ‖ and ⊥, respectively, to the C_3 axis.

Furthermore, the CD measurement of a single crystal of

$$Na(+)\text{-}[Co(en)_3]_2Cl_7 \cdot 6H_2O$$

confirms the interpretation that the strong (+) band of this complex ion is due to the transition ${}^1A_1 \rightarrow {}^1E_a$. The C_3 axis of each complex ion is parallel to the optical axis of the crystal [136], and the measurement was performed with the radiation directed along the optical axis [123]. Under these conditions the transition ${}^1A_1 \rightarrow {}^1A_2$ is forbidden, as Table 3-3 shows. The absolute configuration of the (+) isomer of $[Co(en)_3]^{3+}$ has been determined as Δ [137, 155] (see Sec. 3-2A). From the general theory of optical rotatory power it is difficult at present to assign the absolute configuration from the sign of the CD components.

The pioneer work by Moffitt [132] succeeded in obtaining the approximate selection rules, but Sugano [170] has pointed out that the model adopted by Moffitt to calculate the rotational strength is incorrect. It has been shown that the rotational strengths of the D_3 complexes cannot be predicted with the Moffitt model. Shinada [167] and several other authors [68, 148 149, 152] have modified the Moffitt model in order to predict the rotational strength of a complex. Piper and Karipides [148] showed that the sign of rotatory power of the split states which arise on lowering the symmetry from O_h to D_3 is determined by the sign of a coefficient of the potential function added as a perturbing term, and that the sign of that coefficient is governed by the absolute configuration of the complex and by the angle of the chelate ring ∠L-M-L. Their results are $R(A_1 \rightarrow A_2) > 0$ and $R(A_1 \rightarrow E_a) < 0$ for the ion Δ-$[Co(en)_3]^{3+}$, which contradict the experimental results. Bürer [27] has pointed out that this contradiction is due to an error in their calculations. Attempts to calculate the rotatory power of metal complexes have also been made by several other authors, particularly some molecular orbital treatments by Karipides and Piper [89] and by Liehr [102]. Nevertheless, no theory has been found satisfactory.

3-2 APPLICATIONS

A. Optical Activity and Absolute Configuration

A number of chemical and physical methods have been developed for determining the absolute configuration of optically active metal complexes. The x-ray technique developed by Bijvoet et al. [21] determines the absolute configuration directly (see Sec. 1-1A). Theoretical analysis of the ORD or CD spectra has also been very useful for this problem. Most of the latter works, however, have been made by comparing the spectrum of a molecule of unknown configuration with that of a complex whose absolute configuration has been determined by the x-ray method. The several attempts to obtain a relationship between the sign of rotation and the absolute configuration of an optically active complex should probably be considered preliminary at present.

Saito et al. [155] have shown by an x-ray analysis of

$$Na(+)\text{-}[Co(en)_3]_2Cl_7 \cdot 6H_2O$$

that the complex ion, which is dextrorotatory at the sodium D line, corresponds to the structure (a) in Fig. 3-12. For this configuration they used the symbol D. The antipode should then be structure (b), and it should be L. Piper [146] suggested the symbols Λ and Δ for D and L, respectively. In this chapter we use the symbols Δ and Λ in the reverse sense of Piper's designation, in accord with the recommendation of IUPAC. As will be seen in Sec. 3-2D, there is another kind of optical activity which arises from the optically active ligands coordinated with the metal ion. For optically active ligands of known absolute configuration, we use the symbols D and L. For other ligands the signs of rotation at the sodium D line, (+) or (−), will be used. When the rotation was measured at a wavelength different from that of the sodium D line, we indicate the wavelength in angstroms as a subscript, for example, $(+)_{5461}$. The use of small letters *d* and *l*, which correspond to (+) and (−) respectively, are avoided in this chapter, because of confusion with the symbols of absolute configuration, D and L.

As seen in Fig. 3-11, $\Delta\text{-}[Co(en)_3]^{3+}$ shows two Cotton effects in the region of the I band, one strong and positive at the longer wavelength, and the other weak and negatvie at the shorter wavelength. $\Delta\text{-}[Co(L\text{-}pn)_3]^{3+}$ has the same absolute configuration as the $\Delta\text{-}[Co(en)_3]^{3+}$ ion, and these two ions show almost similar ORD and CD curves. Iwasaki and Saito [80] determined by x-ray analysis the absolute configuration of $\Lambda\text{-}[Co(D\text{-}pn)_3]^{3+}$ which is enantiomorphous to $\Delta\text{-}[Co(L\text{-}pn)_3]^{3+}$. As mentioned before, it is almost impossible to correlate the sign of the Cotton effect with the absolute configura-

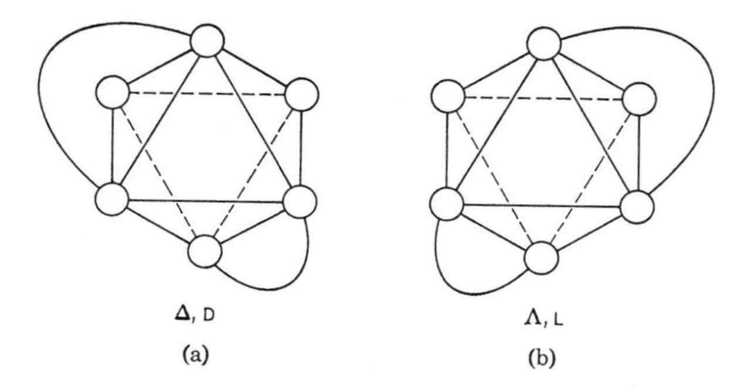

Fig. 3-12. Absolute configurations of the tris- and bis-chelate complexes.

tion of a complex. On the basis of the two experimental results mentioned above, however, the absolute configuration of analogous complexes can be safely assigned from the analysis of their ORD or CD spectra. The most useful and reliable method is to compare the sign of the Cotton effect of a band with the same electronic origin as that of a standard complex. Consequently, most studies have been made on the signs of the Cotton effects associated with $d \rightarrow d$ absorption bands. The ORD and CD of a number of tris-chelate complexes of Co(III), Cr(III), Rh(III), etc., have been measured and their absolute configuration assigned by comparing their spectra with that of Δ-$[Co(en)_3]^{3+}$ [22, 44, 45, 62, 82, 83, 115, 118, 119, 123, 127, 130].

It is also possible to assign the absolute configuration of a bis-diamine complex, cis-$[M(diamine)_2(X)(Y)]^{n+}$ from its ORD or CD spectrum, because the structures of the two optical isomers of this type are related to the Δ and Λ configurations of $[M(diamine)_3]^{n+}$, as Fig. 3-12 shows.

Since the replacement of one chelate ligand in $[M(diamine)_3]^{n+}$ by X or Y results in the shifting and the splitting of the absorption bands, the corresponding ORD or CD curve varies in appearance with the nature of the substituents X and Y. For example, the CD spectrum of

$$cis\text{-}(+)\text{-}[Co(NH_3)_2(en)_2]^{3+}$$

is very similar to that of Δ-$[Co(en)_3]^{3+}$, and it can be assigned the Δ structure (Fig. 3-13a). On the other hand, some complexes exhibit CD curves very

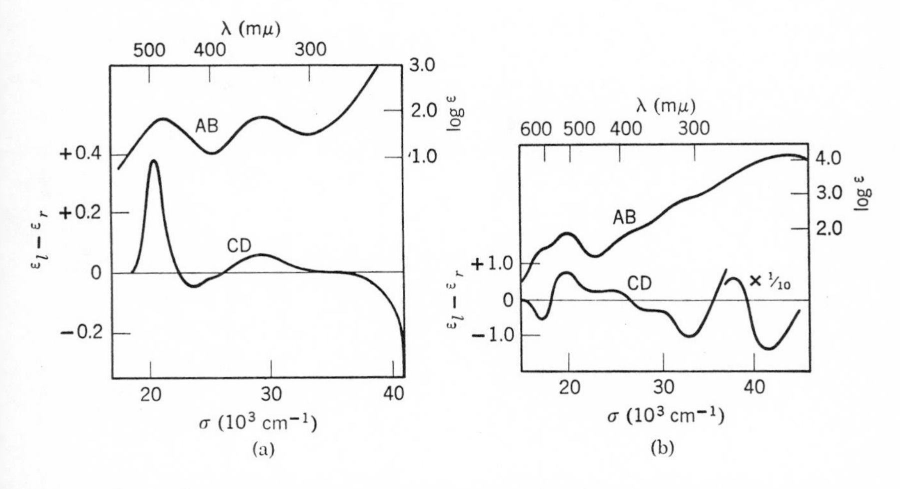

Fig. 3-13. The CD and absorption spectra (AB) of (a) *cis*-(+)-$[Co(NH_3)_2(en)_2](ClO_4)_3$ in water, (b) *cis*-(+)-$[CoCl_2(en)_2]ClO_4$ in water. After McCaffery, Mason, and Norman [128].

different from that of $[Co(diamine)_3]^{n+}$. Figure 3-13b shows an example, *cis*-(+)-$[CoCl_2(en)_2]^+$. Although current theories do not predict precisely the sign of the Cotton effect for each component of the split bands, the absolute configurations of a number of complexes of this type have been assigned from the comparison of their ORD or CD spectra [28, 54, 60, 61, 63, 110, 119, 128].

Also, for more complicated complexes such as $[Co(trien)(X)_2]^{n+}$, the ORD or CD spectra provide very useful information about the absolute configuration. In $[Co(trien)(X)_2]^{n+}$, three geometrical isomers, one *trans* and two *cis* (α and β), are possible. Each *cis* form has its optical isomers. The various structures are shown in Fig. 3-14, which may be compared with Fig. 3-12.

Sargeson and Searle [159] have assigned the absolute configurations of a number of $[Co(trien)(X)(Y)]^{n+}$ complexes from a comparison of their ORD and CD spectra with those of Δ-$[Co(en)_3]^{3+}$ and the related bis-ethylenediamine complexes. Figure 3-15 shows that the CD spectra show a net positive sign in their I-bands. This is consistent with their being assigned the Δ configuration. These results are also consistent with the conclusions obtained from kinetic studies [96, 158].

According to the "octant theory" for octahedral complexes, the Δ configurations of the various types mentioned above correspond to the octant

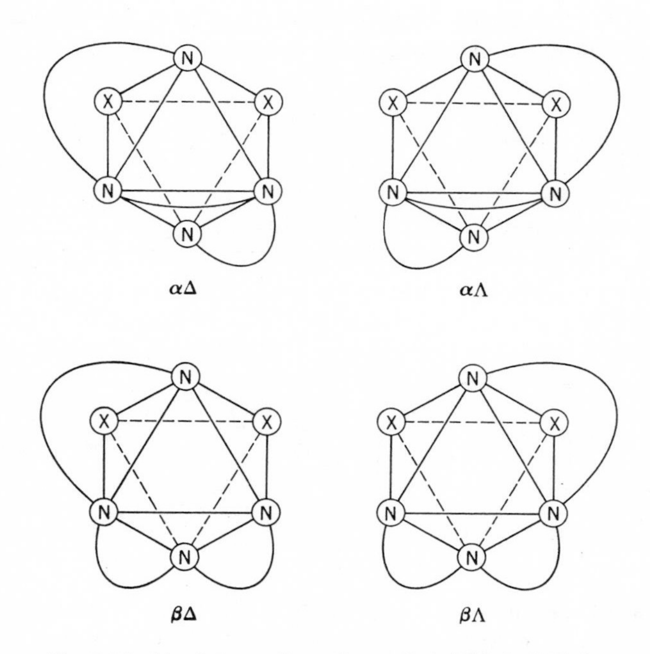

Fig. 3-14. Absolute configurations of *cis*-$[M(trien)(X)_2]$.

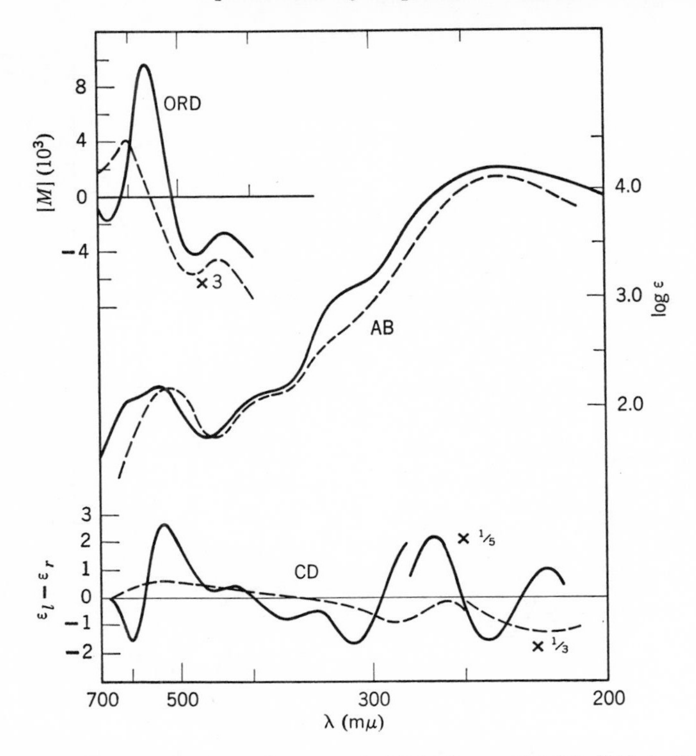

Fig. 3-15. CD, ORD, and absorption spectra (AB) of α- and β-[CoCl$_2$(trien)]$^+$: (+)-α-isomer (———); (+)-β-isomer (----). After Sargeson and Searle [159].

plus structures. Hawkins and Larsen [70] have defined the absolute configurations of metal complexes in terms of an "octant sign" and have shown that a relationship exists between the octant sign and the sign of the CD bands of the $d \rightarrow d$ transitions. Full details of the definition of the "octant sign" are obtainable from the original paper [70]. Eyring and collaborators [174] also discussed the "octant sign" of metal complexes of amino acids.

Although similar work has been reported for various complexes, the ORD or CD spectra seem at present to be useful in determining the absolute configuration for a limited number of complexes analogous to [Co(en)$_3$]$^{3+}$. Figure 3-16 shows the CD spectra of (+)-[Fe(bipy)$_3$]$^{2+}$ and (+)-[Fe(phen)$_3$]$^{2+}$. Hidaka and Douglas [75] assigned to them the same absolute configuration because their CD curves are almost identical. On the other hand, McCaffery et al. [124, 125] concluded that these two ions are enantiomorphous with one another, because their CD peaks at ca. 36,000 cm^{-1}, which are associated with E components of $\pi \rightarrow \pi^*$ ligand absorption bands, have opposite signs.

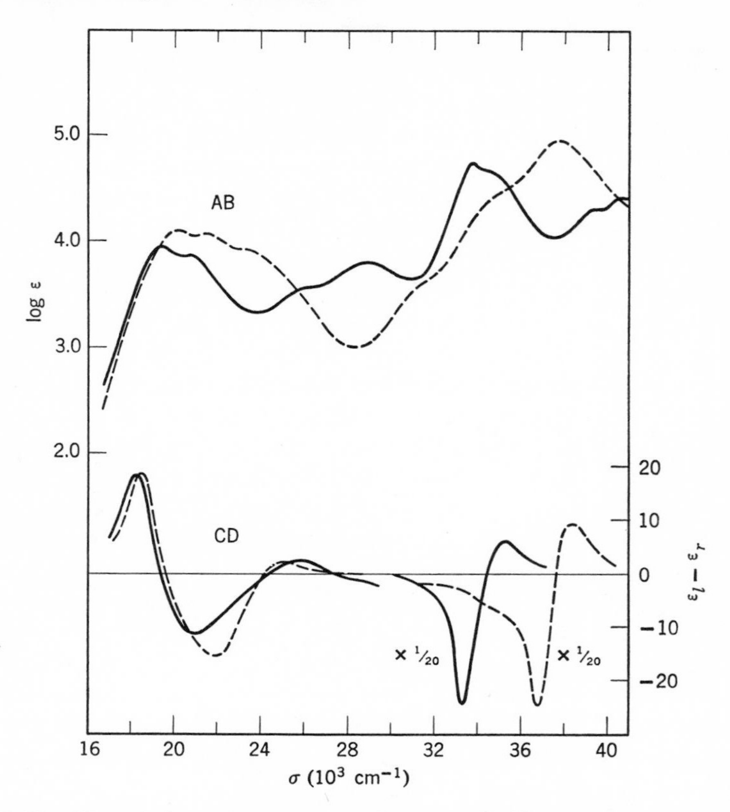

Fig. 3-16. The CD and absorption spectra (AB) of (+)-[Fe(bipy)$_3$](ClO$_4$)$_2$ (———) and (+)-[Fe(phen)$_3$](ClO$_4$)$_2$ (-----). After Hidaka and Douglas [75].

A similar difficulty is seen in the CD spectra of (+)-[Cr(ox)$_3$]$^{3-}$ and (−)-[Cr(mal)$_3$]$^{3-}$. In these complexes the I band splits into two components, $^4A_1$ and ^4E. The polarized spectra of the crystals show that the former component is at the longer wavelength in [Cr(ox)$_3$]$^{3-}$ [147], but that they are reversed in [Cr(mal)$_3$]$^{3-}$ [69]. McCaffery et al. [127] have stated that both ions have a positive sign in the E component, and hence have the same absolute configuration. The CD curves are apparently enantiomeric; further studies are clearly needed.

B. Isomerism Due to Non-planarity of Chelate Rings

It is well known that a chelate ligand such as ethylenediamine or propylenediamine coordinates with the metal ion in a *gauche* form and brings about a non-planar chelate ring. Kobayashi [92] suggested this non-planarity of the ethylenediamine chelate ring from the ORD spectrum of Δ-$[Co(en)_3]^{3+}$. Later Nakahara et al. [135] confirmed it by an x-ray analysis of *trans*-$[CoCl_2(en)_2]Cl \cdot HCl \cdot 2H_2O$ (see Sec. 1-2A).

Figure 3-17 shows that ethylenediamine molecules in a *gauche* form exist in two rotational isomers which are enantiomorphous with each other. In this chapter we use the symbols δ and λ to designate these two optical isomers as suggested by IUPAC. Some authors use the symbols k and k' [36, 52, 53]. Two such isomers of ethylenediamine cannot be resolved into optically active forms, since the activation energy for isomerization, $\delta \rightarrow \lambda$, is smaller than kT at room temperatures in a free ligand. In fact, an attempt to resolve the two isomers of $[Co(NH_3)_4en]^{3+}$ was unsuccessful [165].

In octahedral complexes containing two or more ethylenediamine ligands with the same configuration, δ for example, the spatial conformations around the metal ion of two isomers, Δ and Λ, are somewhat different. The two isomers are not enantiomorphous, but are diastereomeric with each other. In Fig. 3-18 are illustrated the two isomers of $[M(en)_3]^{n+}$ with all the ethylenediamines in the δ form. The carbon-carbon bonds in the Δ complex, denoted by $\Delta(\delta\delta\delta)$, are approximately parallel to the trigonal (C_3) axis, whereas in the $\Lambda(\delta\delta\delta)$ complex the carbon-carbon bonds are slanted obliquely to the trigonal (C_3) axis. Thus the former is designated the "lel" form and the latter the "ob" form [36]. These two isomers, lel and ob, may have different free energies and hence different stabilities, because the repulsions between ligands are different in the two forms. According to Corey and Bailar [36] the $\Delta(\delta\delta\delta)$ is more stable than the $\Lambda(\delta\delta\delta)$ by about 1.8 kcal/mole; similarly the $\Lambda(\lambda\lambda\lambda)$ is more stable than the $\Delta(\lambda\lambda\lambda)$ form by the same amount. If two kinds of ligands, δ and λ, are mixed in a tris-complex, four different forms are possible for a given configuration around the metal ion, $\delta\delta\delta$, $\delta\delta\lambda$, $\delta\lambda\lambda$, and $\lambda\lambda\lambda$. The

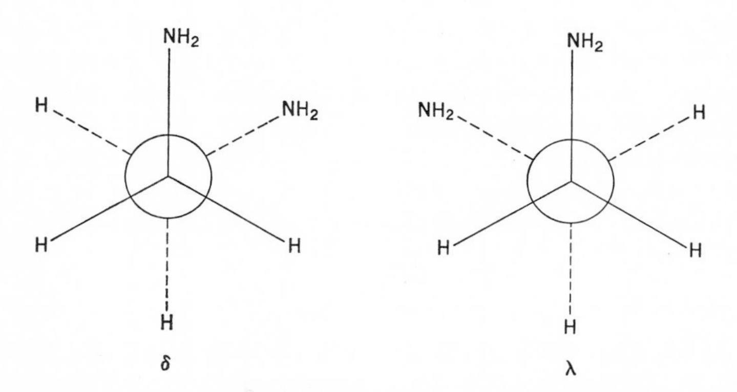

Fig. 3-17. The δ and λ configurations of a *gauche* form of ethylenediamine.

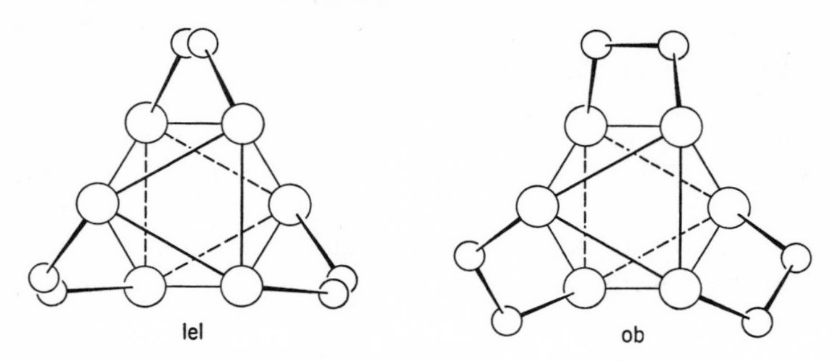

Fig. 3-18. The lel and ob conformations of Δ-[M(en)$_3$].

order of stability for these four structures would be $\Delta(\delta\delta\delta) > \Delta(\delta\delta\lambda) > \Delta(\delta\lambda\lambda) > \Delta(\lambda\lambda\lambda)$ for the Δ configuration. Obviously this order would be reversed for the Λ configuration. Although these isomers could not be resolved, because of the small energy differences between them, it has been shown by x-ray analysis that the most stable form of Δ-$[Co(en)_3]^{3+}$ is lel-$\Delta(\delta\delta\delta)$ in the solid state [155].

In propylenediamine complexes another isomerism due to the substituent —CH_3 group exists. In $[M(pn)_3]^{n+}$, for example, two geometrical isomers "fac" and "mer," are possible; the —CH_3 groups are in *cis-cis* positions in the former, and in *cis-trans* positions in the latter. The structure of Λ-$[Co(\text{D-pn})_3]^{3+}$ determined by x-ray analysis is of the "fac" type (see Chapter 1). So far no report has been published on the "mer" isomers.

The propylenediamine molecule has an asymmetric carbon atom and can be resolved into optically active forms. The absolute configuration of (+)-pn has been determined to be L from a comparison with that of L-alanine [153, 156]; the —CH_2NH_2 group of propylenediamine corresponds to the —COOH group of L-alanine. When the L-(+)-pn (IX) forms a chelate ring on coordination, two isomers are possible, depending on whether the methyl group lies in an axial or an equatorial position (see Fig. 3-19). In these two isomers the ligand has λ and δ configurations.

$$\begin{array}{c} CH_2NH_2 \\ \vdots \\ NH_2\blacktriangleright C \blacktriangleleft H \\ \vdots \\ CH_3 \end{array} \qquad \text{L-(+)-pn}$$

IX

The chelate ring of the equatorial type is estimated to be more stable than that of the axial type by about 2 kcal/mole for an octahedral complex, since the repulsions between the —CH_3 and the —NH_2 groups are less in the

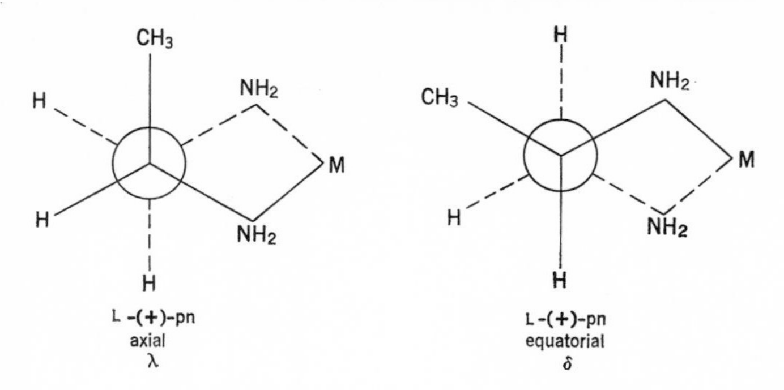

Fig. 3-19. The axial and equatorial conformations of a *gauche* chelate ring of L-(+)-propylenediamine.

equatorial than in the axial type [36]. Thus the stable form of coordinated L-(+)-pn will be δ and that of D-(−)-pn will be λ.

Now we consider the structure of $[M(L\text{-}pn)_3]^{n+}$ with all three chelate rings of the same configuration. The most stable form will be Δ(δδδ) (lel), the same as for $[M(en)_3]^{n+}$, and then Λ(δδδ) (ob) with the equatorial —CH_3 groups will be next in stability. The two other forms, Δ(λλλ) (ob) and Λ(λλλ) (lel) should be unstable, because the —CH_3 groups in these forms are of the axial type. At first, only one species, of the Δ type, was prepared for the tris-L-(+)-pn complex of Co(III) [168]. Later Dwyer et al. [50] isolated the following four optical isomers:

	$[\alpha]_D$	$[\alpha]_{5461}$
1. Δ-$[Co(L\text{-}pn)_3]I_3 \cdot H_2O$	+24°	+184°
2. Λ-$[Co(L\text{-}pn)_3]I_3 \cdot H_2O$	−214	−404
3. Λ-$[Co(D\text{-}pn)_3]I_3 \cdot H_2O$	−24	−184
4. Δ-$[Co(D\text{-}pn)_3]I_3 \cdot H_2O$	+214	+404

Of these four isomers the structure of the bromide corresponding to (3) was determined by x-ray analysis to be fac-Λ(λλλ) (lel), as mentioned above. Thus the structure of (1) should be fac-Δ(δδδ) (lel), enantiomorphous with (3). The ORD and CD curves of (1) and (3) are perfectly enantiomeric with each other. X-ray analyses of compounds (2) and (4) are now in progress [157], but they could be assigned to Λ(δδδ) and Δ(λλλ), respectively, on the basis of the ORD and CD analysis, and the discussion on stabilities above.

Figure 3-20 shows the CD spectra of (a) Δ-$[Co(en)_3]^{3+}$; (b) Δ-$[Co(L\text{-}pn)_3]^{3+}$; (c) Λ-$[Co(L\text{-}pn)_3]^{3+}$ [126]. The asymmetries of these compounds arise from the following: for compound (a), the Δ(δδδ) (lel) structure; for compound (b), the Δ(δδδ) (lel) structure and the asymmetric carbon atoms of three

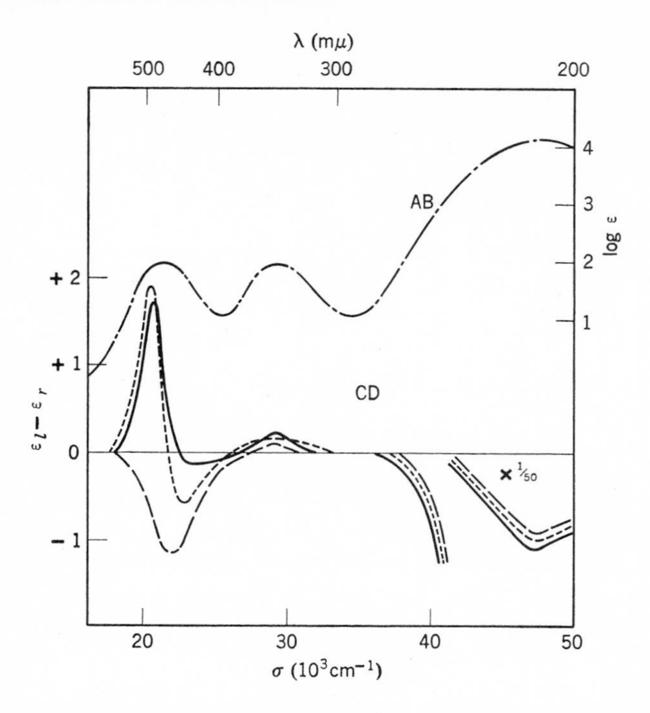

Fig. 3-20. The CD spectra of Δ-$[Co(en)_3]^{3+}$ (———), Δ-[Co(L-pn)$_3$]$^{3+}$ (· · · · ·), and Λ-[Co(L-pn)$_3$]$^{3+}$ (- - - - -). The AB curve is the absorption spectra of the latter two. After McCaffery, Mason, and Norman [126].

L-(+)-pn; for compound (3), the Λ(δδδ) (ob) structure and the same asymmetric carbon atoms as for compound (b). The CD spectra of compounds (a) and (b) are similar in type and have a positive Cotton effect in the I band, indicating the same configuration, Δ(lel). On the other hand, the CD of compound (c) with the Λ configuration has a negative Cotton effect in the I band, but the curve is not enantiomeric with those of the other two. All these compounds have a strong CD peak in the ultraviolet region which is probably associated with a charge transfer transition between the ligands and the metal ion. For all three molecules the bands have the same CD sign and almost the same frequency and intensity, indicating that the diamines have the δ configuration in all three ions [126].

The CD of *trans*-$[CoCl_2\{(-)\text{-chxn}\}_2]^+$ is a good example of the activity due to the non-planarity of a chelate ring, because this ion has no asymmetry around the metal ion. As Fig. 3-21 illustrates, the CD of this ion shows a positive sign in the *E* component of the I band due to the vicinal effect of

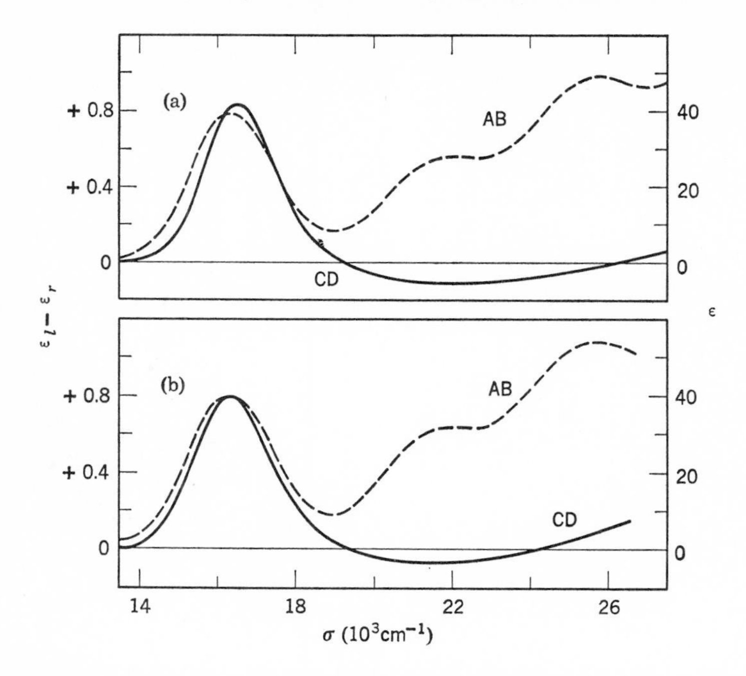

Fig. 3-21. The CD (———) and absorption spectra (AB, - - - -) of (a) *trans*-$[CoCl_2$-(D-pn$)_2]$Cl in methanol, (b) *trans*-$[CoCl_2\{(-)$-chxn$\}_2]$Cl in methanol. After Hawkins, Larsen, and Olsen [71].

$(-)$-chxn in the λ form. This spectrum, being very similar to that of *trans*-$[CoCl_2\{$D$(-)$-pn$\}_2]^+$ with the same λ ligands, indicates that the CD activity of the $d \rightarrow d$ absorption region does not arise from the asymmetric carbon atom, but mostly from the non-planar chelate ring. The CD spectra of $[Co(NH_3)_4$(D-pn$)]^{3+}$ and of *trans*-$[Co(NH_3)_2$(D-pn$)_2]^{3+}$ also exhibit clearly the vicinal effects of the non-planarity of the chelate ring [71].

The different stabilities in complex formation result from the differences in the configuration of ligands in complexes, such as the δ and λ or axial and equatorial forms of diamines. The idea forwarded by Corey and Bailar [36] is very useful for assigning the absolute configuration of related complexes; for example, of four possible structures, *a*(1), *a*(2), *e*(1), and *e*(2) (see Fig. 3-22), of the Co(III) complex which contains the D$(-)$-propylenediaminetetra-acetate ion (PDTA), the most stable is the *e*(1) form, as the first two forms are of the unstable axial type, and the *e*(2) form suffers appreciable steric distortion. In fact, the Co(III) ion forms with D-PDTA only one species, $(+)$-[Co(D-PDTA)$]^-$, which can be assigned the *e*(1) configuration on the basis

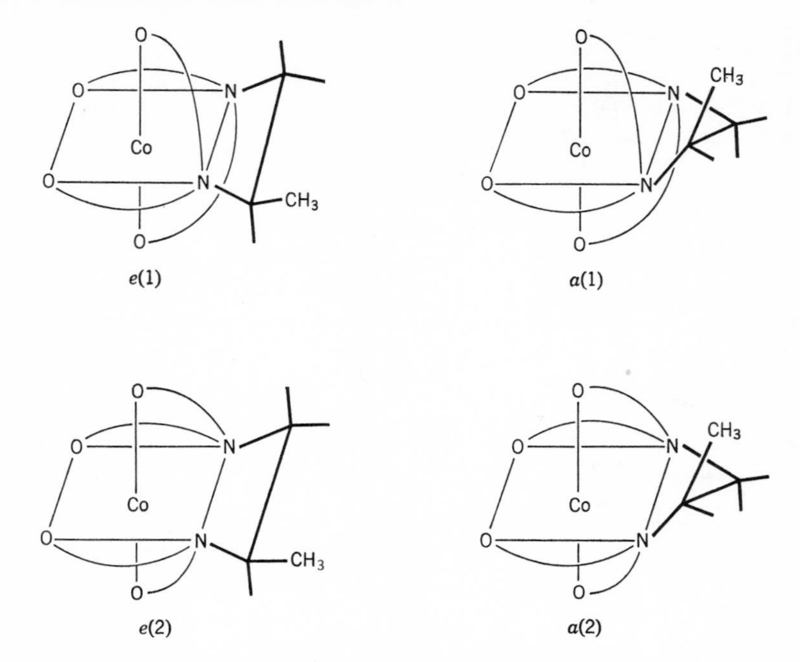

Fig. 3-22. Four possible structures of [Co(D-PDTA)]⁻.

of the preceding discussion [30, 31, 51, 78]. Taking this complex as a standard, the absolute configurations of a number of EDTA complexes such as

$$(+)_{5461}\text{-}[Co(EDTA)]^- \quad \text{and} \quad (-)_{5461}\text{-}[Co(EDTA)X]^{2-} \quad (X = Cl^-, NO_2^-)$$

and of related complexes were assigned from their ORD or CD [39, 63, 67, 110]. On the other hand, the absolute configuration of various optically active ligands can also be assigned by a consideration of the stereospecificity of complex formation [14, 66, 150].

C. Isomers of Amino Acid Complexes

It is well known that amino acid ions form stable chelate compounds with various metals. Amino acid complexes are especially useful for studying the stereochemistry of metal complexes, because optically active forms of various amino acids of known absolute configuration are available. Only recently, however, was the stereochemistry of amino acid complexes established.

First let us consider the isomers of a tris-(amino acidato) complex, $[M(am)_3]$. Since the donor atoms of an amino acid are nitrogen and oxygen, two geometrical isomers are possible, and each of these isomers has its optical enantiomer. These four isomers can be classified as fac-Δ, fac-Λ, mer-Δ, and mer-Λ, as shown in Fig. 3-23. The symbols "fac" (facial) and "mer" (meridional) are used, respectively, to designate an isomer in which three N atoms (or O atoms) occupy the corners of one of the octahedral faces (*cis-cis*), and one in which they do not (*cis-trans*) [172].

Two geometrical isomers of $[Co(gly)_3]$ were prepared many years ago. The one (the β form) is red and sparingly soluble in water, and the other (the α form) is purple and fairly soluble in water [101]. That the β form has the "fac" structure and the α form the "mer" structure could be easily determined from their visible absorption spectra [19, 164]. The I band of the "mer" form, having lower symmetry (C_1), is split to a much greater extent

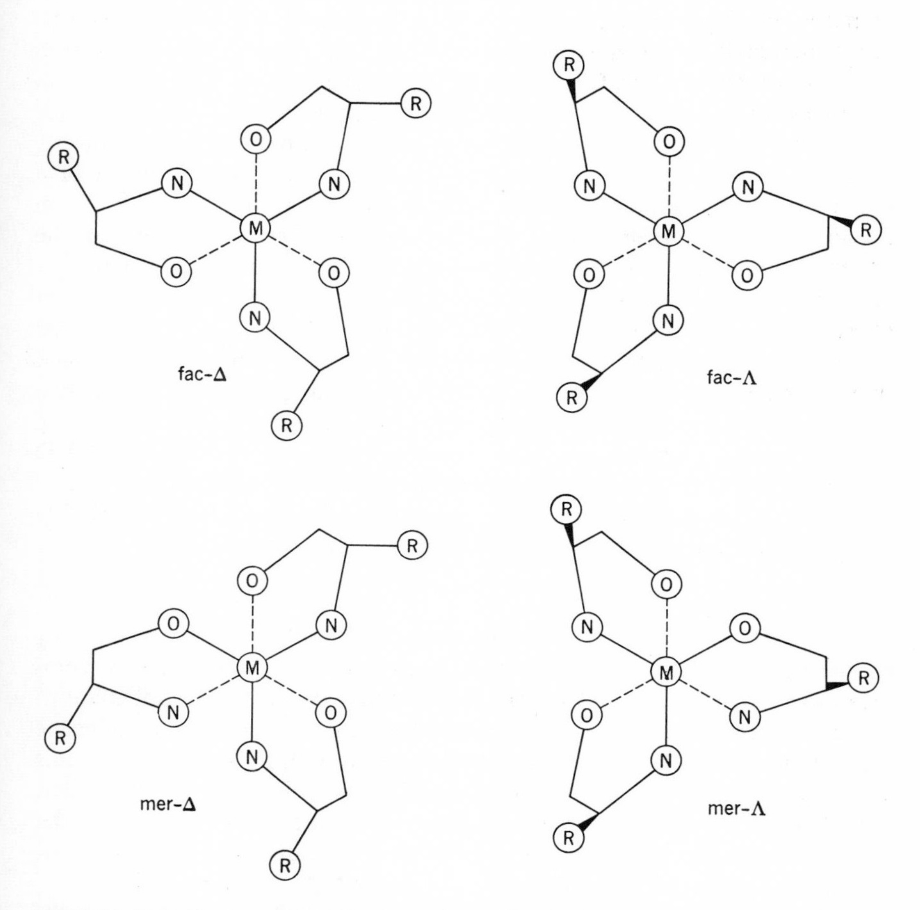

Fig. 3-23. Four isomers of a tris-(L-α-amino acidato) complex. The substituents R are equatorial in the fac-Δ and mer-Δ isomers, and axial in the fac-Λ and mer-Λ isomers.

than that of the "fac" form, having higher symmetry (C_3). Although these geometrical isomers can be isolated easily, the optical resolution of each of them has been made only partially, because they are non-charged complexes and hence the usual chemical methods for resolution could not be applied. Krebs and Rasche [93], Douglas and Yamada [40], and Dunlop and Gillard [46] have resolved partially each fac and mer form of [Co(gly)$_3$] into optically active forms by means of starch column chromatography.

On the other hand, four isomers of [Co(L-ala)$_3$] have been isolated from one another perfectly. For [Co(gly)$_3$], the Δ and Λ forms are enantiomorphous with one another, whereas they are diastereoisomeric for [Co(L-ala)$_3$]. When the L-alaninate ion forms an octahedral complex, the —CH_3 group has two different conformations, *a* type (parallel to the C_3 axis of the complex) and *e* type (perpendicular to the same axis). As Fig. 3-23 shows, the —CH_3 group of a Λ complex will be forced to take the *a* configuration and that of a Δ complex, the *e* configuration. Recently it has been shown that α-amino acids also form non-planar chelate rings on coordination, but their non-planarity seems to be somewhat different from that of a typical *gauche* form of a diamine ring. As mentioned in Sec. 3-2B, the formation of [Co(L-ala)$_3$] from cobalt ion and L-alanine is stereospecific, since the interactions between the —CH_3 groups and the other atoms are different in the *a*- and *e*-type complexes, and consequently the two forms have different stabilities. Four isomers of [Co(L-ala)$_3$] with different stabilities also have different solubilities and therefore can be isolated by the solubility differences. Lifschitz [103] prepared three isomers of this complex. Later Douglas and Yamada [40] and then Larsen and Mason [98] and Dunlop and Gillard [46] succeeded in isolating all four isomers and assigning their absolute configurations from the ORD and CD spectra. Recently Denning and Piper [38] have studied this problem very carefully and reached the same conclusion. Figure 3-24 shows the CD and the visible-ultraviolet absorption spectra of all four isomers of

[Co(L-ala)$_3$].

The assignment of the absolute configuration of each isomer was made by comparing the Cotton effects in the I band with those of tris-diamine complexes mentioned in Sec. 3-2A. That the CD curve of a Δ type is not enantiomeric with that of a Λ type may be due to the vicinal effect of the asymmetric carbon atom of L-alanine. Douglas and Yamada [40] have shown that this is a reasonable explanation in a curve analysis in which they assumed that the CD curve which was obtained by subtracting the vicinal effect from the experimental curve should be the same as that of [Co(gly)$_3$]. The same relationship is seen in the CD of Δ- and Λ-[Co(L-pn)$_3$]$^{3+}$ (see Sec. 3-2B).

According to the x-ray analysis of mer-(+)-[Co(L-ala)$_3$] [42], the conformation of all the —CH_3 groups, being of the *e* type, indicates that this complex

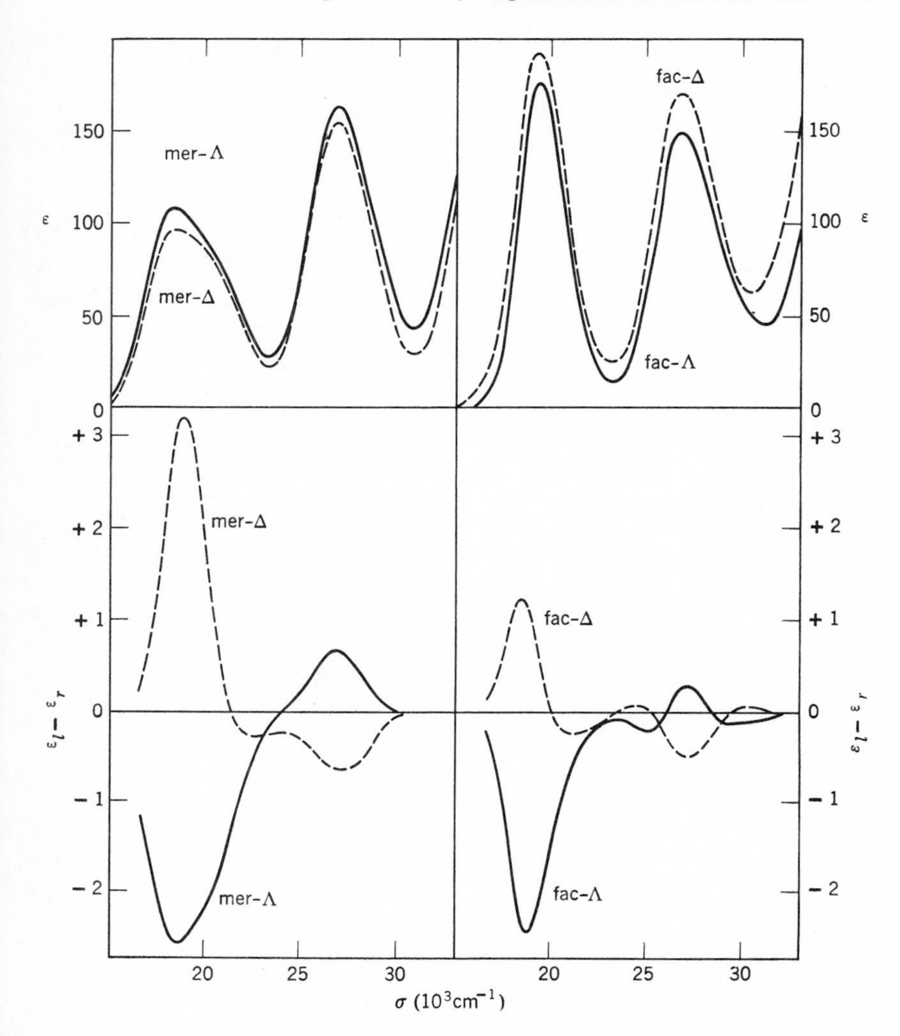

Fig. 3-24. The CD and absorption spectra of four isomers of [Co(L-ala)$_3$]. After Denning and Piper [38].

takes the Δ form. From this result the absolute configuration of [Co(am)$_3$] could be determined by comparing their ORD or CD spectra with that of mer (+)-[Co(L-ala)$_3$].

The assignment of the absolute configuration of each isomer of

[Co(L-ala)$_3$]

is consistent with that obtained from studies of [Co(L-prol)$_3$] (prol = prolinate ion) for which appreciable stereospecificity is expected. The sterically most stable form of four probable isomers of [Co(L-prol)$_3$] can be predicted to be fac-Δ by the construction of molecular models. Yasui et al. [189] prepared purplish pink crystals of [Co(L-prol)$_3$] almost quantitatively by the reaction of [Co(NH$_3$)$_4$(H$_2$O)$_2$](ClO$_4$)$_3$ and L-proline in the presence of activated charcoal. This isomer was assigned the fac-Δ structure from ORD and CD measurements. The Cotton effects of the I band of this complex have the same signs as those of fac Δ-[Co(L-ala)$_3$]. Denning and Piper [38] have isolated three isomers which have been assigned from their CD spectra to the fac-Δ, mer-Λ, and fac-Λ configurations. The mer-Δ isomer, which would be most unstable, has not been obtained.

The L-proline ligand has another characteristic property: the donor N atom of L-proline coordinated as a bidentate ligand becomes an asymmetric atom, and is forced to take a configuration antipodal to that of the similarly coordinated D-proline. The higher CD intensity of the I band of fac-Δ-[Co(L-prol)$_3$] compared with that of fac-Δ-[Co(L-ala)$_3$] could be explained by considering the vicinal effect of this asymmetric nitrogen atom.

Studies have been reported on the ORD and CD of various tris-complexes of the amino acids: L-valine [162], L-leucine [38], L-aspartic acid [105, 163], and L-glutamic acid [98, 104]. ORD and CD work on mixed complexes containing diamines and amino acids has also been published [107, 120, 166].

D. Vicinal Effect

As mentioned before, some complexes in which the asymmetry is not around the metal ion but on an atom of the ligand show the Cotton effect in the ligand field bands. In these complexes the ligand field bands are said to be affected by a vicinal effect. The ORD or CD due to a vicinal effect is usually weak compared with that due to asymmetry around the metal ion. However, the study of vicinal effects seems to be very useful for elucidating the origin of optical activity in metal complexes, as it is relatively easy to prepare complexes with desired symmetries or those containing various ligands of known absolute configuration.

It is known that the Cu(II) complexes of natural α-amino acids (L form) show a negative Cotton effect in the visible absorption band region [143]. By making use of this effect, the absolute configuration of some α-amino acids have been determined [81].

Yasui et al. [188] measured the CD of a number of Cu(II) L-amino acid (L-am) complexes in the region of their *d-d* absorption bands. As Fig. 3-25 illustrates, the [Cu(L-am)$_2$] complexes show four CD components of the $d \rightarrow d$ transitions, (+), (+), (−), and (−), respectively, from longer to shorter wavelengths, although in some of the complexes one or two of the

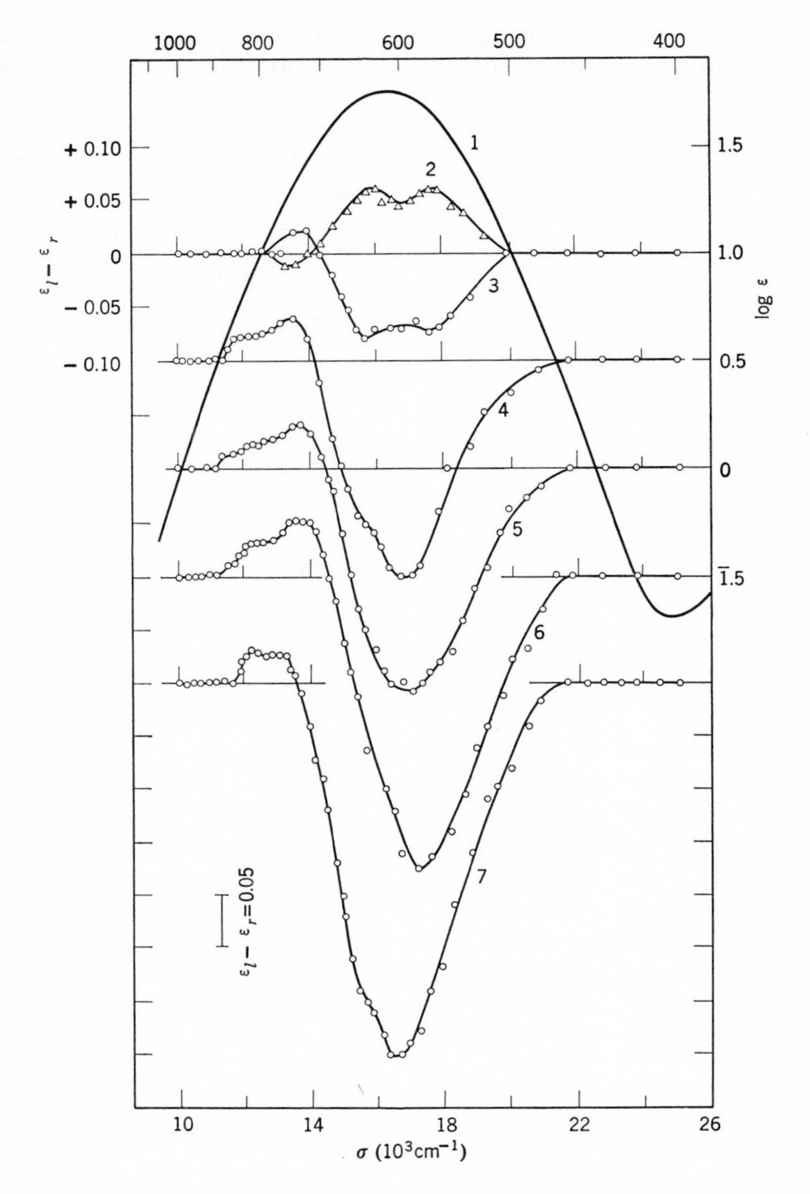

Fig. 3-25. The absorption curve of (1) [Cu(L-ala)$_2$] and the CD curves of (2) [Cu(D-ala)$_2$], (3) [Cu(L-ala)$_2$], (4) [Cu(L-serinate)$_2$], (5) [Cu(L-threoninate)$_2$]·H_2O, (6) [Cu(L-valinate)$_2$]·H_2O, and (7) [Cu(L-allothreoninate)$_2$]·H_2O in aqueous solutions.

components are hidden by an intense component. It is interesting that all four of the possible $d \rightarrow d$ transitions of a Cu(II) complex were observed. The net CD strength is negative for all the complexes of this type of L-amino acid.

As Fig. 3-26 shows, the CD of [Cu(L-prol)$_2$] is quite different from those of the complexes in Fig. 3-25. This variation may be attributed more to the stronger vicinal effect of the asymmetric nitrogen atom attached directly to the copper atom than to that of the α-asymmetric carbon atom. As mentioned before for [Co(L-prol)$_3$], the nitrogen atom of coordinated L-prolinate ion takes a forced antipodal configuration to that of the similarly coordinated D-prolinate ion. The same CD behavior is noted in the [Cu(L-hydroxyprolinate)$_2$] complex (see Fig. 3-26).

Several studies have been made of the vicinal effect for planar-type complexes such as Cu(II)- and Ni(II)-amino acids [151], Cu(II)-Schiff bases [144],

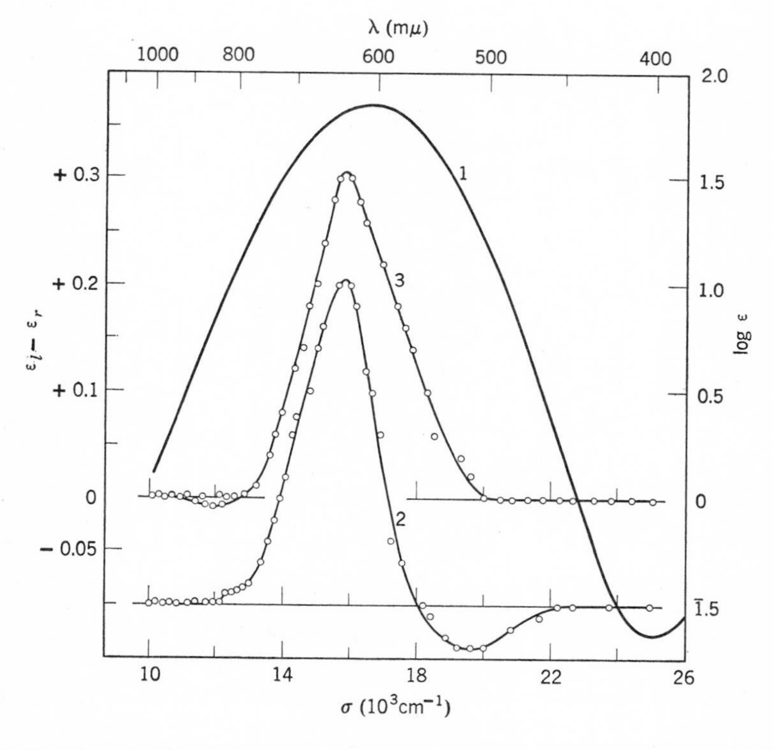

Fig. 3-26. The absorption curve of (1) [Cu(L-prol)$_2$]$\cdot$2H_2O and the CD curves of (2) [Cu(L-prol)$_2$]$\cdot$2H_2O and (3) [Cu(L-hydroxyprolinate)$_2$]$\cdot$3H_2O in aqueous solutions.

Cu(II)-peptides [25], Cu(II)-L-tartrate [64], $[Cu(D\text{-}pn)_2]^{2+}$ [64], Cu(II)-mandelic acid derivatives [55], and Ni(II)-thiosemicarbazides [85].

Jaeger [84] prepared $[Co(NO_2)_3(NH_3)\{(-)\text{-chxn}\}]$, the first example of an octahedral, optically active complex with no dissymmetry around the metal ion. Later Mathieu [121] prepared the same type of complex,

$$[Co(NO_2)_3(NH_3)(L\text{-pn})].$$

It was shown that both complexes exhibit anomalous ORD in the region of their first absorption bands. Shimura [166] prepared $[Co(L\text{-leu})(NH_3)_4]^{2+}$, which has a simpler structure than the complexes above, and showed by curve analysis that the Cotton effect of the first absorption band of this complex consists of two components with different signs. Since then, a number of complexes of this type have been prepared, and their ORD and CD have been measured: $[Co(ch)(NH_3)_4]^{n+}$ (ch = L-phenylalaninate ion [107], L-ala [43], L-tart [20, 43]).

Yasui et al. [190] have shown from the ORD and CD curves of

$$[Co(L\text{-am})(NH_3)_4]^{2+}$$

that the I bands of the complexes actually consist of three components, although the appearances of the CD curves differ according to the relative strengths of the components. The CD of $[Co(L\text{-leu})(NH_3)_4]^{2+}$ shows three components clearly in this region (Fig. 3-27). From the experimental behavior of the CD curves in the first absorption band region, the complexes

$$[Co(L\text{-am})(NH_3)_4]^{2+}$$

can be divided into three groups:

1. L-am = L-ala, L-leu, and L-phenylalaninate ion; these complexes exhibit three CD bands with the signs (−), (+), and (−) from longer to shorter wavelengths.
2. L-am = L-valinate and L-isoleucinate ion; these complexes exhibit two CD bands of negative sign.
3. L-am = L-serinate and L-threoninate ion; these complexes exhibit only one CD band of negative sign.

The differences among these groups are less explicit in their ORD curves than in their CD curves.

The differences in the behavior of CD curves can be explained by supposing that the intramolecular interactions between a coordinated ammonia molecule and the substituents on the β-carbon atom are different, depending on the nature of the substituents. The difference in the type of interaction (repulsive or attractive, depending on the substituents) may result in non-planar chelate rings with different conformations, which will be related to the

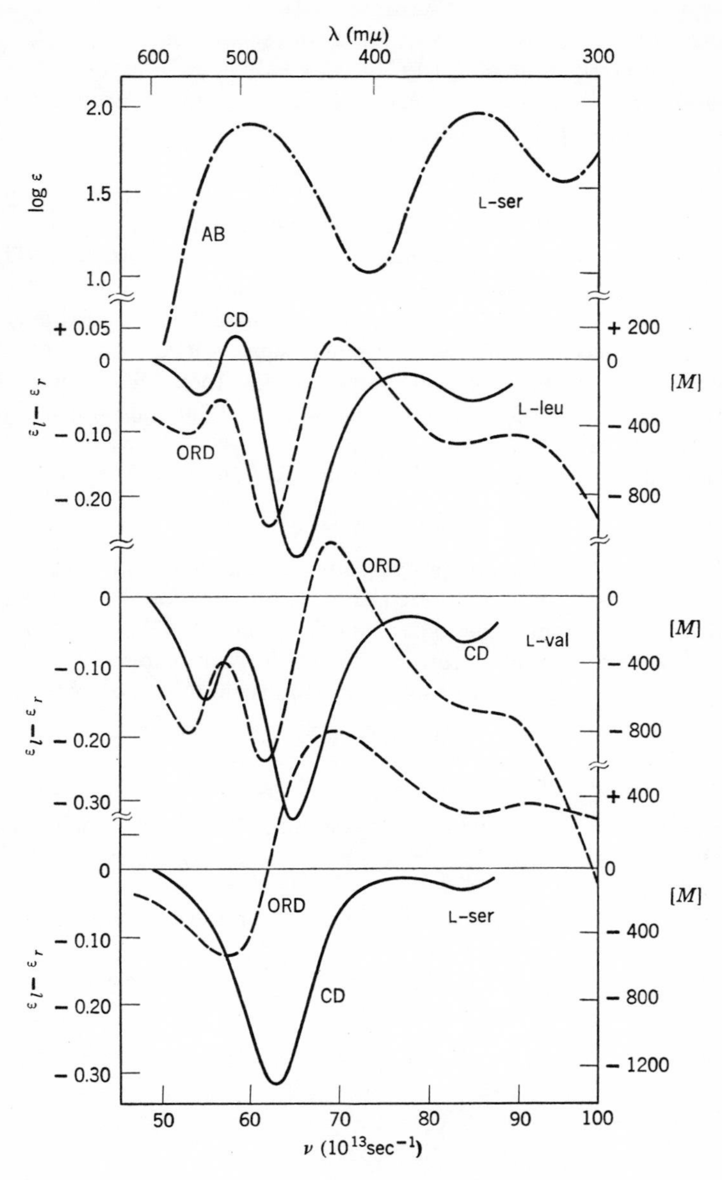

Fig. 3-27. The CD (———), ORD (– – –), and absorption spectra (AB, –·–·–) of $[Co(L\text{-}am)(NH_3)_4]^{2+}$ ions (L-am = L-valinate, L-leucinate, and L-serinate).

rotational strength. At present, however, it is almost impossible to correlate the behavior of CD curves with the stereochemical differences. The curve of $[Co(\text{L-prol})(NH_3)_4]^{2+}$ shows a behavior similar on the whole to that of the L-leucinate complex, but the positive CD component of the former is more intense than that of the latter. This difference can be attributed to the asymmetric nitrogen atom of the coordinated L-prolinate ion, as pointed out in this section and in Sec. 3-2C.

A few studies of the vicinal effect in diamine complexes have been made. For example, the ORD or CD of *trans*-$[CoX_2(\text{D-pn})_2]^+$ ($X = Cl^-$ [121, 175], NO_2^- [175]) have been measured.

Hawkins et al. [71] studied the vicinal effect of coordinated D-propylenediamine for a number of mono- and *trans*-bis-D-propylenediamine-Co(III) complexes. They concluded that the CD of the I band actually consists of three components, although they overlap one another in some of the complexes. The Cotton effects of the components with the same symmetry have the same sign. As mentioned in Sec. 3-2B, the vicinal effect of the coordinated D-propylenediamine originates possibly in its *gauche* configuration with the λ form stabilized by a $—CH_3$ group of type *e*.

It is interesting to note that the sign of the net rotational strength of the I band is negative for the vicinal effect of an L-amino acid, whereas it is positive for D-propylenediamine. From these experimental results it may be suggested that the formation of an asymmetric chelate ring is responsible for the vicinal effect. It will be interesting therefore to see whether complexes containing optically active unidentate ligands show the Cotton effect or not. Larsen and Olsen [97] observed no detectable Cotton effect in the *d-d* absorption bands of Cu(II) and the Co(III) complexes containing the unidentate, optically active ligands (+)-2-aminobutane and (+)-α-methylbutyrate. From this result they proposed tentatively that the optically active, unidentate ligand could not induce CD in the *d-d* absorption bands of a metal complex. On the other hand, the Co(III) complexes containing unidentate, optically active amino acids show Cotton effects in the *d-d* absorption bands, although the strength is considerably weaker than those of the chelate amino acid complexes. In Co(III) complexes the unidentate amino acid is coordinated with the metal ion through the oxygen atom of the carboxylate group. Fujita et al. [59] have measured the ORD in the *d-d* absorption band region of Co(III)-ammine complexes containing unidentate L-alanine. As Fig. 3-28 illustrates, the Cotton effects of the unidentate complex (1, and 3–5) are generally weak compared with that of the chelate complex (2). Of these unidentate complexes, those containing at least one active ligand and another large ligand in a position *cis* to it (4, 5) show stronger Cotton effects than do those of the other complexes (1, 3). From these observations it was suggested that the strength of the Cotton effect of the I band induced by the vicinal

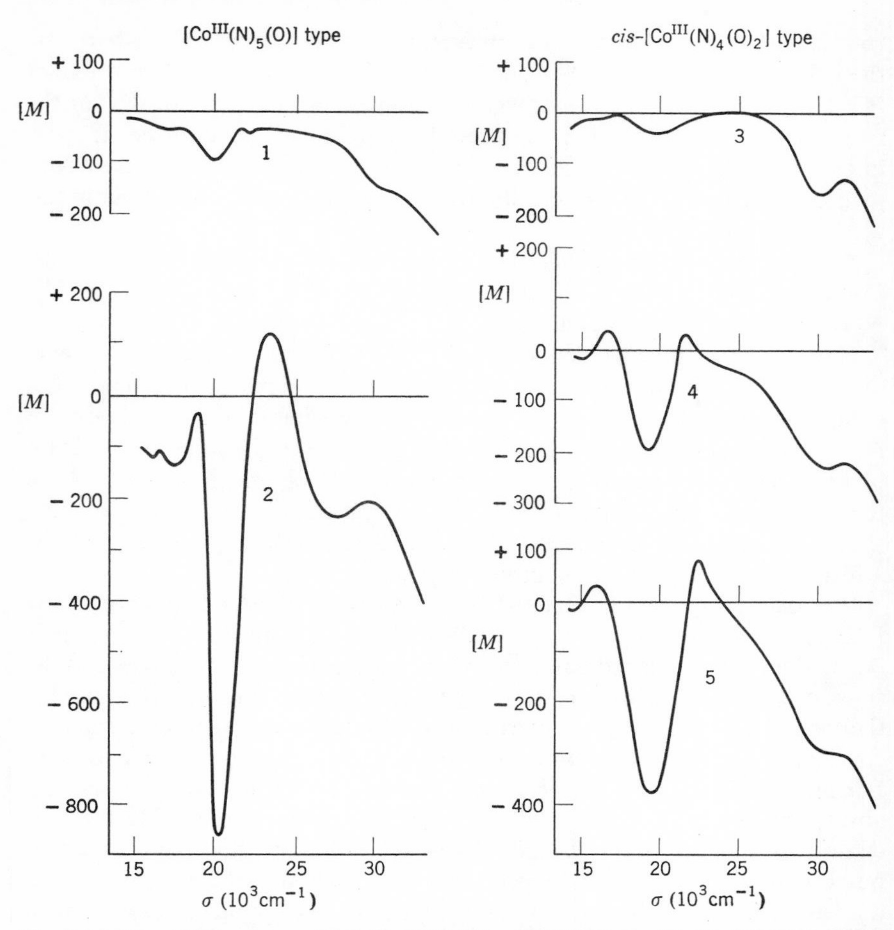

Fig. 3-28. The ORD curves of (1) $[Co(NH_3)_5(\text{L-alaH})](ClO_4)_3$; (2) $[Co(NH_3)_4(\text{L-ala})]SO_4$; (3) *cis*-$[Co(NH_3)_4(OH_2)(\text{L-alaH})](ClO_4)_3$; (4) *cis*-$[Co(NH_3)_4(\text{glyH})(\text{L-alaH})](ClO_4)_3$; (5) *cis*-$[Co(NH_3)_4(\text{L-alaH})_2](ClO_4)_3$.

effect of the coordinated L-alanine depends mainly on the conformational mobility of the ligand. In the chelate complexes the conformation of the coordinated L-alanine is almost fixed by the formation of a chelate ring with the metal ion. A similar observation was made for the complexes

$$[Co(\text{L-tart})(NH_3)_4]^+ \quad \text{and} \quad [Co(\text{L-tart})(NH_3)_5]^+ \text{ [20]}.$$

Yasui et al. [190] have measured the ORD and CD of a number of unidentate L-amino acid complexes of the type

$$[Co(NH_3)_5(\text{L-amH})]^{3+} \quad \text{and} \quad trans\text{-}[Co(en)_2(\text{L-amH})_2]^{3+}.$$

The ORD and CD of these complexes show a behavior similar to those of the L-alanine complexes above in the $d \rightarrow d$ transition band region.

For complexes such as $[Co(L\text{-}am)(en)_2]^{2+}$ or $[Co(L\text{-}am)_3]$ there are two contributions to the optical activity; one is from the asymmetry around the metal ion (configurational effect), and the other from the vicinal effect of the optically active L-amino acid. Liu and Douglas [107] have shown from the CD studies of $[Co(am)(en)_2]^{2+}$ (am = D- or L-ala, D- or L-leu, and L-phenylalaninate ion) that the contributions from these two effects are essentially additive. Douglas and Yamada [40] have analyzed the CD curve for the configurational and vicinal effects for the isomers of $[Co(L\text{-}ala)_3]$. The CD curves of the configurational effects for Δ- and Λ-$[Co(L\text{-}ala)_3]$ are almost mirror images. Therefore the two effects should be additive for these isomers.

E. Ion-Pair Formation

Werner [181] observed that the molar rotation of an optically active complex in solution is affected by the counter ions. Pfeiffer et al. [142] found that the equilibrium ($\Delta \rightleftharpoons \Lambda$) in aqueous solution containing racemic $[Zn(phen)_3]SO_4$ or $[Cd(phen)_3]SO_4$ is shifted toward one side by adding (+)-camphorsulfonic acid, and that the concentration of one of the antipodes increases. Recent studies [16, 90, 99, 100, 117, 169] of these effects have shown that the change in ORD or CD of an optically active complex ion caused by adding counter ions is due mainly to ion-pair (outer sphere complex) formation. The CD is usually more sensitive to this effect than is the ORD.

A most remarkable CD change was observed when phosphate ion was added to an aqueous solution of Δ-$[Co(en)_3]^{3+}$ [100, 117]. As Fig. 3-29 shows, the rotational strengths (measured by the CD band areas) of the lower (E_a) and the higher (A_2) frequency components of the I band are decreased and enhanced, respectively. Another change in the CD occurs in the ultraviolet region with the appearance of a new CD band. The changes in the CD bands depend on the concentration of the phosphate ion, but no appreciable changes in the CD occur when phosphate ions over a certain amount are added. Mason and colleagues [100, 117] have also studied the changes in the CD of Δ-$[Co(en)_3]^{3+}$ (= M) due to ion-pair formation with various counter ions (L). For ion-pair formation an equilibrium of the general form

$$M + nL \rightleftharpoons ML_n$$

is established. The major change in CD is induced, however, by 1:1 ($n = 1$) ion-pair formation with L = PO_4^{3-}. Similar CD behavior was observed for the 1:1 ion pair with L = $[Fe(CN)_6]^{4-}$. But in this case the CD spectrum reverts toward that of the free complex ion M on formation of the 1:2 ($n = 2$) ion pair. For L = SeO_3^{2-} the CD spectra of the 1:1 and the 1:2 ion pairs do not differ greatly from that of the free complex ion, but major changes in

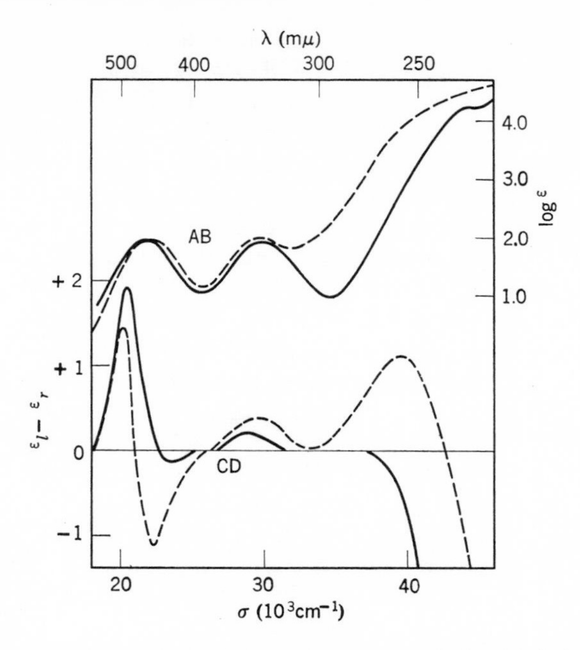

Fig. 3-29. Ion-pairing effects; CD and absorption spectra (AB) of Δ-(+)-$[Co(en)_3]^{3+}$: in aqueous solution (———), in 0.05*F* PO_4^{3-} solution (– – –). After Mason and Norman [117].

the CD spectra occur with the formation of the 1:3 and the 1:4 ion pairs. The effect of the formation of ion pairs on the CD spectrum of Δ-$[Co(en)_3]^{3+}$ is considered to correlate with the appearance of a new CD band in the ultraviolet region, which originates in the charge transfer transition between the counter ion and the metal ion. Since it is unlikely, at least in 1:1 ion pairs, that the charge transfer transition has an intrinsic magnetic dipole moment, the appearance of a CD band associated with the charge transfer transition indicates that the magnetic moment is borrowed from another transition with the same symmetry. Since $R[E_a]$, the rotational strength of the E_a component of the I band, was diminished in the 1:1 and the 1:3 ion pairs, the E_a component of the octahedral $A_{1g} \rightarrow T_{1g}$ transition of the Co(III) ion is the probable source of the borrowed magnetic moment. (In these ion pairs it is probable that the charge transfer transition has E symmetry). A decrease in $R[E_a]$ results in a commensurate increase in $R[A_2]$, since in the free complex ion $R[E_a]$ and $R[A_2]$ overlap with opposed signs. This assumption, proposed by Mason and co-workers, accounts for the equal changes observed in $R[E_a]$ and $R[A_2]$. The same effect on the CD spectra is seen in the Δ-$[Co(\text{L-pn})_3]^{3+}$

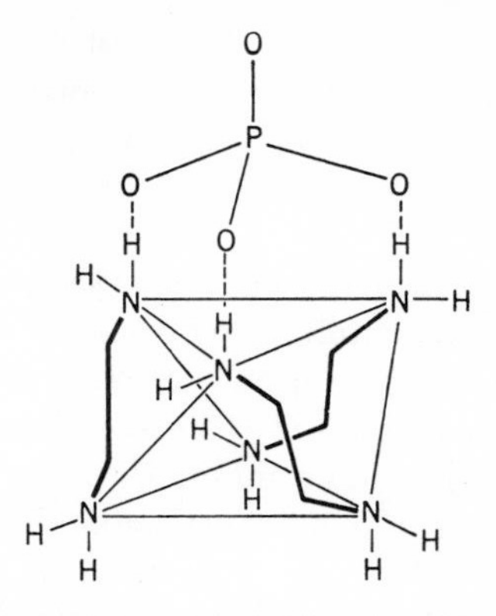

Fig. 3-30. A proposed structure of 1:1 ion pair of Δ-$[Co(en)_3]^{3+}$ and PO_4^{3-}.

complex ion (lel form). The Λ-$[Co(\text{L-pn})_3]^{3+}$ ion, which has the ob form, does not show such distinct changes in the CD with change in counter ion as does the lel form of the complex. From these observations Mason [117] suggested that the PO_4^{3-} and the complex ion have a preferred mutual orientation in the ion pair. As Fig. 3-30 shows, the tris-diamine complex with the lel form can form a mutually oriented ion pair with the PO_4^{3-} ion with hydrogen bonding between the oxygen atoms of the anion and one set of three N—H bonds parallel to the C_3 axis of the complex cation. In the ob form of complex no set of N—H bonds has a suitable orientation to form an ion pair of the lel type. This model implies that hydrogen bonding plays an important role in the ion-pair formation between the PO_4^{3-} and the tris-diamine complex. Smith and Douglas [169] have found that the effects of PO_4^{3-} ion on the CD of $[Co(gly)(en)_2]^{2+}$ and *cis*-$[Co(NO_2)_2(en)_2]^+$, which have less favorable structures for the hydrogen-bond formation with the anion, are not so great as in the tris-diamine complexes. The protonation of PO_4^{3-} resulted also in smaller effects on the CD of tris-diamine complexes, as might be expected.

Mason and Norman [116] have reported that $[Co(NH_3)_6]^{3+}$ in an aqueous solution containing an excess of an optically active substance, such as L-tartrate or L-diethyltartrate, exhibits a marked Cotton effect in the region of the I band. This might be due to the formation of an ion pair with a preferred mutual orientation of the two species. Bosnich [23] has observed a similar effect for the $[PtCl_4]^{2-}$ ion in the optically active solvent D(−)-2,3-butanediol.

F. Assignment of Transitions

It is well known that an ORD or CD study is very useful for observing the absorption components overlapped in ordinary absorption spectra. In fact, it is almost impossible to observe all the absorption components expected theoretically in a complex by analyzing only the ordinary absorption spectra. In general CD is much superior to ORD for the separation of bands, because the CD falls off more rapidly than the ORD in regions away from the band center.

As mentioned in Sec. 3-2D, Yasui et al. [188] have studied the splitting of the *d* orbitals in $[Cu(\text{L-am})_2]$ complexes by means of the CD spectra. As Fig. 3-25 illustrates, the ordinary absorption spectrum of any of these complexes shows a single broad band, whereas the CD curve indicates four bands whose signs are (+), (+), (−), and (−), beginning from the longer wavelength side. Thus the four bands expected for the Cu(II) complex with its low symmetry were first observed by the CD study. Similarly, the *d-d* bands of the VO^{2+} ion in the tartrate and related complexes have been examined by CD [86]. Ito et al. [79] have measured the CD of $[Pt(\text{D-pn})_2]^{2+}$ and related complexes in order to assign the absorption bands of planar Pt(II) complexes.

It is almost impossible to determine the trigonal splitting of the *d* orbitals in a tris-chelate complex from visible-ultraviolet absorption spectra. This splitting can be observed easily by CD [89, 123, 127], but ORD is not always useful.

The CD has also been utilized to separate the bands for various complexes: the separation of three components for each I and II band of a low symmetry d^6 complex [39, 71, 138], the observation of spin-forbidden bands hidden under strong spin-allowed bands [91, 122], and the finding of *d-d* bands overlapped by strong charge transfer bands [76]. For determining the assignment of components thus separated, the most reliable method is the CD measurement of a single crystal of the complex. As mentioned in Sec. 3-1D2, the assignment for the E_a and the A_2 components of the I band of $[Co(en)_3]^{3+}$ was made on the basis of the CD spectrum of a single crystal of

$$Na\text{-}(+)\text{-}[Co(en)_3]_2Cl_7 \cdot 6H_2O$$

[123]. Crystal CD spectra have been measured for only a very few complexes [127] because of technical difficulties in the measurements on microcrystals.

Although current theories on optical activity cannot predict the sign or the strength of a CD band, the assignment of CD bands [127, 128] has been made by comparing the CD data with those of complexes of known absolute configurations or those of structurally related complexes. At the same time, other information, such as linearly polarized spectra of single crystals [112, 178, 186], is useful for assignments.

Wentworth and Piper [175] and Yasui et al. [190] have observed two CD components of the I band of *trans*-$[CoCl_2(\text{D-pn})_2]^+$ and

$$trans\text{-}[Co(en)_2(\text{L-amH})_2]^{3+},$$

which are approximately of D_{4h} symmetry, and have assigned the longer and the shorter wavelength components to the $E_g(D_{4h})$ and the $A_{2g}(D_{4h})$ transitions, respectively. In these complexes the rotational strength of $A_{1g} \rightarrow A_{2g}(D_{4h})$ was usually much weaker than that of $A_{1g} \rightarrow E_g(D_{4h})$. The frequencies of these two components are reversed in *trans*-$[Co(NO_2)_2(D\text{-}pn)_2]^+$; the result coincides with a theoretical prediction.

According to Hawkins et al. [71], the *trans*-$[CoCl_2(D\text{-}pn)_2]^+$ ion shows another very weak negative component on the longer wavelength side of the $A_{1g} \rightarrow E_g(D_{4h})$ band (see Table 3-4). This was assigned to one component resulting from the splitting of $A_{1g} \rightarrow E_g(D_{4h})$ due to the lowering of symmetry (the true symmetry of this ion should be below C_2).

Table 3-4 The CD Data of the I Band of *trans*-Bis(D-propylenediamine)-Co(III) Complexes

Complexes	CD		Assignment of Excited State	
	cm^{-1}	$(\varepsilon_l - \varepsilon_r)_I^0$		
trans-$[CoCl_2(D\text{-}pn)_2]Cl$	14,000	−0.012	B_3 or B_2	$E_g(D_{4h})$
(in 4*M* HCl)	16,450	+0.770	B_2 or B_3	
	21,800	−0.126	B_1	$A_{2g}(D_{4h})$
trans-$[Co(NO_2)_2(D\text{-}pn)_2]ClO_4$	21,700	−0.62		$A_{2g}(D_{4h})$
	24,500	+0.44		$E_g(D_{4h})$

McCaffery et al. [128] have analyzed the CD bands of the complexes *cis*-$[Co(en)_2X_2]^{n+}$ and *cis*-$[Co(en)_2(NH_3)X]^{n+}$, which belong to approximate symmetry C_{2v} and C_{4v}, respectively. They obtained a relationship between the frequencies of the ordinary absorption bands and those of the CD bands (see Fig. 3-31).

Ohkawa et al. [138] have measured the CD of the complexes

$$cis\text{-}[Co(en)_2(CN)X]^+ \quad (X = Cl^-, Br^-)$$

and estimated the σ- and the π-antibonding contribution of the ligands to the metal *d* orbitals. The energies of the I and II bands for a complex of the type *cis*-$[CoN_4XY]$ can be expressed on the basis of Yamatera's theoretical treatment [187] as follows:

$$\mathrm{I}\begin{cases} A' & E_{\mathrm{I}}^0 + \frac{1}{4}\delta\sigma(\mathrm{X}) + \frac{1}{4}\delta\pi(\mathrm{X}) + \frac{1}{4}\delta\sigma(\mathrm{Y}) + \frac{1}{4}\delta\pi(\mathrm{Y}) \\ A'' & E_{\mathrm{I}}^0 + \frac{1}{4}\delta\sigma(\mathrm{Y}) + \frac{1}{4}\delta\pi(\mathrm{Y}) \\ A' & E_{\mathrm{I}}^0 + \frac{1}{4}\delta\sigma(\mathrm{X}) + \frac{1}{4}\delta\pi(\mathrm{X}) \end{cases}$$

$$\mathrm{II}\begin{cases} A' & E_{\mathrm{II}}^0 + \frac{1}{12}\delta\sigma(\mathrm{X}) + \frac{1}{4}\delta\pi(\mathrm{X}) + \frac{1}{12}\delta\sigma(\mathrm{Y}) + \frac{1}{4}\delta\pi(\mathrm{Y}) \\ A'' & E_{\mathrm{II}}^0 + \frac{1}{3}\delta\sigma(\mathrm{X}) + \frac{1}{12}\delta\sigma(\mathrm{Y}) + \frac{1}{4}\delta\pi(\mathrm{Y}) \\ A' & E_{\mathrm{II}}^0 + \frac{1}{12}\delta\sigma(\mathrm{X}) + \frac{1}{3}\delta\sigma(\mathrm{Y}) + \frac{1}{4}\delta\pi(\mathrm{X}) \end{cases}$$

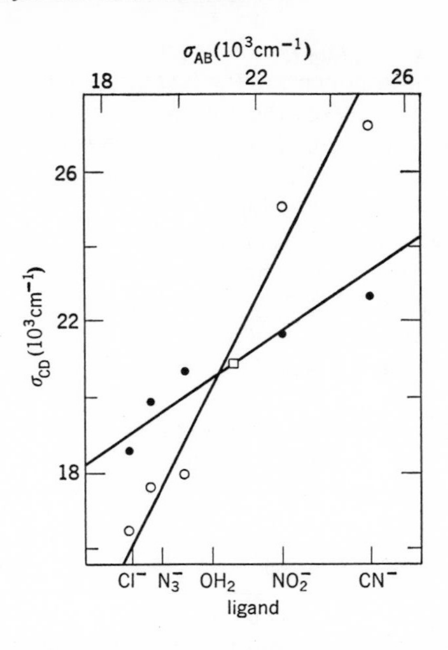

Fig. 3-31. The relationship between the frequency of the ordinary absorption band, σ_{AB}, and the frequency of the major (○) and the minor (●) CD bands, σ_{CD}, for the complexes *cis*-$[Co(en)_2X_2]^{n+}$ or *cis*-$[Co(en)_2(X)(Y)]^{n+}$. The point □ refers to the mean frequency of the A_2 and the E_a CD bands and the absorption frequency of the parent complex, $[Co(en)_3]^{3+}$, obtained from the crystal spectra. After McCaffery, Mason, and Norman [128].

where, $E_I{}^0$ and $E_{II}{}^0$ denote the energy of the first and the second bands of $[Co(en)_3]^{3+}$, respectively, and $\delta\sigma$ and $\delta\pi$ are parameters for the measure of contributions from the σ- and the π-bonding, respectively. As Fig. 3-32 illustrates, the I band of the *cis*-$[Co(en)_2(CN)Cl]^+$ ion shows three CD components at 544, 482, and 431 mμ and the II band, at 379, 353, and 312 mμ. From an analysis of these observed values with some appropriate assumptions, the following parameters were estimated:

$$\delta\sigma(CN) = 36.5, \qquad \delta\pi(CN) = -8.4,$$
$$\delta\sigma(Cl) = -30.5, \qquad \delta\pi(Cl) = -7.4 \quad (\text{in } 10^{13}\ \text{sec}^{-1}).$$

These values indicate that the σ-bonding contributions of the CN^- and Cl^- ions relative to ethylenediamine are almost the same in absolute value, but of opposite sign, whereas the π-bonding contributions of both ions have almost equal values with the same (−) sign.

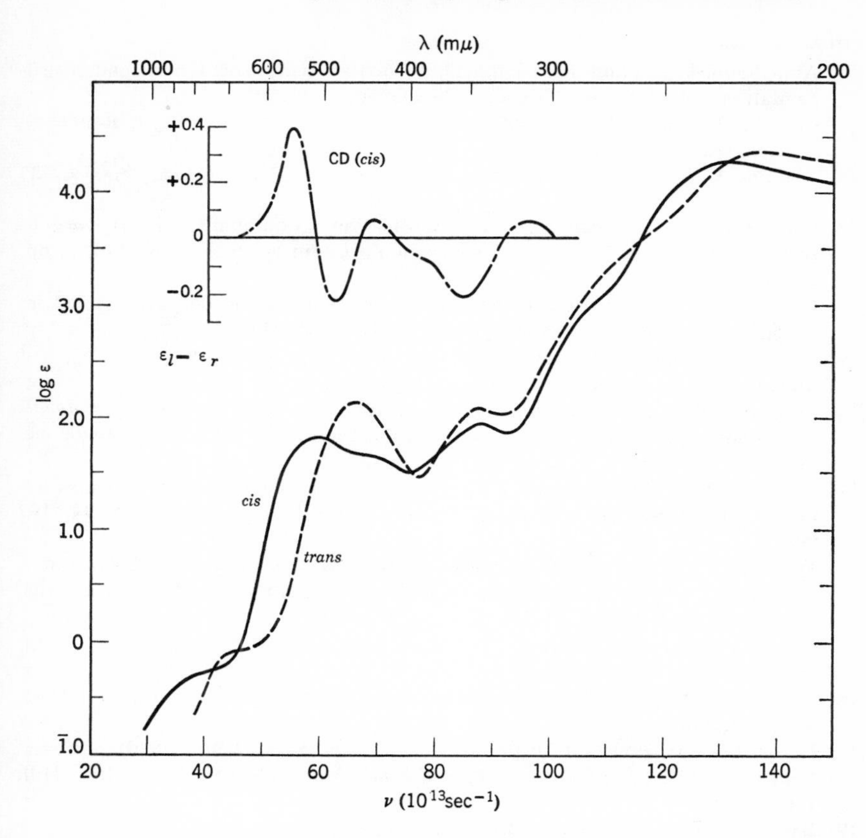

Fig. 3-32. The spectra of isomers of $[Co(en)_2(CN)Cl]^+$: absorption spectra of the *cis* isomer (———) and of the *trans* isomer (– – – –), and the CD spectrum (–––) of the optically active isomer, (−)-*cis*-$[Co(en)_2(CN)Cl]I$.

REFERENCES

Books

[1] Djerassi, C., Optical Rotatory Dispersion. New York: McGraw-Hill, 1960, 293 pp.
[2] Jaeger, F. M., Optical Activity and High Temperature Measurements. New York: McGraw-Hill, 1930, 450 pp.
[3] Lowry, T. M., Optical Rotatory Power. London: Longmans, Green, 1935, 483 pp.
[4] Mathieu, J.-P., Les Theories Moleculaires du Pouvoir Rotatoire Naturel. Paris: Gauthier-Villars, 1946, 243 pp.
[5] Velluz, L., M. Legrand, and M. Grosjean, Optical Circular Dichroism. New York: Academic Press, 1965, 247 pp.

Review Articles

[6] Abu-shumays, A., and J. J. Duffield, "Circular dichroism, theory and instrumentation," *Anal. Chem.*, **38**, No. 7, 29A (1966).

[7] Carroll, B., and I. Blei, "Measurement of optical activity, new approaches," *Science*, **142**, 200 (1963).

[8] Condon, E. U., "Theories of optical rotatory power," *Rev. Mod. Phys.*, **9**, 423 (1937).

[9] Kirschner, S., "Optically Active Coordination Compounds," in *Preparative Inorganic Reactions*, W. L. Jolly, ed. New York: Interscience, 1964, Vol. 1, pp. 29–57.

[10] Kuhn, W., "Eigenschaften, Konfiguration und Korrelation der Unterschied im Verhalten optische aktiver Stoffe," *Angew. Chem.*, **68**, 93 (1956).

[11] Kuhn, W., "Optical rotatory power," *Ann. Rev. Phys. Chem.*, **9**, 417 (1958).

[12] Mason, S. F., "Optical rotatory power," *Quart. Rev.*, **17**, 20 (1963).

[13] Sargeson, A. M., "Optical Phenomena in Metal Chelates," in *Chelating Agents and Metal Chelates*, F. P. Dwyer, and D. P. Mellor, eds. New York: Academic, 1964, pp. 183–235.

[14] Woldbye, F., "Optical rotatory dispersion and circular dichroism of transition metal complexes," *Record Chem. Progr.* (*Kresge-Hooker Sci. Libr.*), **24**, 197 (1963).

[15] Woldbye, F., "Technique of Optical Rotatory Dispersion and Circular Dichrosim," *in Technique of Inorganic Chemistry*, H. B. Jonassen, and A. Weissberger, eds. New York: Interscience, 1965, Vol. 4, pp. 249–368.

Other References

[16] Albinak, M. J., D. C. Bhatnagar, S. Kirschner, and A. J. Sonnessa, *Can. J. Chem.*, **39**, 2360 (1961).

[17] Bailar, J. C., Jr., and D. F. Peppard, *J. Am. Chem. Soc.*, **62**, 105 (1940).

[18] Ballhausen, C. J., *Introduction to Ligand Field Theory*. New York: McGraw-Hill, 1962, 298 pp.

[19] Basolo, F., C. J. Ballhausen, and J. Bjerrum, *Acta Chem. Scand.*, **9**, 810 (1955).

[20] Bhatnagar, D. C., and S. Kirschner, *Inorg. Chem.*, **3**, 1256 (1964).

[21] Bijvoet, J. M., A. F. Peerdman, and A. J. Van Bommel, *Nature*, **168**, 271 (1951).

[22] Billardon, M., *Compt. Rend.*, **251**, 2320 (1960).

[23] Bosnich, B., *J. Am. Chem. Soc.*, **88**, 2606 (1966).

[24] Brushmiller, J. G., E. L. Amma, and B. E. Douglas, *J. Am. Chem. Soc.*, **84**, 3227 (1962).

[25] Bryce, C. F., J. M. H. Pinkerton, L. K. Steinrauf, and F. R. N. Gurd, *J. Biol. Chem.*, **240**, 3829 (1965). C. F. Bryce and F. R. N. Gurd, *J. Biol. Chem.*, **241**, 122, 1439 (1966).

[26] Bürer, T., *Helv. Chim. Acta*, **46**, 2388 (1963).

[27] Bürer, T., *Mol. Phys.*, **6**, 541 (1963).

[28] Bürer, T., *Helv. Chim. Acta*, **46**, 242 (1963).

[29] Busch, D. H., and J. C. Bailar, Jr., *J. Am. Chem. Soc.*, **75**, 4574 (1953).

[30] Busch, D. H., and D. W. Cooke, *J. Inorg. Nucl. Chem.*, **23**, 145, 150 (1961).

[31] Busch, D. H., K. Swaminathan, and D. W. Cooke, *Inorg. Chem.*, **1**, 260 (1962).

[32] Chernyaev, I. I., *Ann. Inst. Platine* (*U.S.S.R*), **6**, 40 (1928); I. I. Chernyaev, T. N. Fedotova, and O. N. Adrianova, *Zh. Neorgan. Khim.*, **10**, 154 (1965).

[33] Chernyaev, I. I., and O. N. Adrianova, *Izv. Akad. Nauk SSSR, Otd. Khim. Nauk*, 204 (1953).

[34] Chernyaev, I. I., T. N. Fedotova, and O. N. Adrianova, *Zh. Neorgan. Khim.*, **10,** 1541 (1965).
[35] Cooley, W. E., C. F. Liu, and J. C. Bailar, Jr., *J. Am. Chem. Soc.*, **81**, 4189 (1959).
[36] Corey, J., and J. C., Bailar, Jr., *J. Am. Chem. Soc.*, **81**, 2620 (1959).
[37] Cotton, A., *Ann. Chim. Phys.*, [7], **8**, 347 (1896).
[38] Denning, R. G., and T. S. Piper, *Inorg. Chem.*, **5**, 1056 (1966).
[39] Douglas, B. E., R. A. Haines, and J. G. Brushmiller, *Inorg. Chem.*, **2**, 1194 (1963).
[40] Douglas, B. E., and S. Yamada, *Inorg. Chem.*, **4**, 1561 (1965).
[41] Douglas, B. E., *Inorg. Chem.*, **4**, 1813 (1965).
[42] Drew, M. G. B., J. H. Dunlop, R. D. Gillard, and D. Rogers, *Chem. Commun.*, 42 (1966).
[43] Dunlop, J. H., and R. D. Gillard, *J. Chem. Soc.*, 2822 (1964).
[44] Dunlop, J. H., R. D. Gillard, and G. Wilkinson, *J. Chem. Soc.*, 3160 (1964).
[45] Dunlop, J. H., and R. D. Gillard, *J. Inorg. Nucl. Chem.*, **27**, 361 (1965).
[46] Dunlop, J. H., and R. D. Gillard, *J. Chem. Soc.*, 6531 (1965).
[47] Dwyer, F. P., and F. Lions, *J. Am. Chem. Soc.*, **72**, 1545 (1950).
[48] Dwyer, F. P., N. S. Gill, E. C. Gyarfas, and F. Lions, *J. Am. Chem. Soc.*, **75**, 1526 (1953).
[49] Dwyer, F. P., E. C. Gyarfas, and D. P. Mellor, *J. Phys. Chem.*, **59**, 296 (1955).
[50] Dwyer, F. P., F. L. Garvan, and A. Schulman, *J. Am. Chem. Soc.*, **81**, 290 (1959).
[51] Dwyer, F. P., and F. L. Garvan, *J. Am. Chem. Soc.*, **81**, 2955 (1959).
[52] Dwyer, F. P., T. E. MacDermott, and A. M. Sargeson, *J. Am. Chem. Soc.*, **85**, 2913 (1963).
[53] Dwyer, F. P., A. M. Sargeson, and L. B. James, *J. Am. Chem. Soc.*, **86**, 590 (1964).
[54] Dwyer, F. P., I. K. Reid, and A. M. Sargeson, *Australian J. Chem.*, **18**, 1919 (1965).
[55] Emerson, T. R., D. F. Weing, W. Klyne, D. G. Nielsen, L. H. Roach, and R. J. Swan, *J. Chem. Soc.*, 4007 (1965).
[56] Eyring, H., J. Walter, and G. E. Kimball, *Quantum Chemistry*. New York: Wiley, 1944, p. 346.
[57] Fraser, R. T. M., *Proc. Chem. Soc.*, 262 (1963).
[58] Fujita, J., and Y. Shimura, *Bull. Chem. Soc. Japan*, **36**, 1281 (1963).
[59] Fujita, J., T. Yasui, and Y. Shimura, *Bull. Chem. Soc. Japan*, **38**, 654 (1965).
[60] Garbett, K., and R. D. Gillard, *J. Chem. Soc.*, 6084 (1965).
[61] Garbett, K., R. D. Gillard, and P. J. Staples, *J. Chem. Soc.* (A), 201 (1966); K. Garbett, and R. D. Gillard, *J. Chem. Soc.* (A), 204 (1966).
[62] Gillard, R. D., and G. Wilkinson, *J. Chem. Soc.*, 3193 (1963).
[63] Gillard, R. D., and G. Wilkinson, *J. Chem. Soc.*, 1368 (1964); *J. Chem. Soc.*, 4271 (1963).
[64] Gillard, R. D., *J. Inorg. Nucl. Chem.*, **26**, 1455 (1964).
[65] Gillard, R. D., and H. M. Irving, *Chem. Rev.*, **65**, 603 (1965).
[66] Gillard, R. D., *Tetrahedron*, **21**, 503 (1965).
[67] Haines, R. A., and B. E. Douglas, *Inorg. Chem.*, **4**, 452 (1965).
[68] Hamer, N. K., *Mol. Phys.*, **5**, 339 (1962).
[69] Hatfield, W. E., *Inorg. Chem.*, **3**, 605 (1964).
[70] Hawkins, C. J., and E. Larsen, *Acta Chem. Scand.*, **19**, 185, 1969 (1965).
[71] Hawkins, C. J., E. Larsen, and I. Olsen, *Acta Chem. Scand.*, **19**, 1915 (1965).
[72] Hidaka, J., S. Yamada, and R. Tsuchida, *Bull. Chem. Soc. Japan*, **31**, 921 (1958).
[73] Hidaka, J., Y. Shimura, and R. Tsuchida, *Bull. Chem. Soc. Japan*, **33**, 847 (1960).

[74] Hidaka, J., Y. Shimura, and R. Tsuchida, *Bull. Chem. Soc. Japan*, **35**, 567 (1962).
[75] Hidaka, J., and B. E. Douglas, *Inorg. Chem.*, **3**, 1180 (1964).
[76] Hidaka, J., and B. E. Douglas, *Inorg. Chem.*, **3**, 1724 (1964).
[77] Hidaka, J., and Y. Shimura, *Bull. Chem. Soc. Japan*, **40**, 2312 (1967).
[78] Irving, H., and R. D. Gillard, *J. Chem. Soc.*, 5266 (1960).
[79] Ito, H., J. Fujita, and K. Saito, *Bull. Chem. Soc. Japan*, **39**, 2056 (1966); **40**, 2584 (1967).
[80] Iwasaki, H., and Y. Saito, *Bull. Chem. Soc. Japan*, **39**, 92 (1966).
[81] Izumiya, N., M. Winitz, S. M. Birnbaum, and J. P. Greenstein, *J. Am. Chem. Soc.*, **78**, 1602 (1956).
[82] Jaeger, F. M., and H. B. Blumendal, *Z. Anorg. Allgem. Chem.*, **175**, 161 (1928).
[83] Jaeger, F. M., and L. Bijkerk, *Z. Anorg. Allgem. Chem.*, **233**, 97 (1937).
[84] Jaeger, F. M., *Bull. Soc. Chim. France*, [5] **4**, 1201 (1937).
[85] Jensen, K. A., S. Burmester, G. Cederberg, R. B. Jensen, C. Th. Pedersen, and E. Larsen, *Acta Chem. Scand.*, **19**, 1239 (1965).
[86] Jones, K. M., and E. Larsen, *Acta Chem. Scand.*, **19**, 1210 (1965).
[87] Jørgensen, C. K., *Acta Chem. Scand.*, **8**, 1495 (1954).
[88] Jørgensen, C. K., *Absorption Spectra and Chemical Bonding in Complexes*. Oxford: Pergamon Press, 1962, 352 pp.
[89] Karipides, A. G., and T. S. Piper, *J. Chem. Phys.*, **40**, 674 (1964).
[90] Kirschner, S., D. C. Bhatnagar, M. J. Albinak, and A. J. Sonnessa, in *Theory and Structure of Complex Compounds*, B. Jezowska-Trzebiatowska, ed. 1964, pp. 63–69.
[91] Kling, O., and F. Woldbye, *Acta Chem. Scand.*, **15**, 704 (1961).
[92] Kobayashi, M., *J. Chem. Soc. Japan, Pure Chem. Sect.* (*Nippon Kagaku Zasshi*), **64**, 648 (1943).
[93] Krebs, H., and R. Rasche, *Z. Anorg. Allgem. Chem.*, **276**, 236 (1954).
[94] Kuebler, J. R., Jr., and J. C. Bailar, Jr., *J. Am. Chem. Soc.*, **74**, 3535 (1952).
[95] Kuhn, W., and E. Braun, *Z. Physik. Chem.* (*Leipzig*), **B8**, 281 (1930).
[96] Kyuno, E., L. J. Boucher, and J. C. Bailar, Jr., *J. Am. Chem. Soc.*, **87**, 4458 (1965); E. Kyuno and J. C. Bailar, Jr., *J. Am. Chem. Soc.*, **88**, 1121, 1125 (1966).
[97] Larsen, E., and I. Olsen, *Acta Chem. Scand.*, **18**, 1025 (1964).
[98] Larsen, E., and S. F. Mason, *J. Chem. Soc.* (A), 313 (1966).
[99] Larsson, R., *Acta Chem. Scand.*, **16**, 2267, 2305 (1962).
[100] Larsson, R., S. F. Mason, and B. J. Norman, *J. Chem. Soc.* (A), 301 (1966).
[101] Ley, H., and H. Winkler, *Chem. Ber.*, **42**, 3894 (1909); **45**, 3894 (1909); **45**, 372 (1912).
[102] Liehr, A. D., *J. Phys. Chem.*, **68**, 665, 3629 (1964).
[103] Lifschitz, I., *Proc. Acad. Sci. Amsterdam*, **27**, 721 (1924); *Z. Physik. Chem.*, (*Leipzig*), **114**, 485 (1925).
[104] Lifschitz, I., *Proc. Roy. Acad. Amsterdam*, **39**, 1192 (1936).
[105] Lifschitz, I., and W. Forentjes, *Rec. Trav. Chim.*, **60**, 225 (1941).
[106] Linhard, M., and M. Weigel, *Z. Physik. Chem.*, *N.F.*, **11**, 308 (1957).
[107] Liu, C. T., and B. E. Douglas, *Inorg. Chem.*, **3**, 1356 (1964).
[108] Liu, J. C. I., and J. C. Bailar, Jr., *J. Am. Chem. Soc.*, **73**, 5432 (1951).
[109] Lowry, T. M., and T. W. Dickson, *J. Chem. Soc.*, **103**, 1067 (1913); T. M. Lowry, and H. H. Abram, *J. Chem. Soc.*, **115**, 300 (1919).
[110] MacDermott, T. E., and A. M. Sargeson, *Australian J. Chem.*, **16**, 334 (1963).
[111] Marchi, L. E., and J. P. McReynolds, *J. Am. Chem. Soc.*, **65**, 333 (1943).
[112] Martin, D. S., Jr., and C. A. Lenhardt, *Inorg. Chem.*, **3**, 1368 (1964); D. S. Martin, Jr., M. A. Tucker, and A. J. Kassman, *Inorg. Chem.*, **4**, 1682 (1965).
[113] Mann, F. G., *J. Chem. Soc.*, 1224 (1927); 890 (1928).

[114] Mann, F. G., *J. Chem. Soc.*, 1745 (1930).
[115] Mason, S. F., and B. J. Norman, *Chem. Commun.*, 73 (1965).
[116] Mason, S. F., and B. J. Norman, *Chem. Commun.*, 335 (1965).
[117] Mason, S. F., and B. J. Norman, *J. Chem. Soc.* (A), 307 (1966); *Proc. Chem. Soc.*, 339 (1964).
[118] Mathieu, J.-P., *J. Chim. Phys.*, **33**, 78 (1936).
[119] Mathieu, J.-P., *Bull. Soc. Chim. France*, [5] **3**, 476 (1936).
[120] Mathieu, J.-P., *Bull. Soc. Chim. France*, [5] **6**, 873 (1939).
[121] Mathieu, J.-P., *Ann. Phys.*, [11] **19**, 335 (1944).
[122] McCaffery, A. J., and S. F. Mason, *Trans. Faraday Soc.*, **59**, 1 (1963).
[123] McCaffery, A. J., and S. F. Mason, *Mol. Phys.*, **6**, 359 (1963).
[124] McCaffery, A. J., and S. F. Mason, *Proc. Chem. Soc.*, 211 (1963).
[125] McCaffery, A. J., S. F. Mason, and B. J. Norman, *Proc. Chem. Soc.*, 259 (1964).
[126] McCaffery, A. J., S. F. Mason, and B. J. Norman, *Chem. Commun.*, 49 (1965).
[127] McCaffery, A. J., S. F. Mason, and R. E. Ballard, *J. Chem. Soc.*, 2883 (1965).
[128] McCaffery, A. J., S. F. Mason, and B. J. Norman, *J. Chem. Soc.*, 5094 (1965).
[129] Meisenheimer, J., L. Angerman, and H. Holsten, *Ann.*, **438**, 261 (1924).
[130] Michelson, K., *Acta Chem. Scand.*, **19**, 1175 (1965).
[131] Mills, W. H., and T. H. H. Quibell, *J. Chem. Soc.*, 839 (1935).
[132] Moffitt, W., *J. Chem. Phys.*, **25**, 1189 (1956).
[133] Moffitt, W., and A. Moscowitz, *J. Chem. Phys.*, **30**, 648 (1955).
[134] Moscowitz, A., ref. [1], pp. 150–177.
[135] Nakahara, A., Y. Saito, and H. Kuroya, *Bull. Chem. Soc. Japan*, **25**, 331 (1952).
[136] Nakatsu, K., M. Shiro, Y. Saito, and H. Kuroya, *Bull. Chem. Soc. Japan*, **30**, 158 (1957).
[137] Nakatsu, K., *Bull. Chem. Soc. Japan*, **35**, 832 (1962).
[138] Ohkawa, K., J. Hidaka, and Y. Shimura, *Bull. Chem. Soc. Japan*, **39**, 1715 (1966.)
[139] Paiaro, G., P. Corrandini, R. Palumbo, and A. Panunzi, *Makromol. Chem.*, **71**, 184 (1964).
[140] Paiaro, G., and A. Panunzi, *J. Am. Chem. Soc.*, **86**, 5148 (1964).
[141] Paiaro, G., and A. Panunzi, *Tetrahedron Letters*, 441 (1965).
[142] Pfeiffer, P., and K. Quehl, *Chem. Ber.*, **65**, 560 (1932); P. Pfeiffer and Y. Nakatsuka, *Chem. Ber.*, **66**, 415 (1933).
[143] Pfeiffer, P., and W. Christeleit, *Z. Physiol. Chem.*, **245**, 197 (1937); **247**, 262 (1937).
[144] Pfeiffer, P., W. Christeleit, T. Hesse, H. Pfitzner, and H. Thielert, *J. Prakt. Chem.*, **150**, 261 (1938).
[145] Pfeiffer, P., and S. Saure, *Chem. Ber.*, **74**, 935 (1941).
[146] Piper, T. S., *J. Am. Chem. Soc.*, **83**, 3908 (1961).
[147] Piper, T. S., and R. L. Carlin, *J. Chem. Phys.*, **33**, 608 (1961).
[148] Piper, T. S., and A. G. Karipides, *Mol. Phys.*, **5**, 475 (1962).
[149] Piper, T. S., *J. Chem. Phys.*, **36**, 2224 (1962).
[150] Piper, T. S., and A. G. Karipides, *J. Am. Chem. Soc.*, **86**, 5039 (1964).
[151] Plicque, F., *Compt. Rend.*, **262**, 381 (1966).
[152] Poulet, H., *J. Chim. Phys.*, **59**, 584 (1962).
[153] Reihlen, H., E. Weinbrenner, and G. V. Hessling, *Ann.*, **494**, 143 (1932); S. Schnell and P. Karrer, *Helv. Chim. Acta*, **38**, 2036 (1955).
[154] Rosenfeld, L., *Z. Physik.*, **52**, 161 (1928).
[155] Saito, Y., K. Nakatsu, M. Shiro, and H. Kuroya, *Bull. Chem. Soc. Japan*, **30**, 795 (1957).

[156] Saito, Y., and H. Iwasaki, *Bull. Chem. Soc. Japan*, **35**, 1131 (1962).
[157] Saito, Y., private communication.
[158] Sargeson, A. M., and G. H. Searle, *Nature*, **200**, 356 (1963).
[159] Sargeson, A. M., and G. H. Searle, *Inorg. Chem.*, **4**, 45 (1965).
[160] Sarma, B. D., and J. C. Bailar, Jr., *J. Am. Chem. Soc.*, **77**, 5480 (1955).
[161] Schäffer, C. E., and C. K. Jørgensen, *Kgl. Danske Videnskab. Selskab, Mat.-Fys. Medd.*, **34**, No. 13 (1965).
[162] Shibata, M., H. Nishikawa, and Y. Nishida, *Bull. Chem. Soc. Japan*, **39**, 2310 (1966).
[163] Shibata, M., *J. Chem. Soc. Japan, Pure Chem. Sect.* (*Nippon Kagaku Zasshi*), **87**, 771 (1966).
[164] Shimura, Y., and R. Tsuchida, *Bull. Chem. Soc. Japan*, **29**, 311 (1956); N. Matsuoka, Y. Shimura, and R. Tsuchida, *J. Chem. Soc. Japan, Pure Chem. Sect.* (*Nippon Kagaku Zasshi*), **82**, 1637 (1961).
[165] Shimura, Y., *Bull. Chem. Soc. Japan*, **31**, 311 (1958).
[166] Shimura, Y., *Bull. Chem. Soc. Japan*, **31**, 315 (1958).
[167] Shinada, M., *J. Phys. Soc. Japan*, **19**, 1607 (1964).
[168] Smirnoff, A. P., *Helv. Chim. Acta*, **3**, 177 (1920).
[169] Smith, H. L., and B. E. Douglas, *Inorg Chem.*, **5**, 784 (1966).
[170] Sugano, S., *J. Chem. Phys.*, **33**, 1883 (1960).
[171] Tanabe, Y., and S. Sugano, *J. Phys. Soc. Japan*, **9**, 753 ,766 (1954).
[172] Trimble, R. F., Jr., *J. Chem. Educ.*, **31**, 176 (1954).
[173] Tsuchida, R., *Bull. Chem. Soc. Japan*, **13**, 388, 436 (1938).
[174] Urry, D. W., and H. Eyring, *J. Am. Chem. Soc.*, **86**, 4574 (1964); D. W. Urry, D. Miles, D. J. Caldwell, and H. Eyring, *J. Phys. Chem.*, **69**, 1603 (1965).
[175] Wentworth, R. A. D., and T. S. Piper, *Inorg. Chem.*, **4**, 202 (1965).
[176] Wentworth, R. A. D., and T. S. Piper, *Inorg. Chem.*, **4**, 709 (1965).
[177] Wentworth, R. A. D., and T. S. Piper, *Inorg. Chem.*, **4**, 1524 (1965).
[178] Wentworth, R. A. D., *Inorg. Chem.*, **5**, 496 (1966).
[179] Werner, A., *Z. Anorg. Chem.*, **3**, 267 (1893).
[180] Werner, A., *Chem. Ber.*, **44**, 1887 (1911).
[181] Werner, A., *Chem. Ber.*, **45**, 121 (1912).
[182] Werner, A., and T. P. McCutcheon, *Chem. Ber.*, **45**, 3281 (1912).
[183] Werner, A., and Y. Shibata, *Chem. Ber.*, **45**, 3287 (1912).
[184] Werner, A., *Chem. Ber.*, **46**, 3674 (1913).
[185] Werner, A., *Chem. Ber.*, **47**, 3087 (1914).
[186] Yamada, S., A. Nakahara, Y. Shimura, and R. Tsuchida, *Bull. Chem. Soc. Japan*, **28**, 221 (1955).
[187] Yamatera, H., *Bull. Chem. Soc. Japan*, **31**, 95 (1958).
[188] Yasui, T., *Bull. Chem. Soc. Japan*, **38**, 1746 (1965); T. Yasui, J. Hidaka, and Y. Shimura, *J. Am. Chem. Soc.*, **87**, 2762 (1965).
[189] Yasui, T., J. Hidaka, and Y. Shimura, *Bull. Chem. Soc. Japan*, **38**, 2025 (1965).
[190] Yasui, T., J. Hidaka, and Y. Shimura, *Bull. Chem. Soc. Japan*, **39**, 2417 (1966).

Additional References

[191] Treptow, R. S., "Optical activity and electronic spectra. Pseudotetragonal *l*-cyclohexanediamine complexes of cobalt(III)," *Inorg. Chem.*, **5**, 1593 (1966).
[192] Kirschner, S., and K. H. Pearson, "The Cotton effect in coordination compounds containing monodentate ligands," *Inorg. Chem.*, **5**, 1614 (1966).

[193] Buckingham, D. A., S. F. Mason, A. M. Sargeson, and K. R. Turnbull, "The stereospecific coordination of sarcosine," *Inorg. Chem.*, **5**, 1649 (1966).
[194] Gillard, R. D., H. M. Irving, R. M. Parkins, N. C. Payne, and L. D. Pettit, "The isomers of complexes of α-amino-acids with copper(II)," *J. Chem. Soc.* (A), 1159 (1966).
[195] Dunlop, J. H., D. F. Evans, R. D. Gillard, and G. Wilkinson, "Optically active co-ordination compounds. Part VI. Stereoselective effects in alkaline tartrato-complexes of transition metals," *J. Chem. Soc.* (A), 1260 (1966).
[196] Bosnich, B., "The circular dichroism of complexes of palladium containing an optically active unidentate ligand," *J. Chem. Soc.* (A), 1394 (1966).
[197] Corradini, P., G. Paiaro, A. Panunzi, S. F. Mason, and G. H. Searle, "Induction of assymmetry in *cis*-dichloro(olefin)(amine)platinum(II) complexes," *J. Am. Chem. Soc.*, **88**, 2863 (1966).
[198] Legg, J. I., and B. E. Douglas, "A general method for relating the absolute configurations of octahedral chelate complexes," *J. Am. Chem. Soc.*, **88**, 2697 (1966).
[199] Halpern, B., A. M. Sargeson, and K. R. Turnbull, "Racemization and deuteration at an asymmetric nitrogen center," *J. Am. Chem. Soc.*, **88**, 4630 (1966).
[200] Boudreaux, E. A., O. E. Weigang, Jr., and J. A. Turner, "On solvent-induced change in intensity of electronic absorption and circular dichroism," *Chem. Commun.*, 378 (1966).
[201] Barclay, G. A., E. Goldschmied, N. C. Stephenson, and A. M. Sargeson, "The absolute configuration of the (+)-*cis*-dinitro-bis(−)-propylenediamine-cobalt(III) ion," *Chem. Commun.*, 540 (1966).
[202] McCaffery, A. J., S. F. Mason, and B. J. Norman, "The circular dichroism of dissymmetric cobalt(III) complexes in the solid state," *Chem. Commun.*, 661 (1966).
[203] Larsen, E., S. F. Mason, and G. H. Searle, "The absorption and the circular dichroism spectra and the absolute configuration of the tris-acetylacetonato silicon(IV) ion," *Acta Chem. Scand.*, **20**, 191 (1966).
[204] Woldbye, F., and S. Bagger, "Circular dichroism and the Bouguer-Lambert-Beer law," *Acta Chem. Scand.*, **20**, 1145 (1966).
[205] Bürer, T., and L. I. Katzin, "Optical rotatory dispersion data and the Drude equation," *J. Phys. Chem.*, **70**, 2663 (1966).
[206] Larsson, R., and B. Norman, "Circular dichroism of the (+)-tris-ethylenediamine cobalt(III) ion and the stepwise formation of outer-sphere complexes," *J. Inorg. Nucl. Chem.*, **28**, 1291 (1966).
[207] Sargeson, A. M., "Conformations of Coordinated Chelates," in *Transition Metal Chemistry*, R. L. Carlin, ed. New York: Marcel Dekker, 1966, Vol. 3. pp. 303–343.
[208] House, D. A., and C. S. Garner, "Transition Metal Complexes of Tetraethylenepentamine. I. Preparation, Properties, and Geometric Configuration of α- and β-Chlorotetraethylenepentaminecobalt(III) Tetrachlorozincate(II) and the α Chromium(III) Analog," *Inorg. Chem.*, **5**, 2097 (1966).
[209] Yoshikawa, S., T. Sekihara, and M. Goto, "Stereochemical Studies of Metal Chelates. I. Cobalt(III) Complexes Containing an Optically Active Tetramine," *Inorg. Chem.*, **6**, 169 (1967).
[210] House, D. A., and C. S. Garner, "Transition Metal Complexes of Tetraethylenepentamine. II. Some Acidotetraethylenepentamine Complexes of Cobalt(III) and Chromium(III)," *Inorg. Chem.*, **6**, 272 (1967).
[211] Legg, J. I., D. W. Cooke, and B. E. Douglas, "Circular Dichroism of *trans*-N,N′-Ethylenediaminediacetic Acid Cobalt(III) Complexes," *Inorg. Chem.*, **6**, 700 (1967).

[212] Kern, R. D., and R. A. D. Wentworth, "Circular Dichroism of the Tris[di-μ-hydroxo-bis(ethylenediamine)cobalt(III)]cobalt(III) Ion," *Inorg. Chem.*, **6**, 1018 (1967).

[213] Buckingham, D. A., P. A. Marzilli, and A. M. Sargeson, "Stereochemistry and Rearrangement in Some Triethylenetetramine Disubstituted Cobalt(III) Ions," *Inorg. Chem.*, **6**, 1032 (1967).

[214] Marzilli, L. G., and D. A. Buckingham, "The Stereochemistry of Some Cobalt(III) Triethylenetetramine Complexes of Glycine and Sarcosine," *Inorg. Chem.*, **6**, 1042 (1967).

[215] Dunlop, J. H., R. D. Gillard, and R. Ugo, "Optically Active Co-ordination Compounds. Part VII. Tris-Complexes of (+)-Hydroxymethylenecamphor," *J. Chem. Soc.* (A), 1540 (1966).

[216] Carter, O. L., and A. T. McPhail, "Optically Active Organometallic Compounds. Part I. Absolute Configuration of (−)-1,1′-Dimethylferrocene-3-carboxylic Acid by X-ray Analysis of Its Quinidine Salt," *J. Chem. Soc.* (A), 365 (1967).

[217] Gillard, R. D., P. M. Harrison, and E. D. McKenzie, "Optically Active Co-ordination Compounds. Part IX. Complexes of Dipeptides with Cobalt(III)," *J. Chem. Soc.* (A), 618 (1967).

[218] Kyuno, E., and J. C. Bailar, Jr., "The Stereochemistry of Complex Inorganic Compounds XXXIII. Reactions of Optically Active α-Dichlorotriethylenetetramine Cobalt(III) Cation with Optically Active Propylenediamine," *J. Am. Chem. Soc.*, **88**, 5447 (1966).

[219] Asperger, R. G., and C. F. Liu, "Correlations of NMR and ORD Spectra for Establishing the Absolute Configurational Assignment of Cobalt(III)-Chelated Optically Active Triethylenetetramine Homolog-α-Amino Acid Adducts," *J. Am. Chem. Soc.*, **89**, 708 (1967).

[220] Buckingham, D. A., L. G. Marzilli, and A. M. Sargeson, "Racemization and Deuteration at the Asymmetric Nitrogen Center of the N-Methylethylenediamine-tetrammine Cobalt(III) Ion," *J. Am. Chem. Soc.*, **89**, 825 (1967).

[221] Panunzi, A., and G. Paiaro, "Molecular Asymmetry in the Coordination of Olefins to Transition Metals. IV. *cis*-Dichloro(olefin)(amine) Platinum(II) Complexes," *J. Am. Chem. Soc.*, **88**, 4843 (1966).

[222] Buckingham, D. A., L. G. Marzilli, and A. M. Sargeson, "Racemization and Proton Exchange in the *trans*, *trans*-Dinitrobis(N-methylethylenediamine) cobalt(III) Ion," *J. Am. Chem. Soc.*, **89**, 3428 (1967).

[223] Wellman, K. M., T. G. Mecca, W. Mungall, and C. Hare, "Optical Rotatory Dispersion Spectra of Bis- and Mono(α-substituted glycino)copper(II) Complexes," *J. Am. Chem. Soc.*, **89**, 3646 (1967).

[224] Wellman, K. M., W. Mungall, T. G. Mecca, and C. R. Hare, "Relationship of Ring Conformation to Rotational Strengths of *d-d* Transitions in Amino Acid-Copper(II) Complexes," *J. Am. Chem. Soc.*, **89**, 3647 (1967).

[225] Gollogly, J. R., and C. J. Howkins, "The Absolute Configuration of NNN′N′-Tetrakis-(2′-aminoethyl)-1,2-Diaminoethanecobalt(III) and Related Complexes," *Chem. Commun.*, 873 (1966).

[226] Dunlop, J. H., R. D. Gillard, N. C. Payne, and G. B. Robertson, "A Case of Kinetic Stereoselectivity and Its Origin," *Chem. Commun.*, 874 (1966).

[227] Denning, R. G., "Vibronic Structure in Circular Dichroism," *Chem. Commun.*, 120 (1967).

[228] Mason, S. F., and J. W. Wood, "The Circular Dichroism of the Dodecammine Hexa-μ-hydroxotetracobalt(III) Ion," *Chem. Commun.*, 209 (1967).

[229] Blount, J. F., and H. C. Freeman, "Absolute Configuration and Stereospecificity of the $(-)_{589}$-Sarcosinatobis(ethylenediamine) cobalt(III) Ion," *Chem. Commun.*, 325 (1967).

[230] Liu, C. F., "Optical Isomers of the *cis*-Dichlorobisethylenediamineplatinum(IV) and Its Stereospecific Reaction with Ethylenediamine," *Chem. Commun.*, 412 (1967).

[231] Buckingham, D. A., P. A. Marzilli, A. M. Sargeson, and S. F. Mason, "The Absolute Configuration of the $(+)_{589}$-*trans*-Dichlorotriethylenetetraminecobalt-(III) Ion," *Chem. Commun.*, 433 (1967).

[232] Larsson, R., "Studies on Cobaltammines X. Some Reactions of the (+)-Tris-ethylenediaminecobalt(III) Ion with the Hexacyanoferrate(II) Ion," *Acta Chem. Scand.*, **21**, 257 (1967).

[233] Stephens, P. J., "Faraday Effect of Vibronically Allowed Transitions: *d–d* Transitions in Cobalt(III) Complexes," *J. Chem. Phys.*, **44**, 4060 (1966).

[234] Schatz, P. N., A. J. McCaffery, W. Suetaka, G. N. Henning, A. B. Ritchie, and P. J. Stephens, "Faraday Effect of Charge-Transfer Transitions in $Fe(CN)_6^{3-}$, MnO_4^-, and CrO_4^{2-}," *J. Chem. Phys.*, **45**, 722 (1966).

[235] Denning, R. G., "Magnetic Circular Dichroism in Tetrahalo Cobaltate Ions," *J. Chem. Phys.*, **45**, 1307 (1966).

[236] Mason, S. F., A. M. Sargeson, R. Larsson, B. J. Norman, A. J. McCaffery, and G. H. Searle, "The Influence of Ring Conformation on the Optical Activity of Tris-diamine Cobalt(III) and Platinum(IV) Complexes," *Inorg. Nucl. Chem. Letters*, **2**, 333 (1966).

[237] Dunlop, J. H., and R. D. Gillard, "Cotton Effect and Configuration in Acido-pentammine-Cobalt(III) Complexes of Asymmetric Acids," *Tetrahedron*, **23**, 349 (1967).

[238] Velluz, L., and M. Legrand, "Circular Dichroism and Configuration in the Sugar Series (particularly of molybdenum-sugar compounds)," *Compt. Rend.*, **263**, 1429 (1966).

[239] Kida, S., T. Isobe, and S. Misumi, "Circular Dichroism of Lanthanide(III) Complexes," *Bull. Chem. Soc. Japan*, **39**, 2786 (1966).

[240] Shibata, M., H. Nishikawa, and K. Hosaka, "The Isolation of the Four Isomers of Trihydrogen Tris-(L(+)-aspergato) Cobaltate(III)," *Bull. Chem. Soc. Japan*, **40**, 236 (1967).

[241] Kaizaki, S., J. Hidaka, and Y. Shimura, "Stereospecific Formation of Optically Active Chromium(III) Complexes of a Novel Binuclear Type Containing Two L-Tartrate Bridges," *Bull. Chem. Soc. Japan*, **40**, 2207 (1967).

[242] Sasaki, Y., J. Fujita, and K. Saito, "Circular Dichroism of the Paramagnetic μ-amido-μ-peroxo-tetrakis-*l*-propylenediamine-dicobalt Ion," *Bull. Chem. Soc. Japan*, **40**, 2206 (1967).

[243] Zand, R., and S. Vinogradov, "Circular Dichroism of Ferricytochrom C," *Biochem. Biophys. Res. Comm.*, **26**, 121 (1967).

[244] Gillard, R. D., "Optical Conformations of Complexes of Aminoacids with Cobalt(III)," *Nature*, **214**, 168 (1967).

[245] Mason, J., and S. F. Mason, "The Electronic Absorption and Circular Dichroism Spectra and the Absolute Stereochemistry of the Tris-Catechylarsenate(V) ion," *Tetrahedron*, **23**, 1919 (1967).

[246] Liehr, A. D., "Optical Activity in Inorganic and Organic Compounds." In R. L. Carlin, ed., *Transition Metal Chemistry*, New York: Marcel Dekker, 1966, pp. 165–338.

4

INFRARED SPECTROSCOPY

KAZUO NAKAMOTO
Illinois Institute of Technology

Vibrational spectra originate in the vibrations of the nuclei constituting a molecule, and are observed both as the infrared and the Raman spectra in the frequency region between approximately 10^4 cm^{-1} (1 μ) and 10 cm^{-1} (10^3 μ). The frequencies of the vibrational transitions are determined by the masses of constituent atoms, the molecular geometry (bond distances, bond angles, and angles of internal rotation), and the interatomic forces, whereas the intensities of infrared and Raman spectra are related to the changes in the dipole moment and polarizability, respectively, caused by molecular vibrations. Thus studies of vibrational spectra provide valuable information about molecular structure and chemical bonding.

Rapid developments in commercial infrared and Raman instrumentation after World War II have brought a revolution to chemical laboratories. It is now a routine procedure to run infrared spectra in order to characterize compounds. However, measurements of Raman spectra are not yet as popular as those of infrared spectra because of technical inconveniences involved in the measurements. In particular, conventional Raman spectrophotometers are not suitable for the study of colored metal chelate compounds because these compounds absorb the exciting line of mercury (4358 A) normally used as a light source in these instruments. At present an attempt is being made to replace the Hg source by a laser such as the He-Ne gas laser (6328 A). Developments of these and other sources will undoubtedly stimulate Raman studies of colored substances in the future. Although both infrared and Raman data are desirable, the lack of Raman data does not cause serious difficulty in vibrational analysis of metal chelate compounds, because, owing to the relatively low symmetry of these compounds, most vibrational transitions are allowed in the infrared spectrum.

For convenience, vibrational spectra of metal chelate compounds can be divided into the high frequency (4000–650 cm^{-1}) or NaCl region, and the low frequency (650–50 cm^{-1}) or far-infrared region. In general, vibrations which occur in the high frequency region originate in the ligand itself, whereas those in the low frequency region originate in the metal-ligand coordinate bonds. High frequency vibrations are *ligand-sensitive* and have already been the subject of a large number of studies. On the other hand, low frequency vibrations are *metal-sensitive*, and very little work has been done on them. The reasons are that far-infrared spectrophotometers are still not very common, and the spectra in the low frequency region are more difficult to interpret on an empirical basis than are those in the high frequency region. As will be seen later, studies of the low frequency spectra provide direct information about the coordinate bond. It is probable that, with the progress in both experimental techniques and theoretical analyses, vibrational studies of metal chelate compounds will in the future be concentrated more on the low than on the high frequency region.

In this chapter typical applications of infrared spectra to the structural determination of metal chelate compounds are described by means of examples. Some of these applications can be understood without much theoretical background. Other applications, however, can be understood only through a knowledge of symmetry, group theory, selection rules, etc. These subjects are not discussed here, as they have already been described in many other books which are listed in the references at the end of the chapter (e.g., [1, 3, 4, 5, 7, 10, 14]).

4-1 THEORY

A. Empirical Interpretation of Spectra

Before we treat any applications, it is appropriate to discuss the method of assigning the bands in observed spectra. Assignments can be made either empirically or theoretically. The empirical method is simple and straightforward, but its application is limited to a few well-known vibrations, and it often fails to give unambiguous band assignments even in the high frequency region. On the other hand, the theoretical method involves a tedious procedure of calculation, but leads to a quantitative description of each vibration in terms of the bond-stretching and angle-bending coordinates chosen for the calculation.

The principle of the empirical method rests on the idea of "group vibrations," that is, the notion that the vibrations of a particular group in a molecule are relatively independent of those of the rest of the molecule. This idea is approximately correct if a group contains atoms which are relatively light (e.g., hydrogen) or relatively heavy (e.g., a halogen) compared with the other atoms in the molecule. Such groups absorb in narrow frequency ranges regardless of the nature of the rest of the molecule. For example, the methyl group bonded to carbon (C—CH_3) is known to absorb at approximately 2980–2920, 2880–2860, 1480–1430, 1420–1300, and 1200–700 cm^{-1}, and the modes of vibration are known to be those shown in Fig. 4-1.†

Such group frequencies have already been found for a number of organic and inorganic groups, and have been summarized in "group frequency

† Throughout this chapter the following abbreviations and symbols are used to represent the vibrational modes. As shown in Fig. 4-1, the methyl group bonded to the C atom exhibits six group vibrations: ν_a (antisymmetric stretching), ν_s (symmetric stretching), δ_d (degenerate deformation), δ_s (symmetric deformation), ρ_r (rocking) and ν (stretching). In general they are applicable to the pyramidal XY_3 type molecule coordinated with an atom (e.g., NH_3 complexes). The CH_2 group exhibits seven group vibrations:

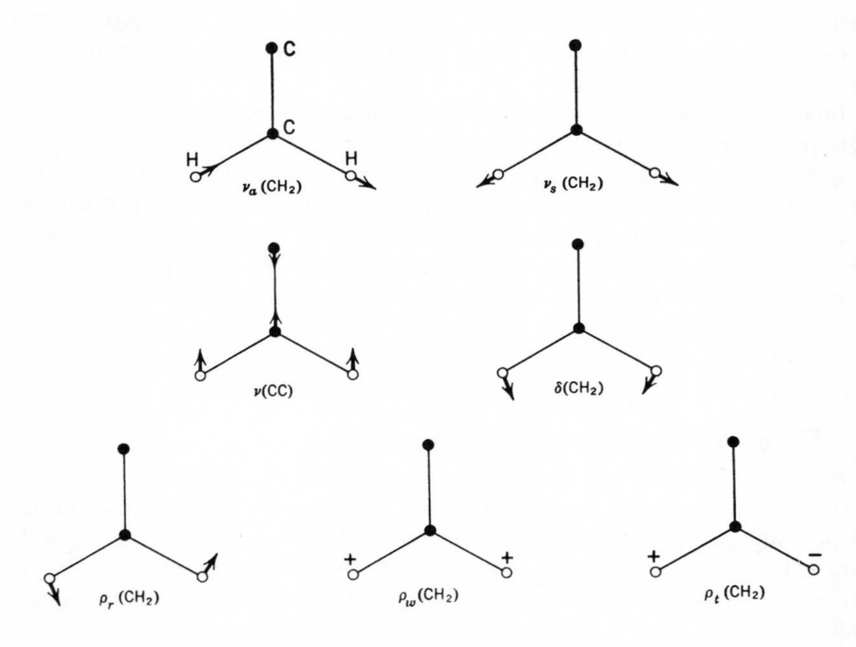

Here + and − indicate the vibrations upward and downward from the plane of the paper, respectively. The four bending modes are designated δ (bending), ρ_r (rocking), ρ_w (wagging), and ρ_t (twisting). In general these designations are applicable to the bent XY_2 molecule coordinated with an atom (e.g., N-bonded NO_2 complexes). In addition, it is convenient to use π (out-of-plane bending) and ring def. (ring deformation) for complex molecules.

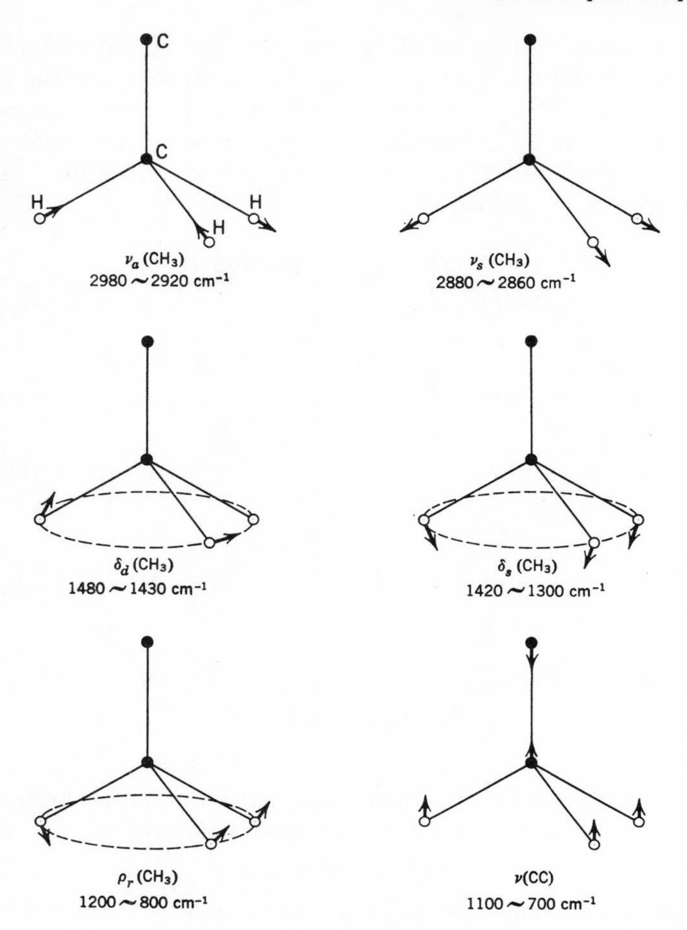

Fig. 4-1. Methyl group vibrations and their band assignments.

charts" prepared by several investigators.† By using these charts it is often possible to identify the bands belonging to a specific group. The band assignments such as those given in Fig. 4-1 are, however, not always listed in these charts. In this case one must resort to the results of normal coordinate analysis carried out on compounds containing the same group as that in question, and assume the same assignments for the bands in similar frequency ranges.

† The most up-to-date group frequency charts are found in books by Nakamoto [14], Colthup et al. [10], and Szymanski [17].

As an example [22] let us discuss the infrared spectrum of potassium bis(oxamido) Ni(II) which is known from microanalysis to have the empirical formula $K_2Ni(C_2H_2N_2O_2)_2$. Several probable structures for this complex ion are shown in I to IV. Although both the tetrahedral and square-planar structures are probable, all the tetrahedral structures can be ruled out, since the Ni(II) complex is diamagnetic.

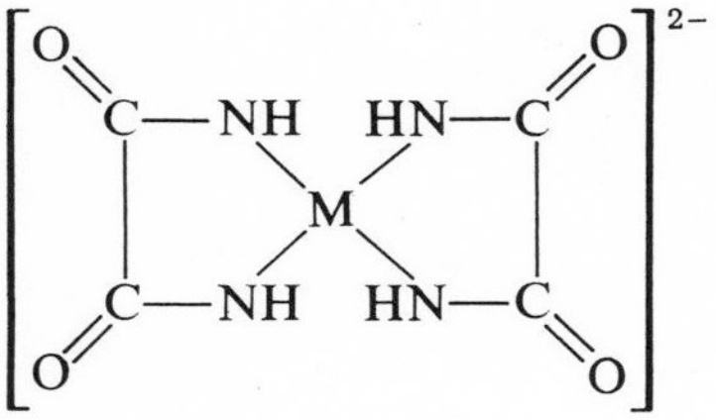

I

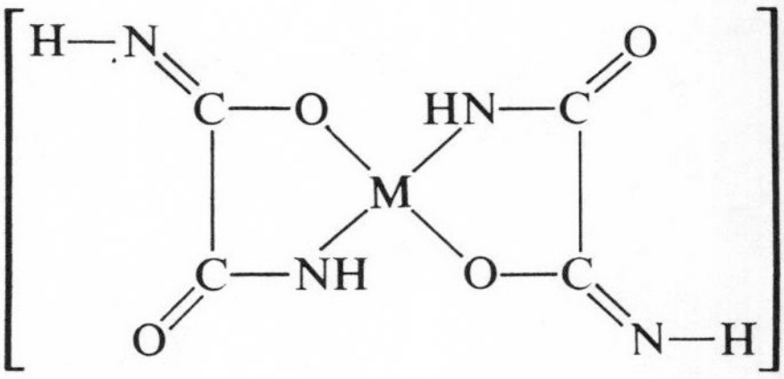

II

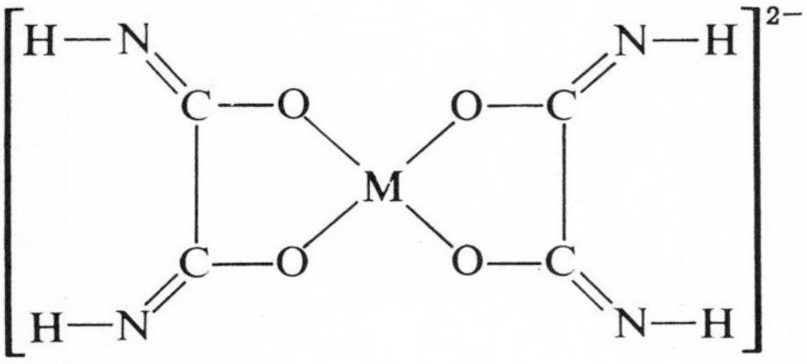

III

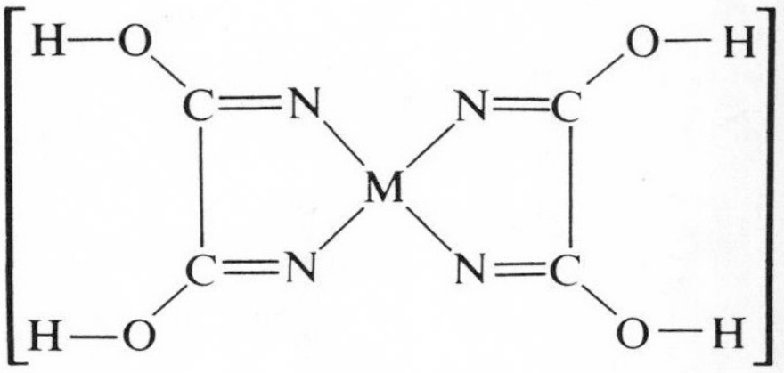

IV

Infrared spectra are very useful in selecting the most probable structure of the four. Figure 4-2 illustrates the infrared spectra of oxamide

$$(NH_2COCONH_2),$$

the potassium bis(oxamido)-Ni(II) and -Cu(II) complexes, and their deutero analogs [22]. The infrared spectrum of oxamide is markedly different from those of the Ni(II) and Cu(II) complexes; the two strong bands of oxamide at 3370 and 3200 cm^{-1} are replaced on chelation by a sharp single band at about 3310 cm^{-1}. It is obvious from group frequency charts and from the spectra of the deuterated species that these bands are due to the NH_2 groups of oxamide and the NH groups of the metal complexes, respectively. This change on coordination to a metal is similar to that observed for the N—H stretching bands in going from primary to secondary amines, and it is interpreted as an indication that one of the hydrogens of each NH_2 group is ionized on coordination. Observation of a single band, or at most a closely spaced doublet, for the N—H stretching mode of the complexes precludes structure II, since substantially different N—H stretching frequencies are expected for such a structure.

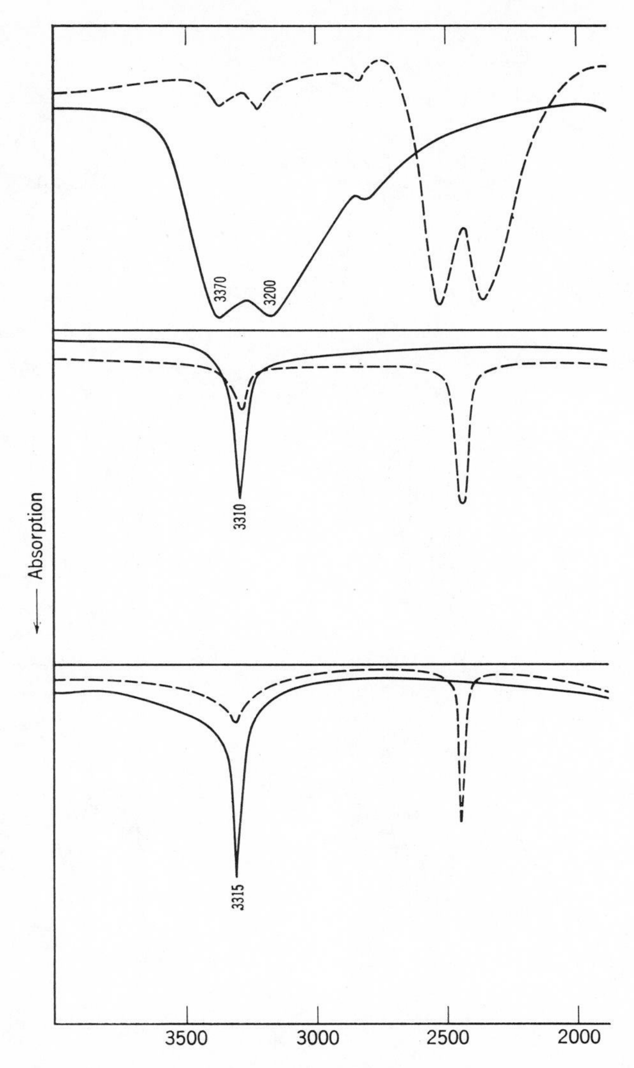

Fig. 4-2. Infrared spectra of oxamide and the bis(oxamido) complexes of Cu(II) and Ni(II) [22]. Broken lines indicate the spectra of deuterated compounds. Frequencies are given for the non-deuterated compounds.

It is also clear from group frequency charts that the strong, broad bands near 1650–1600 cm^{-1} of the free ligand and the complexes are due to the C=O stretching modes; they are too strong for the C=N stretching bands of structures III and IV. Furthermore the stretching mode of the C—O

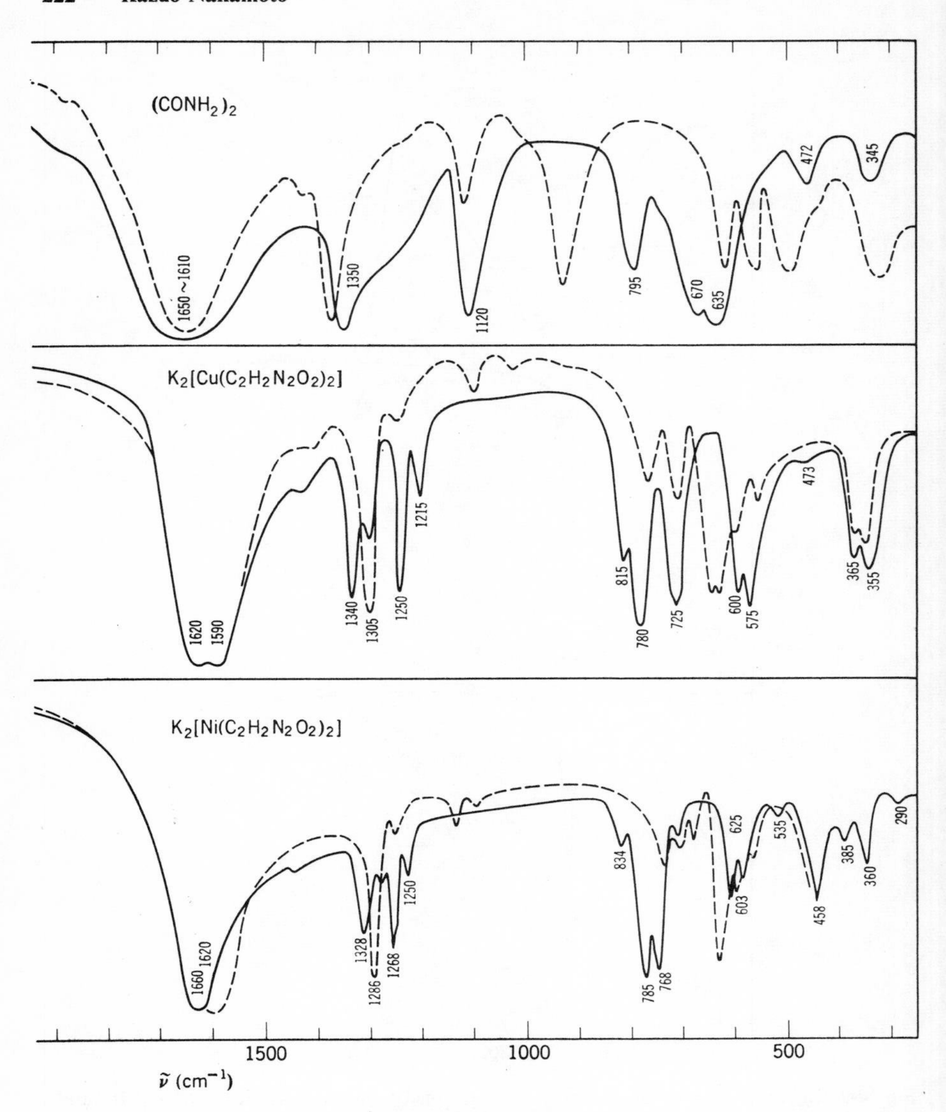

Fig. 4–2. (*Continued*).

single bond which appears in structures III and IV is known to absorb below 1450 cm^{-1}. These results rule out structures III and IV. It is therefore reasonable to conclude that the diimido structure I is the most probable one. Assuming that structure I is correct, we next try to assign the bands below 1600 cm^{-1}.

Figure 4-2 shows that both the Ni(II) and the Cu(II) complexes exhibit four bands between 1350 and 1200 cm^{-1}. The most probable vibrations to be expected in this region are the C—N stretching and N—H in-plane bending modes. On deuteration of the NH hydrogens, two bands at 1286 and 1250 cm^{-1} (a weaker pair) disappear and two other bands at 1328 and 1268 cm^{-1} (a stronger pair) are shifted slightly to lower frequencies. Therefore, the former two bands may be assigned to the N—H in-plane bending modes, and the latter two to the C—N stretching modes.

The bands below 850 cm^{-1} are extremely difficult to assign empirically, because the group frequencies for this region have not yet been established. In order to assign the low frequency region, it is necessary to carry out a normal coordinate analysis of structure I. This analysis not only confirms the empirical band assignments mentioned above, but also provides the force constants which can be used as a measure of the interatomic forces. In principle, the method of normal coordinate analysis is applicable to any compound of known structure. In practice, however, the procedure of calculation becomes more laborious as the molecule becomes larger and more complex, and it is almost impossible to apply it to metal chelate compounds without the aid of electronic computers.

B. Normal Coordinate Analysis

Infrared absorptions and Raman scattering are both due to normal vibrations of a molecule. Before discussing normal coordinate analysis, we must first explain the physical meaning of the normal vibrations and the normal coordinates which form the basis of normal coordinate analysis.

1. Normal Vibrations and Normal Coordinates. The kinetic energy of an N-atom molecule in terms of cartesian coordinates is written

$$2T = \sum_N m_N \left[\left(\frac{d\Delta x_N}{dt} \right)^2 + \left(\frac{d\Delta y_N}{dt} \right)^2 + \left(\frac{d\Delta z_N}{dt} \right)^2 \right], \tag{4-1}$$

where m_N is the mass and Δx_N, Δy_N, and Δz_N are the displacements from the equilibrium position of the Nth atom along the x, y, and z axes, respectively. If the mass-weighted cartesian coordinates such as

$$q_1 = \sqrt{m_1}\, \Delta x_1, \quad q_2 = \sqrt{m_1}\, \Delta y_1, \quad q_3 = \sqrt{m_1}\, \Delta z_1,$$
$$q_4 = \sqrt{m_2}\, \Delta x_2, \quad \ldots \tag{4-2}$$

are used, (4-1) takes a very simple form:

$$2T = \sum_i^{3N} \dot{q}_i^2, \tag{4-3}$$

where $\dot{q}_i$ denotes the first time derivative of q_i. The potential energy is a complex function of all the coordinates and can be expanded as a Taylor series:

$$V = V_0 + \sum_i^{3N} \left(\frac{\partial V}{\partial q_i}\right)_0 q_i + \frac{1}{2}\sum_{ij}^{3N} \left(\frac{\partial^2 V}{\partial q_i\, \partial q_j}\right)_0 q_i q_j + \cdots, \tag{4-4}$$

where the derivatives are evaluated at $q_i = 0$, the equilibrium position. Higher terms in this expansion can be ignored if the displacements are small. The constant term V_0 can be taken as zero if the potential energy at the equilibrium position is taken as a standard. The $(\partial V/\partial q_i)_0$ terms also become zero, since V must be a minimum at $q_i = 0$. Thus V may be written simply

$$2V = \sum_{ij}^{3N} \left(\frac{\partial^2 V}{\partial q_i\, \partial q_j}\right)_0 q_i q_j, \tag{4-5}$$

or

$$2V = \sum_{ij}^{3N} b_{ij} q_i q_j, \tag{4-6}$$

where

$$b_{ij} = b_{ji} = \left(\frac{\partial^2 V}{\partial q_i\, \partial q_j}\right)_0. \tag{4-7}$$

If (4-3) and (4-6) are inserted in the equation of motion,

$$\frac{d}{dt}\left(\frac{\partial T}{\partial \dot{q}_i}\right) + \frac{\partial V}{\partial q_i} = 0, \qquad i = 1, 2, \ldots, 3N, \tag{4-8}$$

there results

$$\ddot{q}_i + \sum_j^{3N} b_{ij} q_j = 0, \qquad i = 1, 2, \ldots, 3N. \tag{4-9}$$

The solutions of these $3N$ homogeneous second-order differential equations are known to take the form

$$q_i = A_i \cos(\sqrt{\lambda}\, t + \delta). \tag{4-10}$$

We now substitute these solutions into (4-9) and obtain the $3N$ equations

$$\sum_j^{3N} b_{ij} A_j - \lambda A_i = 0, \qquad i = 1, 2, \ldots, 3N. \tag{4-11}$$

This is a system of first order simultaneous equations with respect to A_i. In order that they have non-zero A's, the determinant must equal zero.

$$\begin{vmatrix} b_{11} - \lambda & b_{12} & b_{13} & \cdots \\ b_{21} & b_{22} - \lambda & \cdots & \\ \vdots & \vdots & & \cdots \end{vmatrix} = 0. \tag{4-12}$$

Equation (4-12) has $3N$ roots. Any one of these, λ_k, for example, can be substituted into the $3N$ equations (4-11) to give

$$\sum_{j}^{3N} b_{ij}A_{jk} - \lambda_k A_{ik} = 0, \qquad i = 1, 2, \ldots, 3N. \tag{4-13}$$

More explicitly,

$$\begin{aligned} b_{11}A_{1k} + b_{12}A_{2k} + \cdots - \lambda_k A_{1k} &= 0 \\ \vdots \qquad\qquad\qquad\qquad & \vdots \\ b_{3N1}A_{1k} + b_{3N2}A_{2k} \qquad - \lambda_k A_{3Nk} &= 0. \end{aligned} \tag{4-14}$$

These equations can only be solved to give the ratios $A_{1k}:A_{2k}:\cdots$ and not their absolute values. We therefore construct a set of $3N$ coefficients l'_{ik} which are in the same ratio as the A_{ik} and related to them by

$$A_{ik} = K_k l'_{ik}, \qquad i = 1, 2, \ldots, 3N, \tag{4-15}$$

where K_k is a proportionality constant. l'_{ik} can be chosen so that

$$\sum_i l'_{ik} l'_{il} = \delta_{kl}, \tag{4-16}$$

where δ_{kl} is Kronecker's delta (0 for $k \neq l$ and 1 for $k = l$).

The solution of (4-9) for λ_k is now written

$$\begin{aligned} q_{ik} &= A_{ik} \cos(\sqrt{\lambda_k}\, t + \delta_k) \\ &= K_k l'_{ik} \cos(\sqrt{\lambda_k}\, t + \delta_k), \end{aligned} \tag{4-17}$$

where K_k and δ_k are determined by the initial condition of the vibration. This equation indicates that all the N atoms in the molecule vibrate with the same frequency and in phase, and also that the ratios of the vibrational amplitudes between each pair of atoms are constant. This vibration is called a *normal vibration.* It is clear that there are $3N$ λ's similar to λ_k in a N-atom molecule. However, our treatment includes three translational and three (or two for linear molecules) rotational motions of the molecule as a whole, which gives six (or five) zero frequencies. Therefore, an N-atom molecule has $3N - 6$ (or $3N - 5$ if linear) normal vibrations which are represented by (4-17). The most general solution of (4-8) may be written

$$q_i = \sum_k l'_{ik} K_k \cos(\sqrt{\lambda_k}\, t + \delta_k). \tag{4-18}$$

In other words, a complicated vibration performed by a molecule can be resolved into k normal vibrations, as this equation shows.

Equation (4-18) can be written

$$q_i = \sum_k l'_{ik} Q_k, \tag{4-19}$$

where

$$Q_k = K_k \cos(\sqrt{\lambda_k}\, t + \delta_k). \tag{4-20}$$

Q_k is called a *normal coordinate*, and (4-19) gives the relationship between normal coordinates and mass-weighted cartesian coordinates. Using normal coordinates, the kinetic and potential energies can be written simply

$$2T = \sum_k \dot{Q}_k^2, \tag{4-21}$$

$$2V = \sum_k \lambda_k Q_k^2. \tag{4-22}$$

Equation (4-21) can be derived by combining (4-3) with (4-19) and considering the relation given by (4-16). Similarly (4-22) can be obtained by combining (4-6) with (4-19) and considering the relations given by (4-13) and (4-16). Neither the kinetic nor the potential energy in terms of normal coordinates includes any cross terms, such as $Q_k Q_l$. In other words, all the normal vibrations are independent of each other. The purpose of normal coordinate analysis is to calculate the frequencies of these normal vibrations (or to calculate a set of force constants which gives the best fit to the observed frequencies of these normal vibrations), and to describe their modes of vibration in terms of internal coordinates such as increments of the bond length and bond angle. The relationships between the internal and the normal coordinates are shown later [in (4-30)].

2. The GF Matrix Method. As we have just stated, the purpose of normal coordinate analysis is to calculate the frequencies of normal vibrations and to describe the modes of vibrations in terms of proper coordinates. Instead of using $3N$ cartesian coordinates, we can write the kinetic and potential energies in terms of internal coordinates. The advantages of using internal coordinates are twofold: (1) the translational and rotational motions of the molecule as a whole are eliminated initially from the calculation; and (2) a normal mode of vibration can be interpreted in terms of the stretching vibration of a particular bond or the bending vibration of a particular angle in a molecule. In 1939 Wilson [137] developed the GF matrix method, which has been used almost exclusively for the normal coordinate analysis of a large number of molecules. The detailed description of his method is given in several other publications [5, 7, 14], and examples of the application of it to H_2O [14], CH_3Cl [10], benzene [7], and acetylacetonate complexes [14] are also available. The purpose of this section is to give a condensed outline of the GF matrix method.

As a first step we choose a set of $3N - 6$ (or $3N - 5$) internal coordinates to describe both the potential and the kinetic energy of a molecule. If more than $3N - 6$ (or $3N - 5$) coordinates are chosen, these extra coordinates give zero frequencies in the final results and are therefore called *redundant coordinates*. Using these internal coordinates, the potential energy of a molecule can be written in matrix form:†

$$2V = \tilde{\boldsymbol{R}}\boldsymbol{F}\boldsymbol{R}, \tag{4-23}$$

where $\boldsymbol{R}$ is a column matrix whose elements are internal coordinates, and $\tilde{\boldsymbol{R}}$ is the transpose of $\boldsymbol{R}$. $\boldsymbol{F}$ is a matrix whose components are the force constants. The $\boldsymbol{F}$ matrix is normally written in terms of the generalized valence force (GVF) field, which consists of stretching and bending constants (diagonal elements), and constants which express interactions between stretching and stretching, between stretching and bending, and between bending and bending vibrations (off-diagonal elements) [7]. As a molecule becomes larger, the number of force constants in the GVF field becomes too large to allow any reliable evaluation. On the other hand, the Urey-Bradley force (UBF) field [121] consists of stretching (K) and bending (H) constants, as well as repulsive constants (F) between non-bonded atoms. Relative to the GVF field, it has the advantages that (1) the number of force constants is much smaller, (2) the force constants have clearer physical meaning, and (3) they are transferable between similar molecules. Thus the UBF field is used almost exclusively for the normal coordinate analysis of complex molecules such as metal chelate compounds. The GVF force constants are readily converted into the corresponding UBF constants through the general relationships given by Shimanouchi [5, 121]. The UBF field does not include any interactions between non-neighboring stretching, or between bending, vibrations. Ignorance of these interaction constants causes difficulty in fitting the calculated to the observed frequencies for some molecules [123]. For such molecules it is necessary to modify the force field by introducing new interaction constants.

Using the same set of internal coordinates, the kinetic energy of a molecule can be written [7]

$$2T = \tilde{\dot{\boldsymbol{R}}}\boldsymbol{G}^{-1}\dot{\boldsymbol{R}}, \tag{4-24}$$

where $\dot{\boldsymbol{R}}$ is a column matrix whose elements are the time derivatives of the internal coordinates used, and $\tilde{\dot{\boldsymbol{R}}}$ is the transpose of $\dot{\boldsymbol{R}}$. $\boldsymbol{G}^{-1}$ is the reciprocal of the $\boldsymbol{G}$ matrix which is defined by

$$\boldsymbol{G} = \boldsymbol{B}\boldsymbol{M}^{-1}\tilde{\boldsymbol{B}}. \tag{4-25}$$

Here $\boldsymbol{M}^{-1}$ is a diagonal matrix whose components are the reciprocals of the masses of the constituent atoms, and $\boldsymbol{B}$ is a rectangular matrix which relates the internal coordinates to the cartesian coordinates:

$$\boldsymbol{R} = \boldsymbol{B}\boldsymbol{X}. \tag{4-26}$$

† For matrix theory see for example [5].

$\boldsymbol{X}$ denotes a column matrix whose elements are $3N$ cartesian coordinates. The method of calculating the $\boldsymbol{G}$ matrix elements is not described here because it is given in other books [7, 14]. The tables developed by Decius [46] and Shimanouchi [122] are very useful for this purpose.

If the kinetic and potential energies are expressed in the matrix form given in (4-23) and (4-24) and are inserted in the equation of motion (4-8), there results a matrix secular equation of the form

$$\begin{vmatrix} \sum G_{1t}F_{t1} - \lambda & \sum G_{1t}F_{t2} & \cdots \\ \sum G_{2t}F_{t1} & \sum G_{2t}F_{t2} - \lambda & \cdots \\ \cdots & \cdots & \cdots \end{vmatrix} \equiv |\boldsymbol{GF} - \boldsymbol{E}\lambda| = 0, \qquad (4\text{-}27)$$

where $\boldsymbol{E}$ is the unit matrix, and each eigenvalue, λ, is related to a vibrational frequency (wave number, $\tilde{\nu}$) by the relation $\lambda = 4\pi^2C^2\tilde{\nu}^2$, where C is the velocity of light. The final equation (4-27) involves the $\boldsymbol{G}$ matrix rather than its reciprocal, which appeared in (4-24). It is necessary therefore only to evaluate the $\boldsymbol{G}$ matrix and not its reciprocal. The $\boldsymbol{G}$ matrix elements involve the masses of individual atoms, bond distances, and bond angles, and they can be evaluated readily if the structural data are available. On the other hand, the $\boldsymbol{F}$ matrix elements include various types of force constants which are not known for the molecule being studied. In general, then, we set up a secular equation (4-27) with a fixed $\boldsymbol{G}$ matrix and an $\boldsymbol{F}$ matrix which consists of an assumed set of force constants. The force constants (or the $\boldsymbol{F}$ matrix) are refined until good agreement between calculated and observed frequencies is obtained.

If good agreement is obtained, the next step is to make theoretical band assignments. As shown in (4-19), the mass-weighted cartesian coordinates are related to normal coordinates by the relation

$$q_i = \sum_k l'_{ik} Q_k.$$

In matrix form this is written

$$\boldsymbol{q} = \boldsymbol{L_q Q}. \qquad (4\text{-}28)$$

It can be shown that the internal coordinates are related to normal coordinates by an equation similar to (4-28):

$$\boldsymbol{R} = \boldsymbol{LQ}. \qquad (4\text{-}29)$$

Equation (4-29) is written more explicitly as

$$\begin{aligned} R_1 &= l_{11}Q_1 + l_{12}Q_2 + \cdots + l_{1N}Q_N \\ R_2 &= l_{21}Q_1 + l_{22}Q_2 + \cdots + l_{2N}Q_N \\ &\cdots\cdots\cdots\cdots\cdots\cdots \\ R_i &= l_{i1}Q_1 + l_{i2}Q_2 + \cdots + l_{iN}Q_N. \end{aligned} \qquad (4\text{-}30)$$

In a normal vibration in which the normal coordinate Q_N changes with frequency λ_N, all the internal coordinates, $R_1, \ldots, R_i$ change with the same frequency. However, the amplitude of oscillation is different for each internal coordinate, and the relative ratio of the amplitudes is given by

$$l_{1N}:l_{2N}:\cdots:l_{iN}. \tag{4-31}$$

If one of these elements is relatively large compared with the others, this normal vibration is said to be due predominantly to the vibration caused by the change in this internal coordinate. It can be shown [7] that the ratio of l's (4-31) can be obtained as a column matrix which satisfies the relation

$$\boldsymbol{GFl}_N = \lambda_N \boldsymbol{l}_N. \tag{4-32}$$

Here $\boldsymbol{l}_N$ consists of i elements, $l_{1N}, l_{2N}, \ldots, l_{iN}$, i being the number of internal coordinates.

Such a comparison of l values may not be suitable for band assignments, since the dimension of l for a stretching coordinate is different from that for a bending coordinate. It is therefore more appropriate to compare the potential energy distribution defined in the following manner [96]. As shown in (4-23), the potential energy of a molecule is written

$$V = (\tfrac{1}{2})\tilde{\boldsymbol{R}}\boldsymbol{F}\boldsymbol{R}.$$

Using the relation given by (4-29), this can be written

$$V = (\tfrac{1}{2})\tilde{\boldsymbol{Q}}\tilde{\boldsymbol{L}}\boldsymbol{F}\boldsymbol{L}\boldsymbol{Q}. \tag{4-33}$$

On the other hand, the potential energy is given by

$$V = (\tfrac{1}{2})\tilde{\boldsymbol{Q}}\boldsymbol{\Lambda}\boldsymbol{Q}. \tag{4-34}$$

This is the matrix representation of (4-22), and $\boldsymbol{\Lambda}$ is a diagonal matrix whose elements consist of the λ values. A comparison of (4-33) and (4-34) gives

$$\tilde{\boldsymbol{L}}\boldsymbol{F}\boldsymbol{L} = \boldsymbol{\Lambda}. \tag{4-35}$$

If this relation is written for one normal vibration whose frequency is λ_N, there results

$$\lambda_N = \sum_{ij} \tilde{l}_{Ni} F_{ij} l_{jN} = \sum_{ij} F_{ij} l_{iN} l_{jN}. \tag{4-36}$$

Thus the potential energy due to this particular vibration is given by

$$V(\lambda_N) = (\tfrac{1}{2}) Q_N^2 \sum_{ij} F_{ij} l_{iN} l_{jN}, \tag{4-37}$$

where $F_{ij} l_{iN} l_{jN}$ indicates the distribution of the potential energy in each internal coordinate. Since, in general, this term is large when $i = j$, a comparison of $F_{ii} l_{iN}^2$ terms may be useful for band assignments. If the $F_{ii} l_{iN}^2$ term is exceedingly large for R_i, this normal vibration can be assigned to the group vibration represented by R_i. If both $F_{ii}l_{iN}^2$ and $F_{jj}l_{jN}^2$ are relatively large compared with those for the other internal coordinates, this normal vibration is assigned to a coupled vibration between those represented by internal coordinates R_i and R_j (vibrational coupling).

A more accurate method of determining the vibrational mode may be to draw the actual displacements of individual atoms in terms of cartesian coordinates. For this purpose it is necessary to calculate the $\boldsymbol{L}_x$ matrix defined by

$$\boldsymbol{X} = \boldsymbol{L}_x\boldsymbol{Q}. \tag{4-38}$$

The $\boldsymbol{L}_x$ matrix is obtainable from the relation [44a]

$$\boldsymbol{L}_x = \boldsymbol{M}^{-1}\tilde{\boldsymbol{B}}\boldsymbol{G}^{-1}\boldsymbol{L}. \tag{4-39}$$

As noted in Sec. 4-1A, the idea of group vibrations is applicable if a group contains atoms which are relatively light (e.g., OH, NH, NH_2, CH, CH_2, CH_3) or relatively heavy (e.g., CCl, CBr, CI) compared with the other atoms in the molecule. Vibrations of groups having multiple bonds (e.g., C≡C, C≡N, C═O) may also be relatively independent of the rest of the molecule, if the group does not belong to a conjugated system. If atoms of similar mass are connected by bonds of similar strength, the amplitude of oscillation is similar for each atom of the system. For example, a pure C—C or C—N stretching vibration is unlikely to occur in the ethylenediamine molecule (V).

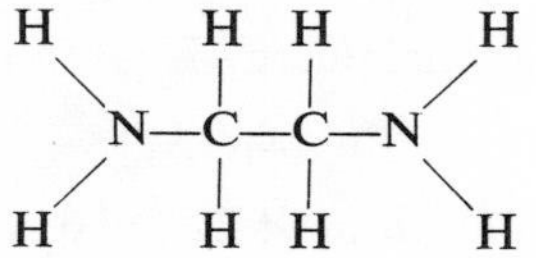

V

A similar situation may occur in a system in which resonance effects average out the single and double bonds by conjugation. This is observed in metal chelates of acetylacetone (VI) (see Sec. 4-2C). In these systems, band assignments can be made only by comparing the potential energy distribution (or similar quantity) in each internal coordinate.

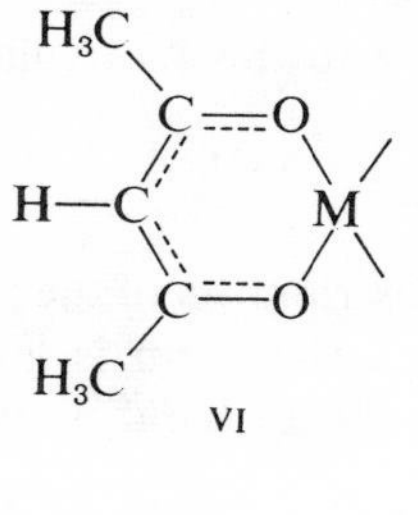

VI

3. An Example: The Bis(oxamido)Ni(II) Complex In Sec. 4-1A we discussed the infrared spectrum of the bis(oxamido)Ni(II) complex on an empirical basis. The purpose of this section is to show how the GF matrix method described in the preceding section is applied to the same compound [22]. As stated previously, the diimido structure I (p. 222) is most probable for the bis(oxamido)Ni(II) ion. Although this ion is relatively simple, it still possesses 45 ($3 \times 17 - 6$) normal vibrations. It is desirable therefore to simplify the calculation as much as possible by making reasonable assumptions.

Selection of Molecular Models. The ratio of metal to bidentate ligands in metal chelate compounds may be 1:1, 1:2 (square-planar and tetrahedral), 1:3 (octahedral and trigonal prismatic), 1:4 (cubic and Archimedean-antiprism), and so on. According to the theory of normal vibrations [7], vibrational couplings between different chelate rings in a complex are not appreciable except for the stretching and bending vibrations involving the coordinate bonds around the central metal. If we approximate the 1:2, 1:3, and 1:4 complexes by the 1:1 model, the results may not be accurate for the coordinate bond stretching and bending vibrations in the low frequency region. However, the 1:1 model provides satisfactory results in the high frequency region where the ligand absorptions appear. In fact, Fig. 4-3 shows that the infrared spectra of $K[Pt(acac)Cl_2]$ and $[Pt(acac)_2]$ are similar in the high frequency region [29]. In some cases the spectra of 1:1 and 1:2 complexes can be markedly different because of the effect of intermolecular interactions in the crystalline state.

It can be shown, in general, that vibrational coupling between different chelate rings decreases as the coordinate bond becomes weaker (or its force constant becomes smaller). Chemical evidence shows that the Ni—N bond in the bis(oxamido)Ni(II) complex is fairly weak. Thus the 1:1 model of this complex shown in Fig. 4-4 is expected to give a fairly accurate result even in the low frequency region.

Coordinate bond stretching and bending (skeletal) vibrations of the whole molecule can be calculated very easily if we consider only the metal and the atoms bonded directly to it. Such an approximation would provide satisfactory results only when the vibrational coupling between the coordinate bond and the bond next to it is small.

*Construction of the **G** and **F** Matrices.* The 1:1 complex model shown in Fig. 4-4 has 21 ($3 \times 9 - 6$) normal vibrations. Since the symmetry of this model is $\mathbf{C}_{2v}$, they are classified into four species: $8A_1 + 3A_2 + 3B_1 + 7B_2$. The 15 in-plane vibrations ($8A_1 + 7B_2$) are separable from the 6 out-of-plane vibrations ($3A_2 + 3B_1$), since they are independent of each other. The procedure of calculating only the 15 in-plane vibrations is described here. At first we choose the 9 stretching and 13 angle bending coordinates shown in Fig. 4-4. This set of internal coordinates includes 7 redundant coordinates,

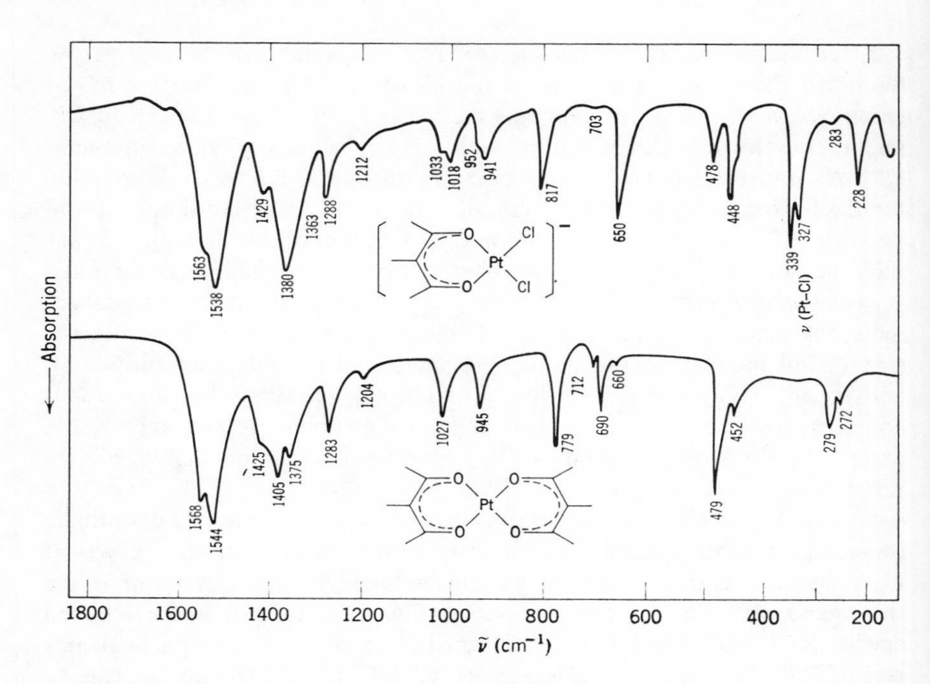

Fig. 4-3. Infrared spectra of 1:1 and 1:2 Pt(II) acetylacetonate complexes [29].

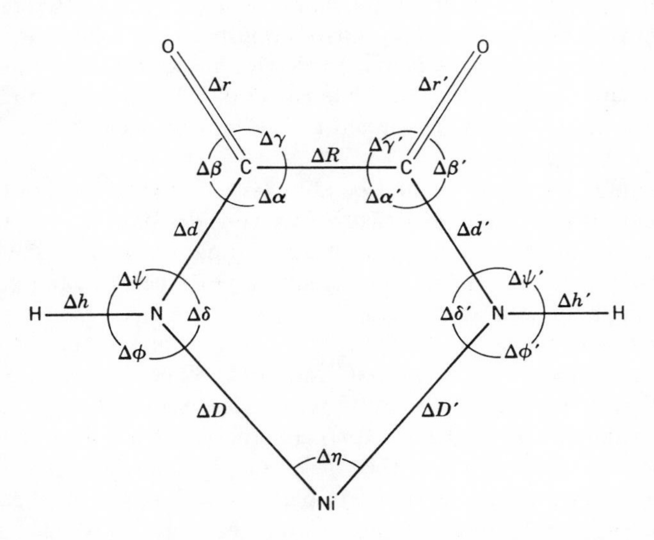

Fig. 4-4. 1:1 complex model and internal coordinates of bis(oxamido)Ni(II) ion.

as the number of normal vibrations is only 15. As a second step we construct the G and F matrices using these 22 internal coordinates. The explicit forms of these matrices are not given here because they are similar to those already given for metal acetylacetonate complexes [14]. Although the order of these matrices is twenty-second, it can be reduced through coordinate transformations. In order to eliminate simple angle redundancies such as

$$\Delta\alpha + \Delta\beta + \Delta\gamma \equiv 0,$$
$$\Delta\alpha' + \Delta\beta' + \Delta\gamma' \equiv 0,$$
$$\Delta\psi + \Delta\delta + \Delta\phi \equiv 0,$$
$$\Delta\psi' + \Delta\delta' + \Delta\phi' \equiv 0,$$

we transform the G and F matrices by the relations

$$G_{\mathrm{I}} = TG\tilde{T} \tag{4-40}$$
$$F_{\mathrm{I}} = \tilde{T}^{-1}FT^{-1},$$

where T is a transformation matrix and T^{-1} is its reciprocal matrix. The explicit form of the T matrix for the oxamidoNi(II) complex is not shown here, as it is similar to that used for a metal acetylacetonate [14]. Through this transformation the four rows and four columns of the G matrix corresponding to the four redundant conditions listed above vanish. Although the corresponding rows and columns of the F matrix may not be completely zero, any non-zero terms can be omitted from the calculation at this stage because they will vanish when G_{I} and F_{I} are multiplied later. This transformation thus reduces the order of the matrices by four. It should be noted that the λ values of (4-27) are not altered by such coordinate transformations, since the product $G_{\mathrm{I}}F_{\mathrm{I}}$ gives the same roots as $TGFT^{-1}$ (similarity transformation).

The next step is to factor the matrices into the A_1 (symmetric) and B_2 (antisymmetric) species and to eliminate any remaining redundancies whenever possible. This process can be accomplished by the second coordinate transformation:

$$G_{\mathrm{II}} = UG_{\mathrm{I}}\tilde{U} \tag{4-41}$$
$$F_{\mathrm{II}} = UF_{\mathrm{I}}\tilde{U}.$$

The U matrix must be so constructed as to meet $\mathbf{C}_{2v}$ symmetry requirements as well as to eliminate the redundant condition:

$$\Delta\alpha + \Delta\alpha' + \Delta\delta + \Delta\delta' + \Delta\eta \equiv 0.$$

The explicit form of the U matrix for the oxamidoNi(II) complex is similar to that given for a metal acetylacetonate [14]. The U matrix is different from the T matrix in that the former can easily be constructed to meet the orthogonality condition. Through this second transformation the row and column corresponding to the redundancy vanish, and both the G_{II} and F_{II} matrices are factored into one ninth order (A_1) and one eighth order (B_2) matrix.

Since the number of normal vibrations is 8 for the A_1 and 7 for the B_2 species, one redundancy is still involved in each species. These two redundancies cannot be removed easily, as they are complicated functions involving bond distances and angles. Hence the ninth order A_1 and the eighth order B_2 secular equations are solved with them included. This does not affect the final result, because the redundancies simply give two "zero frequencies." Their inclusion provides a good check of the $\boldsymbol{G}$ elements, since any mistakes in the original $\boldsymbol{G}$ matrix cause deviation from the exact zero frequencies.

Both the $\boldsymbol{G}_{\mathrm{II}}$ and $\boldsymbol{F}_{\mathrm{II}}$ matrices thus obtained are expressed in terms of the symmetry coordinates listed in Table 4-1, rather than in terms of individual internal coordinates used originally. The calculated frequencies and the band assignments to be discussed later must be interpreted in terms of these symmetry coordinates.

The last step of the calculation is to evaluate the $\boldsymbol{G}_{\mathrm{II}}$ and $\boldsymbol{F}_{\mathrm{II}}$ matrix elements thus obtained. The $\boldsymbol{G}$ elements consist of bond distances, bond angles, and the masses of constituent atoms. They can be evaluated immediately if such information is available from x-ray and other diffraction studies. Small errors in these values do not affect the calculated frequencies seriously, since the

Table 4-1 Symmetry Coordinates for the In-Plane Vibrations of the Oxamido Ni(II) Complex [22]

	Symmetry Coordinate[a]	Vibrational Mode
A_1 species	$S_1 = (1/\sqrt{2})(\Delta h + \Delta h')$	ν_s(N—H)
	$S_2 = (1/\sqrt{2})(\Delta r + \Delta r')$	ν_s(C=O)
	$S_3{}^* = (1/\sqrt{2})(\Delta\varphi - \Delta\psi + \Delta\varphi' - \Delta\psi')$	δ_s(N—H)
	$S_4 = (1/\sqrt{2})(\Delta d + \Delta d')$	ν_s(C—N)
	$S_5 = \Delta R$	ν(C—C)
	$S_6{}^* = (1/\sqrt{2})(\Delta\beta - \Delta\gamma + \Delta\beta' - \Delta\gamma')$	δ_s(C=O)
	$S_7 = (1/\sqrt{2})(\Delta D + \Delta D')$	ν_s(Ni—N)
	$S_8 = (\frac{1}{2})(\Delta\delta + \Delta\delta' - \Delta\alpha - \Delta\alpha')$	ring def.
	$S_9 = (1/\sqrt{20})(4\Delta\eta - \Delta\delta - \Delta\delta' - \Delta\alpha - \Delta\alpha')$	ring def.
	$S \equiv (1/\sqrt{5})(\Delta\delta + \Delta\delta' + \Delta\alpha + \Delta\alpha' + \Delta\eta)$	redundant
B_2 species	$S_{10} = (1/\sqrt{2})(\Delta h - \Delta h')$	ν_{as}(N—H)
	$S_{11} = (1/\sqrt{2})(\Delta r - \Delta r')$	ν_{as}(C=O)
	$S_{12}{}^* = (1/\sqrt{2})(\Delta\varphi - \Delta\psi - \Delta\varphi' + \Delta\psi')$	δ_{as}(N—H)
	$S_{13} = (1/\sqrt{2})(\Delta d - \Delta d')$	ν_{as}(C—N)
	$S_{14}{}^* = (1/\sqrt{2})(\Delta\beta - \Delta\gamma - \Delta\beta' + \Delta\gamma')$	δ_{as}(C=O)
	$S_{15} = (1/\sqrt{2})(\Delta D - \Delta D')$	ν_{as}(Ni—N)
	$S_{16} = (\frac{1}{2})(\Delta\delta - \Delta\delta' + \Delta\alpha - \Delta\alpha')$	ring def.
	$S_{17} = (\frac{1}{2})(\Delta\delta - \Delta\delta' - \Delta\alpha + \Delta\alpha')$	ring def.

[a] Symmetry coordinates with asterisks are not normalized. This is due to the choice of a non-orthogonal $\boldsymbol{T}$ matrix.

frequencies are much more sensitive to the force constants than to the geometrical parameters. Because no structural data are available for the bis(oxamido)Ni(II) ion, most of the bond distances and angles were estimated from the x-ray analysis data on metal oxalato complexes [133]. Other distances such as the N—H and Ni—N bond distances were estimated from data on compounds containing these bonds.

The ***F*** elements consist of various types of force constants. As stated before, the UBF field is more convenient than the GVF field for calculations on a molecule such as the oxamidoNi(II) complex. Using the UBF field, it is possible to transfer the UBF force constants obtained for glycino [42] and oxalato [60] complexes to the oxamido complex, as all these molecules have somewhat similar structures. The remaining force constants must, however, be estimated, because no values have been reported. With this initial set of force constants the first set of frequencies is calculated. Then the force constants are refined until the differences between observed and calculated frequencies are minimized. In carrying out this refinement it should be borne in mind that all the values of the force constants used should be reasonable in view of chemical structure and bonding. A number of force constants have already been obtained for the UBF field [5]. Drastic deviation from these values cannot be justified unless the structure or bonding changes drastically. It is also desirable to check the validity of the force constants by calculating the vibrational frequencies of all the possible isotopic species (D, N^{15}, O^{18} species, etc.). This is particularly important when the number of force constants used exceeds the number of observed fundamentals. Table 4-2 lists the best set of force constants thus obtained for the bis(oxamido)Ni(II) ion.

Table 4-3 compares the calculated frequencies with those observed for the $[Ni(C_2H_2N_2O_2)_2]^{2-}$ and $[Ni(C_2D_2N_2O_2)_2]^{2-}$ ions. The agreement is quite satisfactory; the average error for the observed frequencies is 1.5% and the maximum error is 4.1% (ν_{15} of the D compound). Therefore the set of force constants listed in Table 4-2 can be regarded as a good representation of the interatomic forces in the bis(oxamido)Ni(II) complex.

Table 4-4 gives the potential energy distribution for the B_2 species of the non-deuterated oxamido complex of Ni(II) whose frequencies are listed in Table 4-3. All the vibrations except ν_{11} are "pure vibrations" which can be represented by changes in single internal (symmetry) coordinates. However, the contributions of the ν(C—N) and δ(N—H) coordinates to the potential energy distribution are almost equivalent in ν_{11} (1272 cm^{-1}). Therefore this mode must be assigned to a coupled vibration between these two coordinates. The last column of Table 4-3 gives the theoretical band assignments thus obtained. The results not only confirm the previous empirical band assignments in the high frequency region, but also provide a method of assigning low frequency bands which are difficult to interpret on an empirical basis.

Table 4-2 The UBF Force Constants of the Bis(oxamido)Ni(II) Complex (mdyn/A) [22]

Stretching	Bending	Repulsive
K(Ni—N) = 0.73	H(N—Ni—N) = 0.02	F(N···N) = 0.02
	H(C—C—N) = 0.30	F(C···N) = 0.05
K(C═0) = 8.80		
	H(O═C—C) = 0.55	F(O···C) = 0.55
K(N—H) = 5.83	H(O═C—N) = 0.50	F(O···N) = 1.50
	H(Ni—N—C) = 0.05	F(Ni···C) = 0.02
K(C—N) = 4.50		
	H(H—N—C) = 0.10	F(H···C) = 0.53
K(C—C) = 2.50	H(H—N—Ni) = 0.23	F(H···Ni) = 0.08

Table 4-3 Comparison of Calculated and Observed Frequencies for $K_2[Ni(C_2H_2N_2O_2)_2]$ and $K_2[Ni(C_2D_2N_2O_2)_2]$ (cm^{-1}) [22]

		$K_2[Ni(C_2H_2N_2O_2)_2]$		$K_2[Ni(C_2D_2N_2O_2)_2]$		Band Assignment[a]
		Obs.	Calc.	Obs.	Calc.	
A_1	ν_1	3315	3352	2470	2446	ν(N—H)
	ν_2	1620	1594	1620	1593	ν(C═O)
	ν_3	1328	1342	1305	1337	ν(C—N)
	ν_4	1286	1260	1075	1035	δ(N—H)
	ν_5	834	842	755	740	ν(C—C)
	ν_6	458	460	458	449	δ(C═O) + ν(Ni—N)
	ν_7	360	355	358	352	ν(Ni—N) + δ(C═O)
	ν_8	—	199	—	197	ring def.
B_2	ν_9	3315	3352	2470	2446	ν(N—H)
	ν_{10}	1660	1695	1660	1694	ν(C═O)
	ν_{11}	1268	1272	1250	1264	δ(N—H) + ν(C—N)
	ν_{12}	1250	1251	1040	1000	δ(N—H)
	ν_{13}	768	765	700	692	δ(C═O)
	ν_{14}	535	555	530	546	ring def.
	ν_{15}	290	281	290	278	ν(Ni—N)

[a] Band assignment is given for the H compound.

4-2 APPLICATIONS

A. Conformational Isomerism

Chelating ligands such as ethylenediamine and 1,2-dithiocyanatoethane may take either the *cis*, *trans*, or *gauche* conformation because of internal rotation about the C—C bond (structures VII to IX).

Table 4-4 Potential Energy Distribution for the B_2 Vibrations of Bis(oxamido)Ni(II) [22]

Symmetry Coordinate	Band[a]						
	ν_9	ν_{10}	ν_{11}	ν_{12}	ν_{13}	ν_{14}	ν_{15}
S_{10}	*1.00*	0.00	0.00	0.00	0.00	0.00	0.00
S_{11}	0.00	*1.00*	0.18	0.01	0.01	0.02	0.01
S_{12}	0.00	0.00	*1.00*	*1.00*	0.04	0.00	0.01
S_{13}	0.00	0.14	*0.73*	0.36	0.04	0.04	0.00
S_{14}	0.00	0.01	0.02	0.10	*1.00*	0.31	0.16
S_{15}	0.00	0.00	0.00	0.00	0.06	0.35	*1.00*
S_{16}	0.00	0.01	0.00	0.00	0.15	*1.00*	0.36
S_{17}	0.00	0.00	0.00	0.00	0.00	0.00	0.00
Band assignment[b]	ν(N—H)	ν(C=O)	δ(N—H) + ν(C—N)	δ(N—H)	δ(C=O)	ring def.	ν(Ni—N)

[a] These numbers indicate relative ratios of $F_{ii}l_{iN}^2$ values.
[b] Only the term which contributes more than 0.5 was used to represent vibrational coupling.

The *cis* form may not be stable in the free ligand because of steric repulsion between two X groups. The *trans* form belongs to point group C_{2h}, in which only the *u* vibrations (antisymmetric with respect to center of symmetry) are infrared active. On the other hand, both *gauche* forms† belong to point group C_2, in which all the vibrations are infrared active. It is therefore anticipated that the *gauche* form will exhibit more bands than the *trans* form. Mizushima et al. [98] have shown that 1,2-dithiocyanatoethane in the crystalline state definitely exists in the *trans* form, because no infrared frequencies coincide with Raman frequencies (mutual exclusion rule). By comparing the spectrum of the crystal with that of a $CHCl_3$ solution, they have concluded that the several extra bands observed in solution can be attributed to the *gauche* form. Table 4-5 summarizes the infrared frequencies and band assignments obtained by Mizushima et al. Although theory predicts more bands for the *gauche* form than they observed, this difference may be due to overlapping of bands between similar vibrational modes. Table 4-5 shows that the CH_2 rocking vibration provides the most clear-cut diagnosis of conformation: one band (A_u species) at 749 cm^{-1} for the *trans* form, and two bands (*A* and *B* species) at 918 and 845 cm^{-1} for the *gauche* form.

† These two *gauche* forms are mirror images (optical isomers) and cannot be distinguished by means of their IR spectra. For δ and λ notation see p. 26.

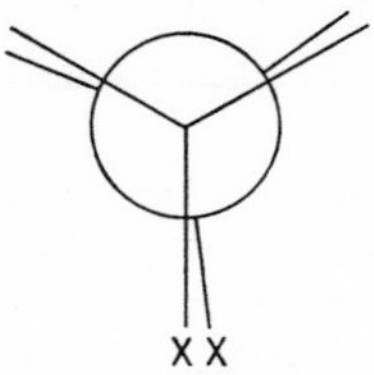

X X
cis ($\mathbf{C}_{2v}$)
VII

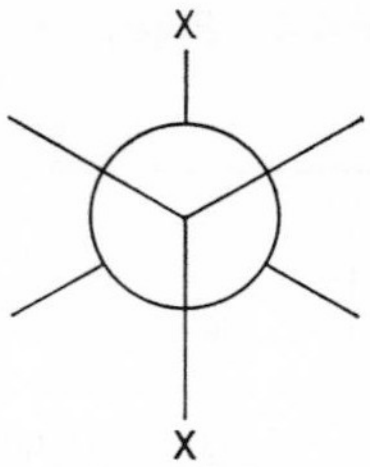

X
X
trans ($\mathbf{C}_{2h}$)
VIII

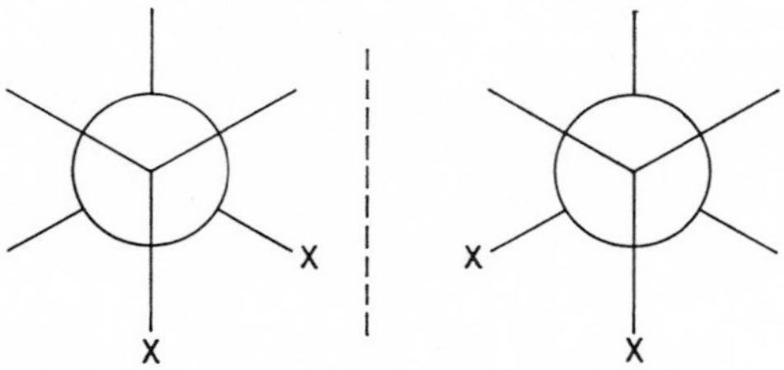

X X
gauche (λ)

X X
gauche (δ)

$\mathbf{C}_2$
IX

1,2-Dithiocyanatoethane may take the *cis* or *gauche* form when it coordinates with a metal. The chelate ring will be completely planar in the *cis*, and puckered in the *gauche*, conformation. It is easy to distinguish *cis* and *gauche* forms by comparing the spectrum of a metal chelate with that of the ligand in $CHCl_3$ solution (*gauche* + *trans*). Table 4-5 compares the infrared spectrum of 1,2-dithiocyanatoethane-dichloroplatinum(II) with that of the free ligand in a $CHCl_3$ solution [98]. Only the bands characteristic of the *gauche* form are observed in the Pt(II) complex. This result provides evidence for the *gauche* conformation of the chelate ring. The method described above has also been applied to the metal complexes of 1,2-dimethylmercaptoethane by Sweeny et al. [126]. In this case the free ligand exhibits one CH_2 rocking band at 735 cm^{-1} in the crystalline state (*trans*), whereas the metal complexes always exhibit two CH_2 rocking modes at 920 ~ 890 and 855 ~ 825 cm^{-1} (*gauche*).

The conformation in metal complexes of common chelating agents such as ethylenediamine is of particular interest in coordination chemistry. Unfortunately, the CH_2 rocking mode discussed above cannot provide a clear-cut diagnosis in this case, because it couples strongly with the NH_2 rocking and C—N stretching modes which appear in the same frequency region. However, x-ray analysis of $[Co(en)_3]Cl_3 \cdot 3H_2O$ [110] indicates that all the ethylenediamines in this complex ion take the *gauche* conformation. Furthermore, the

Table 4-5 Infrared Spectra of 1,2-Dithiocyanatoethane and Its Pt(II) Complex (cm^{-1}) [98]

Ligand			
Crystal (*trans*)	$CHCl_3$ Solution (*gauche* + *trans*)	Pt Complex (*gauche*)	Assignment
—	2170 (*g*)	2165 (*g*)	
2155 (*t*)	2170 (*t*)	—	$\nu(C{\equiv}N)$
1423 (*t*)	1423 (*t*)	—	
—	1419 (*g*)	1410 (*g*)	$\delta(CH_2)$
1291[a] (*t*)	—	—	
—	1285 (*g*)	1280 (*g*)	$\rho_w(CH_2)$
1220 (*t*)	1215 (*t*)	—	
1145 (*t*)	1140 (*t*)	—	$\rho_t(CH_2)$
—	1100 (*g*)	1110 (*g*)	
—	[b] (*g*)	1052 (*g*)	$\nu(C{-}C)$
1037[a]	—	—	
—	918 (*g*)	929 (*g*)	
—	845 (*g*)	847 (*g*)	$\rho_r(CH_2)$
749 (*t*)	[b]	—	
680 (*t*)	677 (*t*)	—	$\nu(C{-}S)$
660 (*t*)	660 (*t*)	—	

[a] Raman frequencies in the crystalline state.
[b] Hidden by $CHCl_3$ absorption.

whole complex ion is in the "lel" form rather than in the "ob" form (see p. 28). Hence the overall symmetry of the $[Co(en)_3]^{3+}$ ion is $\mathbf{D}_3$.

Perhaps the most interesting information obtained from infrared studies is that ethylenediamine takes the *trans* conformation when it functions as a bridging group between two metals. Powell and Sheppard [115] were the first to suggest that ethylenediamine in $(C_2H_4)Cl_2Pt(en)PtCl_2(C_2H_4)$ is likely to be *trans*, since the infrared spectrum of this complex is simpler than that of other complexes in which ethylenediamine is definitely in the *gauche* form (chelated). Figure 4-5 compares the infrared spectrum of $[Pt(en)Cl_2]$ with that of $(C_2H_4)Cl_2Pt(en)PtCl_2(C_2H_4)$ obtained by Powell and Sheppard.

Similar results have been obtained for polymeric chains of $[Hg(en)Cl_2]_\infty$ [33], $[Zn(en)Cl_2]_\infty$ [111], $[Cd(en)Cl_2]_\infty$ [111], and for the binuclear complex

$[Ag(en)Ag]Cl_2$ [112]. The structure of these complex ions may be as shown in X.

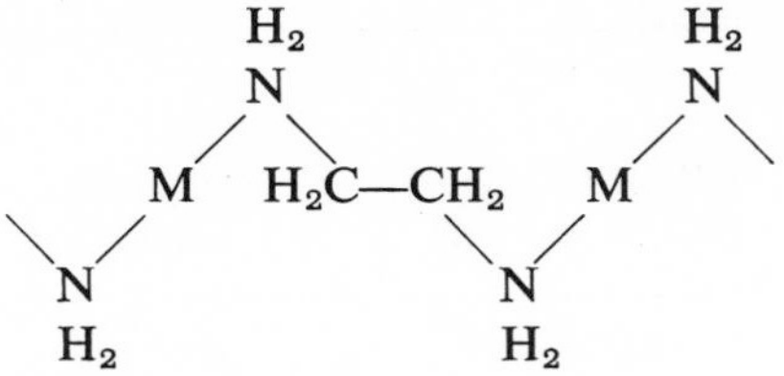

```
              H2                    H2
              N                     N
            /   \                 /   \
          M    H2C—CH2          M
        /            \        /
   \  /                \    /
    N                    N
    H2                   H2
                X
```

Complexes of aliphatic dinitriles provide additional examples of conformational isomerism. According to x-ray analysis the complex ion in bis-(succinonitrilo)Cu(I) nitrate, $[Cu(N{\equiv}C{-}CH_2{-}CH_2{-}C{\equiv}N)_2]NO_3$ [81], has a polymeric chain structure (XI), in which the ligand takes the *gauche* conformation.

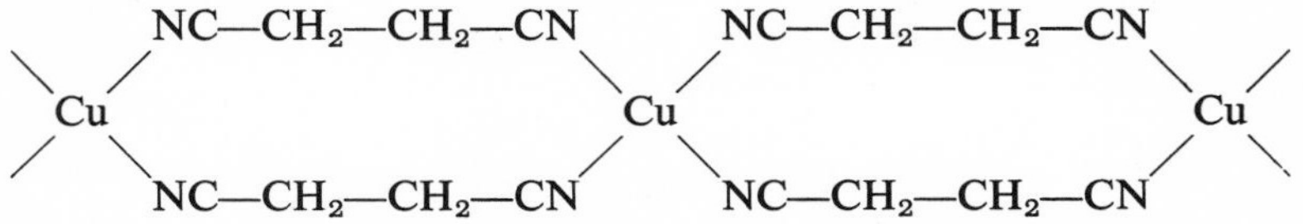

```
      NC—CH2—CH2—CN         NC—CH2—CH2—CN
 \   /              \      /              \   /
  Cu                  Cu                    Cu
 /   \              /      \              /   \
      NC—CH2—CH2—CN         NC—CH2—CH2—CN
                    XI
```

Infrared and Raman studies of succinonitrile [65, 90] indicate that the *trans* and *gauche* forms coexist in the liquid and solid states, and that only the *gauche* form remains when the solid is cooled to $-50°$. Table 4-6 lists the CH_2 rocking frequencies of succinonitrile and those of its Cu(I) complex [90, 65]. Again the similarity of the CH_2 rocking frequencies in the free ligand (solid at $-50°$) and in the complex confirms the *gauche* conformation of the former. On the other hand, the Ag(I) complexes of succinonitrile, such as $Ag(sn)NO_3$ and $Ag(sn)ClO_4$ (sn = succinonitrile), exhibit the CH_2 rocking bands near 760 cm^{-1} which are characteristic of the *trans* form. It has therefore been suggested that in these Ag(I) complexes succinonitrile acts as a bridge between two Ag atoms and assumes the *trans* form [84]. It is interesting that, in this and other aliphatic dinitrile complexes, the $C{\equiv}N$ stretching band is shifted by $25 \sim 35$ cm^{-1} to a higher frequency on coordination. Similar observations had been made previously on the $C{\equiv}N$ stretching bands of metal cyano complexes [14, 47].

There are four possible rotational isomers for glutaronitrile,

$$NC{-}CH_2{-}CH_2{-}CH_2{-}CN,$$

which are spectroscopically distinguishable. They are designated *tt*, *tg*, *gg*, and *gg'*, according to the nomenclature developed by Mizushima [13]. Figure

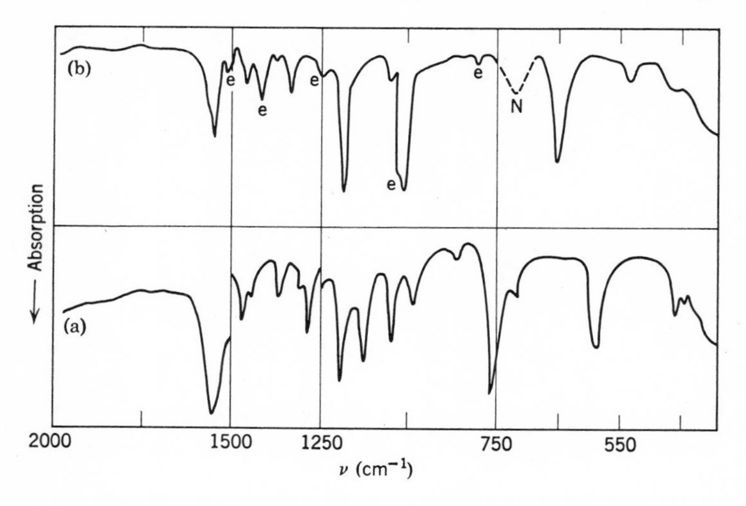

Fig. 4-5. Infrared spectra of (*a*) $Pt(en)Cl_2$ and (*b*) $(C_2H_4)PtCl_2(en)PtCl_2(C_2H_4)$ [115]. e, ethylene band; N, nujol band.

Table 4-6 Infrared Active CH_2 Rocking Frequencies of Succinonitrile and Its Cu(I) Complex (cm^{-1})

Succinonitrile Crystal [90]		Cu(I) Complex [90, 84]
Low Temp. (*gauche*)	Room Temp. (*gauche* + *trans*)	(*gauche*)
963 (*g*)	962 (*g*)	966 (*g*)[a]
820 (*g*)	818 (*g*)	826 (*g*)[a]
—	761 (*t*)	—

[a] Frequencies in nujol mull. In alkali halide pellets the complex decomposes to give free succinonitrile [90].

4-6 illustrates the conformation of these four isomers, their symmetries, and the number of the IR active CH_2 rocking modes predicted from the selection rules. According to the x-ray analysis of bis(glutaronitrilo)Cu(I) nitrate [82], the ligand in the complex definitely takes the *gg* form. The infrared spectrum of this complex is very similar to that of solid glutaronitrile in the stable form. Matsubara [91] therefore concluded that the latter also takes the *gg* form. On the other hand, the spectrum of solid glutaronitrile in the metastable form (produced by rapid cooling) is different from that of the *gg* form. It could be *tt*, *tg*, or *gg'*. However, the *tt* form was excluded because of the absence of the 730 cm^{-1} band characteristic of the *trans*-planar methylene chain [34], and the *gg'* form was considered to be improbable because of steric repulsion

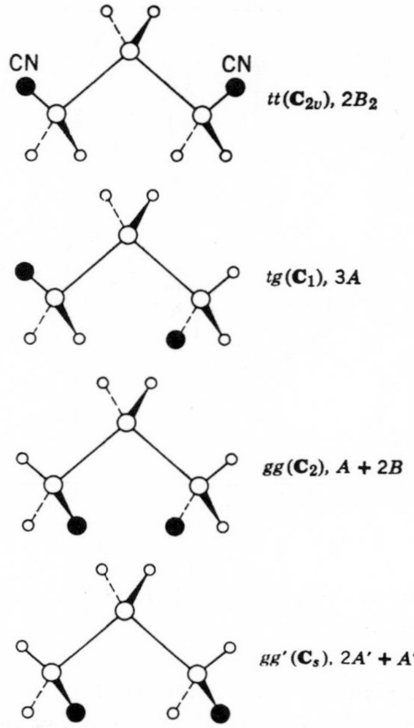

CN CN

$tt(\mathbf{C}_{2v})$, $2B_2$

$tg(\mathbf{C}_1)$, $3A$

$gg(\mathbf{C}_2)$, $A + 2B$

$gg'(\mathbf{C}_s)$, $2A' + A''$

Fig. 4-6. Conformation isomers of glutaronitrile.

between two CN groups. This left only the *tg* form for the metastable solid. The complicated spectrum of liquid glutaronitrile is explainable on the assumption that it is a mixture of the *tg*, *gg*, and *tt* forms. Recently Kubota and Johnston [83] have reported similar studies of the glutaronitrile complexes of other metals. Table 4-7 summarizes the CH_2 rocking frequencies of

Table 4-7 Infrared Active CH_2 Rocking Frequencies of Glutaronitrile and Its Metal Complexes (cm^{-1}) [91, 83]

Liquid[a]	945 (*tg*)	904 (*gg*)	835 (*tg*, *gg*)	757 (*tg*, *gg*)	737 (*tt*)[b]
Solid[a] (metastable)	943 (*tg*)	—	839 (*tg*)	757 (*tg*)	—
Solid[a] (stable)	—	903 (*gg*)	837 (*gg*)	768 (*gg*)	—
$Cu(gn)_2NO_3$[a]	—	913 (*gg*)	830 (*gg*)[c]	778 (*gg*)	—
$Cu(gn)_2ClO_4$[d]	—	908 (*gg*)	875 (*gg*)	767 (*gg*)	—
$Ag(gn)_2ClO_4$[d]	—	904 (*gg*)	872 (*gg*)	772 (*gg*)	—
$SnCl_4(gn)$[d]	—	—	—	—	733 (*tt*)
$TiCl_4(gn)$[d]	—	—	—	—	730 (*tt*)

[a] Ref. [91].

[b] The *tt* form should exhibit two IR active CH_2 rocking. The other one is not known, however.

[c] Overlapped with an NO_3^- absorption.

[d] Ref. [83].

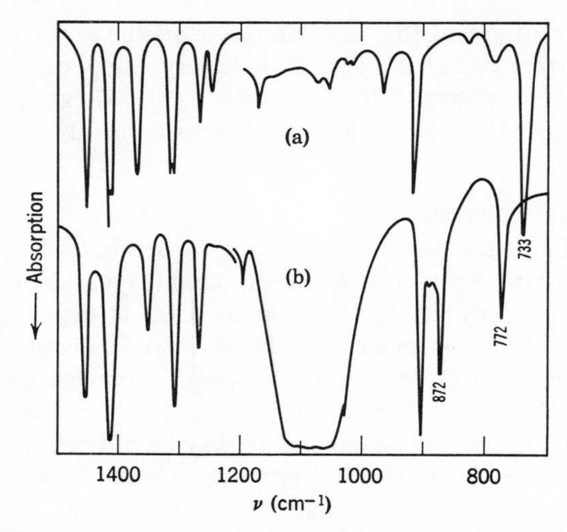

Fig. 4-7. Infrared spectra of complexes of glutaronitrile (gn): (*a*) $SnCl_4 \cdot (gn)$; (*b*) $Ag(gn)_2$-ClO_4 [83]. Frequencies are given only for CH_2 rocking bands.

glutaronitrile and its metal complexes [91, 83]. From these data and other evidence, Kubota and Johnston [83] concluded that glutaronitrile (abbreviated gn) takes the *gg* form in $Ag(gn)_2ClO_4$ and $Cu(gn)_2ClO_4$, and the *tt* form in $TiCl_4(gn)$ and $SnCl_4(gn)$. Figure 4-7 compares the infrared spectrum of $SnCl_4(gn)$ with that of $Ag(gn)_2ClO_4$.

The situation is much more complicated for adiponitrile,

$$NC—CH_2—CH_2—CH_2—CH_2—CN,$$

because many rotational isomers are probable. However, the x-ray analysis of bis(adiponitrilo)Cu(I) nitrate [80] indicates that the ligand acts as a bridge between two metal atoms (diagram XI) and definitely takes the *ttt* form ($\mathbf{C}_{2h}$ symmetry). The infrared spectrum of the Cu(I) complex exhibits two CH_2 rocking bands at 895 (weak) and 737 cm^{-1} (strong), in accordance with the selection rule of the $\mathbf{C}_{2h}$ point group [92]. As stated previously, the strong band is characteristic of the methylene chain of the planar, all-*trans* configuration [34]. The conformation of free adiponitrile can now be elucidated by comparing its spectrum with that of the Cu(I) complex [92]. Among a number of spectroscopically distinguishable rotational isomers, those having a nonplanar C—C—C—C skeleton can be ruled out because of steric repulsion between two CN groups. Thus, only the four forms, *ttt*($\mathbf{C}_{2h}$), *gtg'*($\mathbf{C}_i$), *gtg*($\mathbf{C}_2$), and *ttg*($\mathbf{C}_1$) were regarded as probable. According to the selection rules, both the *ttt* and *gtg'* forms exhibit two IR active CH_2 rocking modes, whereas the *gtg* and *ttg* forms exhibit four IR active CH_2 rocking modes each. The solid adiponitrile (−20°) exhibits two CH_2 rocking modes at 900 and 757–751 cm^{-1}.

It was therefore suggested that adiponitrile takes the *gtg′* form in the solid state. The spectrum of liquid adiponitrile is very complicated; it exhibits at least four CH_2 rocking bands: 900 (*gtg′*), 891 (*ttt*), 752 (*gtg′*), and 734 cm^{-1} (*ttt*). This seems to indicate that the liquid consists mainly of the *ttt* and *gtg′* forms.

B. Geometrical Isomerism

A number of investigations have been made to distinguish geometrical isomers by their infrared spectra. It has been found (1) that the isomer of lower symmetry generally exhibits more bands than that of higher symmetry, and (2) that the *cis* and *trans* isomers absorb radiation of different frequencies for some ligand vibrations. Rule (1) must be applied with caution to the spectra of compounds obtained in the crystalline state, because the symmetry of a molecule or an ion in the crystalline state may be different from that in the free (gaseous) state. Furthermore, non-fundamental vibrations involving lattice modes may complicate the spectra of substances in the crystalline state. Most investigations have been limited to the high frequency region where the vibrations due to the ligands appear. Since the essential difference between *cis* and *trans* structures is in the spatial arrangement of the coordinate bonds, the spectra of the low frequency region, which involves the coordinate bond stretching and bending bands, often show more marked differences than do those of the high frequency region.

Octahedral complexes such as tris(glycino)cobalt(III) may exist in the *cis-cis* and *cis-trans* forms (structures XII and XIII). In fact, two isomers are

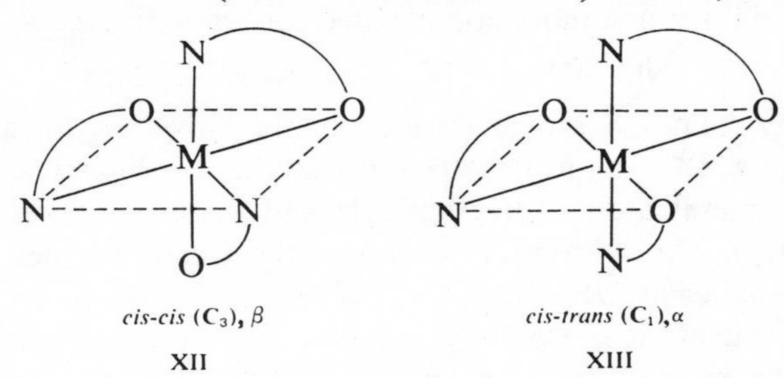

cis-cis (C_3), β — XII   *cis-trans* (C_1), α — XIII

known for $Co(gly)_3$: purple crystals, $Co(gly)_3 2H_2O$ (α form), and red crystals, $Co(gly)_3 H_2O$ (β form). It has been suggested [26, 124] that the α form is *cis-trans* ($\mathbf{C}_1$) and the β-form *cis-cis* ($\mathbf{C}_3$), since the *d—d** band in the ultraviolet region splits into two bands in the α form. Figure 4-8 illustrates the infrared spectra of both forms [76]. In accordance with the UV spectra the IR spectrum of the α form exhibits more bands than that of the β form, which accordingly has a higher symmetry. It is interesting that the Co—N stretching (near 540 cm^{-1}) and Co—O stretching (near 360 cm^{-1}) bands split into two peaks in the α form.

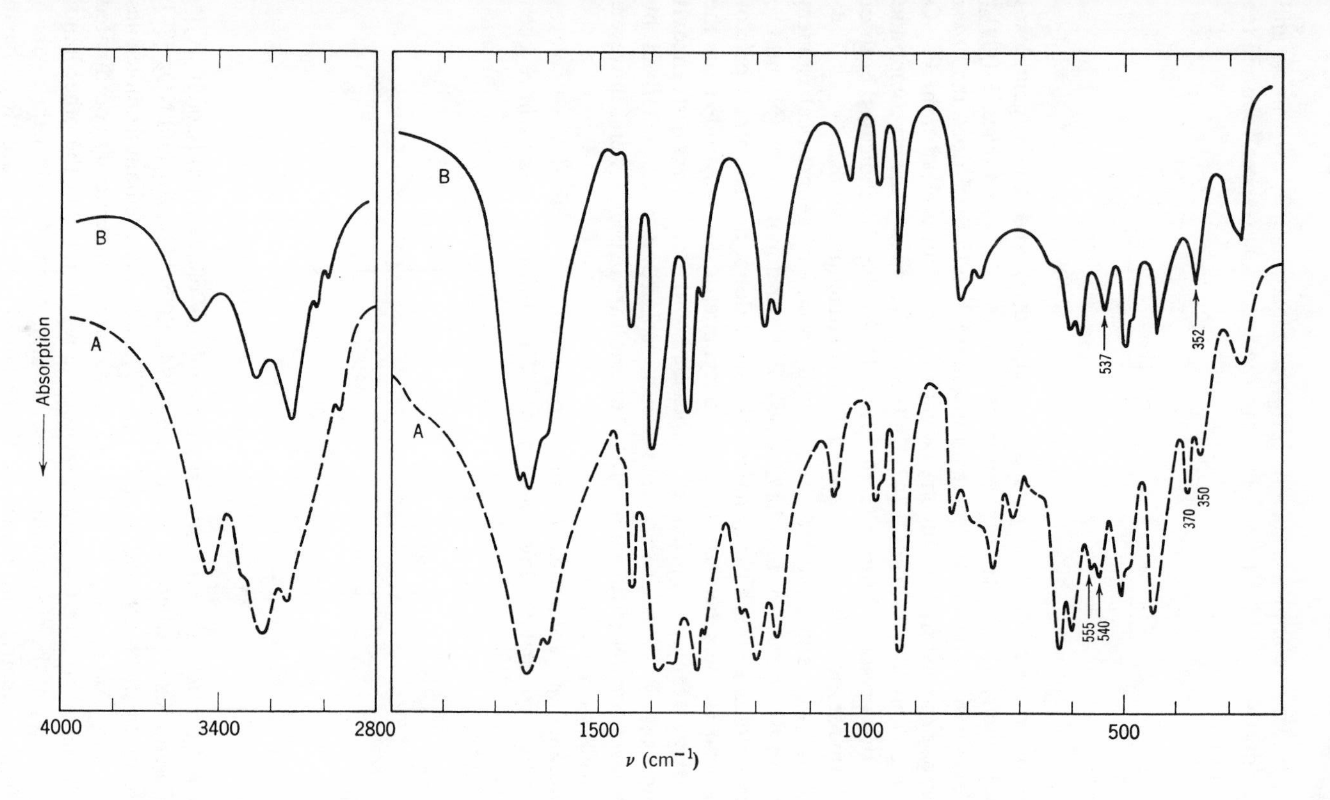

Fig. 4-8. Infrared spectra of tris(glycino)Co(III) complexes [76]; A, *cis-trans* (α) form; B, *cis-cis* (β) form.

Square-planar complexes such as bis(glycino)platinum(II) exist in *cis* and *trans* forms (XIV and XV). Thus far, infrared spectra have been reported for

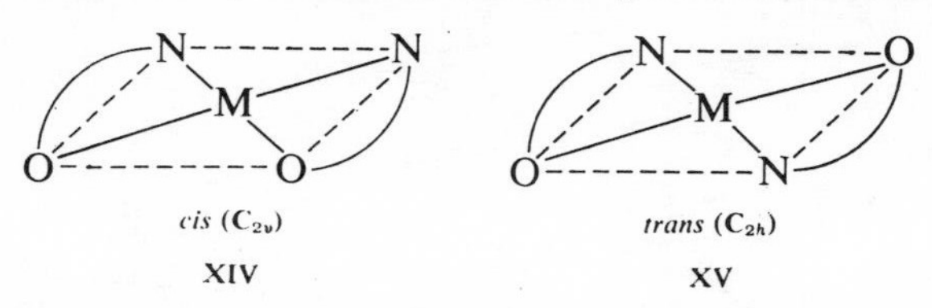

cis (C_{2v}) XIV

trans (C_{2h}) XV

the *cis* and *trans* pairs of bis(glycino)-Pt(II), -Pd(II), and -Cu(II) complexes [42, 85]. Figure 4-9 illustrates the infrared spectra of *cis* and *trans* [$Pt(gly)_2$] complexes [76]. The *cis* isomer exhibits more bands than does the *trans* isomer, as expected from symmetry considerations. This is clear in the Pt—O stretching bands near 400 cm^{-1} [42]. In this particular case the *cis* compound exists in two crystalline forms (α and β) [134]. As Fig. 4-9 indicates, both of these forms show the splitting in their Pt—O stretching bands.

Thus far we have discussed the infrared spectra of geometrical isomers which consist of chelate rings having two different donor atoms (N and O). A number of chelating ligands having the same donor atoms form a pair of geometrical isomers because they are unsymmetrically substituted on the ring. A typical example is provided by the *cis-cis* and the *cis-trans* tris(benzoylacetono)Co(III). These isomers have been isolated and characterized by NMR spectroscopy [50] (see p. 318). Their infrared spectra were, however, found to be very similar.

A number of geometrical isomers are known for octahedral complexes containing two chelate rings (XVI and XVII). Infrared spectra have been

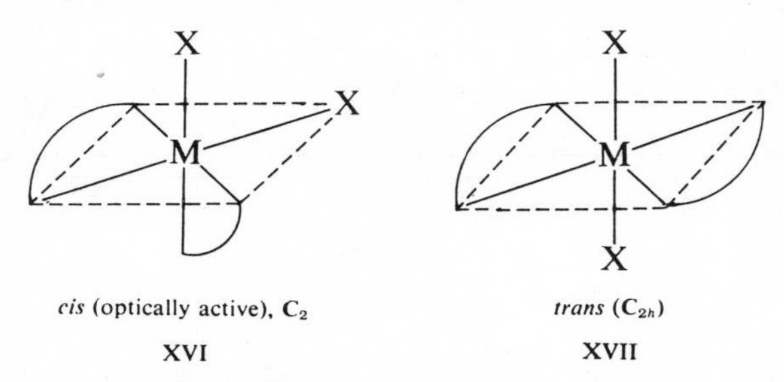

cis (optically active), C_2 XVI

trans (C_{2h}) XVII

compared for *cis* and *trans* isomers of complexes of the types $[Co(en)_2X_2]^+$ and $[Co(en)_2XY]^+$, where X, Y = Cl^-, Br^- [94], SCN^- [24, 36], and NO_2^- [89]. According to the x-ray analysis [100] (p. 27), two ethylenediamine molecules in the *trans*-$[Co(en)_2Cl_2]Cl \cdot HCl \cdot H_2O$ crystal are symmetrically coordinated with the metal to give $\mathbf{C}_{2h}$ symmetry for the whole ion. On the other hand, the symmetry of the *cis* structure cannot be higher than $\mathbf{C}_2$. Only the *u*-vibrations (antisymmetric with respect to the center of symmetry) are IR active in

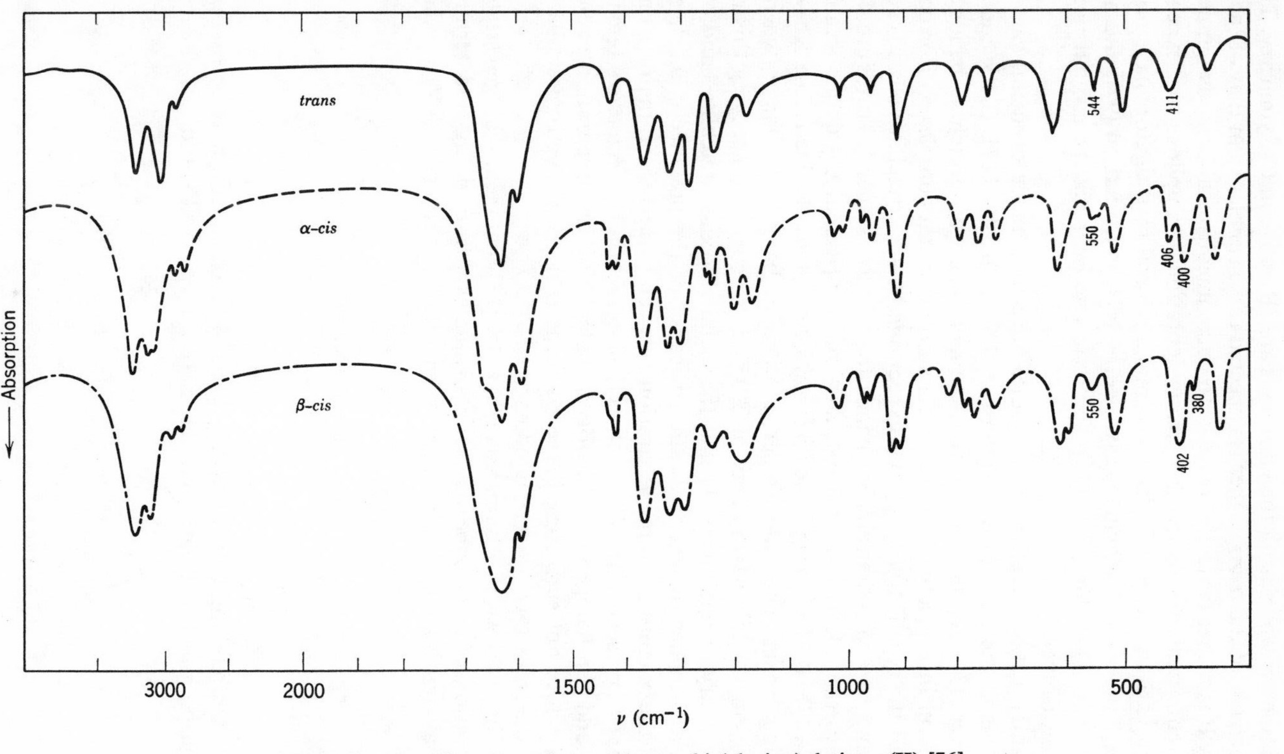

Fig. 4-9. Infrared spectra of *cis*- and *trans*-bis(glycino)platinum(II) [76].

$\mathbf{C}_{2h}$ symmetry, whereas all the vibrations are IR active in $\mathbf{C}_2$ symmetry. Thus the spectrum of the *trans* isomer is expected to be simpler than that of the *cis* isomer. It is rather difficult, however, to make unequivocal band assignments on an empirical basis, because strong vibrational couplings are expected between various modes of ethylenediamine. Several investigators have noted that the bands near 1600 cm^{-1} (asymmetric NH_2 deformation [97]) and near 900 cm^{-1} (CH_2 rocking [24, 116]) split into two bands in the *cis*, but not in the *trans*, isomers.

Recently a number of infrared studies have been made on metal-halogen vibrations in the far-infrared region [19, 20, 38, 40, 41, 117]. Clark and his co-workers [37, 39] have shown that the metal-halogen stretching bands are very useful in predicting the stereochemistry of metal complexes. For complexes of the type $[M(en)_2X_2]^+$ we expect one M—X stretching band for the *trans*, and two M—X stretching bands for the *cis*, isomer [99]. Figure 4-10 compares the infrared spectra of *trans*- and *cis*-$[Rh(en)_2Cl_2]Cl$ as obtained by Kida [75a]. The strong bands near 275 cm^{-1} may be due to the Rh—Cl stretching modes. The *trans* isomer exhibits one band at 270 cm^{-1}, whereas the *cis* isomer exhibits two bands at 290 and 260 cm^{-1}. In applying this rule it is necessary to remember (1) that the metal-chlorine stretching band may exhibit a shoulder for each vibration because of the presence of Cl^{37}, and (2) that the magnitude of the separation between two bands in the *cis* isomer may depend on several factors, such as the interaction between two stretching modes and the halogen-metal-halogen angle (minimum separation at 90°). It is therefore not surprising if the *cis* form exhibits a spectrum similar to that of the *trans* form in certain compounds.

It has been shown by Adams and Chandler [18] that metal-halogen stretching vibrations can also be useful in distinguishing bridging and terminal metal-halogen bonds. For example, compounds of the type XVIII exhibit

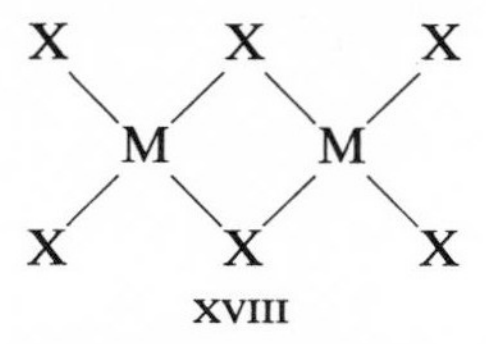

XVIII

the M—X stretching bands listed in Table 4-8. This result may be useful in distinguishing metal chelate compounds of the types XIX and XX.

terminal MX group

XIX

bridging MX group

XX

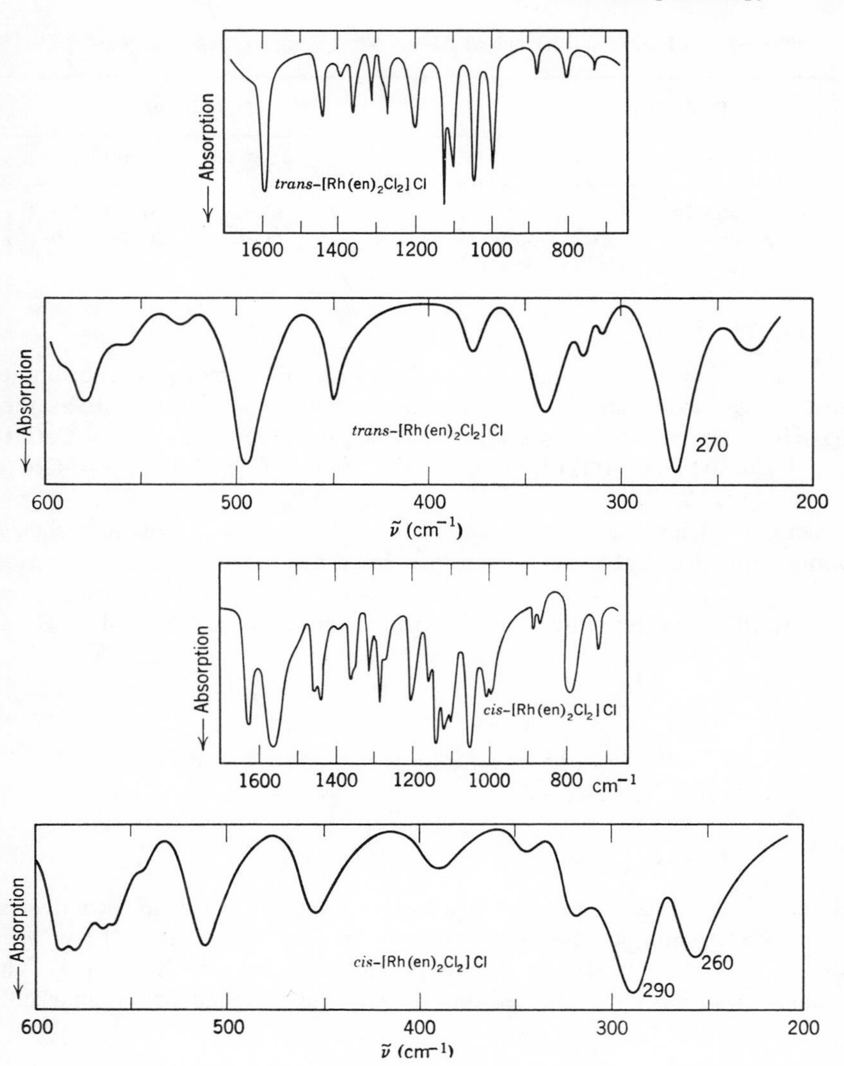

Fig. 4-10. Infrared spectra of *trans*- and *cis*-$[Rh(en)_2Cl_2]Cl$.

C. Linkage Isomerism

Infrared spectroscopy has been very successful in distinguishing linkage isomers involving relatively simple ligands, such as NO_2^-, CN^-, SCN^-, and urea [14]. In most cases, however, only the stabler form has been prepared for a given metal, because the energy difference between the two isomers

Table 4-8 Bridging and Terminal M—X Stretching Frequencies (cm^{-1}) [18]

	Pt(II) Complexes			Pd(II) Complexes	
	Terminal	Bridging		Terminal	Bridging
X = Cl	365–340	335–310, 295–250	X = Cl	370–345	310–300, 280–250
Br	260–235	230–210, 190–175	Br	285–265	220–185, 200–165
I	200–170	190–150, 150–135			

is too large. Thus "true" linkage isomers in which the same ligand coordinates with the same metal through different donor atoms are very rare. So far, they have been reported only for $[M(NH_3)_5(—NO_2$ or $—ONO)]Cl_{2,3}$ [M = Co(III), Rh(III), Ir(III), and Pt(IV)] [114, 28] and *trans*-$[Pd\{As(C_6H_5)_3\}_2(—SCN$ or $—NCS)_2]$ [27].

Only a very few examples are known for linkage isomerism in metal chelate compounds. Again, true linkage isomers are difficult to obtain for the reason mentioned above. An example is seen in platinum(II) acetylacetonate complexes, which may have the types of coordination shown in XXI and XXII.

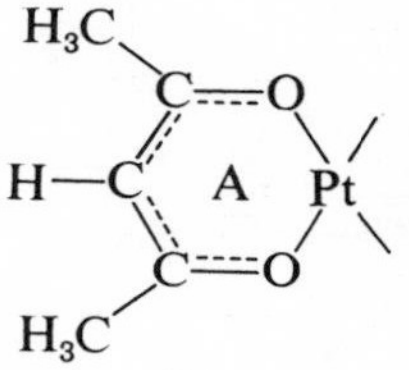

XXI and XXII

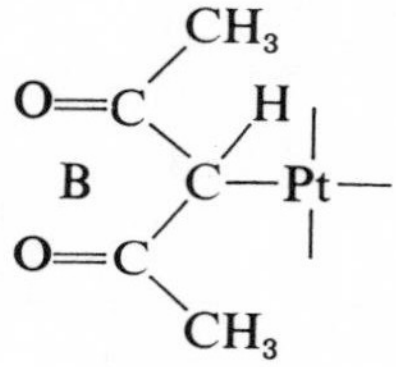

Bidentate coordination through two oxygen atoms (XXI) is almost universal in acetylacetonate complexes. Recently, however, $K[Pt(acac)_2Cl]$ has been shown by x-ray analysis (p. 43) to possess the structure XXIII [55]. The same analysis has also indicated that the CO and CC distances are markedly

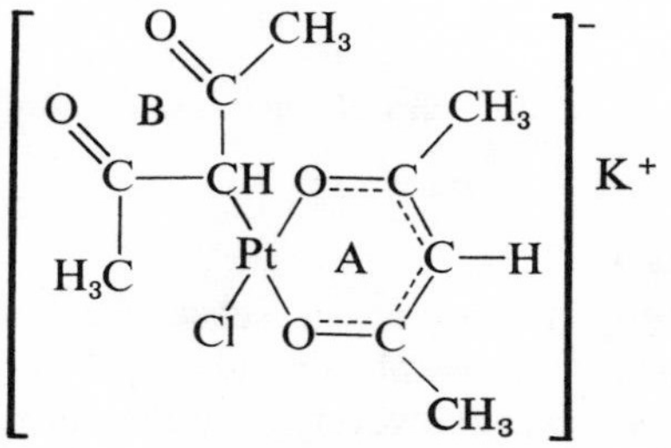

XXIII

different between type A and B coordination; those of type B are close to the pure C—C and C=O distances, whereas those of type A are between the pure single- and double-bond distances. In other words, the π electrons of ring A are delocalized in the OCCCO skeleton, as illustrated in structure XXIII by semidouble bonds (⋯). Such differences in structure are expected to cause significant differences in their infrared spectra. The spectrum of compound XXIII, however, is probably too complicated to allow any reliable interpretation. We therefore begin with compounds having only type A or type B acetylacetonate groups.

Figure 4-11 illustrates the infrared spectra of compounds XXIV and XXV, which have only one type of acetylacetonate group [29]. The spectra of these

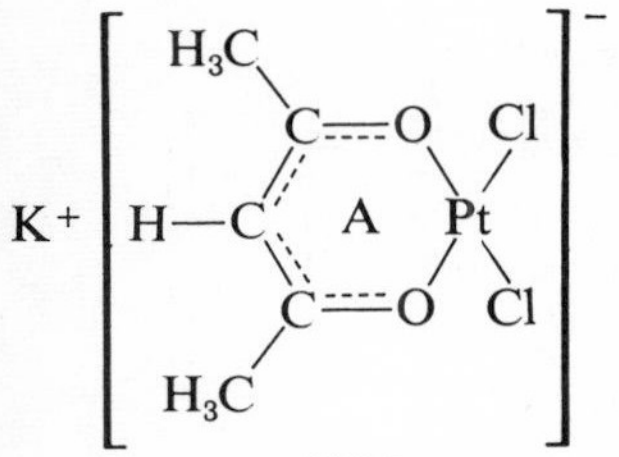

$K^+[Pt(acac\text{-}O,O')Cl_2]^-$ (ring A)

XXIV

$[PtCl_2(acac\text{-}C)_2]^{2-}(Na^+)_2 \cdot 2H_2O$ (groups B)

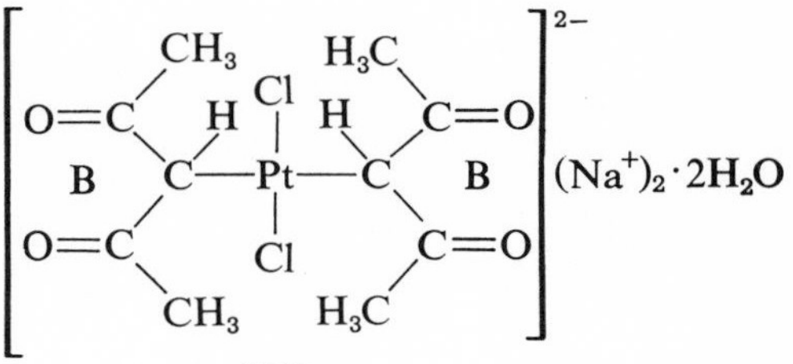

XXV

compounds have been obtained previously by Lewis et al. [88]. However, no theoretical band assignments were available. Behnke and Nakamoto [29] have prepared several deutero analogs of these compounds to carry out normal coordinate analyses on both structures. Table 4-9 lists the stretching force constants thus obtained. Both the CO and CC stretching force constants are considerably different in the two structures, while the Pt—O and Pt—C

Table 4-9 Urey-Bradley Stretching Force Constants for O-Bonded and C-Bonded AcetylacetonoPt(II) Complexes (mdyn/A) [29]

O-Bonded Complex		C-Bonded Complex	
K(C⋯O)	= 6.50	K(C=O)	= 8.84
K(C⋯C)	= 5.23	K(C—C)	= 2.52
K(C—CH_3)	= 3.58	K(C—CH_3)	= 3.85
K(Pt—O)	= 2.46	K(Pt—C)	= 2.50
K(C—H)	= 4.68	K(C—H)	= 4.48
$\rho = 0.43$			

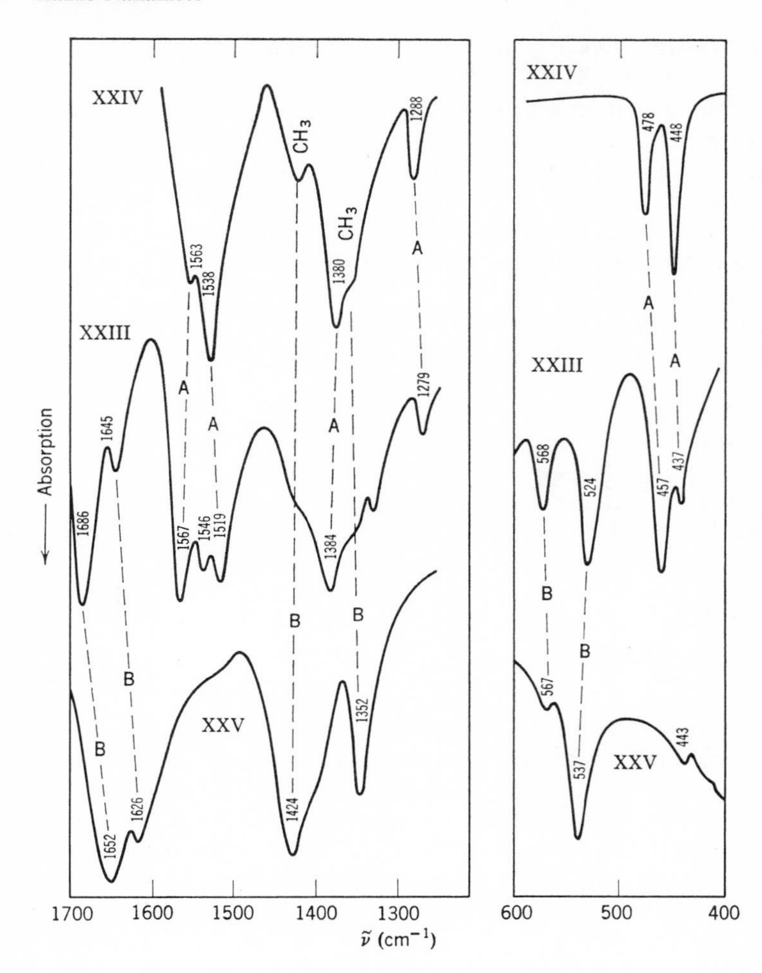

Fig. 4-11. Infrared spectra of Pt(II) acetylacetonate complexes [29].

stretching force constants are almost the same. This indicates that the strength of the latter bonds is quite similar. The introduction of a stretching-stretching interaction constant (ρ) in the OCCCO skeleton was necessary in order to explain the spectra of the type A ring and its deutero analogs. This is similar to the situation found in benzene [118]. The necessity for such an interaction constant in structure A is also consistent with the results of x-ray analysis [55] noted above.

According to the calculation of the potential energy distribution the two

C⋯O stretching bands of the $[Pt(acac)Cl_2]^-$ ion (XXIV) are at 1563 and 1380 cm^{-1}, whereas the two C═O stretching bands of the

trans-$[Pt(acac)_2Cl_2]^{2-}$

ion (XXV) are at 1652 and 1626 cm^{-1}. On the other hand, the C⋯C stretching bands of the former are at 1538 and 1288 cm^{-1}, whereas the C—C stretching bands of the latter are at 1352 and 1193 cm^{-1}. Thus these two types of coordination can be readily distinguished by means of their infrared spectra in the high frequency region. In addition, the two C—H bending bands (1212 and 817 cm^{-1} for XXIV; 1193 and 852 cm^{-1} for XXV) and the coordinate bond stretching bands (650 and 478 cm^{-1} for XXIV; 567 cm^{-1} for XXV) are also useful for the same purpose. These two types of coordination are mixed in the $[Pt(acac)_2Cl]^-$ ion (XXIII). As a result, its infrared spectrum is interpreted as a superposition of the bands characteristic of each type of coordination [29]. This is clearly demonstrated in Fig. 4-11.

Recently Allen et al. [21] have found a third type (type C) of coordination in which acetylacetone coordinates to the platinum atom through the C═C bond (structure XXVIa). This compound was obtained when a solution of

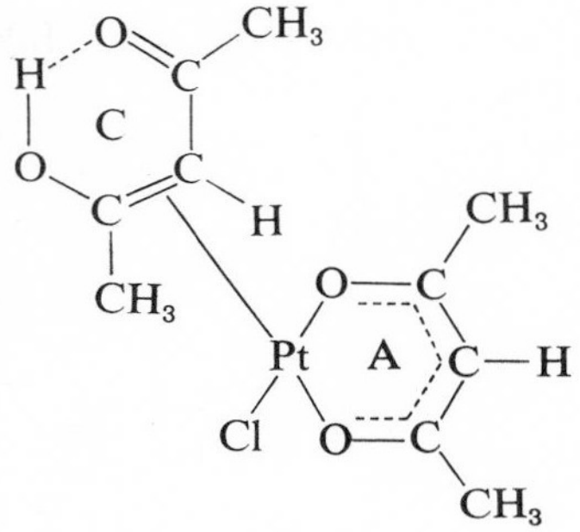

XXVIa

$K[Pt(acac)_2Cl]$ was acidified. Evidently the carbon-bonded acetylacetonate ion of the latter compound was converted into structure XXVIa by the addition of a proton to the oxygen atom, while the oxygen-bonded acetylacetone remained unchanged.

Figure 4-12 compares the infrared spectra (1700–1150 and 500–300 cm^{-1}) of compound XXVIa (types A and C) with that of compound XXIV (type A). The bands characteristic of the type-C ring can be identified by subtracting the type-A spectrum from the spectrum of compound XXVIa. In addition to the C═O stretching at 1627 cm^{-1} and the C—H in-plane bending at 1163 cm^{-1}, the type-C ring exhibits a number of characteristic bands such as the O—H stretching at ca. 2900 and the O—H—O in-plane bending at 1450 cm^{-1}.

The middle curve in Fig. 4-12 shows the infrared spectrum of [Pt(acac)-(C_2H_4)Cl] whose structure is given below (XXVIb).

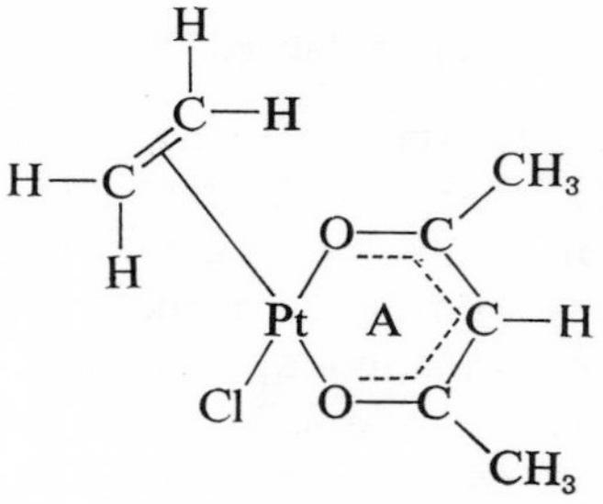

XXVIb

The bands characteristic of the coordinated ethylene are marked by asterisks. In the low-frequency region it exhibits the Pt—C_2H_4 stretching band at 405 cm^{-1}. This frequency is almost the same as that of the same mode in Zeise's salt (407 cm^{-1}) [68].

There are several other types of bonding between the metal and β-diketone. Unidentate coordination such as that in XXVII has been suggested for bis(dipivaloylmethanido)mercury(II) (p. 325), and bridging coordination such as that in XXVIII has been found from the x-ray analysis of Ni(acac)$_2$ (trimer) and Co(acac)$_2$ (tetramer) (p. 46).

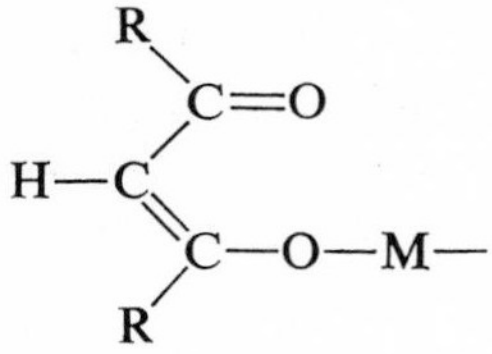

XXVII

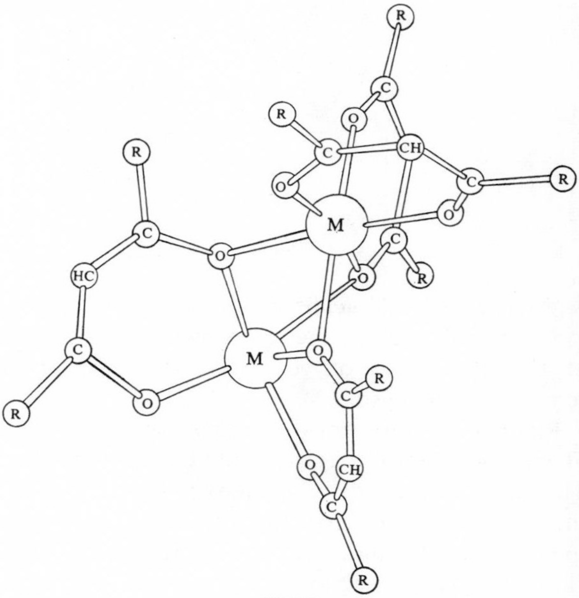

XXVIII

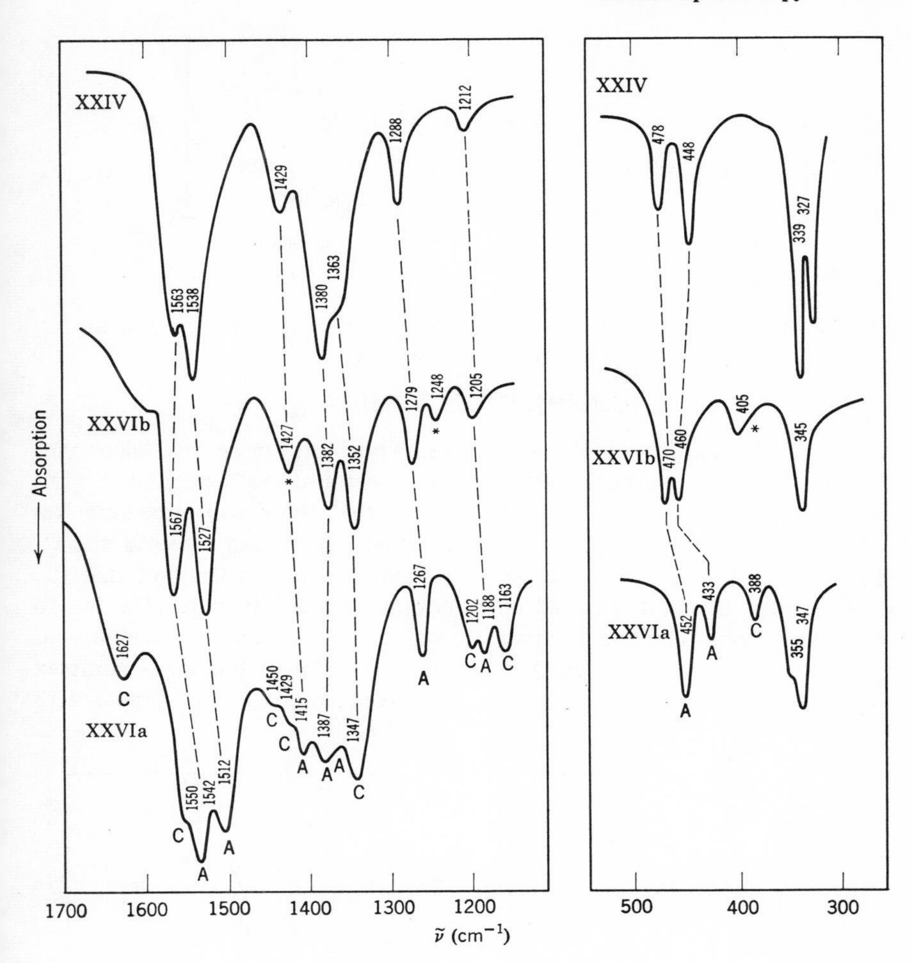

Fig. 4-12. Infrared spectra of Pt(II) acetylacetonate complexes. * = bands expected for coordinated ethylene.

Complexes XXIII and XXVIa are not true linkage isomers, since a proton is attached to the acac anion in the latter. If we relax the definition of linkage isomerism so that compounds like XXIII and XXVIa are included, we can find more examples of linkage isomerism in metal chelate compounds. For example, biuret ($NH_2CONHCONH_2$) reacts with Cu(II) to form linkage isomers XXIX and XXX [57] (p. 42). Violet crystals of composition

$$K_2[Cu(NHCONHCONH)_2]\cdot 4H_2O$$

violet

XXIX

blue-green

XXX

are obtained when the Cu(II) ion is added to an alkaline solution of biuret, whereas pale blue-green crystals of composition

$$[Cu(NH_2CONHCONH_2)_2]Cl_2,$$

are obtained when the Cu(II) ion is mixed with biuret in neutral (alcoholic) solution. Figure 4-13 compares the infrared spectra of the N-bonded (XXIX) and O-bonded (XXX) complexes. It is seen that the C=O stretching frequencies of the N-bonded complex are rather lower than those of the O-bonded complex. This unexpected result may be due to the fact that the C=O groups of the N-bonded complex are strongly hydrogen-bonded to the crystal water. The two compounds can, however, be easily distinguished by comparing the absorption due to the NH group of the violet complex with that of the NH_2 group of the blue-green complex [75]. Figure 4-13

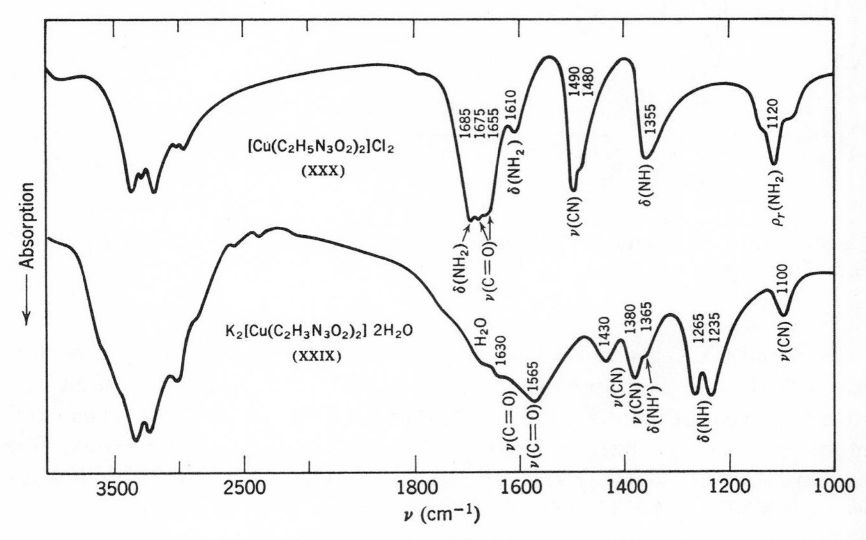

Fig. 4-13. Infrared spectra of O-bonded and N-bonded biuret complexes of Cu(II) [75].

shows that the NH_2 bending and rocking modes (1685 and 1120 cm^{-1}) of XXX are completely missing in XXIX. Instead, the N—H (coordinated) in-plane bending bands appear at 1265 and 1235 cm^{-1} in XXIX. (All these NH and NH_2 group frequencies were confirmed by deuterium substitution.) Using this criterion, it is easy to see that Co(II) coordinates with biuret through the N atoms and Zn(II) coordinates through the O atoms. Infrared spectra also indicate that biuret complexes of Ni(II) exist in two forms having N-coordinated (XXIX) and O-coordinated (XXX) structures [75]. $CdCl_2$ and $HgCl_2$ also form complexes having compositions similar to that of the pale blue-green Cu(II) complex. According to x-ray analysis (p. 43) [35], the neutral biuret molecule in the Cd(II) complex coordinates to the metal through one of the carbonyl oxygens, as shown in XXXI. The structure of the

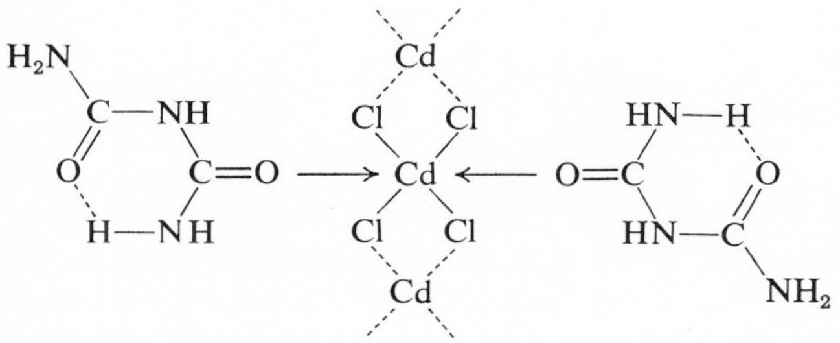

H_2N—C(=O)—NH—C=O → Cd ← O=C—NH—C(=O)—NH_2 (with bridging Cl—Cd—Cl)

XXXI

biuret molecule in this compound is almost the same as that of free biuret [71]. As a result, the spectrum of the Cd(II) complex is similar to that of biuret itself [75].

The definition of linkage isomerism may be expanded further so that the three types of coordination (structures XXXII to XXXIV) in platinum(II) glycino complexes are included. The infrared spectra of $K[Pt(gly)Cl_2]$ (XXXII) and *trans*-$[Pt(glyH)_2Cl_2]$ (XXXIII) are compared in Fig. 4-14.

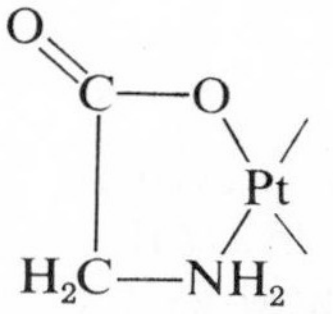

O=C—O—Pt—NH_2—H_2C (chelate ring)

XXXII

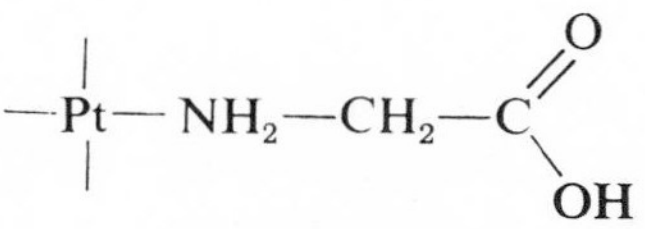

—Pt—NH_2—CH_2—C(=O)OH

XXXIII

—Pt—NH_2—CH_2—C($O^{-\frac{1}{2}}$)($O^{-\frac{1}{2}}$)

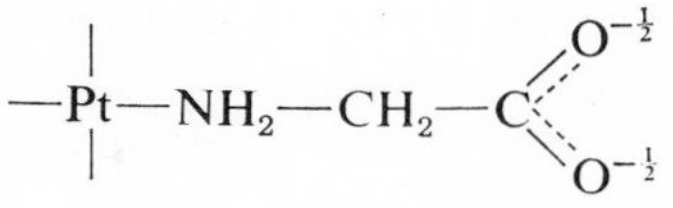

XXXIV

These two structures can be distinguished in the C=O stretching region; XXXII absorbs at 1645 cm^{-1}, whereas XXXIII absorbs at 1710 cm^{-1}. The most notable difference, however, is in the Pt—O stretching region near 400 cm^{-1}; XXXII exhibits a strong Pt—O stretching band at 388 cm^{-1}, whereas this band is completely missing in XXXIII [76, 42]. Figure 4-14 also shows that the spectrum of the complex XXXV, which involves two types of

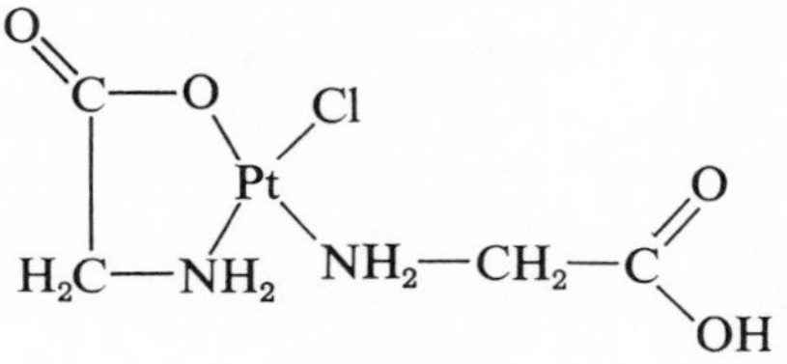

XXXV

coordination, can be interpreted roughly as a superposition of the spectra of types XXXII and XXXIII.

D. Unidentate and Bidentate Coordination

Simple inorganic anions such as SO_4^{2-} and CO_3^{2-} belong to the high symmetry point groups, $\mathbf{T}_d$ and $\mathbf{D}_{3h}$, respectively. On coordination to a metal the symmetry of these ions will be lowered, because the oxygen atom bonded to a metal is different from the other oxygen atoms. For example, the symmetry of the sulfate ion in unidentate coordination (XXXVI) becomes $\mathbf{C}_{3v}$, whereas in bidentate coordination it becomes $\mathbf{C}_{2v}$ both for chelating (XXXVII) and for bridging (XXXVIII) coordination. Accordingly, the IR and Raman

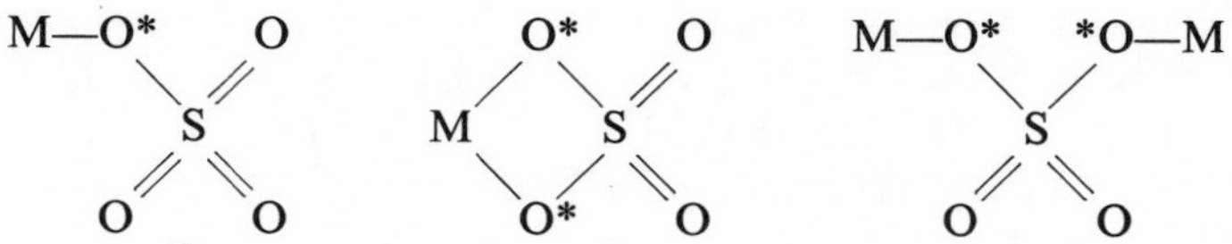

$\mathbf{C}_{3v}$ XXXVI — $\mathbf{C}_{2v}$ XXXVII — $\mathbf{C}_{2v}$ XXXVIII

selection rules are changed, as shown in Table 4-10 [102]. In the free ion, only one stretching (ν_3) and one bending (ν_4) mode are IR active (both triply degenerate, F_2). In unidentate complexes each of these vibrations is split into two bands (A_1 and E), and the ν_1 and ν_2 vibrations, which are only Raman active in the free ion, become IR active. Thus three stretching and three bending vibrations are expected in the infrared spectrum. In bidentate complexes the symmetry is further lowered to $\mathbf{C}_{2v}$, and all the degenerate vibrations are split completely. Thus we expect four stretching and four bending vibrations in the infrared spectrum.

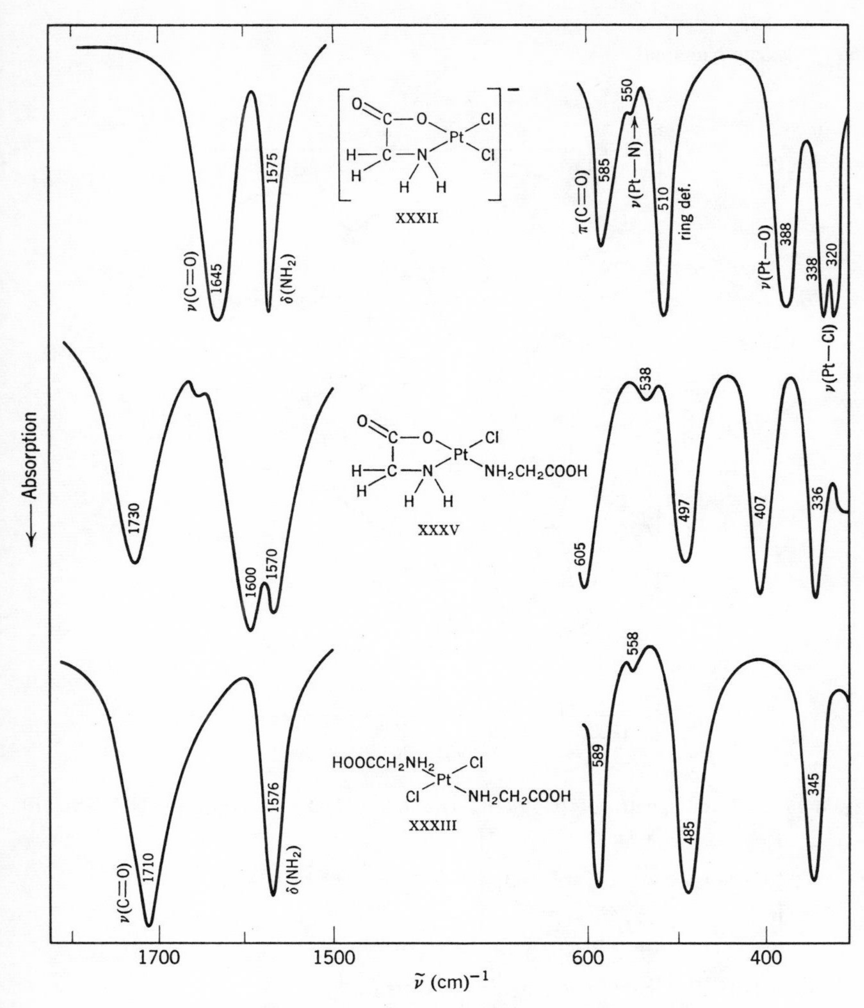

Fig. 4-14. Infrared spectra of Pt(II) glycine complexes.

Table 4-10 Correlation between T_d, C_{3v}, and C_{2v}

	ν_1(str)	ν_2(bend)	ν_3(str), ν_4(bend)
Free ion ($\mathbf{T}_d$)	A_1(R)	E(R)	F_2(IR, R)
Unidentate coordination ($\mathbf{C}_{3v}$)	A_1(IR, R)	E(IR, R)	A_1(IR, R) + E(IR, R)
Bidentate coordination ($\mathbf{C}_{2v}$) (chelating and bridging)	A_1(IR, R)	A_1(IR, R) + A_2(R)	A_1(IR, R) + B_1(IR, R) + B_2(IR, R)

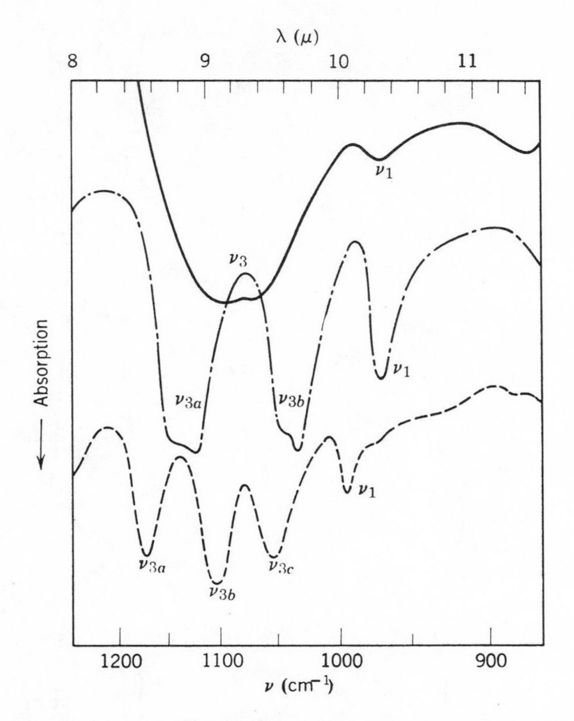

Fig. 4-15. Infrared spectra of $[Co(NH_3)_6]_2(SO_4)_3 \cdot 5H_2O$ (———); $[Co(NH_3)_5SO_4]Br$ (-·-·-); $\left[(NH_3)_4Co \begin{smallmatrix} NH_2 \\ \diagup \quad \diagdown \\ \diagdown \quad \diagup \\ SO_4 \end{smallmatrix} Co(NH_3)_4\right](NO_3)_3$ (- - - - -) [102].

Figure 4-15 illustrates the infrared spectra of

$$[Co(NH_3)_6]_2(SO_4)_3 \cdot 5H_2O, \quad [Co(NH_3)_5SO_4]Br,$$

and

$$[(NH_3)_4Co \begin{smallmatrix} NH_2 \\ \diagup \quad \diagdown \\ \diagdown \quad \diagup \\ SO_4 \end{smallmatrix} Co(NH_3)_4](NO_3)_3$$

in the SO stretching region [102]. The infrared spectrum of the first compound exhibits only one stretching band (ν_3) at 1140 ~ 1130 cm^{-1}, which is indicative of $\mathbf{T}_d$ symmetry. Although ν_1 appears weakly at 973 cm^{-1}, this may be attributed to weak perturbation of the sulfate ion by neighboring ions in the crystal lattice. In $[Co(NH_3)_5SO_4]Br$, three stretching bands are clearly seen at 1143–1117 (ν_{3a}), 1044 ~ 1032 (ν_{3b}), and 970 cm^{-1} (ν_1). This result indicates that the symmetry is lowered to $\mathbf{C}_{3v}$ by unidentate coordination. Finholt et al. [56] have found that the unidentate sulfate group in

$$[Cr(H_2O)_5(SO_4)]Cl \cdot \frac{1}{2}H_2O$$

also exhibits three bands at 1118, 1068, and 1002 cm^{-1}.

The binuclear complex involving a bridging SO_4 group,

$$[(NH_3)_4Co\overset{NH_2}{\underset{SO_4}{\diamond}}Co(NH_3)_4](NO_3)_3,$$

exhibits four stretching bands at 1170 (ν_{3a}), 1105 (ν_{3b}), 1060 ~ 1050 (ν_{3c}), and 995 cm^{-1} (ν_1), because the symmetry of the sulfate group is further lowered to $\mathbf{C}_{2v}$. A mononuclear complex involving a genuine chelating SO_4^{2-} group, $[Co(en)_2SO_4]Br$, was first prepared by Barraclough and Tobe [25]. As the sulfato group of this complex also belongs to $\mathbf{C}_{2v}$, it exhibits four SO stretching bands at 1211, 1176, 1075, and 993 cm^{-1}. It is therefore impossible to distinguish the bridging and chelating sulfato groups from the number of the SO stretching bands. However, the SO stretching frequencies of the chelating sulfato group obtained by Barraclough and Tobe are in general higher than those of the bridging sulfato group obtained by Nakamoto et al. Recently Eskenazi et al. [49] have confirmed this trend in the palladium sulfato complexes. Table 4-11 summarizes their results. Thus a comparison of the SO stretching frequencies, with an examination of solubility of the compound, may be useful in distinguishing the bridging from the chelating sulfato group; the latter type is expected to be more soluble.

Table 4-11 SO Stretching Frequencies of Bridging and Chelating Sulfato Complexes (cm^{-1}) [49]

A. Bridging Sulfato Complexes	
$Pd(NH_3)_2SO_4$	1195, 1110, 1030, 960
$PdSO_4$	1160, 1105, 1035, 996
B. Chelating Sulfato Complexes	
$[Pd(phen)SO_4]$	1240, 1125, 1040–1015, 955
$[Pd(py)_2SO_4]H_2O$	1235, 1125, 1020, 930
$[PdSO_4(H_2O)_2]$	1230, 1089, 995, 960

Differences between unidentate and bidentate coordination are also seen in the OSO bending region between 650 and 450 cm^{-1}. $[Co(NH_3)_5SO_4]Br$ exhibits three bands at 647, 612–610, and 438 cm^{-1}, and the binuclear sulfato complex exhibits four bands at 653, 641, 613, and 454 cm^{-1}. In order to confirm these band assignments, Tanaka et al. [127] have carried out normal coordinate analyses of unidentate and bidentate sulfate ions.

Farraro and Walker [52] have compared the infrared spectra of several metal salts in various states of hydration, and have shown that it is possible to determine the symmetry of the anions from the spectra. Table 4-12 indicates their results for cupric sulfate.

Table 4-12 Infrared Frequencies of Cupric Sulfate in Various Hydrated States (cm^{-1}) [52]

	ν_1	ν_2	ν_3	ν_4	Probable Symmetry of SO_4^{2-} Group
$CuSO_4 \cdot 5H_2O$	975	—	1080	625	Almost $\mathbf{T}_d$
$CuSO_4 \cdot H_2O$	1015	485	1190 1100 1075	675 632 598	$\mathbf{C}_{2v}$
$CuSO_4$	962	500 395	1215 1153 1085	704 613 593	Lower than $\mathbf{C}_{2v}$

It has long been thought that the ClO_4^- ion has little tendency to coordinate with a metal. Recent infrared studies have revealed, however, that both unidentate and bidentate perchlorato complexes can be prepared for Cu(II) and Ni(II). Figure 4-16 illustrates the infrared spectra (ClO stretching region) of $Cu(ClO_4)_2 \cdot 6H_2O$, $Cu(ClO_4)_2 \cdot 2H_2O$, and $Cu(ClO_4)_2$ obtained by Hathaway and Underhill [69]. By the argument used for the sulfate ion, it is concluded that the symmetries of the perchlorate group in these three compounds are $\mathbf{T}_d$, $\mathbf{C}_{3v}$, and $\mathbf{C}_{2v}$, respectively. Wickenden and Krause [136] have found from infrared spectra that the perchlorate groups in $[Ni(CH_3CN)_4(ClO_4)_2]$ act as unidentate ligands, whereas those in $[Ni(CH_3CN)_2(ClO_4)_2]$ act as bidentate ligands. Moore [95] has also concluded that the perchlorate groups in a series of Ni(II) complexes of the type [Ni(substituted-py)$_4$(ClO_4)$_2$] coordinate with the Ni(II) as unidentate ligands, since they exhibit three ClO stretching bands. Siebert [125] reports that four PO stretching bands (1109, 1043, 918, and 895 cm^{-1}) are seen in the infrared spectrum of $[Co(NH_3)_4PO_4] \cdot 2H_2O$, in which the PO_4^{3-} group acts as a bidentate ligand.

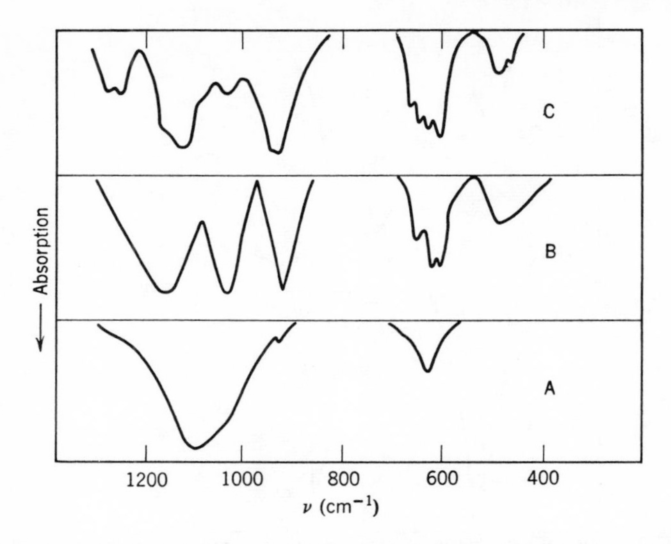

Fig. 4-16. Infrared spectra of (A) $Cu(ClO_4)_2 \cdot 6H_2O$, (B) $Cu(ClO_4)_2 \cdot 2H_2O$, and (C) $Cu(ClO_4)_2$ [69].

Trigonal-planar inorganic ions such as CO_3^{2-} and NO_3^-, ($\mathbf{D}_{3h}$ symmetry) can also coordinate with a metal as unidentate (XXXIX) or as bidentate (XL) ligands. The symmetry of the ion is lowered to $\mathbf{C}_{2v}$ in both types of coordination. It is, accordingly, not possible to distinguish these two modes of co-

M—O* O / C / O — $\mathbf{C}_{2v}$ — XXXIX

O* / M C=O / O* — $\mathbf{C}_{2v}$ — XL

ordination by means of the selection rules alone. Table 4-13 shows the relationship between the $\mathbf{D}_{3h}$ and $\mathbf{C}_{2v}$ point groups. Owing to the lowering of the symmetry, the ν_1 vibration becomes infrared active and both the ν_3 and the ν_4 (doubly degenerate) vibrations split into two components. Figure 4-17 illustrates the infrared spectra of $[Co(NH_3)_5CO_3]Br$ and $[Co(NH_3)_4CO_3]Cl$ in the CO stretching region [102], and Table 4-14 lists the observed frequencies of the same compounds. The splitting of the ν_3 band is much larger in bidentate than in unidentate coordination. It is therefore possible to distinguish them on this basis. Gatehouse et al. [66] observed that the splitting of the ν_3

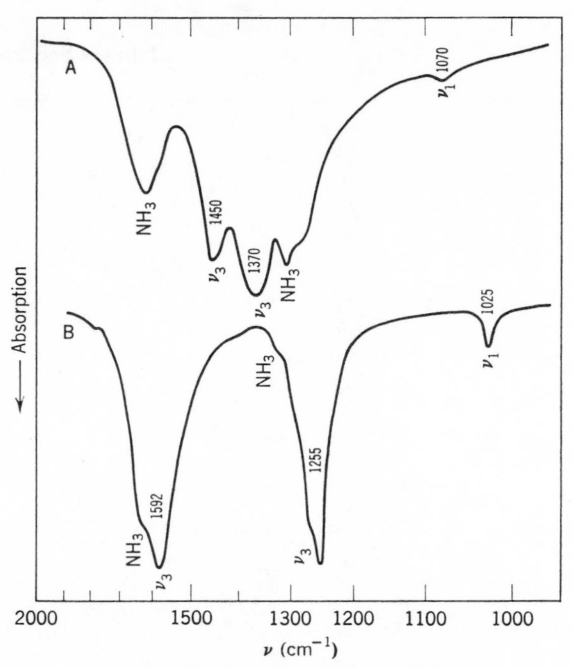

Fig. 4-17. Infrared spectra of (A) $[Co(NH_3)_5CO_3]Br$ and (B) $[Co(NH_3)_4CO_3]Cl$ [102].

Table 4-13 Correlation between D_{3h} and C_{2v}

	ν_1 (str)	ν_2 (bend)	ν_3 (str), ν_4 (bend)
Free ion (D_{3h})	A_1'(R)	A_2''(IR)	E'(IR, R)
Unidentate and bidentate coordination (C_{2v})	A_1(IR, R)	B_1(IR, R)	A_1(IR, R) + B_2(IR, R)

Table 4-14 Infrared Frequencies of Carbonato Ammine Complexes of Co(III) (cm^{-1}) [102][a]

	ν_3	ν_1	ν_2	ν_4
Free ion (D_{3h})	1415 (E')	1080[b] (A_1')	879 (A_2'')	680 (E')
$[Co(NH_3)_5CO_3]Br$ (C_{2v})	1453 (B_2) 1373 (A_1)	1070 (A_1)	850 (B_1)	756 (A_1) 678 (B_2)
$[Co(NH_3)_4CO_3]Cl$ (C_{2v})	1593 (A_1) 1265 (B_2)	1030 (A_1)	834 (B_1)	760 (A_1) 673 (B_2)

[a] The assignment of the symmetry property of each mode was based on ref. [62].
[b] Raman active.

band increases in the order

ionic carbonate < carbonato complex < acid carbonate < organic carbonate

Fujita et al. [62] carried out a normal coordinate analysis of both types of coordination including the Co atom, and obtained the UBF stretching force constants (all in units of mdyn/A) shown in XLI, XLII, and XLIII. The equivalence of the three CO bonds of the free ion is lost by coordination, so

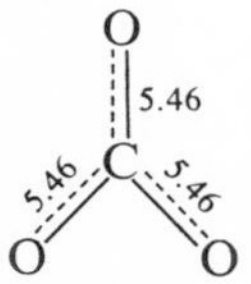

XLI 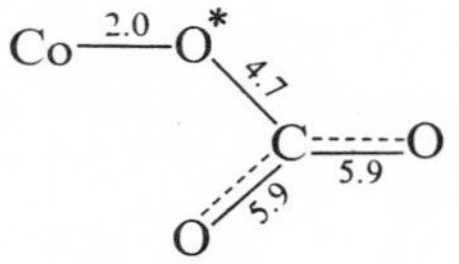 XLII 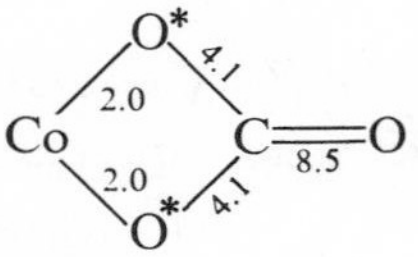 XLIII

that the CO* bond becomes weaker and the CO bonds become stronger than those in the free ion. Evidently this effect is greater in bidentate than in unidentate coordination. Furthermore, they located the Co—O stretching vibrations at 362 cm^{-1} in $[Co(NH_3)_5CO_3]Br$ and at 430 and 395 cm^{-1} in

$$[Co(NH_3)_4CO_3]Cl.$$

Elliott and Hathaway [48] confirmed the band assignments of Fujita et al. [62] by the polarized infrared study of a single crystal of $[Co(NH_3)_4CO_3]Br$.

Apparently the arguments used for carbonato complexes apply also to nitrato complexes. Thus Gatehouse et al. [67] have observed that ν_3 (1390 cm^{-1}) of the free ion splits into two bands (1531 ~ 1481 and 1290 ~ 1253 cm^{-1}), and that ν_1 (1050 cm^{-1}, Raman active) of the free ion appears in the infrared when the nitrate group coordinates with a metal as a unidentate ligand ($\mathbf{C}_{2v}$ symmetry). Bidentate coordination of the nitrate group has been suggested for some uranyl [54] and nickel compounds [87]. No definite evidence was available, however, until Cotton and Soderberg [44] showed by x-ray analysis that both nitrate groups in $[(CH_3)_3PO]_2Co(NO_3)_2$ coordinate as bidentate ligands. Recently Curtis and Curtis [45] have studied the infrared spectra of a number of nitratotetramminenickel(II) complexes, and have found that the splitting of the ν_3 band is larger in bidentate than in unidentate coordination. For example, the unidentate NO_3 group in $[Ni(en)_2(NO_3)_2]$ exhibits two NO stretching bands at 1420 and 1305 cm^{-1}, whereas the bidentate NO_3 group in $[Ni(en)_2(NO_3)]ClO_4$ exhibits them at 1476 and 1290 cm^{-1}. It is interesting that $[Ni(den)(NO_3)_2]$ exhibits the bands characteristic of both types of coordination: 1440 and 1315 (unidentate), and 1480 and 1300 cm^{-1} (bidentate). One of the probable structures is shown in XLIV.

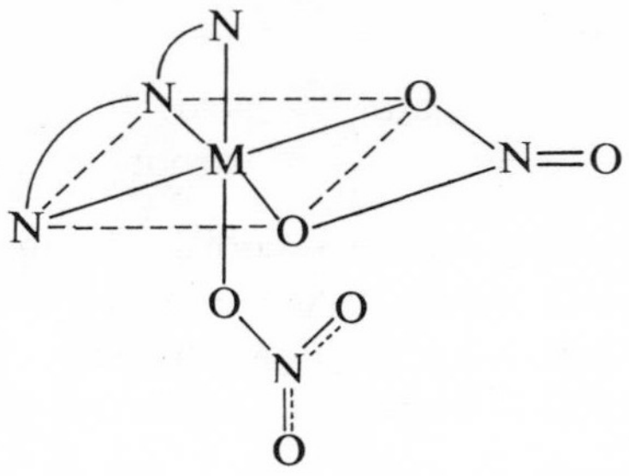

XLIV

Biagetti and Haendler [32] suggested that a similar mixing of unidentate and bidentate coordination occurs in octahedral complexes such as

$$[M(py)_3(NO_3)_2] \quad \text{and} \quad [M(py)_3(NO_3)_2]\cdot 3py,$$

where M is Co(II) and Ni(II). Lever [87] has concluded that the nitrate groups in nitratoamine complexes of the type $[M(A)_2(NO_3)_2]$, where A is isoquinoline, quinoline, and 2-picoline, and M is Co(II), Ni(II), Cu(II), and Zn(II), act as bidentate ligands to complete six-coordination around the metals. Two NO stretching bands (two components of ν_3) were observed at 1517–1484 and at 1305–1258 cm^{-1} for all the compounds studied.

As noted above, both unidentate and bidentate coordination split the ν_3 band of the free ion into A_1 and B_2 species of $\mathbf{C}_{2v}$ symmetry. According to the results of normal coordinate analysis on carbonato complexes [62] (Table 4-14), the higher frequency component of a unidentate complex is of B_2 symmetry, whereas that of a bidentate complex is A_1. Thus the lines corresponding to the higher and lower frequency components may be polarized and depolarized, respectively, in the Raman spectrum of bidentate nitrato complexes.† Ferraro et al. [53] have measured the Raman spectrum of $Th(NO_3)_4{\cdot}2TBP$ (TBP = tributyl phosphate) and have concluded that all the nitrate groups in this compound act as bidentate ligands, since the line at ca. 1550 cm^{-1} is polarized. Therefore a Raman polarization study may provide another tool for distinguishing unidentate from bidentate coordination.

The criterion based on the magnitude of the ν_3 splitting may not be useful if the coordinate bond is very weak (primarily ionic). Topping [130] has pointed out that the unidentate nitrato groups in $Cs_2[U(NO_3)_6]$ give almost the same splitting as the bidentate nitrato groups in $Rb[UO_2(NO_3)_3]$ (1531 and 1274 for the former, and 1536 and 1276 cm^{-1} for the latter). In this case the metal-oxygen stretching bands in the far-infrared region are more useful in distinguishing unidentate from bidentate coordination: 224 cm^{-1} for the former, and 262 and 223 cm^{-1} for the latter. Ferraro and Walker [51] have studied the far-infrared spectra of a number of anhydrous metal nitrates and

† For polarization of Raman lines, see [1, 3, 5, 7].

have shown that covalent nitrates ($\mathbf{C}_{2v}$), such as $Cu(NO_3)_2$ and $Co(NO_3)_2$, exhibit strong metal-oxygen stretching bands between 350 and 250 cm^{-1}, whereas ionic nitrates such as KNO_3 and $NaNO_3$ ($\mathbf{D}_{3h}$) do not exhibit any absorption in the same frequency region. They [135] have also shown that anhydrous rare earth nitrates such as $Pr(NO_3)_3$ and $Nd(NO_3)_3$ show metal-oxygen stretching bands between 270 and 180 cm^{-1} and so must have the nitrate groups covalently bonded to the rare earth metal ions ($\mathbf{C}_{2v}$ symmetry).

The pyramidal sulfite ion, SO_3^{2-}, can coordinate with a metal as a unidentate or a bidentate or a bridging ligand. Thus structures XLV to L are probable [113].

Unidentate Structures

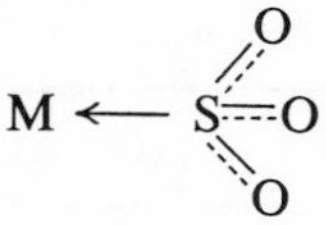

$\mathbf{C}_{3v}$ XLV $\mathbf{C}_s$ XLVI

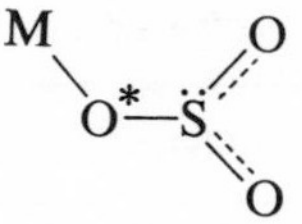

Bidentate Structures

Chelating:

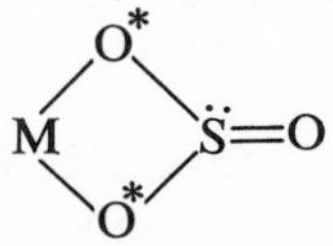

$\mathbf{C}_s$ XLVII $\mathbf{C}_s$ XLVIII

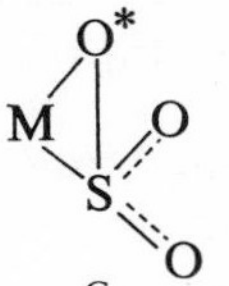

Bridging:

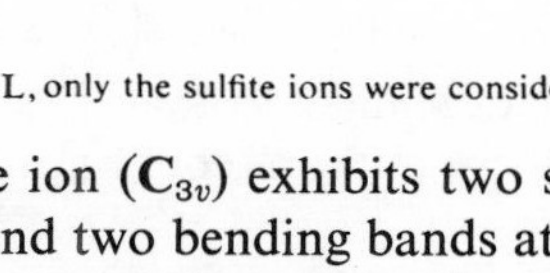

$\mathbf{C}_s$ XLIX $\mathbf{C}_s$ L

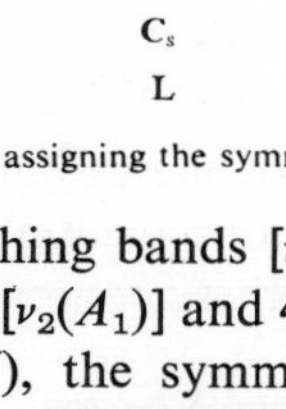

(In XLV-L, only the sulfite ions were considered in assigning the symmetry.)

The free sulfite ion ($\mathbf{C}_{3v}$) exhibits two stretching bands [$\nu_3(E)$ and $\nu_1(A_1)$] near 980 cm^{-1}, and two bending bands at 633 [$\nu_2(A_1)$] and 498 cm^{-1} [$\nu_4(E)$]. If coordination occurs through sulfur (XLV), the symmetry will remain unchanged. If coordination occurs through oxygen (XLVI), the symmetry is lowered to $\mathbf{C}_s$. As a result, both ν_3 and ν_4 are split into two components. Thus it is possible to distinguish S-coordinated and O-coordinated unidentate structures by means of the selection rules. It is also expected [43] that coordination through sulfur will shift the SO stretching bands to higher frequencies, whereas coordination through oxygen will shift them to lower frequencies compared with the free ion. On the basis of these criteria Newman and Powell [113] have shown that the sulfito groups in $[Pt(SO_3)_4]^{6-}$ and $[Co(NH_3)_5SO_3]^+$ are S-bonded, whereas those in $[Cu(SO_3)_2]^{2-}$ are O-bonded (see Table 4-15). The latter is the only known example of this type. Baldwin [23] has also concluded that the sulfito groups in Co(II) complexes of the types $[Co(en)_2(SO_3)_2]^{2-}$ and $[Co(en)_2(SO_3)X]^{0,+}$ ($X = NCS^-$, Cl^-, OH^-, and NH_3) are all S-bonded, since they exhibit the spectra predicted by $\mathbf{C}_{3v}$ symmetry.

Table 4-15 Infrared Spectra of Sulfito Complexes[a] [113]

	Symmetry	$\nu_1(A_1)$	$\nu_3(E)$	$\nu_2(A_1)$	$\nu_4(E)$
Free SO_3^{2-} ion	$\mathbf{C}_{3v}$	1010	961	633	496
$[Co(NH_3)_5SO_3]Cl$	$\mathbf{C}_{3v}$(S-bonded)	1100	985	620	469
$Tl_2[Cu(SO_3)_2]$	$\mathbf{C}_s$(O-bonded)	989	902	673	506
			862		460

[a] These assignments are tentative (made by the author).

There are at least four probable structures (XLVII to L) for bidentate coordination of the sulfite ion. It is, however, impossible to distinguish them by means of the selection rules, because they all belong to the same point group, $\mathbf{C}_s$. Baldwin [23] has prepared Co(III) complexes of the type

$$[Co(en)_2SO_3]X$$

($X = ClO_4^-$, NO_2^-, I^-, or SCN^-), which are monomeric. They exhibit four stretching bands near 1120, 1080, 1030, and 980 cm^{-1}, one of which may be an overtone; they are probably chelated through two oxygen atoms of the sulfito group (XLVII) [113]. Newman and Powell [113] have obtained the infrared spectra of

$$Na_3[Co(SO_3)_3]\cdot 4H_2O, \quad K_2[Pt(SO_3)_2]\cdot 2H_2O,$$
$$K_3[Rh(SO_3)_3]\cdot 2H_2O, \quad \text{and} \quad K_2[Pd(SO_3)_2].$$

All except the last one exhibit spectra similar to that of the $[Co(en)_2SO_3]^+$ ion. For the reasons mentioned, these authors could not, however, deduce the structures of these compounds.

E. Spectra and the Strength of Coordinate Bonds

On coordination to a metal, the ligand bands are shifted to lower or higher frequencies with concomitant variation in intensity. So far, almost no quantitative measurements have been made on the intensity of infrared bands of metal chelate compounds. Therefore this discussion is limited to the band shifts. In a series of metal complexes having the same structure the magnitude of these band shifts becomes larger as the coordinate bond becomes stronger. It is therefore possible to determine the order of the strength of the coordinate bonds by comparing the magnitudes of the band shifts. A number of infrared studies have already been carried out to compare, on the basis of the band shifts observed, the strengths of the coordinate bonds. However, this simple approach needs the cautions described below.

First, the direction of the band shift depends on the structure of the metal complex formed. In the case of the acetate ion the three types of coordination shown in LI to LIV are known to exist. In the free ion both the CO bonds

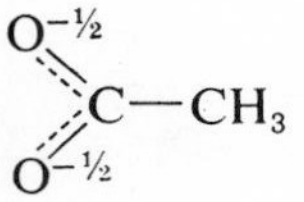

$O^{-1/2}$ / $O^{-1/2}$ C—CH_3 (LI) M—O / O C—CH_3 (LII) M O / O C—CH_3 (LIII) M—O / M—O C—CH_3 (LIV)

LI 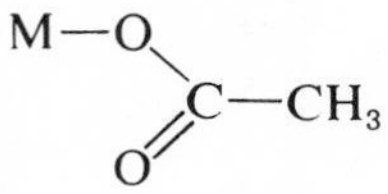 LII 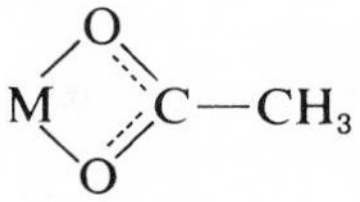 LIII 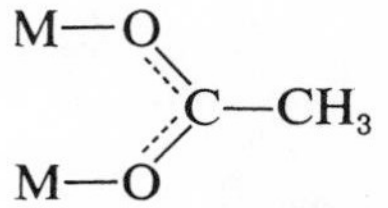 LIV

are equivalent, and the antisymmetric and symmetric COO stretching bands appear at 1582 and 1425 cm^{-1}, respectively [109]. If coordination occurs as shown by structure LII, the antisymmetric and symmetric COO stretching bands will be shifted to higher and lower frequencies, respectively. If coordination occurs symmetrically (structures LIII and LIV), both the COO stretching bands may be shifted in the same direction, since the bond orders of both CO bonds may be changed by the same amount. In fact, this was found to be true for $[Cu_2(CH_3COO)_4]\cdot 2H_2O$ and $[Cr_2(CH_3COO)_4]\cdot 2H_2O$, which have the same structure (LIV); in these compounds both the COO stretching bands are shifted to higher frequencies in going from Cr(II) to Cu(II) [107].

Secondly, the direction of the band shift depends on the nature of the normal vibration and the effect of coordination on it. Consider the chelation of the glycino anion to a metal shown in the accompanying reaction. In this

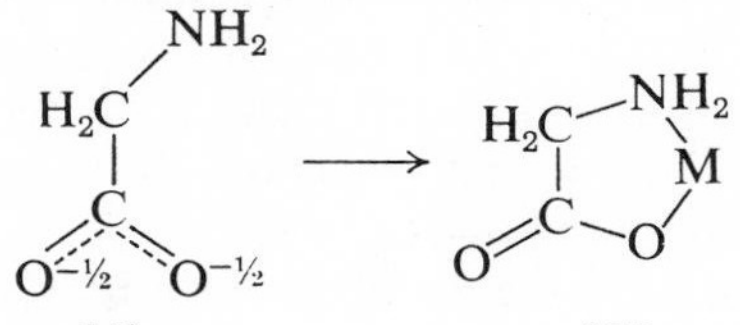

NH_2—H_2C—C($O^{-1/2}$)($O^{-1/2}$) (LV) ⟶ H_2C—NH_2—M—O—C(=O) (LVI)

LV LVI

case the N—H stretching bands will be shifted to lower frequencies because the N—H bond order will be reduced on coordination [42]. The COO stretching bands will be shifted as previously described for the asymmetrical coordination (LII) of the acetate ion. The CH_2 vibrations may not be sensitive to coordination because changes in electronic structure in this group may not be appreciable. Table 4-16 lists all the group frequencies of the NH_2 and COO groups observed for sodium glycinate [131] and *trans*-bis(glycino) complexes of several divalent metals [42]. As Table 4-16 shows, the N—H stretching and the COO stretching are shifted in the directions predicted for each

Table 4-16 The N—H and CO Stretching Frequencies of Sodium Glycinate [131] and *trans*-Bis(glycino) Complexes [42]

	Na(gly)	$Ni(gly)_2$	$Cu(gly)_2$	$Pd(gly)_2$	$Pt(gly)_2$
$\nu(NH_2)$	3330	3340	3320	3230	3230
		3280	3260	3120	3090
$\delta(NH_2)$	1600[a]	1610	1608	1616	1610
$\rho_w(NH_2)$	1005[b]	1038	1058	1025	1023
$\rho_t(NH_2)$	1045[b]	1095	1151	1218	1245
$\rho_r(NH_2)$	—	630	644	771	792
$\nu(C{=}O)$	1600[a]	1589	1593	1642	1643
$\nu(C{-}O)$	1415	1411	1392	1374	1374

[a] Probably overlapped with $\delta(NH_2)$.
[b] Tentative assignments made by the author.

mode, and furthermore the magnitude of the band shifts increases as the metal is changed in the order

$$Ni(II) < Cu(II) < Pd(II) < Pt(II).$$

It is therefore concluded that the strength of the metal-nitrogen and metal-oxygen coordinate bonds becomes stronger in this order of the metals. In general the bending modes show some irregularity in their shifts because their frequencies are sensitive to other factors, such as vibrational coupling and intermolecular interaction. As Table 4-16 shows, the NH_2 rocking and twisting modes are also sensitive to the nature of the metal and are shifted progressively to higher frequencies as the metal is changed in the order given above. Thus these two NH_2 group vibrations serve as a good measure of the strength of the coordinate bonds.

Thus far we have discussed the metal-sensitive bands in the high frequency region. It is anticipated that the coordinate bond stretching vibrations, such as metal-nitrogen and metal-oxygen stretching modes which appear in the low frequency region, should be more sensitive to the nature of the metal than the ligand vibrations in the high frequency region. As stated in Sec. 4-1A, however, these low frequency bands can only be assigned by carrying out a normal coordinate analysis of the whole chelate ring. Normal coordinate analyses have in fact been carried out for the chelate rings of acetylacetone [103, 104, 106], oxalic acid [60, 61], dithiocarbamic acid [101], dithiooxalic acid [63], glycine [42], glycolic acid [105], and oxamide [22]. As an example, Table 4-17 shows the results obtained for the oxalato complexes of divalent metals [60].

Table 4-17 Infrared Spectra of Bis(oxalato) Complexes of Divalent Metals (cm^{-1}) [60]

	$[Zn(ox)_2]^{2-}$	$[Cu(ox)_2]^{2-}$	$[Pd(ox)_2]^{2-}$	$[Pt(ox)_2]^{2-}$	Theoretical Assignment
ν_1	1632	1672	1698	1709	ν(C=O)
ν_2	—	1645	1675, 1657	1674	ν(C=O)
ν_3	1433	1411	1394	1388	ν(C—O) + ν(C—C)
ν_4	1302	1277	1245, 1228	1236	ν(C—O) + δ(OCO)
ν_5	890	886	893	900	ν(C—O) + δ(OCO)
ν_6	785	795	818	825	δ(OCO) + ν(M—O)
ν_7	519	541	556	575, 559	ν(M—O) + ν(C—C)
ν_8	519	481	469	469	ring def. + δ(OCO)
ν_9	428, 419	420	417	405	ν(M—O) + ring def.
ν_{10}	377, 364	382, 370	368	370	δ(OCO) + ν(C—C)

In the free oxalate anion all the four C┄O bonds are equivalent. On coordination to a metal, two CO bonds bonded to it are weakened and the other two CO bonds are rather strengthened as depicted in LVII and LVIII.

LVII LVIII

The results of x-ray analyses on ammonium oxalate monohydrate [73] and *trans*-potassium dioxalatodiaquochromate trihydrate [132] indicate that the C┄O distance in the former is 1.23 ~ 1.25 A, whereas the C—O and C=O distances in the latter are 1.32 ~ 1.30 and 1.19 ~ 1.28 A, respectively. It is therefore reasonable to expect that the C=O stretching bands will be shifted to higher frequencies, and the C—O stretching bands to lower frequencies, as the M—O bonds become stronger. If we ignore the differences in mass of the metal atoms, the strength of the coordinate bond may be approximated by the frequency of the M—O stretching mode. Table 4-17 shows that both the C=O stretchings (ν_1 and ν_2) are shifted to higher frequencies, and both the C—O stretchings (ν_3 and ν_4) are shifted to lower frequencies, as the M—O stretching (ν_7) is shifted to higher frequency in the order of the metals Zn(II) < Cu(II) < Pd(II) < Pt(II). This result clearly indicates that the metal-sensitive bands both in the high and low frequency regions can be used as a measure of the strength of the coordinate bond.

The last column of Table 4-17 shows strong vibrational couplings in almost all the normal modes of oxalato complexes. The results of normal coordinate analyses of other systems indicate that such vibrational couplings are fairly universal among metal chelate compounds, and often occur in the coordinate bond stretching modes. If the frequency of the coordinate stretching vibration is used as a measure of the bond strength, the effect of the mass of the metal as well as of vibrational coupling is not taken into consideration. For this reason the frequency sometimes gives a misleading indication of the order of the strength of coordinate bonds.

Up to this point we have ignored the effect of environment on the band shifts. The spectra of metal complexes are obtained mostly in the crystalline state, as a nujol mull or in a KBr pellet. As stated in Chapter 1, the structure of a metal complex in the crystalline state can be markedly different from that in solution or in the gaseous phase. In the crystalline state the configuration around the metal may be distorted or changed by coordination of neighboring molecules. In extreme cases the complex may be dimerized or polymerized. Even if such changes do not occur, molecules or ions in crystals are under the influence of crystal field effects, which can cause the bands to shift from the positions at which they are found in solution or gaseous spectra. Of the environmental effects, hydrogen bonding is the best-known one, since it causes very appreciable band shifts. The occurrence and the magnitude of crystal field effects are, however, almost unpredictable because they vary from compound to compound. Furthermore, the shifts due to crystal field effects are rather difficult to separate from those due to coordination, which were discussed above [64]. It is desirable therefore to obtain the spectra in other physical states whenever possible. As an example, Table 4-18 lists the antisymmetric and symmetric COO stretching frequencies of metal glycino complexes in D_2O solution, in the hydrated crystal, and in the anhydrous solid [107]. The antisymmetric and symmetric COO stretching bands are shifted to higher and lower frequencies, respectively, in these three physical states, as the metal is changed in the order

$$\text{Ni(II)} < \text{Zn(II)} < \text{Cu(II)} < \text{Co(III)} < \text{Pd(II)} \approx \text{Pt(II)} < \text{Cr(III)}.$$

This result indicates at least that the effect of intermolecular interaction on spectra is smaller than that of coordination.

Previously we discussed a rather qualitative relationship between the band shift and the strength of the coordinate bond. Several attempts have been

Table 4-18 Carboxyl Stretching Frequencies of Metal Glycino Complexes in Various Physical States (cm^{-1}) [107]

	Antisym. COO str.			Sym. COO str.		
	D_2O Sol'n.	Hydrated Crystal	Anhydrous Solid	D_2O Sol'n.	Hydrated Crystal	Anhydrous Solid
$Ni(gly)_2 \cdot 2H_2O$	1589	1609	1583	1413	1408	1400
$Zn(gly)_2 \cdot H_2O$	1594	1598	1603	1407	1400	1384
$Cu(gly)_2 \cdot H_2O$	1604	1593	1607	—	1387	1366
α-$Co(gly)_3 \cdot 2H_2O$	1624	1625	—	1366	1364	—
β-$Co(gly)_3 \cdot H_2O$	—	1636	—			
trans-$Pd(gly)_2$	—	—	1642	—	—	1373
trans-$Pt(gly)_2$	—	—	1643	—	—	1374
$Cr(gly)_3 \cdot H_2O$	—	1659, 1639	1658, 1643	—	1381, 1365	1372

made to make it quantitative. For example, Bellamy and Branch [30] have plotted the C=O stretching frequency against the stability constant (log k_1k_2) for a series of bis(salicylaldehydato) complexes of divalent metals, and have obtained the linear relationship shown in Fig. 4-18. The C=O stretching band is shifted linearly to lower frequency with increase in stability constant in the order of metals

$$Mg(II) < Zn(II) < Co(II) < Ni(II) < Cu(II) < Pd(II).$$

This order is in perfect agreement with the well-known stability order of divalent metal complexes found by Mellor and Maley [93]:

$$Mg(II) < Mn(II) < Cd(II) < Zn(II) < Co(II) < Ni(II) < Cu(II) < Pd(II).$$

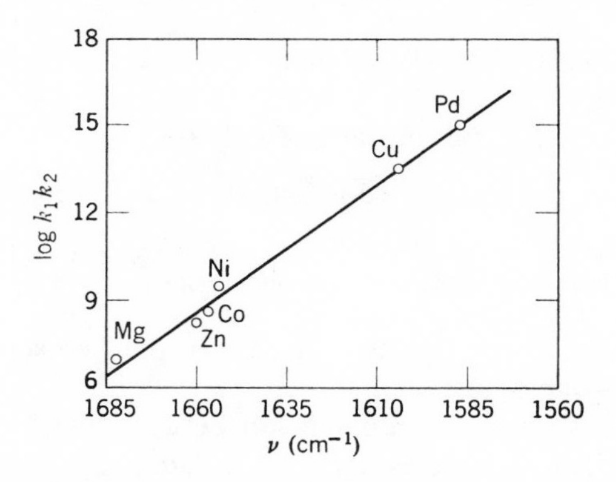

Fig. 4-18. C = O stretching frequencies and stability constants in divalent metal salicylaldehyde complexes [30].

It is evident that the C=O bond order decreases and the C=O stretching frequency decreases as the metal-oxygen bond of salicylaldehyde complexes becomes stronger. Sharma et al. [120] also noted that a linear relationship holds between the stability constant and the N—H stretching frequency of bis(β-alanino) complexes of divalent metals; the N—H stretching frequency (i.e., the highest frequency band among several N—H stretching bands observed) decreases linearly as the stability constant increases in the order Co(II) < Ni(II) < Cu(II) < Pd(II). This result was interpreted as an indication that the N—H bond is weakened as the metal-nitrogen bond is strengthened in this order of metals.

However, the metal-sensitive bands, such as the C=O stretching bands, are not always easy to identify on an empirical basis. This is particularly true for the metal chelates of β-diketones (LIX) and α,β-unsaturated-β-ketoamines (LX), which exhibit both the C⋯O and C⋯C (or C⋯N) stretching bands in the same frequency region. This difficulty has been the origin of a con-

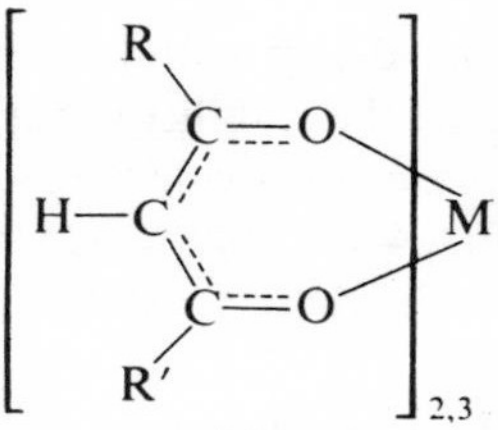

LIX

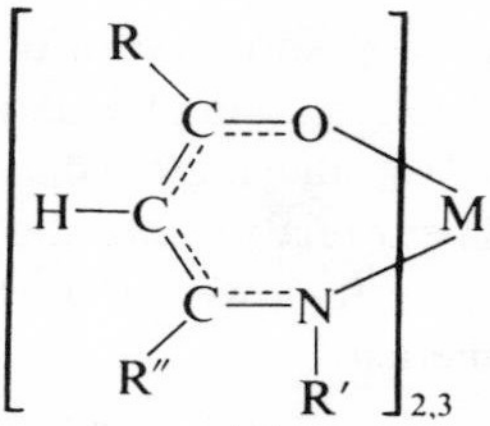

LX

troversy in which the existence of any relationship between the frequency and the stability in these compounds is questioned [31, 70]. Approximate normal coordinate analyses on a series of acetylacetonates of divalent metals [106] give the following metal-oxygen stretching force constants:

$$\begin{array}{ccccccccc} \mathrm{Co(II)} & \sim & \mathrm{Zn(II)} & \ll & \mathrm{Ni(II)} & \sim & \mathrm{Cu(II)} & \ll & \mathrm{Pd(II)} \\ 1.50 & & 1.50 & & 2.05 & & 2.20 & & 2.65 \end{array}$$

(all in units of mdyn/A)

On the other hand, the stability constants ($\log k_1 k_2$) are known to be [72]

$$\begin{array}{ccccccccc} \mathrm{Zn(II)} & < & \mathrm{Co(II)} & < & \mathrm{Ni(II)} & \ll & \mathrm{Cu(II)} & \ll & \mathrm{Pd(II)}. \\ 8.81 & & 9.51 & & 10.38 & & 14.98 & & 27.1 \end{array}$$

Although the order of the metals is similar in the two series, they are not quantitatively in good agreement.

In order to derive a better relationship between the force constant and thermodynamic data, it is necessary to calculate the enthalpy ΔH_{ML} of the reaction

$$[\mathrm{M}^{2+}]_{\mathrm{g}} + 2[\mathrm{L}]_{\mathrm{aq}} \rightleftarrows [\mathrm{M}^{2+}\mathrm{L}_2]_{\mathrm{aq}}.$$

The quantity ΔH_{ML} can be calculated from the relation

$$\Delta H_{ML} = \Delta H_h + \Delta H_c,$$

where ΔH_h is the heat of hydration of the metal and ΔH_c is the enthalpy of the reaction

$$[M^{2+}(H_2O)_x]_{aq} + 2[L]_{aq} \rightleftarrows [M^{2+}L_2]_{aq} + xH_2O_{liq}.$$

The stability constant ($\ln k_1k_2$) of this reaction is related to ΔH_c by

$$-\ln k_1k_2 = (1/RT)(\Delta H_c - T\Delta S_c)$$

It is therefore possible to calculate ΔH_c if the stability constant, the temperature, and ΔS_c are known. These quantities are known for divalent acetylacetonates [67a, 72]. ΔH_{ML} thus calculated takes the values [104]

Co(II) ~	Zn(II) ≪	Cu(II) ~	Ni(II)	[relative to Mn(II)]
45.8	48.3	66.7	68.7	
	(kcal/mole)			

It seems therefore that the M—O stretching force constant can be related better to ΔH_{ML} than to the stability constant. Unfortunately, such an attempt has been made thus far only for a series of divalent acetylacetonate complexes.

F. Equilibrium Studies in Aqueous Solution

Thus far we have discussed the applications of infrared spectroscopy to "static systems," in which the molecules or ions do not undergo structural changes. It is possible, however, to apply infrared spectroscopy to the study of "dynamic systems," such as solution equilibria, and to obtain structural as well as thermodynamic information. Previous equilibrium studies by infrared spectroscopy have been limited largely to aqueous solution, since most coordination compounds are not soluble enough to have their spectra determined in non-aqueous solvents. Using H_2O as a solvent, one can obtain infrared spectra in the regions ca. 2900–1800 and 1500–1000 cm^{-1}. For example, as metal cyano [74] and thiocyanato [59] complexes exhibit their C≡N stretching bands between 2200 and 2000 cm^{-1}, it is possible to study aqueous solution equilibria involving these complexes.

Up to the present time, aqueous (H_2O) infrared spectroscopy has been applied to a very few metal chelate systems. Fronaeus and Larsson [58], for example, have studied the infrared spectra of the Fe(III) and Al(III) oxalato systems in the CO stretching region. They measured the infrared spectra (1500–1200 cm^{-1}) of the Fe(III) oxalato system by changing the molar ratio of $FeCl_3/Na_2C_2O_4$ and obtained the results shown in Fig. 4-19. When this ratio was below 2, only two bands were observed at ca. 1400 and 1260 cm^{-1}. However, two new bands at 1352 and 1309 cm^{-1} appeared and began to

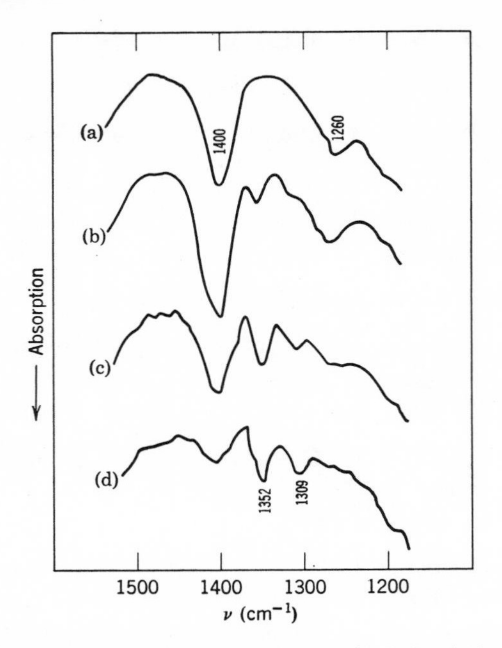

Fig. 4-19. Infrared spectra of the Fe(III) oxalate system in H_2O solutions [58]. (a) Mixture of 0.6$M$$Na_2$(ox) and 1.2$M$$FeCl_3$(1:2); (b) mixture of 1.2M Na_2(ox) and 3.0$M$$FeCl_3$ (1:2.5); (c) mixture of 1.2$M$$Na_2$(ox) and 3.6$M$$FeCl_3$(1:3); (*d*) mixture of 1.0$M$$Na_2$(ox) and 4.1$M$$FeCl_3$(1:4.1).

grow as more $FeCl_3$ was added. These new bands were attributed to the formation of polynuclear complexes, although no detailed information is available about their structures.

Larsson [86] has also studied the infrared spectra of the glycolate

$$(HOCH_2COO^-)$$

complexes with various metal ions in H_2O solution. Through careful studies of the C—OH stretching band near 1060 cm^{-1}, he was able to distinguish three types of complex ions: (1) a species with pure chelated ligand, (2) a species with ligand chelated through a water molecule, and (3) a species with ligand bonded through the carboxylate group only.

If D_2O is used instead of H_2O, it is possible to obtain aqueous infrared spectra in the regions ca. 4000–2900, 2000–1300, and 1100–900 cm^{-1}. As stated in Sec. 4-2E, the un-ionized, ionized, and coordinated carboxyl groups exhibit relatively strong bands between 1750 and 1550 cm^{-1}. Thus, it is possible to study solution equilibria involving carboxylato complexes by observing the CO stretching bands in D_2O solution. Such attempts were first made by

Nakamoto et al. [108] for the equilibrium studies of iminodiacetic acid, nitrilotriacetic acid, ethylenediaminetetraacetic acid, and related chelating agents by changing the pH (pD) of the solution. The following carboxyl group frequencies were established by these investigations:

Type A: un-ionized carboxyl(R_2N—CH_2COOH), 1730–1700 cm^{-1}
Type B: α-ammonium carboxylate(R_2N^+H—CH_2COO^-), 1630–1620 cm^{-1}
Type C: α-aminocarboxylate(R_2N—CH_2COO^-), 1595–1575 cm^{-1}

On the other hand, the coordinated carboxyl group absorbs between 1650 and 1590 cm^{-1}, the exact frequency depending on the nature of the metal. Its frequency becomes higher as the metal-oxygen bond becomes more covalent (LXI and LXII). The COO group absorbs at 1650–1620 cm^{-1} when

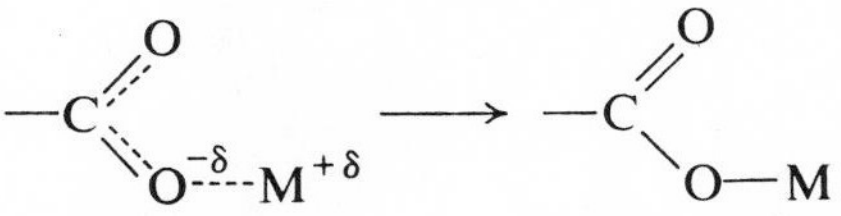

$$-C(\overset{\cdot\cdot}{=}O)\cdots O^{-\delta}\cdots M^{+\delta} \longrightarrow -C(=O)-O-M$$

ionic LXI — covalent LXII

coordinated to such metal ions as Cr(III) and Co(III), whereas it absorbs at 1630–1575 cm^{-1} when coordinated with Cu(II) and Zn(II). It is therefore necessary to choose the metal ions carefully if one intends to distinguish free carboxylate (types B and C) from coordinated carboxylate groups.

Tomita et al. [129] have studied the complex formation of nitrilotriacetic acid, $H_3(NTA)$, with Mg(II) by aqueous infrared spectroscopy. The spectra shown in Fig. 4-20 are of solutions of equimolar amounts of $H_3(NTA)$ and the metal chloride in D_2O at concentrations ca. 5–10% by weight; the pD of the solutions were adjusted to various values by the addition of concentrated NaOD. The spectra of the Mg-NTA chelate from pD 3.2 to 4.2 exhibit a single band at 1625 cm^{-1}, which is identical with that of the free NTA^{3-} ion in the same pD range [128]. This was interpreted as an indication that no complex formation occurs in this pD range, and that the 1625 cm^{-1} band is due to species LXIII. When the pD became higher than 4.2, Tomita et al.

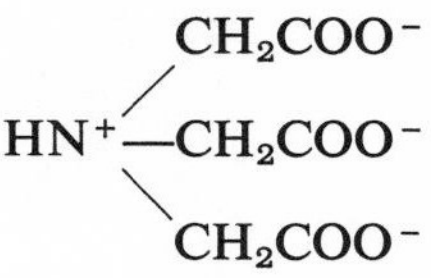

$$HN^{+}(CH_2COO^-)_3$$

LXIII

observed a new band at 1610 cm^{-1}, which was not observed for the free NTA^{3-} ion over the entire pD range investigated. As seen in Fig. 4-20, the

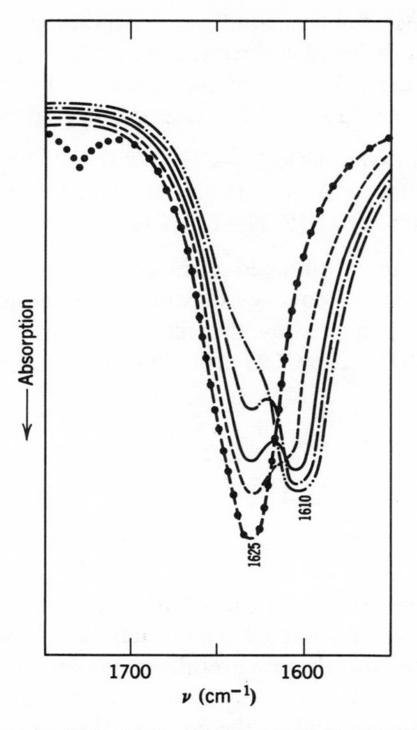

Fig. 4-20. Infrared spectra of Mg-NTA complex in D_2O solutions [129]. (···) pD 3.2; (— — —) pD 4.2; (- - - - -) pD 5.5; (———) pD 6.8; (-·-·-·) pD 10.0; (—- - -—) pD 11.6.

1610 cm^{-1} band becomes more intense and the 1625 cm^{-1} band becomes weaker as the pD increases. It was concluded that this change is due mainly to the shift of the following equilibrium in the direction of complex formation:

$$\underset{\substack{1625\ cm^{-1}\\ \text{LXIII}}}{HN^{\pm}(CH_2COO^-)_3} + Mg^{2+} \rightleftharpoons \underset{\substack{1610\ cm^{-1}\\ \text{LXIV}}}{\left[N(CH_2COO^-)_3 \cdots Mg\right]^-} + H^+$$

By plotting the intensity of these two bands as a function of pD, these authors have calculated the stability constant of the complex ion (log k) to be 5.24. This value is in fairly good agreement with that obtained from potentiometric titration (5.41) [119].

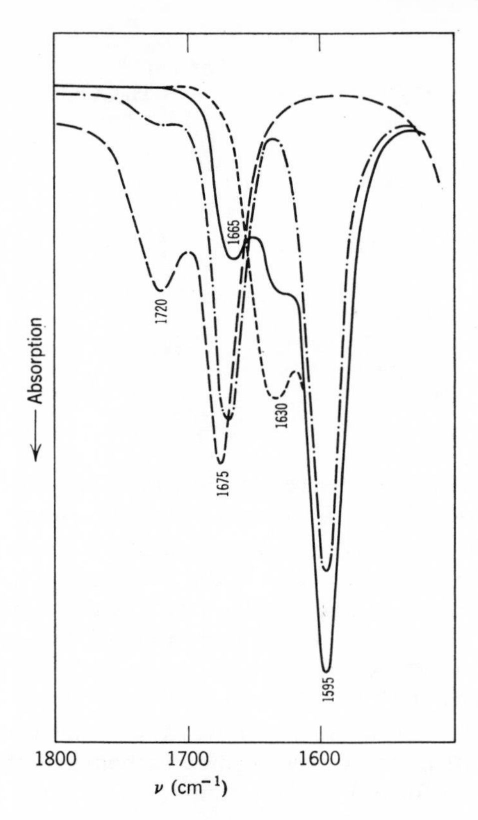

Fig. 4-21. Infrared spectra of glycylglycine in D_2O solution at 0.288*M* concentration and ionic strength 1.0, adjusted with KCl. (— — —) pD 1.75; (·—·—·—) pD 4.31; (———) pD 8.77; (-----) pD 10.29 [77].

Kim and Martell [77] have also studied solution equilibria involving the formation of glycylglycino Cu(II) complexes; they did this by combining aqueous infrared spectroscopy with potentiometric titrations and visible spectroscopy. Figure 4-21 illustrates the infrared spectra of glycylglycine in D_2O solution as a function of the pD. The spectral changes observed were interpreted in terms of the solution equilibria shown on page 280 [78].

At low pD values (1.75) two bands are observed at 1720 and 1675 cm^{-1}; the former is due to the un-ionized carboxyl group and the latter is due to the peptide carbonyl group adjacent to the terminal ammonium group of species LXV. As the pD is raised (4.31), a new band appears at 1595 cm^{-1} which is due to the ionized carboxyl group. Thus the spectrum at this pD value was interpreted as being due to a mixture of species LXV and LXVI. At high

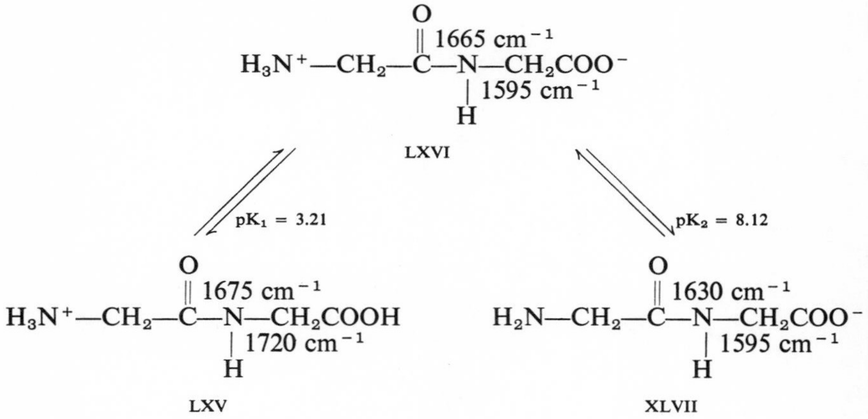

$$\mathrm{H_3N^+{-}CH_2{-}\overset{\overset{\displaystyle O}{\|\ 1665\ cm^{-1}}}{C}{-}\underset{\underset{\displaystyle H}{|\ 1595\ cm^{-1}}}{N}{-}CH_2COO^-}$$

LXVI

$pK_1 = 3.21$ $\quad$ $pK_2 = 8.12$

$$\mathrm{H_3N^+{-}CH_2{-}\overset{\overset{\displaystyle O}{\|\ 1675\ cm^{-1}}}{C}{-}\underset{\underset{\displaystyle H}{|\ 1720\ cm^{-1}}}{N}{-}CH_2COOH} \qquad \mathrm{H_2N{-}CH_2{-}\overset{\overset{\displaystyle O}{\|\ 1630\ cm^{-1}}}{C}{-}\underset{\underset{\displaystyle H}{|\ 1595\ cm^{-1}}}{N}{-}CH_2COO^-}$$

LXV $\qquad$ XLVII

pD (8.77) three bands are observed at 1665, 1630, and 1595 cm^{-1}. The 1665 and 1630 cm^{-1} bands are attributed to the peptide carbonyl groups of species LXVI and LXVII, respectively. At very high pD (10.29) only the two bands (1630 and 1595 cm^{-1}) characteristic of the latter species are observed.

The infrared spectra of glycylglycine mixed with copper chloride in 1:1 molar ratio in D_2O solution (total concentration of the ligand and the metal is 0.2333 *M*) [77] are shown in Fig. 4-22. As shown previously, glycylglycine exhibits three bands at 1720, 1675, and 1595 cm^{-1} at pD = 3.58. However, at the same pD value, the mixture exhibits one extra band at 1625 cm^{-1}. This extra band was attributed to the peptide carbonyl of the metal complex (LXVIII) which was formed by the reaction

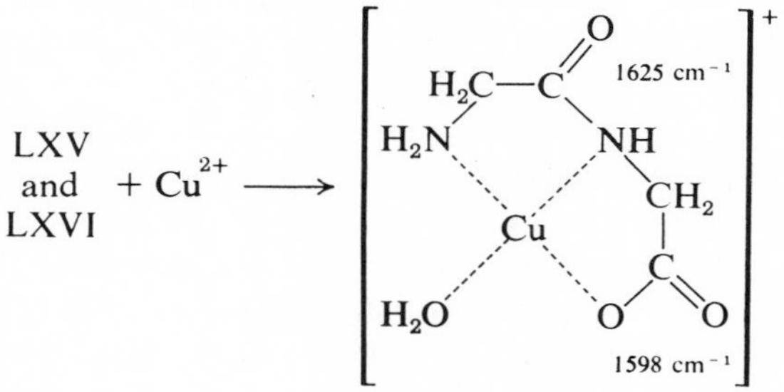

LXV and LXVI $+ Cu^{2+} \longrightarrow$ [H_2C—C(=O, 1625 cm^{-1})—NH—CH_2—C(=O)—O (1598 cm^{-1}); H_2N, NH, H_2O, O coordinated to Cu]$^+$ $+ xH^+$

LXVIII

The same solution exhibits one broad band at ca. 1610 cm^{-1}, if the pD is raised to 5.18. This result was interpreted as an indication that the equilibrium is shifted almost completely to the right-hand side, and that the 1610 cm^{-1} band is an overlap of two bands due to the two carbonyl groups of species LXIX. The shift of the peptide CO stretching band from 1625 (LXVIII) to

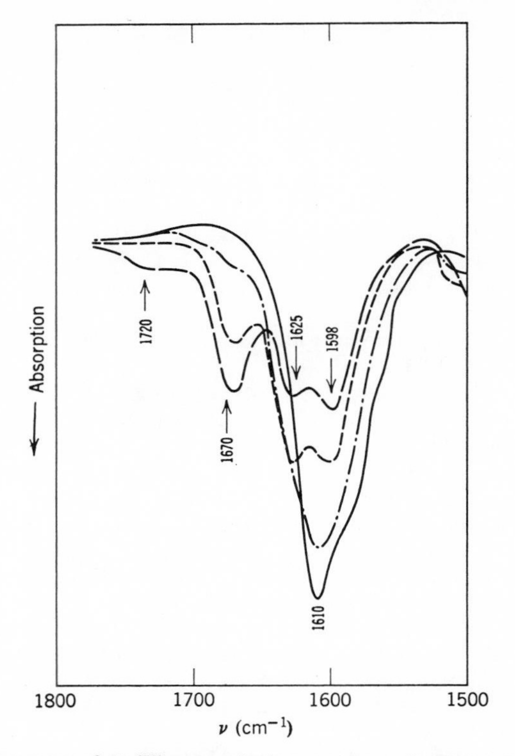

Fig. 4-22. Infrared spectra of Cu(II) glycylglycine complexes in D_2O solutions (1:1); ionic strength 1.0, adjusted with KCl. (———) pD 3.58; (-----) pD 4.24; (·-·-·-) pD 5.18; (———) pD 10.65 [77].

1610 cm^{-1} (LXIX) was interpreted as direct evidence for the ionization of the peptide NH hydrogen, since such an ionization results in the resonance of

LXVI + Cu^{2+} ⟶ [H_2C—C(=O)(1610 cm^{-1})—N—CH_2—C(=O)—O (1598 cm^{-1}); H_2N, N, O and H_2O coordinated to Cu] + $2H^+$

LXIX

the O—C—N system as indicated in structure LXIX. The same authors [79] have also extended their investigations to the triglycine- and tetraglycine-Cu(II) systems.

REFERENCES

Books

[1] Eyring, H., J. Walter, and G. E. Kimball, *Quantum Chemistry*. New York: John Wiley, 1944.

[2] Frazer, R. A., W. J. Duncan, and A.R. Collar, *Elementary Matrices*. Cambridge: University Press, 1960.

[3] Herzberg, G., *Molecular Spectra and Molecular Structure II: Infrared and Raman Spectra of Polyatomic Molecules*. Princeton, N.J.: Van Nostrand, 1945.

[4] King, G. W., *Spectroscopy and Molecular Structure*. New York: Holt, Rinehart and Winston, 1964.

[5] Mizushima, S., and T. Shimanouchi, *Infrared Absorption and Raman Effect*. Tokyo: Kyoritsu, 1958.

[6] Potts, W. J., *Chemical Infrared Spectroscopy*. New York: Wiley, 1963.

[7] Wilson, E. B., J. C. Decius, and P. C. Cross, *Molecular Vibrations*. New York: McGraw-Hill, 1955.

Reviews

[8] Bellamy, L. J., *Infrared Spectra of Complex Molecules*. New York: Wiley, 1957.

[9] Brügel, W., *Einführung in die Ultrarotspektroskopie*. Weinheim: Verlag, 1954.

[10] Colthup, N. B., L. H. Daly, and S. E. Wiberley, *Introduction to Infrared and Raman Spectroscopy*. New York: Academic, 1964.

[11] Jones, R. N., and C. Sandorfy, "The Application of Infrared and Raman Spectrometry to the Elucidation of Molecular Structure," in *Chemical Applications of Spectroscopy*, W. West, ed. New York: Interscience, 1956.

[12] Lecomte, J., "Spectroscopie dans l'Infrarouge," in *Handbuch der Physik*. Berlin: Springer, Vol. 26, 1957.

[13] Mizushima, S., *Structure of Molecules and Internal Rotation*. New York: Academic, 1954.

[14] Nakamoto, K., *Infrared Spectra of Inorganic and Coordination Compounds*. New York: Wiley, 1963.

[15] Nakanishi, K., *Infrared Absorption Spectroscopy*. San Francisco: Holden-Day, 1962.

[16] Siebert, H., *Anwendungen der Schwingungspektroskopie in der Anorganischen Chemie*. Berlin: Springer-Verlag, 1966.

[17] Szymanski, H. A., *IR: Theory and Practice of Infrared Spectroscopy*. New York: Plenum Press, 1964.

Other References

[18] Adams, D. M., and P. J. Chandler, *Chem. Commun.*, **3**, 69 (1966).

[19] Adams, D. M., J. Chatt, J. M. Davidson, and J. Gerratt, *J. Chem. Soc.*, 2189 (1963).

[20] Adams, D. M., J. Chatt, J. Gerratt, and A. D. Westland, *J. Chem. Soc.*, 734 (1964).

[21] Allen, G., J. Lewis, R. G. Long, and C. Oldham, *Nature*, **202**, 580 (1964).

[22] Armendarez, P. X., and K. Nakamoto, *Inorg. Chem.*, **5**, 796 (1966).

[23] Baldwin, M. E., *J. Chem. Soc.*, 3123 (1961).

[24] Baldwin, M. E., *J. Chem. Soc.*, 4369 (1960); 471 (1961).

[25] Barraclough, C. G., and M. L. Tobe, *J. Chem. Soc.*, 1993 (1961).

[26] Basolo, F., C. J. Ballhausen, and J. Bjerrum, *Acta Chem. Scand.*, **9**, 810 (1955).
[27] Basolo, F., J. L. Burmeister, and A. J. Poe, *J. Am. Chem. Soc.*, **85**, 1700 (1963); *Inorg. Chem.*, **3**, 1202 (1964).
[28] Basolo, F., and G. S. Hammaker, *J. Am. Chem. Soc.*, **82**, 1001 (1960); *Inorg. Chem.*, **1**, 1 (1962).
[29] Behnke, G. T., and K. Nakamoto, *Inorg. Chem.*, **6**, 433, 440 (1967).
[30] Bellamy, L. J., and R. F. Branch, *J. Chem. Soc.*, 4487 (1954).
[31] Bellamy, L. J., and R. F. Branch, *J. Chem. Soc.*, 4491 (1954).
[32] Biagetti, R. V., and H. M. Haendler, *Inorg. Chem.*, **5**, 383 (1966).
[33] Brodersen, K., *Z. Anorg. Allgem. Chem.*, **298**, 142 (1959).
[34] Brown, J. K., N. Sheppard, and D. M. Simpson, *Phil. Trans. Roy. Soc. London*, **A247**, 35 (1954).
[35] Cavalca, L., M. Nardelli, and G. Fava, *Acta Cryst.*, **13**, 594 (1960).
[36] Chamberlain, M. M., and J. C. Bailar, Jr., *J. Am. Chem. Soc.*, **81**, 6421 (1959).
[37] Clark, R. J. H., *Spectrochim. Acta*, **21**, 955 (1965).
[38] Clark, R. J. H., and T. M. Dunn, *J. Chem. Soc.*, 1198 (1963).
[39] Clark, R. J. H., and C. S. Williams, *Inorg. Chem.*, **4**, 350 (1965).
[40] Coates, G. E., and C. Perkin, *J. Chem. Soc.*, 421 (1963).
[41] Coates, G. E., and D. Ridley, *J. Chem. Soc.*, 166 (1964).
[42] Condrate, R. A., and K. Nakamoto, *J. Chem. Phys.*, **42**, 2590 (1965).
[43] Cotton, F. A., and R. Francis, *J. Am. Chem. Soc.*, **82**, 2986 (1960).
[44] Cotton, F. A., and R. H. Soderberg, *J. Am. Chem. Soc.*, **85**, 2402 (1963).
[44a] Crawford, B. L., and W. H. Fletcher, *J. Chem. Phys.*, **19**, 141 (1951).
[45] Curtis, N. F., and Y. M. Curtis, *Inorg. Chem.*, **4**, 804 (1965).
[46] Decius, J. C., *J. Chem. Phys.*, **16**, 1025 (1948).
[47] Dows, D. A., A. Haim, and W. K. Wilmarth, *J. Inorg. Nucl. Chem.*, **21**, 33 (1961).
[48] Elliott, H., and B. J. Hathaway, *Spectrochim. Acta*, **21**, 1047 (1965).
[49] Eskenazi, R., J. Raskovan, and R. Levitus, *J. Inorg. Nucl. Chem.*, **28**, 521 (1966).
[50] Fay, R. C., and T. S. Piper, *J. Am. Chem. Soc.*, **84**, 2303 (1962).
[51] Ferraro, J. R., and A. Walker, *J. Chem. Phys.*, **42**, 1273 (1965).
[52] Ferraro, J. R., and A. Walker, *J. Chem. Phys.*, **42**, 1278 (1965).
[53] Ferraro, J. R., A. Walker, and C. Cristallini, *Inorg. Nucl. Chem. Letters*, **1**, 25 (1965).
[54] Field, B. O., and C. J. Hardy, *Quart. Rev. (London)*, **18**, 361 (1964).
[55] Figgis, B. N., J. Lewis, R. F. Long, R. Mason, R. S. Nyholm, P. J. Pauling, and G. B. Robertson, *Nature*, **195**, 1278 (1962).
[56] Finholt, J. E., R. W. Anderson, J. A. Fyfe, and K. G. Caulton, *Inorg. Chem.*, **4**, 43 (1965).
[57] Freeman, H. C., J. E. W. L. Smith, and J. C. Taylor, *Nature*, **184**, 707 (1959); *Acta Cryst.*, **14**, 407 (1961).
[58] Fronaeus, S., and R. Larsson, *Acta Chem. Scand.*, **14**, 1364 (1960).
[59] Fronaeus, S., and R. Larsson, *Acta Chem. Scand.*, **16**, 1433, 1447 (1962).
[60] Fujita, J., A. E. Martell, and K. Nakamoto, *J. Chem. Phys.*, **36**, 324 (1962).
[61] Fujita, J., A. E. Martell, and K. Nakamoto, *J. Chem. Phys.*, **36**, 331 (1962).
[62] Fujita, J., A. E. Martell, and K. Nakamoto, *J. Chem. Phys.*, **36**, 339 (1962).
[63] Fujita, J., and K. Nakamoto, *Bull. Chem. Soc. Japan*, **37**, 528 (1964).
[64] Fujita, J., K. Nakamoto, and M. Kobayashi, *J. Am. Chem. Soc.*, **78**, 3295 (1956).
[65] Fujiyama, T., K. Tokumaru, and T. Shimanouchi, *Spectrochim. Acta*, **20**, 415 (1964).
[66] Gatehouse, B. M., S. E. Livingstone, and R. S. Nyholm, *J. Chem. Soc.*, 3137 (1958).

[67] Gatehouse, B. M., S. E. Livingstone, and R. S. Nyholm, *J. Chem. Soc.*, 4222 (1957); *J. Inorg. Nucl. Chem.*, **8**, 75 (1958).
[67a] George, P., and D. S. McClure, *Prog. Inorg. Chem.*, **1**, 426 (1959).
[68] Grogan, M. J., and K. Nakamoto, *J. Am. Chem. Soc.*, **88**, 5454 (1966).
[69] Hathaway, B. J., and A. E. Underhill, *J. Chem. Soc.*, 3091 (1961).
[70] Holtzclaw, H. F., and J. P. Collman, *J. Am. Chem. Soc.*, **79**, 3318 (1957).
[71] Hughes, E. W., and H. L. Yakel, *Acta Cryst.*, **14**, 345 (1961).
[72] Izatt, R. M., C. G. Hass, B. P. Block, and W. C. Fernelius, *J. Phys. Chem.*, **58**, 1133 (1954).
[73] Jeffrey, G. A., and G. S. Parry, *J. Chem. Soc.*, 4864 (1952).
[74] Jones, L. H., and R. A. Penneman, *J. Chem. Phys.*, **22**, 965 (1954); **24**, 293 (1956); *J. Inorg. Nucl. Chem.*, **20**, 19 (1961).
[75] Kedzia, B., P. X. Armendarez, and K. Nakamoto, to be published.
[75a] Kida, S., *Bull. Chem. Soc. Japan*, **39**, 2415 (1966).
[76] Kieft, J., and K. Nakamoto, *J. Inorg. Nucl. Chem.*, **29**, 2561 (1967).
[77] Kim, M. K., and A. E. Martell, *Biochemistry*, **3**, 1169 (1964).
[78] Kim, M. K., and A. E. Martell, *J. Am. Chem. Soc.*, **85**, 3080 (1963).
[79] Kim, M. K., and A. E. Martell, *J. Am. Chem. Soc.*, **88**, 914 (1966).
[80] Kinoshita, Y., I. Matsubara, T. Higuchi, and Y. Saito, *Bull. Chem. Soc. Japan*, **32**, 1221 (1959).
[81] Kinoshita, Y., I. Matsubara, and Y. Saito, *Bull. Chem. Soc. Japan*, **32**, 741 (1959).
[82] Kinoshita, Y., I. Matsubara, and Y. Saito, *Bull. Chem. Soc. Japan*, **32**, 1216 (1959).
[83] Kubota, M., and D. L. Johnston, *J. Am. Chem. Soc.*, **88**, 2451 (1966).
[84] Kubota, M., D. L. Johnston, and I. Matsubara, *Inorg. Chem.*, **5**, 386 (1966).
[85] Lane, T. J., J. A. Durkin, and R. J. Hooper, *Spectrochim. Acta*, **20**, 1013 (1964).
[86] Larsson, R., *Acta Chem. Scand.*, **19**, 783 (1965).
[87] Lever, A. B. P., *Inorg. Chem.*, **4**, 1042 (1965).
[88] Lewis, J., R. F. Long, and C. Oldham, *J. Chem. Soc.*, 6740 (1965).
[89] Morris, M. L., and D. H. Busch, *J. Am. Chem. Soc.*, **82**, 1521 (1960).
[90] Matsubara, I., *Bull. Chem. Soc. Japan*, **34**, 1710 (1961).
[91] Matsubara, I., *Bull. Chem. Soc. Japan*, **34**, 1719 (1961); *J. Chem. Phys.*, **35**, 37 (1961).
[92] Matsubara, I., *Bull. Chem. Soc. Japan*, **35**, 27 (1962).
[93] Mellor, D. P., and L. E. Maley, *Nature*, **159**, 379 (1947).
[94] Merritt, P. E., and S. E. Wiberley, *J. Phys. Chem.*, **59**, 55 (1955).
[95] Moore, L. E., R. B. Gayhart, and W. E. Bull, *J. Inorg. Nucl. Chem.*, **26**, 896 (1964).
[96] Morino, T., and K. Kuchitsu, *J. Chem. Phys.*, **20**, 1809 (1952).
[97] Morris, M. L., and D. H. Busch, *J. Am. Chem. Soc.*, **82**, 1521 (1960).
[98] Mizushima, S., I. Ichishima, I. Nakagawa, and J. V. Quagliano, *J. Phys. Chem.*, **59**, 293 (1955).
[99] Nakagawa, I., and T. Shimanouchi, *Spectrochim. Acta*, **22**, 759 (1966).
[100] Nakahara, A., Y. Saito, and H. Kuroya, *Bull. Chem. Soc. Japan*, **25**, 331 (1952).
[101] Nakamoto, K., J. Fujita, R. A. Condrate, and T. Morimoto, *J. Chem. Phys.*, **39**, 423 (1963).
[102] Nakamoto, K., J. Fujita, S. Tanaka, and M. Kobayashi, *J. Am. Chem. Soc.*, **79**, 4904 (1957).
[103] Nakamoto, K., and A. E. Martell, *J. Chem. Phys.*, **32**, 588 (1960).
[104] Nakamoto, K., P. J. McCarthy, and A. E. Martell, *J. Am. Chem. Soc.*, **83**, 1272 (1961).

[105] Nakamoto, K., P. J. McCarthy, and B. Miniatas, *Spectrochim. Acta*, **21**, 379 (1965).
[106] Nakamoto, K., P. J. McCarthy, A. Ruby, and A. E. Martell, *J. Am. Chem. Soc.*, **83**, 1066 (1961).
[107] Nakamoto, K., Y. Morimoto, and A. E. Martell, *J. Am. Chem. Soc.*, **83**, 4528 (1961).
[108] Nakamoto, K., Y. Morimoto, and A. E. Martell, *J. Am. Chem. Soc.*, **84**, 2081 (1962); **85**, 309 (1963).
[109] Nakamura, K., *J. Chem. Soc. Japan*, **79**, 1411 (1958).
[110] Nakatsu, K., Y. Saito, and H. Kuroya, *Bull. Chem. Soc. Japan*, **29**, 428 (1956).
[111] Newman, G., and D. B. Powell, *J. Chem. Soc.*, 477 (1961).
[112] Newman, G., and D. B. Powell, *J. Chem. Soc.*, 3447 (1962).
[113] Newman, G., and D. B. Powell, *Spectrochim. Acta*, **19**, 213 (1963).
[114] Penland, R. B., T. J. Lane, and J. V. Quagliano, *J. Am. Chem. Soc.*, **78**, 887 (1956).
[115] Powell, D. B., and N. Sheppard, *J. Chem. Soc.*, 3089 (1959).
[116] Rigg, J. M., and E. Sherwin, *J. Inorg. Nucl. Chem.*, **27**, 653 (1965).
[117] Sabatini, A., and L. Sacconi, *J. Am. Chem. Soc.*, **86**, 17 (1964).
[118] Scherer, J. R., and J. Overend, *Spectrochim. Acta*, **17**, 719 (1961).
[119] Schwarzenbach, G., E. Kampitsch, and R. Stener, *Helv. Chim. Acta*, **28**, 828 (1945).
[120] Sharma, V. S., H. B. Mathur, and A. B. Biswas, *Spectrochim. Acta*, **17**, 895 (1961).
[121] Shimanouchi, T., *J. Chem. Phys.*, **17**, 245, 734 and 848 (1949).
[122] Shimanouchi, T., *J. Chem. Phys.*, **25**, 660 (1956).
[123] Shimanouchi, T., *Pure Appl. Chem.*, **7**, 131 (1963).
[124] Shimura, Y., and T. Tsuchida, *Bull. Chem. Soc. Japan*, **29**, 311 (1956).
[125] Siebert, H., *Z. Anorg. Allgem. Chem.*, **296**, 280 (1958).
[126] Sweeny, D. M., S. Mizushima, and J. V. Quagliano, *J. Am. Chem. Soc.*, **77**, 6521 (1955).
[127] Tanaka, N., H. Sugi, and J. Fujita, *Bull. Chem. Soc. Japan*, **37**, 640 (1964).
[128] Tomita, Y., and K. Ueno, *Bull. Chem. Soc. Japan*, **36**, 1069 (1963).
[129] Tomita, Y., T. Ando, and K. Ueno, *J. Phys. Chem.*, **69**, 404 (1965).
[130] Topping, G., *Spectrochim. Acta*, **21**, 1743 (1965).
[131] Tsuboi, M., T. Onishi, I. Nakagawa, T. Shimanouchi, and S. Mizushima, *Spectrochim. Acta*, **21**, 253 (1958).
[132] van Niekerk, J. N., and F. R. L. Schoening, *Acta Cryst.*, **4**, 35 (1951).
[133] van Niekerk, J. N., and F. R. L. Schoening, *Acta Cryst.*, **4**, 381 (1951).
[134] Varshavaskii, Y. S., E. N. Inkova, and A. A. Grinberg, *Russ. J. Inorg. Chem.* (*English Transl.*), **8**, 1394 (1963).
[135] Walker, A., and J. R. Ferraro, *J. Chem. Phys.*, **43**, 2689 (1965).
[136] Wickenden, A. E., and R. A. Krause, *Inorg. Chem.*, **4**, 407 (1965).
[137] Wilson, E. B., *J. Chem. Phys.*, **7**, 1047 (1939); **9**, 76 (1941).

5

ELECTRON SPIN RESONANCE SPECTROSCOPY

SHIZUO FUJIWARA
The University of Tokyo

Electron spin resonance (ESR) is a branch of spectroscopy which has been developed since 1945 and has proved useful for the investigation of electronic states of atoms and molecules. It relates to the transitions between Zeeman energy levels of any electronic paramagnetic system that carries angular momentum. Systems that show ESR phenomena are atoms having an odd number of electrons, ions having partly filled inner electron shells, such as the transition elements and the lanthanides, and other molecular species that carry angular momentum of electronic origin.

Like other branches of spectroscopy, ESR is experimentally concerned with resonant position, line width, line shape, and intensity. The resonant position of ESR is referred to as the *g* value and is directly determined by the separation of the energy levels of the system under investigation. The variation of the *g* value is interpreted in terms of the first and the second order perturbation by the spin-orbit interactions. The line width is produced by the relaxation of the spin energy state. Several mechanisms contribute to the relaxation of ESR, and it is an interesting problem to elucidate these mechanisms. Investigation of the correlation of line width with the chemical nature of bonding is of special importance. Among the systems that give ESR spectra, chelates of transition elements are discussed in this chapter, mainly because they form most of the common chelated complex compounds. Before proceeding to a discussion of the results of the investigations, we shall briefly summarize several important points which should be remembered in interpreting ESR spectra. For details the reader is referred to the books and reviews listed at the end of the chapter; see especially the excellent recent review by McGarvey.

5-1 THEORY

Chelate compounds whose ESR spectra have been studied in detail are those of Cu(II), VO(IV), and Ti(III) ions. In the first row transition metal ions, the unpaired $3d$ electron(s) is (are), in general, under the influence of a pronounced crystal field. In this discussion we deal only with the first row transition metal ions with non-degenerate orbital ground state. In Cu(II) and Ti(III) ions, the fivefold orbital degeneracy is lifted in tetragonal and trigonal crystal fields, respectively. The effect of the crystal field is felt by the electron spin through spin-orbit interaction. The ESR spectra of the first row transition metal ions of this type are described by using an *effective spin Hamiltonian* [16, 158],

$$\mathscr{H} = \beta_0 \mathbf{S}\cdot\tilde{\mathbf{g}}\cdot\mathbf{H} + \mathbf{S}\cdot\tilde{\mathbf{D}}\cdot\mathbf{S} + \mathbf{S}\cdot\tilde{\mathbf{A}}\cdot\mathbf{I}. \tag{5-1}$$

Here β_0 is the Bohr magneton, $\mathbf{S}$ the electron spin, $\mathbf{I}$ the nuclear spin of the paramagnetic ion, and $\mathbf{H}$ is the magnetic field; $\tilde{\mathbf{g}}$, $\tilde{\mathbf{D}}$, $\tilde{\mathbf{A}}$ are tensors of second rank. In the first term of (5-1) $\tilde{\mathbf{g}}$ is given by

$$g_{ij} = g_e(\delta_{ij} - 2\lambda\Lambda_{ij}), \tag{5-2}$$

where g_e is the g value for the free electron and is equal to 2.0023; δ_{ij} is Kronecker's delta; λ is the spin-orbit coupling constant, which is positive for a $3d$ shell less than half filled and negative for one more than half filled. Λ_{ij} is defined in terms of the matrix elements of the orbital angular momentum operator $\mathbf{L}$ by

$$\Lambda_{ij} = \sum_{n \neq 0} \frac{\langle 0|\, L_i \,|n\rangle\langle n|\, L_j \,|0\rangle}{E_n - E_o}, \tag{5-3}$$

where $|n\rangle$ and E_n denote the nth eigenstate and corresponding energy of the $3d$ electronic state of an ion residing in a crystalline electric potential; 0 refers to the lowest state [2].

In the second term of (5-1) D_{ij} is of second order with respect to λ and is given by

$$D_{ij} = -\lambda^2\Lambda_{ij}. \tag{5-4}$$

Though this term is responsible for the *fine structure* of ESR spectra, this is of no consequence for ions of spin $\frac{1}{2}$, such as Cu(II) and Ti(III), with which we are mainly concerned here.

The third term of (5-1) gives a *hyperfine splitting* of ESR spectra. The hyperfine tensor, $\tilde{\mathbf{A}}$, gives a measure of the interaction between electron spin and nuclear spin. Usually the nucleus involved is that of the metal atom itself, but in some cases hyperfine structure due to the nuclear spin(s) belonging to the ligands is observed. This is usually called *super hyperfine structure.* The hyperfine coupling tensor is determined by two factors: the contact interaction of nuclear and electronic spins, and the ordinary dipole-dipole coupling. Explicit expressions for the hyperfine spin Hamiltonian in a variety of cases are found in the literature [16, 5].

Since $\tilde{\mathbf{g}}$ and $\tilde{\mathbf{A}}$ are tensors, the observed spectra of a single crystal are dependent on the direction of the applied magnetic field. If a spin Hamiltonian contains only the g term, the resonance frequency, ν, is given by

$$\nu = \frac{\beta_0 g^* H}{h}, \tag{5-5}$$

where h is Planck's constant, H the magnitude of the magnetic field, and g^* is given by

$$g^* = [g_{xx}{}^2\alpha^2 + g_{yy}{}^2\beta^2 + g_{zz}{}^2\gamma^2]^{1/2}. \tag{5-6}$$

Here g_{xx}, g_{yy}, and g_{zz} are the principal values of $\tilde{\mathbf{g}}$, and α, β, and γ are the direction cosines of the applied magnetic field with respect to the principal axes. In powdered samples the observed spectra result from the superposition of the spectrum of a single crystal with a magnetic field applied in all directions. In solutions the spectra corresponding to the spin Hamiltonian,

$$\mathscr{H} = \beta_0 g_0 \mathbf{S}\cdot\mathbf{H} + A_0 \mathbf{S}\cdot\mathbf{I}, \tag{5-7}$$

are observed in which g_0 and A_0 are the averages of the principal values of the $\tilde{\mathbf{g}}$ and $\tilde{\mathbf{A}}$ tensors, respectively.

A number of factors influence the line width of the observed spectra. In solids, when the magnetic ions are close to each other, the dipole-dipole interaction usually gives rise to a broad spectrum. As a result, ESR lines are sometimes too broad to be detected. Broadening from this source can usually be avoided by using samples which are diluted magnetically with various diamagnetic ions or molecules. γ-Irradiated polyvinyl alcohol (PVA) gel has been used successfully as a medium for magnetic dilution [72]. The dipolar width is also narrowed through exchange interaction between electrons. For this it is necessary that the electron clouds of the two magnetic ions overlap each other. When exchange interaction is comparable to, or stronger than, the Zeeman term [the first term of (5-1)], the two magnetic ions should be considered a pair. In this case two electron spins form singlet and triplet states. The triplet state is usually the upper of the two and is responsible for the ESR spectrum. The energy separation between these levels determines the intensity of the spectrum. In solutions, on the contrary, the line width is narrowed through the Brownian motion of a complex molecule. Factors possibly influencing the line width are (1) the fluctuating parts of the

$\tilde{\mathbf{g}}$ and $\tilde{\mathbf{A}}$ tensors, (2) dipole and exchange interaction between the magnetic ions in the solution, and (3) the modulating effect of the crystal field through spin-orbit interaction.

5-2 APPLICATIONS

A. ESR Parameters and Chemical Bonding

As a first example of application of ESR to the study of metal chelate compounds, we show in this section how the magnitude of the elements of $\tilde{\mathbf{g}}$ and $\tilde{\mathbf{A}}$ tensors (hereafter referred to as ESR parameters) can be correlated with the nature of the metal-ligand bonds. The method is based on the analyses made by Abragam and Pryce [17], Bleaney, Bowers, and Ingram [44], Bleaney, Bowers, and Trenam [47], Maki and McGarvey [123, 124], and Neiman and Kivelson [102, 147, 148].

The first step is to determine ESR parameters as accurately as possible. A typical example concerns Cu(II) ions. Copper ion possesses a ground state ^2D corresponding to the $3d^9$ configuration. When the copper ion is placed in a cubic crystal field, the energy state is split into two levels, $^2E_g(t_{2g})^6(e_g)^3$, the ground state, and $^2T_{2g}(t_{2g})^5(e_g)^4$, the excited state. These states are further split into finer levels when copper ion is placed in a tetragonal field. The appropriate spin Hamiltonian for a square-planar Cu(II) complex is, in general, given by

$$\mathscr{H} = \beta_0[g_\parallel H_z S_z + g_\perp(H_x S_x + H_y S_y)] + A_\parallel S_z I_z + A_\perp(S_x I_x + S_y I_y), \tag{5-8}$$

where x, y, and z refer to the direction of the three principal axes of the $\tilde{\mathbf{g}}$ and $\tilde{\mathbf{A}}$ tensors, which are assumed to possess common principal axes; the z and $x(y)$ axes are parallel and perpendicular to the symmetry axis of the molecule, respectively; and $g_\parallel = g_{zz}$ and $g_\perp = g_{xx} = g_{yy}$.

A single crystal sample of a cupric compound with the spin Hamiltonian (5-1) will absorb energy, $h\nu$, according to the conditions given in [5-9, 5-10, and 5-11]:

$$h\nu = g\beta_0 H + Am, \tag{5-9}$$

where

$$g = (g_\parallel^2 \cos^2 \xi + g_\perp^2 \sin^2 \xi)^{1/2} \tag{5-10}$$

and

$$A = g^{-1}(A_\parallel^2 g_\parallel^2 \cos^2 \xi + A_\perp^2 g_\perp^2 \sin^2 \xi)^{1/2}. \tag{5-11}$$

In (5-9) H is the magnitude of the magnetic field intensity; m is the eigenvalue of the z component of the copper nuclear spin (i.e., $m = \pm\frac{1}{2}$ and $\frac{3}{2}$); ξ is the angle between the z axis and the applied magnetic field. As noted above, the hyperfine structure is mostly smeared out in the undiluted paramagnetic crystals, and the use of a magnetically diluted single crystal is strongly recommended.

Although a single crystal provides, in principle, the most accurate ESR parameters, a chelated metal compound is usually in a polycrystalline or vitreous state. In such a compound, the molecules are randomly oriented and the spectrum is the sum of the resonances of molecules having all orientations. The spectra of transition metal complexes in glasses and in magnetically diluted polycrystalline samples were discussed first by Pake and Sands [153, 177], and by Malmström and Vänngård [128]. More detailed discussions were given by Neiman and Kivelson [102, 147, 148] for systems in which definite hyperfine structure and super hyperfine structure due to the interaction of the unpaired electron with ligand nuclei are observed.

Gersmann and Swalen [75] obtained ESR spectra of cupric diethyldithiocarbamate in chloroform-toluene and in chloroform-pyridine solutions, and of cupric and vanadyl acetylacetonates in chloroform-toluene solution. From the spectra observed in the vitreous state, they obtained $g_{\parallel}$ and $A_{\parallel}$, and from these values they calculated $g_{\perp}$ and $A_{\perp}$ using g_0 and A_0, which are obtained in solution. They calculated the intensity function from the magnetic parameters and compared it with the spectra observed in the vitreous state. Some of the magnetic parameters obtained are shown in Table 5-1.

A glassy state sample may differ slightly from a polycrystalline one in magnetic properties. In a polycrystalline sample all molecules are assumed to be in identical environments, and the microcrystalline axes are assumed to be randomly oriented. In a glassy sample, on the other hand, the environment of each molecule, and hence its magnetic parameters, may differ from that of the others. The parameters which are obtained from magnetically diluted samples have been conventionally taken as applicable for the undiluted sample. Recently, however, evidence has been obtained that this conventional assumption does not necessarily hold. Copper oxinate shows different values of A among samples with different diluents [70].

If a molecule containing Cu(II) has a tetragonal structure stretched along the symmetry axis (taken as the z axis), ${}^2B_{1g}(d_{x^2-y^2})$ is the ground state, and $g_{\parallel}$ and $g_{\perp}$ values are related to the energy scheme [29, 2] as shown in (5-12) and (5-13):

$$g_{\parallel} = 2\left(1 - \frac{4\lambda}{\Delta_1}\right), \tag{5-12}$$

$$g_{\perp} = 2\left(1 - \frac{\lambda}{\Delta_2}\right). \tag{5-13}$$

Here

$$\Delta_1 = E(B_{2g}) - E(B_{1g}) \tag{5-14}$$

and

$$\Delta_2 = E(E_g) - E(B_{1g}), \tag{5-15}$$

Table 5-1 ESR Parameters of Complexes of Cu(II) and VO(IV) [75]

Compound	Solvent[a]	$g_{\parallel}$	$g_{\perp}$	$A_{\parallel} \times 10^4$ (cm^{-1})	$A_{\perp} \times 10^4$ (cm^{-1})	$A_N \times 10^4$[b] (cm^{-1})
Bis(acetylacetono)-copper(II)	A	2.264	2.036	145.5	29	c
Bis(diethyldithio-carbamato)-copper(II)	A	2.098	2.035	154[d]	40	c
				165	43	c
	B	2.121	2.040	134	25	c
				146.5	27	c
Bis(salicylaldoximato)-copper(II)	C	2.171	2.020	183	41	14
Bis(salicylaldimino)-copper(II)	C	2.14	2.08	168	16	e
Bis(8-quinolinolato)-copper(II)	C	2.172	2.042	162	25	10
Dichlorophenanthro-linecopper(II)	C	2.22	2.08	119	29	e
Oxo-(bis-acetylaceto-no)vanadium(IV)	C	1.944	1.996	173.5	63.5	c

[a] Solvent A: 60% toluene, 40% chloroform. Solvent B: 40% pyridine, 60% chloroform. Solvent C: 60% chloroform, 40% toluene.
[b] Hyperfine splitting from ligand nitrogen atoms.
[c] Hyperfine structure from ligands not expected.
[d] Results for the two isotopes of copper, 63 and 65, respectively.
[e] Hyperfine structure not observed.

where E designates the electronic energy of the metal atomic orbital, given in parentheses. If the molecule is compressed along the z axis,

$$g_{\parallel} = 2, \tag{5-16}$$

$$g_{\perp} = 2\left(1 - \frac{3\lambda}{\Delta_3}\right), \tag{5-17}$$

where

$$\Delta_3 = E(E_g) - E(A_{1g}). \tag{5-18}$$

McGarvey [134] first applied molecular orbital theory to the interpretation of the observed parameters, and several papers followed his treatment [23, 51, 88, 163, 192, 207]. One experimental feature which made it necessary to introduce molecular orbital treatment was the observation of super hyperfine interaction between the electronic spin and the nitrogen nucleus of the ligand. Although most papers assume tetragonal symmetry around the z axis, Gersmann and Swalen [75] extended the treatment to the more general case in which a symmetry axis is not present. Here it is assumed that tetragonal symmetry around the z axis is maintained in the coordinated compounds of copper(II).

Gersmann and Swalen [75] combined the proper linear combination of ligand orbitals with the copper d orbitals:

$$\psi_{B_{1g}} = \alpha d_{x^2-y^2} - \tfrac{1}{2}\alpha'[-\sigma_x^{(1)} + \sigma_y^{(2)} + \sigma_x^{(3)} - \sigma_y^{(4)}], \quad (5\text{-}19)$$

$$\psi_{B_{2g}} = \beta d_{xy} - \tfrac{1}{2}(1 - \beta^2)^{1/2}[p_y^{(1)} + p_x^{(2)} - p_y^{(3)} - p_x^{(4)}], \quad (5\text{-}20)$$

$$\psi_{A_{1g}} = \gamma d_{3z^2-r^2} - \tfrac{1}{2}(1 - \gamma^2)^{1/2}[\sigma_x^{(1)} + \sigma_y^{(2)} - \sigma_x^{(3)} - \sigma_y^{(4)}], \quad (5\text{-}21)$$

$$\psi_{E_g} = \begin{cases} \delta d_{xz} - (1 - \delta^2)^{1/2}[p_z^{(1)} - p_z^{(3)}]/\sqrt{2} & (5\text{-}22) \\ \delta d_{yz} - (1 - \delta^2)^{1/2}[p_z^{(2)} - p_z^{(4)}]/\sqrt{2}, & (5\text{-}23) \end{cases}$$

where the notation follows that of Gersmann and Swalen. The overlap integral is involved only in the $\psi_{B_{1g}}$ function, where α and α' are related by

$$\alpha^2 + \alpha'^2 - 2\alpha\alpha' S = 1, \quad (5\text{-}24)$$

where S is the overlap integral. In these wave functions the σ orbitals are hybridized sp orbitals of the type

$$\sigma^{(i)} = np^{(i)} \mp (1 - n^2)^{1/2} s^{(i)}, \quad (5\text{-}25)$$

where the plus sign applies to the ligand atoms on the positive x and y axes, the corresponding minus sign to those on the negative x and y axes and $0 \leqq n \leqq 1$.

The B_{1g} and A_{1g} states account for the σ-bonding to the copper. The B_{2g} state represents the in-plane π bonding and the E_g states represent the out-of-plane π bonding. The smaller the value of α, β, γ, or δ, the more covalent is the bonding of the appropriate type associated with each parameter. The observed magnetic parameters are related to the bonding parameters as follows:

$$g_\parallel = 2.0023 - \left(\frac{8\lambda}{\Delta E_{xy}}\right)[\alpha^2\beta^2 - f(\beta)], \quad (5\text{-}26)$$

$$g_\perp = 2.0023 - \left(\frac{2\lambda}{\Delta E_{xz}}\right)[\alpha^2\delta^2 - g(\delta)], \quad (5\text{-}27)$$

$$A_\parallel = P\left[-\alpha^2(\tfrac{4}{7} + \kappa) - 2\lambda\alpha^2\left(\frac{4\beta^2}{\Delta E_{xy}} + \frac{3}{7}\frac{\delta^2}{\Delta E_{xz}}\right)\right], \quad (5\text{-}28)$$

$$A_\perp = P\left[\alpha^2(\tfrac{2}{7} - \kappa) - \frac{22}{14}\frac{\lambda\alpha^2\delta^2}{\Delta E_{xz}}\right], \quad (5\text{-}29)$$

where

$$f(\beta) = \frac{\alpha\alpha'\beta^2 S + \alpha\alpha'\beta(1 - \beta^2)^{1/2}T(n)}{\sqrt{2}}, \quad (5\text{-}30)$$

$$g(\delta) = \frac{\alpha\alpha'\delta^2 S + \alpha\alpha'\delta(1 - \delta^2)^{1/2}T(n)}{\sqrt{2}}. \quad (5\text{-}31)$$

λ is the spin-orbit coupling constant for the free ion,

$$P = 2\gamma_{cu}\beta_0\beta_N \langle d_{x^2-y^2}|\, 1/r^3 \,|d_{x^2-y^2}\rangle \approx 0.36 \text{ cm}^{-1}, \tag{5-32}$$

γ_{cu} is the gyromagnetic ratio of copper, and β_N is the nuclear magneton. The constant κ, introduced by Abragam and Pryce [16], corrects for the Fermi contact term of excited configurations of copper. The constant $T(n)$ is an integral over the ligand functions and arises from the calculation of the matrix elements of the Hamiltonian with the wave functions.

Kivelson and Neiman [102] give an approximate formula for α:

$$\alpha^2 = -\left(\frac{A_{\|}}{P}\right) + (g_{\|} - 2) + \tfrac{3}{7}(g_{\perp} - 2) + 0.04, \tag{5-33}$$

assuming $\frac{4}{7} + \kappa = 1$. The parameter α' in (5-19) is calculated from (5-24). When the super hyperfine structure is observed as the result of the isotropic Fermi interaction W_L of the unpaired electronic spin with the nuclear spin of the ligands, α' is also calculable according to the expression

$$W_L = \left(\frac{4\pi}{9}\right) \gamma_L \beta_0 \beta_N \alpha'^2 |\rho(0)|^2 S_z I_z, \tag{5-34}$$

where γ_L is the gyromagnetic ratio of the ligand atom's nucleus, and $|\rho(0)|^2$ is the electron density at the ligand nucleus. Maki and McGarvey [123] estimated $|\rho(0)|^2$ for nitrogen to be 33.4×10^{24} cm^{-3}. $A_{\perp}$ is not accurately

Table 5-2 ESR Parameters in Several Copper Salts of Amino Acids [192]

Amino Acid	ESR Parameters						Notes
	g_0	$g_{\|}$	$g_{\perp}$	$\|A_0\|$ ($\times 10^4$ cm^{-1})	$\|A_{\|}\|$	$\|A_{\perp}\|$	
Glycine	2.129	2.252	2.068	69	196	5	In H_2O, 77°K, pH = 11.8
L-Alanine	2.131	2.260	2.067	69	184	12	In (H_2O:EtOH = 1:3), 77°K, pH = 9.8
L-Valine	2.127	2.268	2.056	69	165	21	In (H_2O:EtOH = 1:1), 77°K, pH = 8.3
L-Glutamic acid	2.128	2.254	2.065	69	182	12	In (H_2O:EtOH) = 1:1, 77°K, pH = 8.3
		2.249			178		Cu:Glu = 1:6 (powder)
L-Aspartic acid	2.133	2.262	2.068	65	181	7	In (H_2O:EtOH = 1:1), 77°K, pH = 8.63
		2.244			197		In H_2O, 77°K, pH = 7.8
		2.252			174		Cu:Asp = 1:6 (powder)
L-Threonine	2.125	2.268	2.053	69	171	17	In (H_2O:EtOH = 1:1), 77°K, pH = 8.3

obtained because of poor resolution, and it is common to assume a proper value which gives good correlation between observation and theory.

We now give an example of application of the method of analysis discussed above. Tables 5-2 and 5-3 show how the ESR parameters vary with ligands [192]. It was found by Takeshita and Taminaga [191] that the tetrahydroxocuprate ion, $[Cu(OH)_4]^{2-}$, which is formed in highly alkaline solution of cupric ion, gives a spectrum with four hyperfine lines at room temperature and a spectrum with anisotropy at lower temperatures. The spectrum observed gives the following values: $g_0 = 2.133$, $A_0 = 76 \times 10^{-4}$ cm^{-1}, $g_{\parallel} = 2.272$, $g_{\perp} = 2.047$, $A_{\parallel} = 195 \times 10^{-4}$ cm^{-1}, and $A_{\perp} = 34 \times 10^{-4}$ cm^{-1}. These values give the α^2 value shown in Table 5-4. According to Table 5-4, the tetrahydroxo complex ion gives the highest value of α^2, 0.87, among the compounds investigated. Ionicity decreases from this particular compound to those with amino acids, which carry nitrogen and oxygen as the bonding atoms, to diethyldithiocarbamate. In other words, the ionicity of the copper-ligand bond decreases in the order of the donor atoms: $0 > N > S$.

Table 5-3 ESR Parameters of Several Copper Salts of Sulfur-Containing Ligands [192]

Ligand	ESR Parameters							Notes
	g_0	$g_{\parallel}$	$g_{\perp}$	$\|A_0\|$	$\|A_{\parallel}\|$	$\|A_{\perp}\|$	$\|A_N\|$	
Diethyldithio-carbamic acid	2.047[a]	2.108[b]	2.017[b]	76[a]	139[b]	44[b]		[a] $CHCl_3$ solution [b] Powder, diluted with Zn salt
8-Mercapto-quinoline	2.075[a]	2.158[b]	2.034[b]	63[a]	114[b]	38[b]	15[a] 11[b]	[a] $CHCl_3$ solution [b] powder, diluted with Zn salt
Rubeanic acid	2.05[c] 2.13[c]	2.138[b] 2.167[b]			131[b] 139[b]			[c] powder [b] powder, diluted with Zn salt
Diphenylthio-carbazone	2.042[c]							[c] powder

It must be remembered that the analysis which leads to these results contains several assumptions, and further investigation of the method of analysis is necessary in order to carry out a more precise investigation of the relation between bonding nature and the values of α^2 [112].

The solvent can have a significant effect on ESR parameters. For example, the ESR parameters in the copper oxinate (8-hydroxyquinolinate) are influenced by the diamagnetic oxinates which are added as the magnetic

Table 5-4 The Values of α^2 of Copper Complexes with Various Ligands[a]

Ligand	α^2	Reference
$[Cu(OH)_4]^{2-}$	0.87	192
Glycine	0.86	192
L-Alanine	0.84	192
L-Valine	0.79	192
L-Glutamic acid	0.82	192
L-Aspartic acid	0.83	192
L-Threonine	0.80	192
Diethyldithiocarbamic acid	0.55	192
8-Mercaptoquinoline	0.54	192
Rubeanic acid	0.56	192
Maleonitriledithiol ion	0.57	42
$S_2C_2O_2^{2-}$	0.62	42

[a] Two signals were observed. The α^2 value cited in the table refers to the signal with lower g value.

diluent [70, 71]. The super hyperfine structure, which results from the interaction of unpaired electron spin with the nuclear spin of nitrogen, is observed when pure oxine is used as the diluent, whereas it becomes unobservable when oxinates of zinc(II), mercury(II), or magnesium(II) are used. Furthermore, it is observed that the α^2 values of Cu(II) oxinate vary with the nature of the diluent. For example, $\alpha^2 = 0.85$ and 0.81 for copper oxinate diluted with 6,7-dichloro- and 6,7-dibromooxinates of TiO(IV), respectively. The α^2 values of Cu(II) oxinate are 0.85 and 0.77 when diluted with 6,7-dichlorooxinates of TiO(IV) and Zn(II), respectively. All these results should be taken as an indication that the chemical bonding in a chelated molecule is strongly influenced by the nature of the surroundings of the molecule. Analysis of the details of this effect is a problem to be investigated in the future.

In a single crystal of Cu(II) dimethylglyoximate, it was found [99] that $g_{zz} = 2.136$, $g_{xx} = 2.065$, and $g_{yy} = 2.003 \pm 0.005$, and the d_{xy} orbital of Cu(II) ion was assumed to mix with the p orbital of nitrogen in the ligand, and that $0.80 \leqq \alpha^2 \leqq 0.84$. In a single crystal of bis(acetylacetono)copper(II) it was found [123] that the in-plane σ- and π-bondings are covalent, whereas the out-of-plane π-bonding is ionic, and the following values were obtained: $g_{\parallel} = 2.266$, $g_{\perp} = 2.053_5$, $A_{\parallel} = -1.60 \times 10^{-2}\ cm^{-1}$, and $A_{\perp} = 0.195 \times 10^{-2}\ cm^{-1}$. This investigation reached the conclusions that $\alpha^2 = 0.81$, $\beta^2 = 0.85$, $\delta^2 = 0.99$, and $\kappa = 0.33$.

Bis(salicylaldimino)Cu(II) also gave results [124] which suggest the covalent nature of the metal-ligand bond, where $\alpha^2 = 0.83$, $\beta^2 = 0.72$, $\delta^2 \geqq 0.91$, and $\kappa = 0.34$. Viscous media were used [151] in order to obtain well-resolved hyperfine spectra of VO(IV) etioporphyrin I, where $g_{\parallel} = 1.948 \pm 0.009$, $g_{\perp} = 1.987 \pm 0.005$, $|A_{\parallel}| = 0.0159 \text{ cm}^{-1}$, and $|A_{\perp}| = 0.0052 \text{ cm}^{-1}$. An ESR study of bis(cyclopentadienyl)vanadium(II) has shown [132] that $g_0 = 2.00$ and $A_0 = 77$ mc in benzene, and that these values and the spectral pattern are much changed by oxidation. A theoretical investigation has also been carried out for this type of compound [171].

B. ESR Relaxation and Nature of Chemical Bonding

As mentioned above, chelate compounds of transition elements, such as Cu(II) and Ti(III), present ESR spectra with varieties of line widths. Highly acidic solutions of Ti(III) chelates show no observable spectra at room temperature [67]. In this case the chelation is broken and Ti(III) ion, which may exist as the hexaquo complex, possesses a rather short relaxation time. As the pH value is increased, the Ti(III) ions are chelated strongly and the absorption line appears. The spectrum observed is a single line or a line with hyperfine structure. This situation may be interpreted as follows [184]. Since the relaxation of paramagnetic resonance in Ti(III) chelates is, as will be shown later, determined by the vibrational motion of chelating bonds which modulates the spin-orbit interactions of metal ions, we can assume that, as the metal-ligand bond becomes more covalent, the vibrational frequency of the bond becomes higher and less effective for relaxation. Thus the line widths or the spin-lattice relaxation time of the ESR spectrum of Ti(III) ion can be a good criterion of the ionic nature of the metal-ligand bonds. To some extent, similar trends are evidenced with Cu(II) ions [69].

Investigation of the mechanism of ESR relaxation of paramagnetic ions in solution is, hence, very profitable for a consideration of the ionic states in solution. For example, cupric ion presents a single line with a width of about 140 gauss in aqueous solution at room temperature. This width is maintained constant in a lower concentration range, and is independent of concentration and of anion species which have little coordinating ability. Hence it is assumed to refer to the solvated ion, $[Cu(H_2O)_6]^{2+}$ [69].

Of the sources of relaxation [28, 101, 208–210], namely, (1) rotational motion of spins with anisotropic $\tilde{\mathbf{g}}$ and $\tilde{\mathbf{A}}$ tensors, (2) dipolar interactions, (3) quadrupolar interactions, (4) perturbation of spin-orbit interactions, only (4) is taken as effective in producing the observed line width for Cu(II) in aqueous solution [69]. Source (2) is eliminated by the investigation of very dilute solutions, and (3) is omitted, because the Ti(III) and Cu(II) ions belong to a ^2D state. If source (1) is operative, the line width $1/T_2$, which should be taken as equal to $1/T_1$ in a liquid system, should show dependence on viscosity, temperature, and magnetic field according to the formula [101, 131]:

$$\frac{1}{T_1} = \frac{1}{T_2} = (\Delta g \beta_0 H + \Delta A m_I)^2 \tau_c, \qquad (5\text{-}35)$$

where Δg refers to the anisotropy and is expressed by $\Delta g = g_{\parallel} - g_{\perp}$; H is the magnetic field strength, $\Delta A = A_{\parallel} - A_{\perp}$, and τ_c is the correlation time of rotation, which may be roughly formulated as $\tau_c = 4\pi\eta a^3/3kT$, with viscosity, η, the ionic radius, a, and temperature, T. Experimental results confirm that (1) is not the main source of the relaxation observed. Hence mechanism (4) is assumed to dominate the observed line width. The same conclusion is reached for Ti(III) chelates, namely, that the observed relaxation is determined by the modulation effect of spin-orbit interactions [184]. This feature is interpreted in terms of the crystal field theory as follows. Suppose that the ground state of Ti(III), $\{|+2\rangle - |-2\rangle\}/\sqrt{2}$, is admixed with excited states $|+1\rangle$, $|-1\rangle$, $\{|+2\rangle + |-2\rangle\}/\sqrt{2}$, and $|0\rangle$ as the result of the perturbation effect by the spin-orbit interactions. The latter is represented by $\mathscr{H}_1(t) = \lambda \boldsymbol{L}(t)\cdot\boldsymbol{S}$ which is a random perturbation. Spin-lattice relaxation time T_1 is related to the perturbation energy, $f_n(t)$, in such a way that the orbital angular momentum, $\boldsymbol{L}(t)$, is perturbed through vibrational motion of the ligands or the rotational motion of the spin. The latter refers to the rotational motion of the spin itself or of a microcrystal which consists of ligands and a central metal ion. The spin-lattice relaxation time determined by the vibrational perturbation of the spin-orbit interaction will be expressed as T_1(VSO), and the perturbation due to the rotational motion, as T_1(RSO).

After defining the correlation time, τ_c, for either one of these motions, T_1 is formulated as

$$\frac{1}{T_1} = 2\hbar^{-2} \sum_n \overline{|f_n(t)|^2} \frac{2}{\omega_n{}^2 \tau_c}, \tag{5-36}$$

where ω_n is the resonant frequency, n denotes the excited levels, and

$$f_{(n)}(t) = \langle n|\, \mathscr{H}_1(t)\, |0\rangle. \tag{5-37}$$

Equations (5-36) and (5-37) are calculated for the excited levels of Ti(III) as

$$\frac{1}{T_1(\mathrm{RSO})} = 4\tau_c{}^{-1}\left(\frac{\lambda^2}{2\delta'^2} + \frac{\lambda^2}{\Delta'^2}\right), \tag{5-38}$$

where the eigenfunctions are those for a cubic field with small contributions from tetragonal and rhombic fields. In (5-38) δ' and Δ' designate differences in energy between the ground state, $^2B_{2g}$, and ^2E_g and $^2B_{1g}$, respectively. If we take λ as equal to 100 cm^{-1}, $\delta' < 5000$ cm^{-1}, and $\Delta' > 14{,}000$ cm^{-1}, the second term of (5-38) is neglected in comparison to the first term and hence

$$\frac{1}{T_1(\mathrm{RSO})} = \frac{2\tau_c{}^{-1}\lambda^2}{\delta'^2}. \tag{5-39}$$

Similarly perturbation of the spin-orbit interaction produced by vibrational motion is calculated and related to T_1 as shown:

$$\frac{1}{T_1(\mathrm{VSO})} = \frac{42\lambda^2\tau_v^{-1}\overline{Q^2}}{\delta'^4}, \tag{5-40}$$

where τ_v is the correlation time of vibrational motion, and $\overline{Q^2}$ is the average of the square of the deviation of the coordinates from the equilibrium position. On the other hand,

$$g_\parallel = 2 - \frac{8\lambda}{\Delta'} - \frac{\lambda^2}{\delta'^2} \tag{5-41}$$

and

$$g_\perp = 2 - \frac{2\lambda}{\delta'} - \frac{2\lambda^2}{\Delta'^2}. \tag{5-42}$$

Hence we obtain

$$g_0 = 2 - \frac{4}{3}\frac{\lambda}{\delta'} - \frac{8\lambda}{3\Delta'} - \frac{\lambda^2}{3\delta'^2}. \tag{5-43}$$

This becomes

$$2 - g_0 \approx \frac{4}{3}\left(\frac{\lambda}{\delta'} + \frac{2\lambda}{\Delta'}\right). \tag{5-44}$$

If $2\delta' < \Delta'$,

$$(2 - g_0)^2 \cong \frac{\frac{16}{9}\lambda^2}{\delta'^2}. \tag{5-45}$$

Consequently,

$$\frac{1}{T_1(\mathrm{RSO})} \propto (2 - g_0)^2 \tag{5-46}$$

and

$$\frac{1}{T_1(\mathrm{VSO})} \propto (2 - g_0)^4 \quad \text{for } \lambda \approx \text{const.}, \tag{5-47a}$$

$$\propto (2 - g_0)^2 \quad \text{for } \delta' \approx \text{const.} \tag{5-47b}$$

As we measure the line-widths of the central peaks of chelated Ti(III) and correlate them with the g shift values, a linear relation is obtained between $\log \Delta H$ and $\log (2 - g_0)$. A fourth power relation, $\Delta H \propto (2 - g_0)^4$, holds for chelates with $g_0 < 1.96$, and deviation from the fourth power relation becomes appreciable for chelates with $g_0 > 1.96$.

Figures 5-1 and 5-2 reproduce the ESR spectra of Ti(III)-EDTA at pH 6.9 and 8, respectively [68]. The heavy line, 1, refers to the observed spectrum which is decomposed into two components, 2 and 3. The two components, being different in line width from each other, show constant values of g which do not vary with pH. It is assumed that these components refer to the molecular species II and III shown on page 300. The equilibrium constants for this system at several pH values are given in Table 5-5.

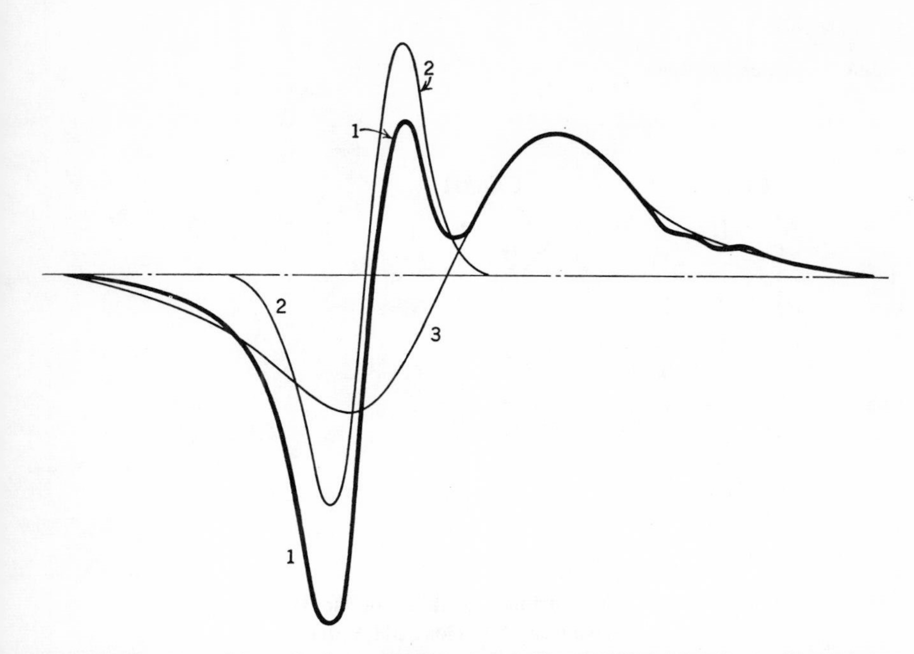

Fig. 5-1. Decomposition of observed ESR signal (1) of Ti(III)-EDTA at pH 6.9 into components (2) and (3).

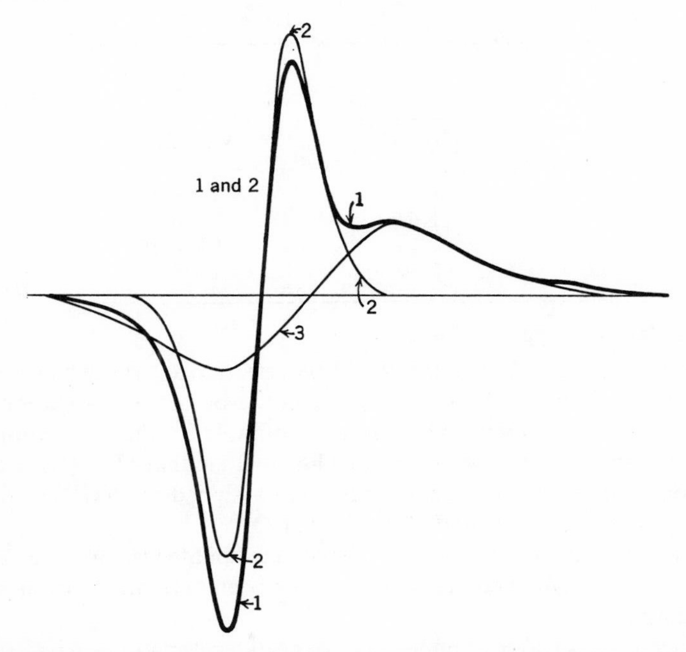

Fig. 5-2. Decomposition of observed ESR signal (1) of Ti(III)-EDTA at pH 8 into components (2) and (3).

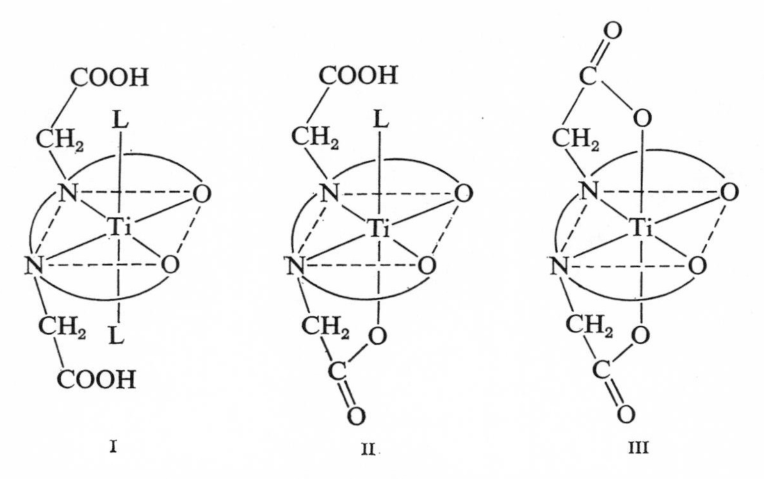

Table 5-5 Equilibrium Constants of the Ti(III)-EDTA Chelates at Various pH Values

$$\text{Ti(III)EDTAH} \rightleftharpoons \text{Ti(III)EDTA}^- + \text{H}^+$$

$$K_e = \frac{(\text{TiEDTA}^-)(\text{H}^+)}{(\text{TiEDTAH})}$$

pH	Ratio of Relative Area (II):(III)	K_e(m/l), Calc.
6.9	1:1.8	2.1×10^{-7}
7.7	1:6.8	1.4×10^{-7}
8.2	1:24	1.5×10^{-7}
8.6	1:30	0.7×10^{-7}

C. Miscellaneous Applications

Several examples of ESR studies will be referred to here which relate to elements other than Cu(II) or Ti(III). As noted before, investigation of the ESR spectrum of metal ions in solution is profitable for the elucidation of the ionic state in solution. Studies along this line include that of Fe(III) extracted by isopropyl ether [91]; different spectra are assigned to Fe(III) ions in the forms of anhydrous $FeCl_3$ and of $FeCl_3 \cdot 6H_2O$.

The ESR spectrum of chromium(III) acetylacetonate was observed at room temperature [185]. Magnetic parameters have been determined using a single crystal sample.

Acetylacetonates and tri- and hexafluoro acetylacetonates of trivalent metal ions, such as Ti(III), Cr(III), Fe(III), Mo(III), and Ru(III), have been treated to a first approximation by crystal field theory [98]; the ESR signal was unobservable for $V(acac)_3$, $Mn(acac)_3$, and $Fe(hfac)_3$. It is pointed out in this experiment that the oxygen atoms form a regular octahedron about a metal ion, and hence point charges at the oxygen positions would give only a cubic component in the crystal field interaction energy. The axial field may arise from a dipole moment of the CO groups or of the entire ligand. The d_ε orbital is superposed by the extension of the ligand orbital, and the metal-oxygen bond takes a distance intermediate between covalent and ionic.

The copper, nickel, cobalt, and rhodium complexes of maleonitriledithiolate, $[S_2C_2(CN)_2]^{2-}$, have been investigated [125]. This ligand is of interest because both of its donor atoms are sulfur. With the interpretation of the ESR spectra, the ground states of $[CuS_4C_4(CN)_4]^{2-}$, $[NiS_4C_4(CN)_4]^{-}$, $[CoS_4C_4(CN)_4]^{2-}$, and $[RhS_4C_4(CN)_4]^{2-}$ were assigned as $|\varepsilon\rangle$, $|\varepsilon^2y\rangle$, $|\varepsilon^2y\rangle$, and $|\varepsilon^2 0\rangle$, respectively, where $\varepsilon \equiv d_{xy}$, $y \equiv d_{yz}$, and $0 \equiv d_{3z^2-r^2}$.

ESR line widths of dithiocarbamates of Cu, Ag, and Au have been investigated, and dependence on concentration, temperature, and solvents have been reported.

A triplet state was evidenced by the ESR spectrum of copper acetate, and the results obtained were successfully interpreted in terms of the presence of metal-metal interaction in the sample [45, 46]. The cupric ion in PVA gel [72] and DPPH (diphenyl picryl hydrazyl) in tetrahydrofuran [90] also reveal ESR signals at liquid nitrogen temperature which correspond to the $\Delta m = \pm 2$ transitions.

Although a number of papers on metal chelates have been published, only some of them are cited in this chapter. For the remaining papers the reader is referred to the references summarized in Table 5-6. The data of ESR parameters of paramagnetic complexes, including chelate compounds, are compiled in the *New Series of Landolt-Börnstein Tables*, Group II, Volume 2, "Magnetic Properties of Coordination and Organo-metallic Transition Metal Compounds," by E. König, published in 1966.

Table 5-6

ESR of Metal Chelate Compounds

Metal	Oxidation state	References
Ti	Ti(I)	65, 107
	Ti(III)	48, 98, 136, 152
V	V(0)	58, 65, 107
	V(II)	92
	V(IV)	19, 23, 26, 39, 56, 58, 73, 80, 81, 87, 93, 103, 106, 111, 113, 151, 170, 179, 203, 211, 212

Table 5-6. (*Continued*)

Metal	Oxidation state	References
Cr	Cr(I)	65, 107
	Cr(III)	6, 85, 86, 113, 137, 138, 185
	Cr(V)	56, 205
Mo	Mo(I)	98
	Mo(V)	56, 205
W	W(V)	56, 205
Re	Re(VI)	187
Fe	Fe(II)	32, 78, 92, 108, 126, 133, 176
	Fe(III)	4, 5, 34, 35, 37, 38, 63, 74, 76, 77, 82, 93, 94, 96–98, 113, 165, 176, 180, 182, 190, 206
Co	Co(II)	3, 23–25, 57, 77, 92, 93, 125, 200
Rh	Rh(II)	41, 125
Ni	Ni(II)	92, 154
	Ni(III)	54, 55, 57, 66, 84, 125, 188
Pd	Pd(III)	54, 55, 57
Pt	Pt(III)	54, 55, 57
Cu	Cu(II)	1–3, 7–12, 15, 18, 20–22, 27, 30–33, 36, 42, 45, 46, 49, 53, 54, 57, 59, 61, 62, 64, 66, 75, 77, 79, 88, 89, 92, 93, 95, 99, 100, 102, 105, 109, 110, 113–120, 122–125, 127, 128, 134, 139–145, 147, 150, 155–157, 159, 162–164, 166, 168, 169, 172, 175, 178, 181, 183, 186, 189, 193–200, 202, 204, 207, 213–215
Ag	Ag(II)	49, 104, 121, 122, 140, 155, 199
Au	Au(II)	199

REFERENCES

Books

Al'tshuler, S. A., and B. M. Kozyrev, *Electron Paramagnetic Resonance*. New York: Academic, 1964.

Bersohn, M., and J. C. Baird, *An Introduction to Electron Paramagnetic Resonance*. New York: Benjamin, 1966.
Low, W., "Paramagnetic Resonance in Solids," *Solid State Phys. Suppl.*, **2** (1960).
Pake, G. E., *Paramagnetic Resonance*. New York: Benjamin, 1962.

Reviews

Bersohn, M., *Electron Paramagnetic Resonance of the Organometallics*, in *Determination of Organic Structures by Physical Methods*, E. A. Braude and F. C. Nachod, eds. New York: Academic, Vol. 2, Chapter 16, 1962.
McGarvey, B. R., *Electron Spin Resonance of Transition Metal Complexes*, *Transition Metal Chemistry*, R. L. Carlin, ed. New York: Marcel Dekker, Vol. 3, 1966.

Other References

[1] Aasa, R., R. Pettersson, and T. Vänngård, *Nature*, **190**, 258 (1961).
[2] Aasa, R., B. G. Malmström, P. Saltman, and T. Vänngård, *Biochim. Biophys. Acta*, **75**, 203 (1963).
[3] Aasa, R., and T. Vänngård, *Z. Naturforsch.*, **19a**, 1425 (1964).
[4] Aasa, R., and T. Vänngård, *Arkiv Kemi*, **24**, 331 (1965).
[5] Aasa, R., K. E. Carlsson, S. A. Reyes, and T. Vänngård, *Arkiv Kemi*, **25**, 285 (1966).
[6] Aasa, R., K. E. Falk, and S. A. Reyes, *Arkiv Kemi*, **25**, 309 (1966).
[7] Abe, H., *Phys. Rev.*, **92**, 1572 (1953).
[8] Abe, H., and J. Shimada, *Phys. Rev.*, **90**, 316 (1953).
[9] Abe, H., and K. Ono, *J. Phys. Soc. Japan*, **11**, 947 (1956).
[10] Abe, H., and J. Shimada, *J. Phys. Soc. Japan*, **12**, 1255 (1957).
[11] Abe, H., *J. Phys. Soc. Japan*, **13**, 987 (1958).
[12] Abe, H., and H. Shirai, *J. Phys. Soc. Japan*, **16**, 118 (1961).
[13] Abe, H., H. Morigaki, and K. Koga, *Phys. Rev. Letters*, **9**, 338 (1962).
[14] Abe, H., and H. Morigaki, *Paramagnetic Resonance*, W. Low, ed. New York: Academic, Vol. 2, p. 587, 1963.
[15] Ablov, A. V., Yu. V. Yablokov, and I. I. Zheru, *Dokl. Akad. Nauk SSSR*, **141**, 343 (1961).
[16] Abragam, A., and M. H. L. Pryce, *Proc. Roy. Soc. (London)*, **A205**, 135 (1951).
[17] Abragam, A., and M. H. L. Pryce, *Proc. Roy. Soc. (London)*, **A206**, 164 (1951).
[18] Allen, H. C., G. F. Kokoszka, and R. G. Inskeep, *J. Am. Chem. Soc.*, **86**, 1023 (1964).
[19] Anderson, W. A., and L. H. Piette, *J. Chem. Phys.*, **30**, 591 (1959).
[20] Anufrienko, V. F., and A. P. Zeif, *Opt. i Spektroskopiya*, **20**, 652 (1966).
[21] Anufrienko, V. F., E. K. Mamaeva, and N. P. Keier, *Dokl. Akad. Nauk SSSR*, **168**, 116 (1966).
[22] Assour, J. M., and S. E. Harrison, *Phys. Rev.*, **136**, A1368 (1964).
[23] Assour, J. M., *J. Chem. Phys.*, **43**, 2477 (1965).
[24] Assour, J. M., *J. Am. Chem. Soc.*, **87**, 4701 (1965).
[25] Assour, J. M., and W. K. Kahn, *J. Am. Chem. Soc.*, **87**, 207 (1965).
[26] Assour, J. M., J. Goldmacher, and S. E. Harrison, *J. Chem. Phys.*, **43**, 159 (1965).
[27] Atherton, N. M., Q. H. Gibson, and C. Greenwood, *Biochem. J.*, **86**, 554 (1963).
[28] Atkins, P. W., and D. Kivelson, *J. Chem. Phys.*, **44**, 169 (1966).
[29] Ballhausen, C. J., *Introduction to Ligand Field Theory*. New York: McGraw-Hill, p. 134, 1962.

[30] Bates, C. A., W. S. Moore, K. J. Standley, and K. W. H. Stevens, *Proc. Phys. Soc.* (*London*), **79**, 73 (1962).
[31] Beinert, H., D. E. Griffiths, D. C. Wharton, and R. H. Sands, *J. Biol. Chem.*, **237**, 2337 (1962).
[32] Bel'skii, N. K., and V. N. Tsikunov, *Dokl. Akad. Nauk SSSR*, **142**, 380 (1962).
[33] Bennett, J. E., and D. J. E. Ingram, *Nature*, **175**, 130 (1955).
[34] Bennett, J. E., D. J. E. Ingram, P. George, and J. S. Griffith, *Nature*, **176**, 394 (1955).
[35] Bennett, J. E., and D. J. E. Ingram, *Nature*, **177**, 275 (1956).
[36] Bennett, J. E., and D. J. E. Ingram, *Phil. Mag.*, (8) **1**, 970 (1956).
[37] Bennett, J. E., J. F. Gibson, and D. J. E. Ingram, *Proc. Roy. Soc.* (*London*), **A240**, 67 (1957).
[38] Bennett, J. E., J. F. Gibson, D. J. E. Ingram, T. M. Haughton, G. A. Kerkut, and K. A. Munday, *Phys. Med. Biol.*, **1**, 309 (1957); *Proc. Roy. Soc.* (*London*), **A262** 395 (1961).
[39] Bernal, I., and P. H. Rieger, *Inorg. Chem.*, **2**, 256 (1963).
[40] Bijl, D., *Proc. Phys. Soc.* (*London*), **A63**, 405 (1950).
[41] Billig, E., S. I. Shupack, J. H. Waters, R. Williams, and H. B. Gray, *J. Am. Chem. Soc.*, **86**, 926 (1964).
[42] Billig, E., R. Williams, I. Bernal, J. H. Waters, and H. B. Gray, *Inorg. Chem.*, **3**, 663 (1964).
[43] Bleaney, B., and D. J. E. Ingram, *Proc. Phys. Soc.* (*London*), **A63**, 408 (1950).
[44] Bleaney, B., K. D. Bowers, and D. J. E. Ingram, *Proc. Phys. Soc.* (*London*), **A64**, 758 (1951).
[45] Bleaney, B., and K. D. Bowers, *Phil. Mag.*, **43**, 372 (1952).
[46] Bleaney, B., and K. D. Bowers, *Proc. Roy. Soc.* (*London*), **A214**, 451 (1952).
[47] Bleaney, B., K. D. Bowers, and R. S. Trenam, *Proc. Roy. Soc.* (*London*), **A228**, 157 (1955).
[48] Bogle, G. S., and J. Owen, *Rept. Progr. Phys.*, **18**, 304 (1955).
[49] Bowers, K. D., *Proc. Phys. Soc.* (*London*), **A66**, 666 (1953).
[50] Brill, A. S., and J. H. Venable, Jr., *Nature*, **203**, 752 (1964).
[51] Bryce, G. F., *J. Phys. Chem.*, **70**, 3549 (1966).
[52] Codell, M., S. Fujiwara, K. Nagashima, and T. Seki, *Bull. Chem. Soc. Japan*, **38**, 21 (1965).
[53] Date, M., M. Motokawa, and H. Yamazaki, *J. Phys. Soc. Japan*, **18**, 911 (1963).
[54] Davison, A., N. Edelstein, R. H. Holm, and A. H. Maki, *J. Am. Chem. Soc.*, **85**, 2029 (1963).
[55] Davison, A., N. Edelstein, R. H. Holm, and A. H. Maki, *Inorg. Chem.*, **2**, 1227 (1963).
[56] Davison, A., N. Edelstein, R. H. Holm, and A. H. Maki, *J. Am. Chem. Soc.*, **86**, 2799 (1964).
[57] Davison, A., N. Edelstein, R. H. Holm, and A. H. Maki, *Inorg. Chem.*, **3**, 814 (1964).
[58] Davison, A., N. Edelstein, R. H. Holm, and A. H. Maki, *Inorg. Chem.*, **4**, 55 (1965).
[59] Deal, R. M., D. J. E. Ingram, and R. Srinivasan, *Electronic Magnetic Resonance and Solid Dielectrics*, Proc. XII Colloque Ampère. Bordeaux, 1963, p. 239 (publ. 1964).
[60] De Armond, K., B. B. Garrett, and H. S. Gutowsky, *J. Chem. Phys.*, **42**, 1019 (1965).
[61] Degtyarev, L. S., and L. N. Ganyuk, *Vysokomolekul. Soedin.*, **6**, 28 (1964).

[62] Dunhill, R. H., J. R. Pilbrow, and T. D. Smith, *J. Chem. Phys.*, **45**, 1474 (1966).
[63] Ehrenberg, A., *Arkiv Kemi*, **19**, 119 (1962).
[64] Ehrenberg, A., and T. Yonetani, *Acta Chem. Scand.*, **15**, 1071 (1961).
[65] Elschner, B., and S. Herzog, *Arch. Sci. (Geneva)*, **11**, 160 (1958).
[66] Fritz, H. P., B. Golla, and H. J. Keller, *Z. Naturforsch.*, **21b**, 97 (1965).
[67] Fujiwara, S., and M. Codell, *Bull. Chem. Soc. Japan*, **37**, 49 (1964).
[68] Fujiwara, S., K. Nagashima, and M. Codell, *Bull. Chem. Soc. Japan*, **37**, 773 (1964).
[69] Fujiwara, S., and H. Hayashi, *J. Chem. Phys.*, **43**, 23 (1965).
[70] Fujiwara, S., and K. Nagashima, *Anal. Chem.*, **38**, 1464 (1966).
[71] Fujiwara, S., and K. Nagashima, IX ICCC Conference Proceedings, 1966, p. 391.
[72] Fujiwara, S., S. Katsumata, and T. Seki, *J. Phys. Chem.*, **71**, 115 (1967).
[73] Garifyanov, N. S., and B. M. Kozyrev, *Teor. i Eksperim. Khim. Akad. Nauk Ukr. SSR*, **1**, 525 (1965).
[74] George, P., J. E. Bennett, and D. J. E. Ingram, *J. Chem. Phys.*, **24**, 627 (1956).
[75] Gersmann, H. R., and J. D. Swalen, *J. Chem. Phys.*, **36**, 3221 (1961).
[76] Gibson, J. F., D. J. E. Ingram, *Nature*, **180**, 29 (1957).
[77] Gibson, J. F., D. J. E. Ingram, and D. Schonland, *Discussions Faraday Soc.*, **26** 72 (1958).
[78] Gibson, J. F., *Nature*, **196**, 64 (1962).
[79] Gibson, J. F., *Trans. Faraday Soc.*, **60**, 2105 (1964).
[80] Golding, R. M., *Mol. Phys.*, **5**, 369 (1962).
[81] Golding, R. M., *Trans. Faraday Soc.*, **59**, 1513 (1964).
[82] Gordy, W., and H. N. Rexroad, *Free Radicals*, Biol. Systems, Proc. Symp. Stanford, Calif., 1960, p. 263 (publ. 1961).
[83] Gray, H. B., R. Williams, I. Bernal, and E. Billig, *J. Am. Chem. Soc.*, **84**, 3596 (1962).
[84] Gray, H. B., and E. Billig, *J. Am. Chem. Soc.*, **85**, 2019 (1963).
[85] Gregorio, S., and R. Lacroix, *Electronic Magnetic Resonance and Solid Dielectrics*, Proc. XII Colloque Ampère. Bordeaux, 1963, p. 213 (publ. 1964).
[86] Gregorio, S., J. Weber, and R. Lacroix, *Helv. Phys. Acta*, **38**, 172 (1965).
[87] Griffiths, J. H. E., and I. M. Ward, *Rept. Progr. Phys.*, **18**, 304 (1955).
[88] Harrison, S. E., and J. M. Assour, *Paramagnetic Resonance*, W. Low, ed. New York: Academic, Vol. 2, p. 855, 1963.
[89] Harrison, S. E., and J. M. Assour, *J. Chem. Phys.*, **40**, 365 (1964).
[90] Hasegawa, H., and T. Maruyama, International Conference on Chemical Application of ESR, Cardiff, Wales, July, 1966.
[91] Hatel, G. R., and H. M. Clark, *J. Phys. Chem.*, **65**, 1930 (1961).
[92] Ingram, D. J. E., and J. E. Bennett, *J. Chem. Phys.*, **22**, 1136 (1954).
[93] Ingram, D. J. E., and J. E. Bennett, *Discussions Faraday Soc.*, **19**, 140 (1955).
[94] Ingram, D. J. E., and J. C. Kendrew, *Nature*, **178**, 905 (1956).
[95] Ingram, D. J. E., J. E. Bennett, P. George, and J. M. Goldstein, *J. Am. Chem. Soc.*, **78**, 3545 (1956).
[96] Ingram, D. J. E., J. F. Gibson, and M. F. Purutz, *Nature*, **178**, 906 (1956).
[97] Ingram, D. J. E., *Arch. Sci. (Geneva)*, **10**, 109 (1957).
[98] Jarrett, H. S., *J. Chem. Phys.*, **27**, 1298 (1957).
[99] Jarrett, H. S., *J. Chem. Phys.*, **28**, 1260 (1958).
[100] Jesson, J. P., *J. Chem. Phys.*, **45**, 1049 (1966).
[101] Kivelson, D., *J. Chem. Phys.*, **33**, 1094 (1960).
[102] Kivelson, D., and R. Neiman, *J. Chem. Phys.*, **35**, 149 (1961).

[103] Kivelson, D., and S.-K. Lee, *J. Chem. Phys.*, **41**, 1896 (1964).
[104] Kneubühl, F. K., W. S. Koski, and W. S. Caughey, *J. Am. Chem. Soc.*, **83**, 1607 (1961).
[105] Kokoszka, G. F., H. C. Allen, Jr., and G. Gordon, *J. Chem. Phys.*, **42**, 3730 (1965).
[106] Kokoszka, G. F., H. C. Allen, Jr., and G. Gordon, *Inorg. Chem.*, **5**, 91 (1966).
[107] König, E., *Z. Naturforsch.*, **19a**, 1139 (1964).
[108] Korshak, Yu. V., T. A. Pronyuk, and B. E. Davidov, *Neftekhimiya*, **3**, 677 (1963).
[109] Kozyrev, B. M., and A. I. Rivkind, *Dokl. Akad. Nauk SSSR*, **127**, 1044 (1954).
[110] Kuska, H. A., and M. T. Rogers, *J. Chem. Phys.*, **43**, 1744 (1965).
[111] Kuska, H. A., and M. T. Rogers, *Inorg. Chem.*, **5**, 313 (1966).
[112] Kuska, H. A., M. T. Rogers, and R. E. Drillinger, *J. Phys. Chem.*, **71**, 109 (1967).
[113] Lancaster, W., and W. Gordy, *J. Chem. Phys.*, **19**, 1181 (1951).
[114] Larin, G. M., V. M. Dziomko, and K. A. Dunaevskaya, *Zh. Strukt. Khim.*, **5**, 783 (1964).
[115] Larin, G. M., *Zh. Strukt. Khim.*, **6**, 548 (1965).
[116] Larin, G. M., V. M. Dziomko, K. A. Dunaevskaya, and Ya. K. Syrkin, *Zh. Strukt. Khim.*, **6**, 391 (1965).
[117] Lewis, W. B., M. Alei, Jr., and L. O. Morgan, *J. Chem. Phys.*, **44**, 4003 (1966).
[118] Loesche, A., and W. Windsch, *Phys. Status Solidi*, **11**, K55 (1965).
[119] Lohskin, B. V., A. K. Piskunov, L. A. Kazitsyna, and D. N. Shigorin, *Dokl. Akad. Nauk SSSR*, **143**, 867 (1962).
[120] Lokshin, B. V., A. K. Piskunov, L. A. Kazitsyna, and D. N. Shigorin, *Izv. Akad. Nauk SSSR, Ser. Fiz.*, **27**, 75 (1963).
[121] MacCragh, A., and W. S. Koski, *J. Am. Chem. Soc.*, **85**, 2375 (1963).
[122] MacCragh, A., C. B. Storm, and W. S. Koski, *J. Am. Chem. Soc.*, **87**, 1470 (1965).
[123] Maki, A. H., and B. R. McGarvey, *J. Chem. Phys.*, **29**, 31 (1958).
[124] Maki, A. H., and B. R. McGarvey, *J. Chem. Phys.*, **29**, 35 (1958).
[125] Maki, A. H., N. Edelstein, A. Davison, and R. H. Holm, *J. Am. Chem. Soc.*, **86**, 4580 (1964).
[126] Maki, A. H., and T. E. Berry, *J. Am. Chem. Soc.*, **87**, 4437 (1965).
[127] Malmström, B. G., R. Mosbach, and T. Vänngård, *Nature*, **183**, 321 (1959).
[128] Malmström, B. G., and T. Vänngård, *J. Mol. Biol.*, **2**, 118 (1960).
[129] Marriage, A. J., *Australian J. Chem.*, **18**, 463 (1965).
[130] McCain, D. C., and R. J. Meyers, *J. Phys. Chem.*, **71**, 192 (1967).
[131] McConnell, H. M., *J. Chem. Phys.*, **25**, 709 (1956).
[132] McConnell, H. M., W. M. Porterfield, and R. E. Robertson, *J. Chem. Phys.*, **30**, 442 (1959).
[133] McDonald, C. C., W. D. Phillips, and H. F. Mower, *J. Am. Chem. Soc.*, **87**, 3319 (1965).
[134] McGarvey, B. R., *J. Phys. Chem.*, **60**, 71 (1956).
[135] McGarvey, B. R., *J. Chem. Phys.*, **33**, 1094 (1960).
[136] McGarvey, B. R., *J. Chem. Phys.*, **38**, 388 (1963).
[137] McGarvey, B. R., *J. Chem. Phys.*, **40**, 809 (1964).
[138] McGarvey, B. R., *J. Chem. Phys.*, **41**, 3743 (1964).
[139] McMillan, J. A., and B. Smaller, *J. Chem. Phys.*, **35**, 763 (1961).
[140] McMillan, J. A., and B. Smaller, *J. Chem. Phys.*, **35**, 1698 (1961).
[141] Miroshnichenko, I. V., and G. M. Larin, *Teor. i Eksperim. Khim. Akad. Nauk Ukr. SSR*, **1**, 545 (1965).

[142] Miroshnichenko, I. V., G. M. Larin, B. I. Stepanov, and B. A. Korolev, *Teor. i Eksperim. Khim. Akad. Nauk Ukr. SSR*, **2**, 131 (1966).
[143] Miroshnichenko, I. V., G. M. Larin, B. I. Stepanov, and B. A. Korolev, *Teor. i Eksperim. Khim. Akad. Nauk Ukr. SSR*, **2**, 405 (1966).
[144] Miroshnichenko, I. V., G. M. Larin, and E. G. Rukhadze, *Teor. i Eksperim. Khim. Akad. Nauk Ukr. SSR*, **2**, 409 (1966).
[145] Mizuno, Z., O. Matsumura, K. Fukuda, and K. Horai, *Mem. Fac. Sci. Kyushu Univ.*, **B2**, 13 (1956).
[146] Mori, M., and S. Fujiwara, *Bull. Chem. Soc. Japan*, **36**, 1636 (1963).
[147] Neiman, R., and D. Kivelson, *J. Chem. Phys.*, **35**, 156 (1961).
[148] Neiman, R., and D. Kivelson, *J. Chem. Phys.*, **35**, 162 (1961).
[149] Nöth, H., J. Voitländer, and M. Nussbaum, *Naturwissenschaften*, **47**, 57 (1960).
[150] Okamura, T., Y. Torizuka, and M. Date, *Phys. Rev.*, **89**, 525 (1953).
[151] O'Reilly, D. E., *J. Chem. Phys.*, **29**, 1188 (1958).
[152] Ostendorf, H. K., *Phys. Rev. Letters*, **13**, 295 (1964).
[153] Pake, G. E., and R. H. Sands, *Phys. Rev.*, **98**, 226 (1955).
[154] Peter, M., *Phys. Rev.*, **116**, 1432 (1959).
[155] Pettersson, R., and T. Vänngård, *Arkiv. Kemi*, **17**, 249 (1961).
[146] Piskunov, A. K., D. N. Shigorin, V. I. Smirnova, and B. I. Stepanov, *Dokl. Akad. Nauk SSSR*, **130**, 1284 (1960).
[157] Piskunov, A. K., D. N. Shigorin, B. I. Stepanov, and E. R. Klinshpont, *Dokl. Akad. Nauk SSSR*, **136**, 871 (1961).
[158] Pryce, M. H. L., *Proc. Phys. Soc. (London)*, **A63**, 25 (1950).
[159] Rajan, R., *J. Chem. Phys.*, **37**, 460 (1962).
[160] Rajan, R., *J. Chem. Phys.*, **37**, 1901 (1962).
[161] Rajan, R., *Physica*, **29**, 1191 (1963).
[162] Rajan, R., and T. R. Reddy, *J. Chem. Phys.*, **39**, 1140 (1963).
[163] Reddy, T. R., and R. Srinivasan, *J. Chem. Phys.*, **43**, 1404 (1965).
[164] Reimann, C. W., G. F. Kokoszka, and H. C. Allen, Jr., *J. Res. Natl. Bur. Std.*, **A70**, 1 (1966).
[165] Rein, H., O. Ristau, and F. Jung, *Z. Physik. Chem. (Leipzig)*, **228**, 102 (1965).
[166] Rivkind, A. I., *Dokl. Akad. Nauk SSSR*, **135**, 365 (1960).
[167] Rivkind, A. I., *Dokl. Akad. Nauk SSSR*, **158**, 1401 (1964).
[168] Roberts, E. M., and W. S. Koski, *J. Am. Chem. Soc.*, **82**, 3006 (1960).
[169] Roberts, E. M., and W. S. Koski, *J. Am. Chem. Soc.*, **83**, 1865 (1961).
[170] Roberts, E. M., W. S. Koski, and W. S. Caughey, *J. Chem. Phys.*, **34**, 591 (1961).
[171] Robertson, R. E., and H. M. McConnell, *J. Phys. Chem.*, **64**, 70 (1960).
[172] Rode, V. V., L. I. Nekrasov, A. P. Terentev, and E. G. Rukhadze, *Vysokomolekul. Soedin.*, **4**, 13 (1962).
[173] Rogers, R. N., and G. E. Pake, *J. Chem. Phys.*, **33**, 1107 (1960).
[174] Robertson, R. E., H. M. McConnell, and W. W. Porterfield, *J. Chem. Phys.*, **30**, 442 (1959).
[175] Russel, D. B., and S. J. Wyard, *Nature*, **191**, 65 (1961).
[176] Sancier, K. M., G. Freeman, and J. S. Mills, *Science*, **137**, 752 (1962).
[177] Sands, R. H., *Phys. Rev.*, **99**, 1222 (1955).
[178] Sands, R. H., and H. Beinert, *Biochem. Biophys. Res. Commun.*, **1**, 175 (1959).
[179] Saraceno, A. J., D. T. Fanale, and N. D. Coggeshall, Am. Chem. Soc., Div. Petrol. Chem., Preprints 5/3, 141 (1960).
[180] Schoffa, G., O. Ristau, and F. Jung, *Naturwissenschaften*, **47**, 227 (1960).

[181] Schoffa, G., O. Ristau, and B. E. Wahler, *Z. Physik. Chem. (Leipzig)*, **215**, 203 (1960).
[182] Schoffa, G., *Nature*, **203**, 640 (1964).
[183] Schübel, W., and E. Lutze, *Z. Angew. Phys.*, **17**, 332 (1964).
[184] Seki, T., M. S. Thesis, The University of Tokyo, 1966.
[185] Singer, L. S., *J. Chem. Phys.*, **23**, 379 (1955).
[186] Spacu, P., V. Voicu, and I. Pascaru, *J. Chim. Phys.*, **60**, 368 (1963).
[187] Stiefel, E. I., and H. B. Gray, *J. Am. Chem. Soc.*, **87**, 4012 (1965).
[188] Stiefel, E. I., J. H. Waters, E. Billig, and H. B. Gray, *J. Am. Chem. Soc.*, **87**, 3016 (1965).
[189] Sundaramma, K., *Proc. Indian Acad. Sci., Sect. A*, **42**, 292 (1955).
[190] Symmons, H. F., and G. S. Bogle, *Proc. Phys. Soc. (London)*, **82**, 412 (1963).
[191] Takeshita, T., and I. Taminaga, B. S. Thesis, The University of Tokyo, 1965.
[192] Taminaga, I., M. S. Thesis, The University of Tokyo, 1967.
[193] Terentev, A. P., V. V. Rode, and E. G. Rukhadze, *Vysokomolekul. Soedin.*, **4**, 91 (1962).
[194] Terentev, A. P., G. V. Panova, D. N. Shigorin, and E. G. Rukhadze, *Dokl. Akad. Nauk SSSR*, **156**, 1174 (1964).
[195] Tikhomirova, N. N., and D. M. Chernikova, *Zh. Strukt. Khim.*, **3**, 335 (1962).
[196] Tikhomirova, N. N., and K. I. Zamaraev, *Zh. Strukt. Khim.*, **4**, 224 (1963).
[197] Timerov, R. Kh., Yu. V. Yablokov, and A. V. Ablov, *Dokl. Akad. Nauk SSSR*, **152**, 160 (1963).
[198] Toyoda, K., and K. Ochiai, *Proc. Intern. Symp. Mol. Struct. Spectry., Tokyo*, D211 (1962).
[199] Vänngård, T., and S. Åkerström, *Nature*, **184**, 183 (1959).
[200] Vänngård, T., and R. Aasa, in *Paramagnetic Resonance*, W. Low, ed. New York: Academic, Vol. 2, p. 509, 1963.
[201] Wakim, F. G., H. K. Henisch, and H. A. Atwater, *J. Chem. Phys.*, **42**, 2619 (1965).
[202] Walaas, E., O. Walaas, and S. Haavaldsen, *Arch. Biochem. Biophys.*, **100**, 97 (1963).
[203] Walker, F. A., R. L. Carlin, and P. H. Rieger, *J. Chem. Phys.*, **45**, 4181 (1966).
[204] Walsh, W. M., Jr., L. W. Rupp, Jr., and B. J. Wyluda, in *Paramagnetic Resonance*, W. Low, ed. New York: Academic, Vol. 2, p. 836, 1963.
[205] Waters, J. H., R. Williams, H. B. Gray, G. N. Schrauzer, and H. W. Finck, *J. Am. Chem. Soc.*, **86**, 4198 (1964).
[206] Wickman, H. H., M. P. Klein, and D. A. Shirley, *J. Chem. Phys.*, **42**, 2113 (1965).
[207] Wiersema, A. K., and J. J. Windle, *J. Phys. Chem.*, **68**, 2316 (1964).
[208] Wilson, R., and D. Kivelson, *J. Chem. Phys.*, **44**, 154 (1966).
[209] Wilson, R., and D. Kivelson, *J. Chem. Phys.*, **44**, 4440 (1966).
[210] Wilson, R., and D. Kivelson, *J. Chem. Phys.*, **44**, 4445 (1966).
[211] Wuethrich, K., *Helv. Chim. Acta*, **48**, 779 (1965).
[212] Wuethrich, K., *Helv. Chim. Acta*, **48**, 1012 (1965).
[213] Yablokov, Yu. V., and A. V. Ablov, *Dokl. Akad. Nauk SSSR*, **144**, 173 (1962).
[214] Yablokov, Yu. V., *Zh. Strukt. Khim.*, **5**, 222 (1964).
[215] Young, J. E., and R. K. Murmann, *J. Phys. Chem.*, **67**, 2647 (1963).

6

NUCLEAR MAGNETIC RESONANCE SPECTROSCOPY

PAUL J. McCARTHY, S.J.

Canisius College

The use of high resolution nuclear magnetic resonance (NMR) spectroscopy in the study of the bonding and structure of metal chelate compounds has been limited enough that most of the pertinent references through 1966 could be, and we hope have been, included in this review. Broad line NMR studies of metal chelates have been so rare that they have been omitted from the discussion. Although Sec. 6-1 contains a general outline of the theory of NMR spectroscopy, the principles behind contact interaction shifts are, for greater convenience, treated in Sec. 6-2B. For the same reason the applications to biological systems have all been treated together, even though this departs somewhat from the general format of this volume. Although no reference to metal chelates is available for studies of the NMR of liquid crystals, this application has been included because of its potential use on such systems.

6-1 THEORY

A. Basic Principles

1. The NMR Signal. The spin angular momentum of a magnetic nucleus has the value, $\hbar\sqrt{I(I + 1)}$, where I is the spin quantum number (some multiple of $\frac{1}{2}$), and $\hbar$ is Planck's constant, h, divided by 2π. In a magnetic field (H_0) this spin angular momentum vector can be projected along the direction of H_0 and can have values of $m\hbar$ ($m = I, I - 1, \ldots, -I$). The state in which $m = I$ (spin vector most nearly parallel with the field) is the state of lowest energy. Since the spin vector (together with its collinear magnetic moment, μ) lies at an angle to H_0, it precesses about the direction of the field, just as a gyroscope whose rotational axis is not parallel to the earth's gravitational field precesses about the direction of that gravitational field. The frequency of precession, the Larmor frequency (ν), is proportional to the magnetic field strength and can be shown to be

$$\nu = \frac{\mu H_0}{hI} = \frac{\gamma H_0}{2\pi}, \tag{6-1}$$

where γ, the magnetogyric ratio, is $(\mu/I\hbar)$. For example, for H^1, $I = \frac{1}{2}$, $\mu = 2.7927$ nuclear magnetons, and ν will be 60 megacycles per second (mc sec^{-1}) when $H_0 \cong 14{,}100$ gauss. If there is placed perpendicular to H_0 a small alternating magnetic field, H_1, whose frequency is just equal to the precessional frequency of the nuclei (the resonance frequency), some of the precessing nuclei absorb a quantum of energy, $h\nu$, and pass to the next higher energy state. The selection rule governing such transitions is $\Delta m = \pm 1$. This resonance absorption of energy can be detected by a probe perpendicular to both H_0 and H_1, and the signal is usually displayed graphically as a peak on the NMR spectrum.

2. Population of Spin States. Relaxation. Since $I = \frac{1}{2}$ for the proton, two orientations of its spin momentum vector are possible in a magnetic field. If the molecules follow a Boltzmann distribution, it can be shown that in a field of 10,000 gauss the ratio of populations of the higher to the lower energy state is 0.999993. As transitions are induced by the magnetic field, H_1, this ratio approaches unity. Since the appearance of the NMR signal depends on a population difference, the signal decreases in size and then disappears as the populations are equalized. This is obviated if some molecules in the upper state can lose a quantum of energy and revert to the lower state. There are two common mechanisms for this *relaxation*. The energy can be transmitted to the surroundings or "lattice," and the time constant for this process is called the *spin-lattice* or *longitudinal relaxation time*, T_1. It is in reality a half-life for the excited nucleus. For some solutions T_1 can be as short as ca. 10^{-5} sec, and for some crystals at low temperature as long as several hours [34]. In general T_1 must be longer than 0.1 sec if high resolution NMR signals are to be observed [8]. The spin energy can also be dissipated by being transmitted to another magnetic nucleus. This is called *spin-spin* or *transverse relaxation*, and a measure of this spin-spin interaction is the transverse relaxation time, T_2. This is the characteristic time for spins precessing in phase to lose phase when the alternating field, H_1, is removed [34].

3. The Chemical Shift. In a molecule a given isotopic species can give several resonance signals if it exists in several magnetically different environments. Such environmental differences arise because the nuclei are in chemically different groups. The resulting differences in resonance have accordingly been labeled *chemical shifts*. They have their origin in the fact that the atoms in a molecule are surrounded by electrons, which in a magnetic field, H_0, precess about the direction of the field and set up an opposing magnetic field. The field at the nucleus, H, is accordingly different from H_0. The two are related thus:

$$H = (1 - \sigma)H_0, \tag{6-2}$$

where σ is a shielding constant which is proportional to the electron density around a given nucleus. For two chemically different nuclei, i and j, of the same isotopic species their chemical shift difference, δ, at a given field, H_0, is $\delta = \sigma_i - \sigma_j$.

It would be most convenient if chemical shifts could be referred to the resonance of a completely unshielded nucleus. But it has not been possible to calculate exactly the resonance frequency for any unshielded nucleus. Consequently chemical shifts are referred to some standard, usually an internal one added directly to the solution being studied. If an external reference is used, corrections for the difference in bulk diamagnetic susceptibility of the reference and test solutions must be made; see [116, 220]. For proton resonances, tetramethylsilane (TMS) is most often used as an internal reference because of its sharp high field signal, high volatility (b.p. 27°), low reactivity, and reasonable solubility in most organic solvents. Since the chemical shift is dependent on H_0, it is now most common to state it in a dimensionless unit (δ) to avoid confusion:

$$\delta = \frac{10^6(H_c - H_r)}{H_r} \text{ ppm}, \tag{6-3}$$

where H_c is the field at which a given nucleus resonates and H_r is that at which the reference compound resonates. Most protons resonate at lower field than TMS, and so δ is almost always negative. To obviate this, another scale (τ) is often used, where $\tau = 10.00 + \delta$. This scale has the advantage that most resonance positions will be positive values, and increasing τ indicates resonance at increasing field strength because of increased shielding of the proton.

4. Spin-Spin Coupling. In a molecule containing several magnetic nuclei it is possible that the nuclear spins of these nuclei can couple together. One mechanism suggested for this is through the bonding electrons [262]. If, for example, A and X are two magnetic nuclei of the same isotopic species ($I = \frac{1}{2}$) in non-equivalent positions in a molecule, the NMR spectrum can appear as in Fig. 6-1. The A nuclei are coupled with X nuclei in two different energy states, namely, with their spins parallel or antiparallel to the A spins. Consequently two absorption peaks occur for the A resonance. Exactly the same

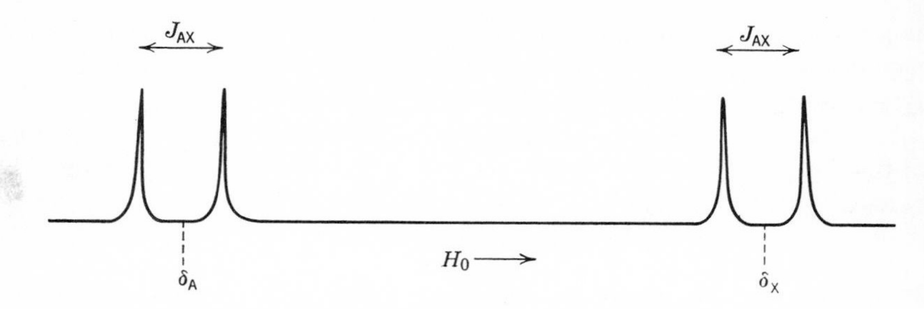

Fig. 6-1. Schematic NMR spectrum of an AX group.

can be said of the X resonance. The chemical shifts for A and X are at the centers of the multiplets. If we have one A and two equivalent X nuclei, A can couple with the two X nuclei existing in four states ↑↑, ↑↓, ↓↑, ↓↓. The middle two states are of equal energy, and so the A resonance appears as a triplet with the relative areas of the peaks equal to 1:2:1. (The relative areas of the components of a regular multiplet are the same as the ordinary binomial coefficients; thus a quartet has areas 1:3:3:1, a quintet, 1:4:6:4:1, etc.) The two equivalent X nuclei couple with A and their resonance appears as a doublet. The relative areas of the doublet and triplet are 2:1, since there are two X and one A nuclei. Again the chemical shifts are at the centers of the multiplets.

In both of these cases the separation of the peaks in the multiplets is a measure of the strength of coupling. The separation is J cycles per second (cps), where J is called the *spin-spin coupling constant*. J, unlike δ, is independent of H_0 and, in general, increases with increase in atomic number of the two coupled nuclei. The examples given are for cases in which the chemical shift is much greater than J, and the coupling schemes are said to be *first order*. Such spectra are always had if A and X are of different nuclear species, but can also be observed if A and X are of the same nuclear species. For such spectra the patterns expected may be determined by the following rule: If n_A equivalent A nuclei of spin I_A are coupled with n_X equivalent nuclei of spin I_X, the A signal has $(2n_XI_X + 1)$ components and the X signal has $(2n_AI_A + 1)$ components.

The term "equivalent" used above should be clarified. Nuclei can be said to have *chemical shift equivalence* if they have the same chemical shift. They are said to be *magnetically equivalent* if, in addition to having the same chemical shift, they also couple to the same extent with all other magnetic nuclei in the molecule. In a general scheme, groups of magnetically equivalent nuclei all separated by chemical shifts much larger than J are designated by letters from different parts of the alphabet, as, for example, A_n, P_m, and X_r;

the subscripts *n*, *m*, and *r* give the number of nuclei in each group. Groups of magnetically equivalent nuclei whose chemical shifts are all of about the same magnitude as *J* are designated by letters from the same part of the alphabet, as, for example, A_n, B_m, and C_r; again the subscripts *n*, *m*, and *r* tell the number of nuclei in each group. In both cases it is supposed that the chemical shift of each group is different from that of any other. Two nuclei which show chemical shift equivalence but which are not magnetically equivalent are designated AA′ or XX′, etc.

While groups of nuclei such as AX_2 give a first order spectrum, the specta of groups such as AB_2 are more complex, and the exact chemical shifts and coupling constants are not immediately evident from the absorption pattern. They can be determined only by quantum mechanical analysis. Resonance patterns for many combinations, $A_nB_m \cdots X_r$, and for various ratios of the chemical shift to the coupling constant have, however, already been analyzed; see especially [16] and also [6, 8, 11, 12, 24, 33]. Comparison of an observed

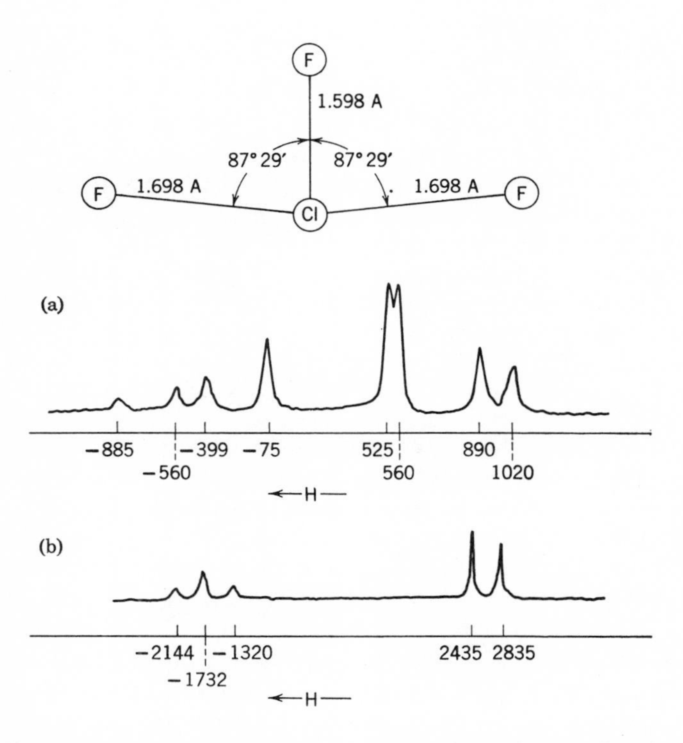

Fig. 6-2. F^{19} NMR spectrum of neat ClF_3 at −60°; shifts are in cps referred externally to SF_6 (a) at 10 mc sec^{-1}; (b) at 40 mc sec^{-1} [244].

spectrum with tabulated spectra can often yield values of the chemical shifts and coupling constants. For values of various coupling constants, see [246] and [8] (passim). Figure 6-2 shows the difference between an AX_2 and an AB_2 pattern [244]. The F^{19} resonance of ClF_3 at $-60°$ and 40 mc sec^{-1} shows two widely separated sets of resonances; the spectrum can be described quite well as an AX_2 pattern. As the field is reduced, the magnitude of the chemical shift becomes more nearly equal to that of J, and at 10 mc sec^{-1} an AB_2 pattern results.

5. ***Simplification of Spectra.*** The analysis of complex spectra can at times be aided by one or more of the following techniques: (a) variation of solvent sometimes separates certain overlapping absorptions; in certain cases the use of preferential solvent shifts can also bring about the coalescence of spin multiplets and the splitting of single lines into doublets [149]; (b) variation of the magnetic field enables one to distinguish chemical shift patterns (field dependent) from spin-spin coupling patterns (field independent); (c) substitution of a non-magnetic isotope for one with a spin can simplify the spin-spin coupling pattern and make it more intelligible; (d) determination of the NMR spectra of two different isotopic species (e.g., H^1 and F^{19} or P^{31}) can provide two complementary approaches to a problem; and (e) double irradiation also can be used to simplify a spectrum. In the last technique there is used an additional radio frequency field whose frequency is equal to the resonance frequency of one group of nuclei in the molecule. This field causes these nuclei to undergo frequent transitions to a higher energy level thus effectively decoupling them from the other magnetic nuclei in the molecule. The spin coupling pattern is thus simplified. Double resonance techniques can also be used to determine the relative signs of the various J's in a molecule (see [3, 8, 11, 20, 27]).

B. An Example

Figure 6-3 shows the proton magnetic resonance (PMR) spectra of two isomeric ligands and their diamagnetic nickel(II) chelates [228]. Assignment of peaks was made on the basis of those made for similar, simpler compounds, and was aided by integrated band intensities. Tables of proton chemical shifts can also be of use if less is known about the molecular structure than was known in the present case. (See [7, 248, 288], pp. 287–291 of [23], and pp. 727 ff. of [8] for tables of proton chemical shifts). Because bonding within the isomeric ligands (or chelates) should be identical, differences in the PMR spectra must be ascribed to magnetic effects arising principally from the positions of the two phenyl groups relative to each other and to the other protons in the molecule. The following assignments may be made for the two isomeric ligands: the two methyl groups adjacent to the bridge resonate at 8.20 and 8.32 τ, and the other two at 7.97 and 7.98 τ, for the *d,l* and *meso* forms, respectively. The former methyls, because of nearness to the π-electron cloud of the phenyl group, are shielded more than the latter, and so resonate at a higher field. The difference between the isomers implies different degrees of diamagnetic shielding by the phenyl groups. The resonances at 5.01 and 5.03 τ are due to the two identical CH groups of the acetylacetone rings, and the complex patterns near 2.8 τ are due to the phenyl hydrogens. The bands at -1.73 and -1.47 τ are typical of hydrogen-bonded protons and so are assigned to the N—H···O groups. The appearance of these bands is an indication that the ligands are largely, if not completely, in an enolized form. The splitting of this resonance peak is due to spin-spin coupling of the NH proton with the CH of the bridging groups. This indicates that the ligands are in the ketamine form shown in Fig. 6-3. In the *d,l* isomer the complex band at 5.28 τ, which is due to the bridge CH—CH group, was found to maintain the same multiplet separation at 100 mc sec^{-1}. This indicated a multiplet due to spin-spin coupling. The pattern was further proved to be due to coupling of the bridge protons with the NH protons by means of double irradiation. When the molecule was irradiated with an additional rf field whose frequency corresponded to that of the NH proton resonance (-1.73 τ), the complex pattern centered at 5.28 τ collapsed to a sharp singlet at the same position. Accordingly, it seems that the complex pattern is due to "virtual coupling" of the bridge CH protons with both NH protons. This phenomenon is discussed in detail by Musher and Corey [247].

In the nickel chelate spectra the separation of the two methyl signals at ca. 8.1 and 8.5 τ is larger than in the free ligands. This is undoubtedly related to the fact that the steric arrangement of the chelates is more rigid. No rotation about the bridge C—C bond can help to relieve the interaction between the methyl and phenyl groups. The absorptions at 5.13 and 5.14 τ are due to the acetylacetone CH group, and those at 5.34 and 5.81 τ to the bridge CH groups. The latter bands are singlets, since coupling with the NH protons has been removed by metal chelation. The rather significant differences between the isomers can be explained thus: in the *meso* isomer the two bridge protons are adjacent to one another and rather removed from any diamagnetic shielding effect of the phenyl groups. In the *d,l* isomer, on the other hand, each proton is rather strongly shielded by the π-electron cloud of the phenyl group on the next carbon. The *d,l* bridge proton resonance is accordingly 0.47 ppm upfield from that of the *meso* isomer. The chelate phenyl resonances (1.7–3.0 τ) are very distinctive. The somewhat unexpected shapes and separations of the multiplets are also most probably due to mutual shielding effects of one phenyl group on the protons of the other. Detailed analysis does not, however, seem feasible without more exact knowledge of the molecular parameters.

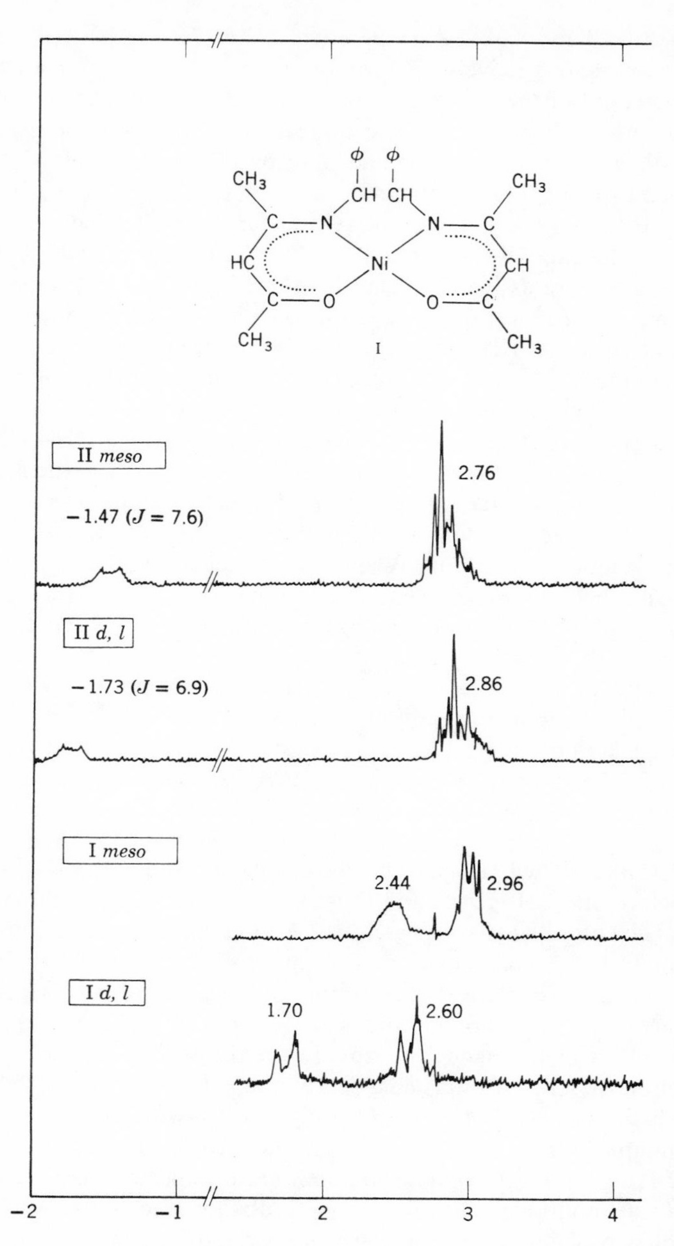

Fig. 6-3. PMR spectra at 35° and 60 mc sec^{-1} of 10% solutions in $CDCl_3$ of bis(acetyl-acetono)-*meso*-stilbenediimine, its nickel(II) chelate, and the analogous *d,l*-ligand and nickel(II) chelate internally referenced to TMS [228].

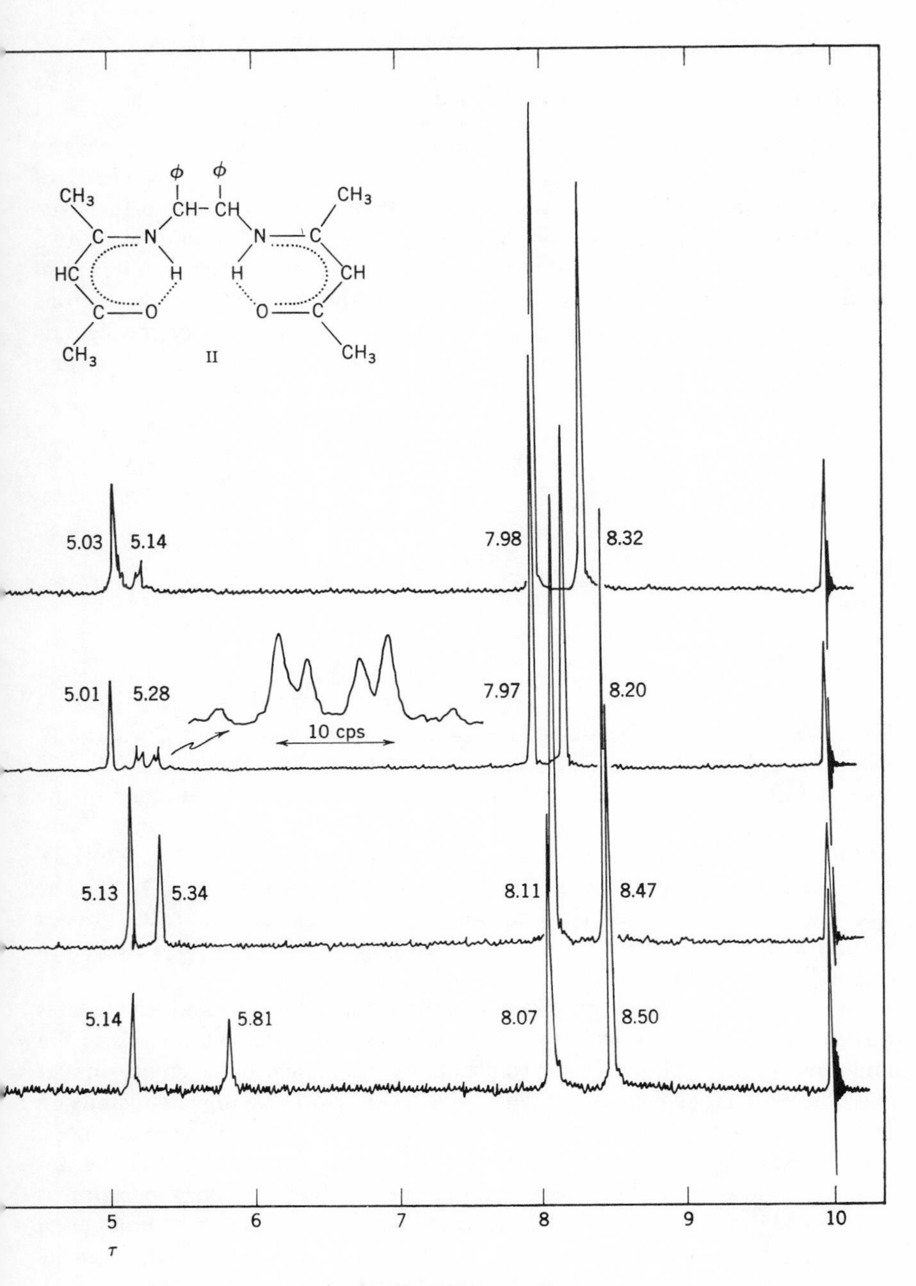

Fig. 6.3. (*Continued.*)

6-2 APPLICATIONS

A. Diamagnetic Chelates: Structure and Bonding

1. Geometrical Isomerism. Octahedral Complexes. In some cases *cis* and *trans* isomers of octahedral complexes can be distinguished by means of their NMR spectra. For example, Fay and Piper [139] have noted that tris-bidentate chelates formed from unsymmetrical ligands should yield two stereoisomers. One of these, (I) (*cis*), contains a threefold rotation axis and has the three ligands in identical environments (the three X's are in positions 1, 2, 3). The other, (II) (*trans*), has no symmetry, and each of the ligands has a different environment (the three X's are in positions 1, 2, 6). Fay and

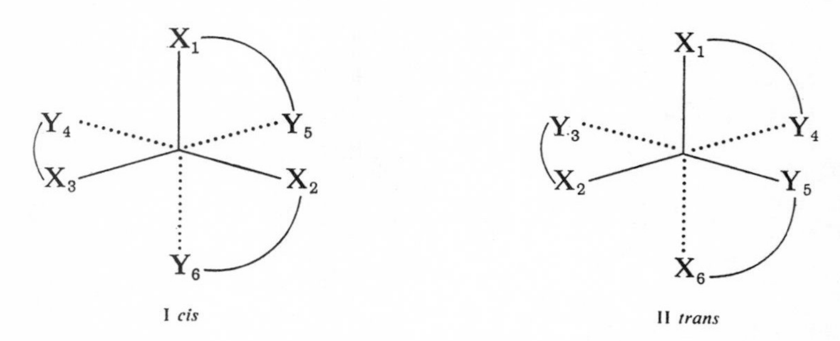

Piper synthesized tris-chelates of benzoylacetone with trivalent Co, Cr, Rh, Al, Mn, and Fe. By making use of solubility differences in ethanol and column chromatography on alumina they separated the two isomers of the first three compounds. The fact that only *one* isomer of the other three compounds was isolated is consistent with the known lability of Al(III), Mn(III), and Fe(III) chelates (see e.g., the exchange studies of Kluiber [191]). And the fact that the one isomer isolated was the *trans* isomer indicates greater thermodynamic stability of this isomer, as might possibly be expected from its smaller dipole moment.

The identification of the two isomers of the diamagnetic Co and Rh chelates was made by PMR spectroscopy. The *cis* form, which has all ligands in the same environment, should give a single methyl resonance and a single, smaller methyne (=CH—) resonance, while the *trans* form, having all ligands in different environments, should show three peaks for each of the resonances, if the chemical shifts are large enough. Figure 6-4 indicates that the *trans* forms do show somewhat broad multiple peaks, but that the separation is often so slight (0.6–1.5 cps at 60 mc sec^{-1}) as to make the exact number of bands indeterminate. The one form of the Al chelate isolated is, from its observed PMR spectrum, clearly the *trans* form. The *cis* chelates, as expected, show single peaks. The methyl resonances occur at about 7.6–7.8 τ, while

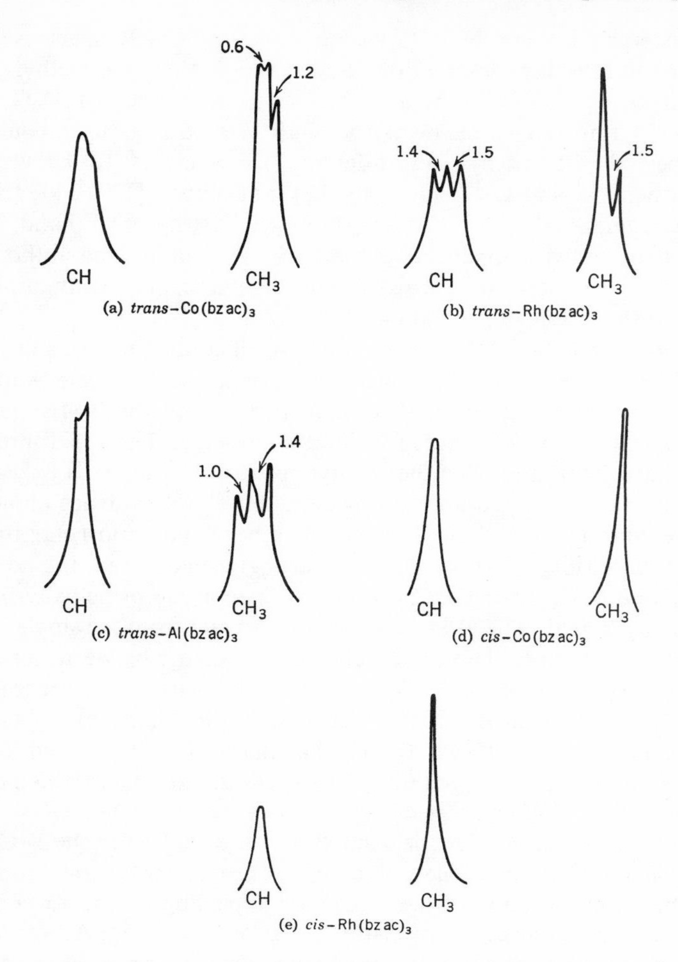

Fig. 6-4. PMR spectra of benzoylacetonates of trivalent metals in $CHCl_3$ solution at 60 mc sec^{-1} and 25° referenced internally to TMS [139].

the =CH— resonances are at about 3.8 τ. Identification of the isomers of the paramagnetic chelates was made on the basis of isomorphism of the crystals of these complexes with those of the diamagnetic complexes. The same authors have also reported a similar study of the isomers of tris-chelate complexes of trifluoroacetylacetone [140].

Archer and Cotsoradis [47] have prepared the complexes

$$cis\text{-}[Co^{(III)}(acac)_2L_2]^{x\pm} \qquad [L_2 = \text{bipy, en, (py)}_2, (NH_3)_2]$$

and have studied them by IR, visible, UV, and PMR spectroscopy. A particularly interesting observation they made is that the methyl protons in $[Co(acac)_2en]^+$ show only a single sharp resonance in D_2O and in CH_3OD. As this ion is of necessity in the *cis* configuration, one might have expected the methyls in different environments to show slightly different chemical shifts, as noted by Fay and Piper [139, 140] for some analogous compounds. *cis*-$[Co(acac)_2(NH_3)_2]I$, on the other hand, has the expected two methyl resonances separated by 0.02 ppm. The authors point out that this indicates the hazard involved in assigning configuration on the basis of the NMR spectra alone.

McGrady and Tobias [236] have studied octahedral complexes of the type $[R_2SnL_2]$ (R = CH_3, C_6H_5; L = acac, bzac, dbm, picolinic acid, 8-hydroxyquinoline, and hfac). The IR and Raman spectra, and the NMR spectra in $CDCl_3$, indicate a *trans* structure for these compounds. There is, for example, a single sharp PMR peak for the methyl groups of acac, as is expected for the *trans* isomer. As noted above, however, the NMR evidence alone is not conclusive. Smith and Wilkins [279], on the other hand, report that the PMR spectra of $[Sn(acac)_2X_2]$ (X = Cl, Br, I) strongly suggest that the *cis* form is the stable one in C_6H_6 or $CDCl_3$ solution. They argue principally from the double methyl signal (0.02–0.11 ppm separation) and from the single —CH= peak, and attempt to explain the difference in structure between this and the former group of compounds in terms of the polarity and strength of the bonds formed. The splitting of the methyl signal in $[Sn(acac)_2X_2]$ (X = Cl, Br, I) was also noted by Kawasaki and Tanaka [187], who proved it to be a chemical shift difference by obtaining the spectrum at 60 and 100 mc sec^{-1}. They also noted that the spectrum of $[Sn(acac)_2(C_6H_5)X]$ (X = Cl, Br) shows a complex quartet for the methyl and a doublet for the —CH= of the acetylacetone groups. They assume a *trans* structure and attempt to explain the added splitting of the signals by appealing to anisotropic effects of the phenyl group joined with distortion in the acac ring. An explanation in terms of a *cis* configuration for all these molecules seems, however, more plausible.

Preferential stability for the *cis* isomer has also been noted for $[M(acac)_2Cl_2]$ [M = Ti(IV), Ge(IV), and Sn(IV)] [280] and for $[Ti(acac)_2X_2]$, where X is an alkoxide group [56]. The presence of the *cis* isomer in the latter compounds [56] was again indicated by the doubling of the methyl resonance of the acetylacetone moiety. Bradley and Holloway [56] suggest as a reason for the greater stability of the *cis* form the fact that in this isomer each of the three titanium d_ε orbitals can be engaged in $d\pi$-$p\pi$ bonding with an oxygen antibonding molecular orbital, while only two of the d_ε orbitals can be so engaged in the *trans* form. Holloway, Luongo, and Pike [171] have also studied the isomerization of $[Si(acac)_2(acetate)_2]$. The PMR spectrum of the compound in $CHCl_3$ or CH_2Cl_2 changed continuously for one-half hour from the time of mixing. By this time a steady state had been reached. On the basis of IR and NMR spectroscopy they interpret the change as isomerization of the initial *trans* isomer to a mixture containing a 1.6:1.0 *cis*:*trans* ratio. The reason given in this case is the greater polarity of the *cis* form, which is favored in the polar solvents used. Partly on the basis of their NMR spectra, a *trans* configuration has been assigned to several nitroamminebis-(acetylacetono)cobalt(III) complexes by Boucher and Bailar [55].

Clifton and Pratt [85] report that the fumaratopentamminecobalt(III) ion, $[(HOOC{-}CH{=}CH{-}COO)Co(NH_3)_5]^{2+}$, has a PMR spectrum consisting of a sharp, low field doublet (300.3 and 313.1 cps downfield from *t*-butyl alcohol in D_2O at pD $\approx$ 1 and 56.45 mc sec^{-1}) and two broad bands at 152 $\pm$ 2 and 94.5 $\pm$ 2 cps below *t*-butyl alcohol. The latter peaks have relative areas of 4:1. The doublet is from the olefinic protons, and the broad lines are from the ammine hydrogens *cis* (4) and *trans* (1) to the fumarato group. The broadening is due to quadrupole relaxation by the N^{14}. Likewise the spectrum of the *cis*-$[Co(en)_2Cl_2]^+$ ion shows a multiplet at 84–96 cps due to the $—CH_2—$ groups, and two equal peaks at 250 and 178 cps. By analogy the latter is assigned to the NH_2 groups *trans* to the chlorides. Once these correlations have been made, the change in spectra on substitution can be interpreted to show whether the new substituent has entered in a *cis* or a *trans* position. For example, in D_2O the broad ammine proton lines become gradually weaker owing to deuterium substitution. But the intensity of the high field line (178 cps) decreases faster than that of the low field line (250 cps). This is an indication of the greater lability of the *trans* protons; the conclusion agrees with those of other rate studies of hydrogen exchange in ammine complexes [252]. See [168] for another study of isomerism in Co(III)-en complexes.

Asperger and Liu [48] have reported the preparation and characterization (largely by NMR spectroscopy) of the three geometric isomers (III, IV, V) of the dichlorodimethyltriethylenetetraminecobalt(III) ion. They made no attempt to isolate pairs of diastereoisomers. And Legg and Cooke [211, 212] used ion exchange chromatography to separate, and PMR and visible absorption spectroscopy to identify, the various isomers of the cobalt(III) complexes, $[Co(en)(L)]^+$, where L is ethylenediamine-N,N′-diacetic acid or its N,N′-dimethyl and -diethyl analogs. They have also studied $[Co(NH_3)_2L]^+$ and the cobalt(III) complexes of mixed tridentate ligands, $[Co(den)(L)]^+$, in which den is diethylenetriamine and L is iminodiacetate (IDA), methyliminodiacetate, or pyridine-2,6-dicarboxylate. They were able to isolate three isomers of $[Co(den)(IDA)]^+$ (VI, VII, VIII). This they report as the first known case of an octahedral chelate containing two tridentate ligands being shown to exist in three isomeric forms. Cooke [100] has also made a

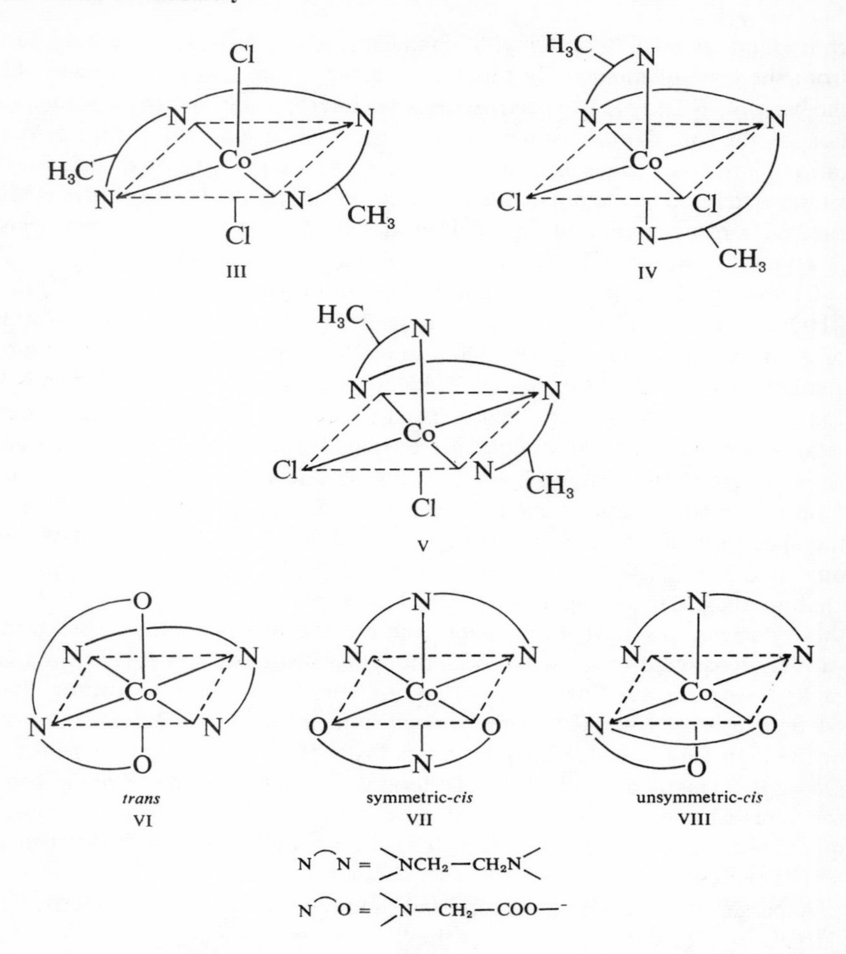

study of the stereochemistry of the bis complexes of Co(III) with iminodiacetic acid and its methyl derivative.

Isomerism in some diamagnetic tris-(N-R-salicylaldimino)cobalt(III) chelates (R = CH_3, C_2H_5, *p*-tolyl) and tris-(N-R-pyrrole-2-aldimino)cobalt(III) complexes (R = CH_3, C_2H_5, *i*-C_3H_7, *sec*-C_4H_9) has been studied by means of PMR spectroscopy by Chakravorty and Holm [73]. The spectra show no evidence of any *cis* compounds, a finding which the authors feel is undoubtedly due to the steric repulsion that would ensue in the *cis* isomers between the bulky R groups.

Gillard and Wilkinson [154] have shown that the presence of an O—H···O hydrogen resonance ($\sim -8.5\ \tau$) in the PMR spectra of cobalt(III) dimethylglyoximato complexes is an indication of the *trans* configuration of the bis-

(dioximato)cobalt(III) moiety. Their observations are confirmed by the IR spectra of the compounds.

Square-Planar Complexes. Few examples of the NMR spectra of square-planar geometrical isomers are available. The spectrum of bis(benzoylacetono)palladium(II) shows in addition to the phenyl resonance an unresolved multiplet at 7.78 τ (due to CH_3) and a clear doublet (separation 1.5 cps; relative areas 2:1) at 3.93 τ (due to =CH—). The unexpected doubling of the latter peak is most probably due to the compound's being a mixture of *cis* and *trans* planar forms [227]. Kluiber [192] has used PMR spectroscopy with many other measurements to show that the copper(II) and nickel(II) chelates of N-alkylthiopicolinamides have *trans*, square-planar structures (IX).

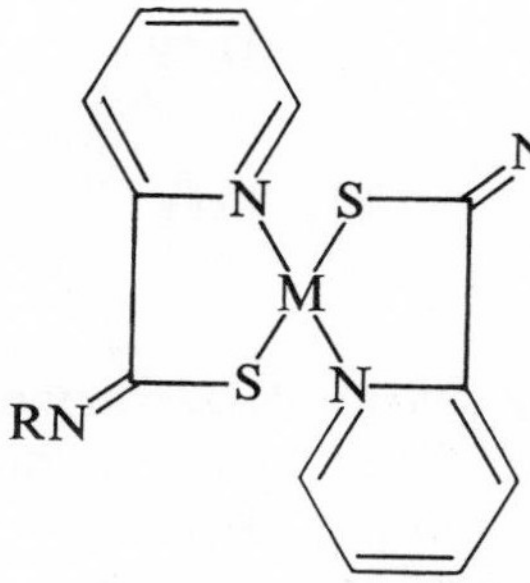

M = Cu(II), Ni(II);
R = alkyl groups

IX

2. *Linkage Isomerism.* When a ligand can be bonded to a metal in several alternative ways, NMR spectroscopy can sometimes show which particular linkage isomer is had. For example, it was used as an aid in elucidating the structure of the unusual chelate $K[Pt(acac)_2Cl]$, which has been shown by Figgis and co-workers [146] to have structure X. When a solution of this is acidified, a yellow, light-sensitive complex separates which is readily soluble

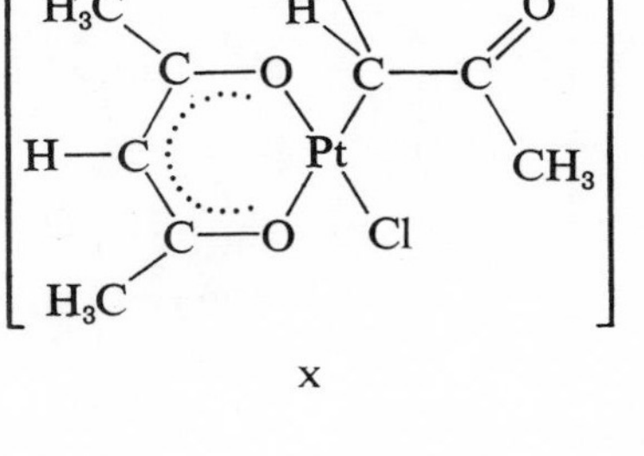

X

in organic solvents. Its PMR spectrum contains methyl signals at 7.37, 7.94, and 8.08 τ in a ratio of 1:1:2, and a peak at -2.47 τ. Because of the position of the latter peak, and because its size decreases on addition of D_2O, it was judged to be due to a hydrogen-bonded proton. These data together with coupling constants and IR information (see p. 253) led Allen and coworkers [44] to assign to the yellow complex formula XI.

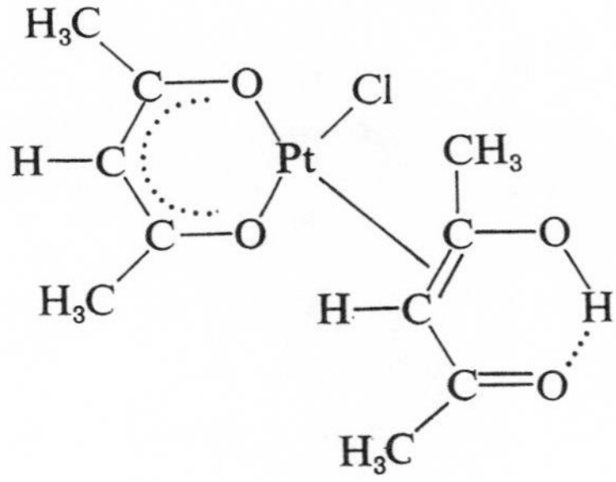

XI

Lewis, Long, and Oldham [214] have also reported the IR (see p. 251) and PMR spectra of the platinum(II) compounds: $[Pt(acac)_2]$, $K[Pt(acac)_nCl_{3-n}]$ ($n = 1, 2, 3$), and $Na_2[Pt(acac)_2X_2]\cdot 2H_2O$ (X = Cl, Br) (see also [152]). They were able to distinguish between chelated acac groups and those attached to the Pt through the γ-carbon. In the former the coupling between platinum and the γ-proton is through four bonds and $J(Pt^{195}—H)$ is 2–11 cps (Pt^{195} has a spin of $\frac{1}{2}$ and a natural abundance of 33.7%). In the latter the coupling is through only two bonds and J is about 120 cps. Thus, for example, the compound $K[Pt(acac)_3]$ has the following absorptions measured in D_2O relative to dioxane referenced to TMS (relative intensities in parentheses): 105 cps (12), $J = 16$ cps; 77 cps (6), $J = 2$ cps; 260 cps (2), $J = 123$ cps; 305 cps(1). These data are consistent with structure XII.

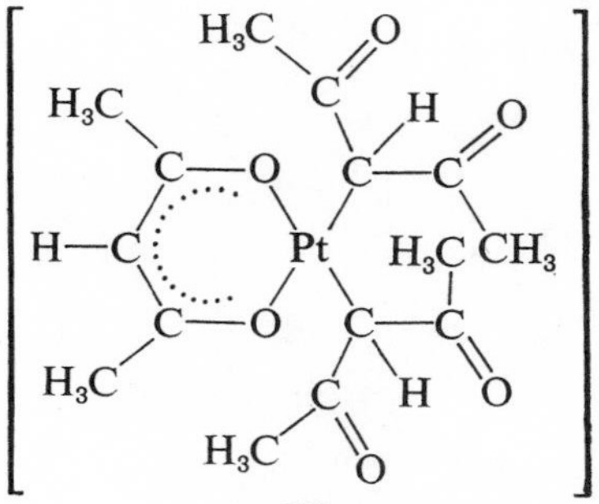

XII

The three types of bonding between Pt and acetylacetone discussed above do not exhaust the known types of linkage isomerism in derivatives of acac and related compounds. Hammond, Nonheber, and Wu [162] have noted that bis(dipivaloylmethanido)mercury(II) shows in $CDCl_3$ a resonance at 5.08 τ. This peak ($\frac{1}{18}$ the size of the methyl peak) is due to the γ-proton, and its position is about 0.5 ppm higher than that in metal chelates. This fact, coupled with UV and IR data, has led them to propose for the compound

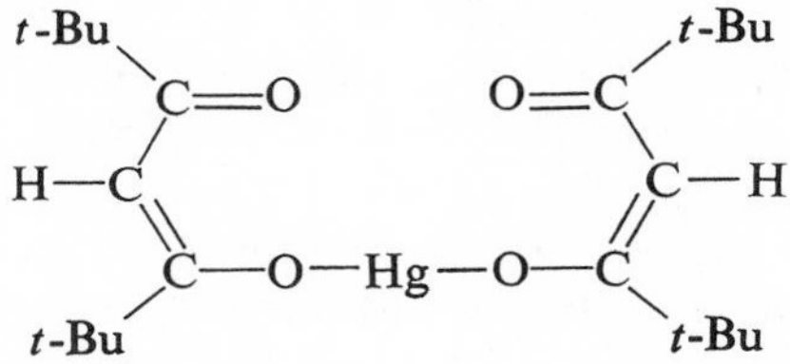

t-Bu t-Bu
C=O O=C
H—C C—H
C—O—Hg—O—C
t-Bu t-Bu

XIII

the non-chelated structure XIII. Still other examples of somewhat unexpected structures found in acetylacetonates are reported in the study of acetylacetone derivatives of S, Se, and Te carried out by Dewar and co-workers [111].

3. Ligand Isomerism. Collman and Jui-Yuan Sun [99] have studied the ligand isomerization and disproportionation of some unsymmetrically substituted metal acetylacetonates. They found, for example, that heating monoformyl cobalt acetylacetonate (XIV) in boiling toluene at 111° for two days caused isomerization of the chelate. Thin layer chromatography showed at least six components in the resulting mixture. Identification of some of the

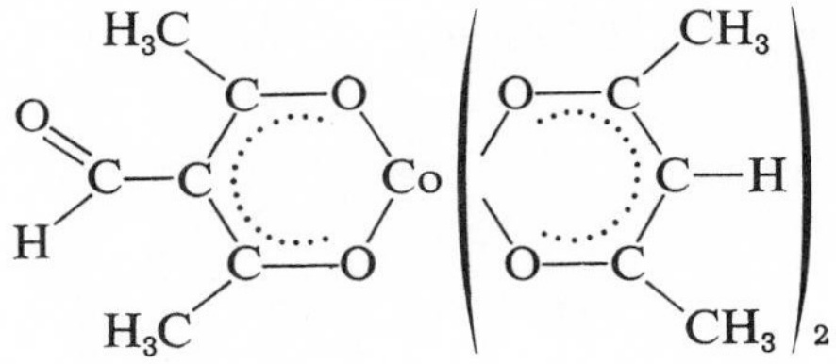

H_3C CH_3
C—O O—C
O
C—C Co C—H
H
C—O O—C
H_3C CH_3 2

XIV

H CH_3
C—O O—C
O
C—C Co C—H
H_3C
C—O O—C
H_3C CH_3 2

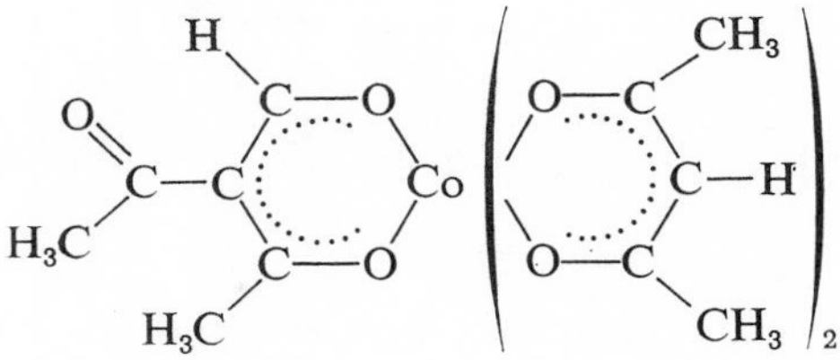

XV

components was aided by comparison of their PMR spectra with those of authentic samples of some cobalt chelates. One of the components definitely identified was the ligand isomer (XV). Their data on this and other mixed chelates suggest without strict proof that ligand isomerization is an intermolecular process. Their results also suggest that the disproportionation of cobalt chelates requires a lower energy of activation than that of similar chromium chelates.

4. Intramolecular Conversion. Much of the evidence we now have for intramolecular tunneling processes, or pseudorotation, comes from NMR spectroscopy. Several years ago Gutowsky, McCall, and Slichter [160] found the F^{19} spectrum of PF_5 to be a 1:1 doublet whose separation was independent of field strength. The five fluorines are accordingly equivalent, their signal being split by coupling with the P^{31} nucleus (spin $\frac{1}{2}$). From the electron diffraction study of Brockway and Beach [60] the compound is also known to be a trigonal bipyramid. Such a structure contains two non-equivalent kinds of fluorine atoms, which might be expected to yield different NMR signals. This rather anomalous result has found a cogent explanation in the possibility of intramolecular rearrangements that are rapid enough to prevent the appearance of separate NMR signals for atoms which are at structurally non-equivalent sites in a molecule. Berry [51] has given a schematic representation for this "pseudorotation" in trigonal bipyramidal molecules (Fig. 6-5). The rearrangement involves only a vibrational process and no bond breaking. This could account both for the equivalence of the five fluorines and for the maintaining of the spin–spin coupling between the P^{31} and F^{19} in PF_5. However, for the ligands to be spectroscopically equivalent, the time that the molecule remains in a given structural arrangement must be shorter than the time scale of the particular spectroscopic technique used.

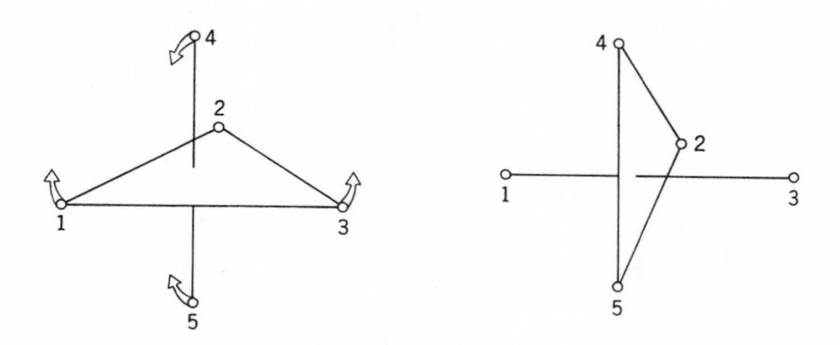

Fig. 6-5. Pseudorotation or intramolecular conversion in a trigonal bipyramidal molecule [51].

Abel and co-workers [39] have noted that in the compound

$$Pt(CH_3SCH_2CH_2SCH_3)Cl_2$$

the methyl signal consists of two triplets. The triplet structure is due to splitting of the methyl hydrogens by Pt^{195} (see p. 324). They attribute the doubling of the triplet to the existence of two forms of the molecule, XVI and XVII. The fact that the two triplets coalesce at 95° indicates a barrier

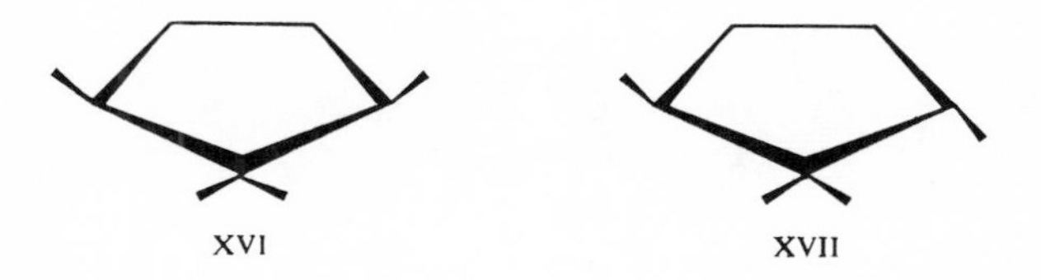

of about 5 kcal to the interconversion. Similar data were obtained for the palladium complex. For several interesting examples of intramolecular conversion in non-chelated systems, see [243] and references cited therein.

5. Bonding Sites in Chelates. Day and Reilly [108] have analyzed the various types of metal-EDTA PMR patterns in terms of the lability of the M—O and/or M—N bonds. If *all these bonds are labile*, one sharp peak should occur for the ethylenic protons and one for the acetate protons. If the metal has an appreciable amount of isotopes of spin $\frac{1}{2}$, splitting of the proton signals will occur and can persist despite lability of the M—N bond. This coupling would not persist in the event of complete metal-ligand exchange, but it does persist when there is continual unmaking and reforming of the M—N bond on the same metal atom. Such a situation was found for Pb-EDTA (Pb^{207} has spin $\frac{1}{2}$ and a natural abundance of 21%) (Figure 6-6a).

If the *M—N bond is non-labile* while the *M—O bond is labile*, structural diagrams show that the ethylenic protons should be equivalent and should therefore give a single peak, if the metal has all isotopes of zero spin. In this situation, however, the two protons of the acetate —CH_2— groups are not equivalent and give an AB pattern. Since the M—O bond is labile and there is free rotation about the C—N bond, this lack of equivalence may be due in part to the proton's spending different amounts of time in different rotational conformations. Cadmium-EDTA is of this type, but its spectrum is further complicated by the fact that 25% of cadmium's nuclei have spin $\frac{1}{2}$ (Fig. 6-6b,c). Kula and co-workers [200] have noted the presence of a single acetate methylene resonance in the spectra of, for example, the alkali and alkaline earth complexes of EDTA. They conclude that this seems to indicate rapid coupling and uncoupling in the metal-oxygen bonds in these compounds.

Cobalt(III) EDTA is an example of a type in which *all the metal-ligand bonds are non-labile.* The acetate groups in this situation are non-equivalent

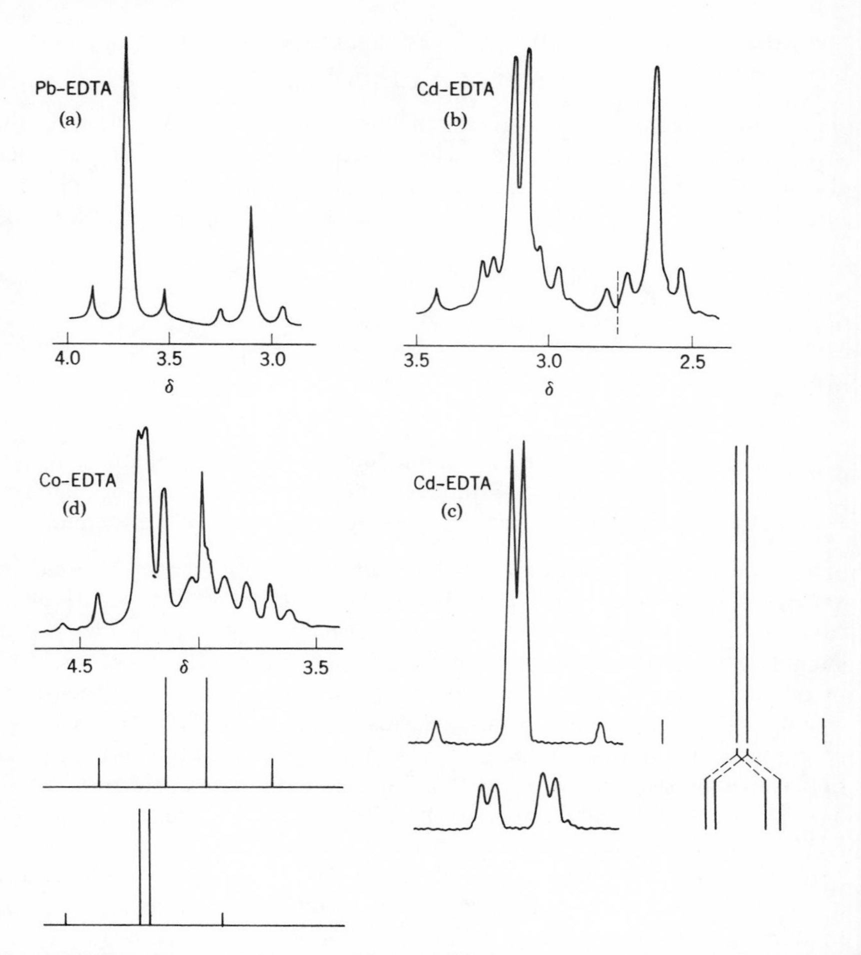

Fig. 6-6. PMR spectra at 60 mc sec^{-1} of 0.5–1.0M aqueous solutions of metal-EDTA complexes [108]; reference: sodium 3-(trimethylsilyl)-1-propanesulfonate. (a) Pb-EDTA; acetate and ethylenic proton resonances at ca. 3.7 and 3.1 δ, respectively. (b) Cd-EDTA; PMR for acetate protons to left of broken line and for ethylenic protons to right. (c) Cd-EDTA; acetate portion of PMR spectrum; on top the AB pattern due to Cd with $I = 0$; on bottom ABX pattern due to Cd with $I = \frac{1}{2}$. (d) Co(III)-EDTA; two AB patterns for the acetate groups shown below spectrum; remaining peaks due to ethylenic protons.

but exist in two pairs. Furthermore, the protons in any one acetate group are non-equivalent. Therefore the spectrum of the acetates may give two overlapping AB patterns. When there is added to this (and superimposed on it) the pattern of the ethylenic protons, which also may be non-equivalent, there results a complex pattern which may effectively resist analysis (Fig. 6-6d). At any rate an inspection of the PMR spectrum of a metal-EDTA complex should allow one to say something about the lability of the metal bonds in the compound. The PMR spectrum of $[Co(EDTA)]^-$ has also been reported by Legg and Cooke [211]. They note, without being able to explain it, that their spectrum is shifted 0.30 ppm upfield from that reported by Day and Reilly [108].

A PMR study by Chan, Kula, and Sawyer [75] of the Mo(VI) complexes of EDTA, NTA (nitrilotriacetic acid), and MIDA (methyliminodiacetic acid) shows that the Mo-carboxylate bonding in these molecules is non-labile. The PMR spectra of the MIDA and EDTA complexes show an AB type of pattern for the —CH_2— of the acetate group, indicating some non-equivalence of protons. The metal-ligand ratios of the complexes were determined by the PMR spectra to be 1:1 for NTA and MIDA and 2:1 for EDTA.

A PMR study of the EDTA complexes of Zr(IV), Hf(IV), and Pd(II) indicates non-labile bonding between the metal and the ligand donor groups in these complexes. It was found also that, with Pd(II), EDTA can act as either a bi- or a tetradentate ligand [46]. Also a study [62] of Mg, Ca, Sr, and Ba complexes with di-(2-aminoethoxy)-ethanetetraacetic acid,

$$[CH_2OCH_2CH_2N(CH_2COO)_2]_2,$$

has indicated that coordination is through the two nitrogens, two ether oxygens, and two carboxyl oxygens; two carboxylate oxygens are accordingly free.

Similar studies in which bonding sites in chelates have been elucidated by NMR spectroscopy have been made by Fritz and co-workers [150], Dudek and Snow [117], Brinckman and co-workers [59], Jensen and Pflaum [184], Sawyer and Brannan [274], and Li and co-workers [216, 226, 300] (see Sec. 6-2C, p. 357).

6. Pseudoaromatic Character in Chelate Rings. Some evidence for pseudo-aromatic character in metal acetylacetonates is furnished by the reactions these chelates undergo. Collman and co-workers among others have studied these reactions [93, 98], and Collman has recently reviewed the work completely [91, 92]. For example, the γ-protons on the chelate rings in several tris-acetylacetonate complexes can be replaced by I, Br, Cl, SCN, NO_2, CHO, etc, if the chelate is treated with the proper electrophilic reagent. Mono-, di-, and trisubstituted products are all possible if the reagent's concentration is properly limited.

The structures of some of these mixed ligand chelates of Co(III) and Rh(III) have been elucidated by PMR spectroscopy [92, 94–97]. As expected, the unsubstituted complexes gave two signals in a ratio of 1:6, the first due to the —CH═ of each chelate ring, the other due to its two equivalent methyl groups. In the tris-γ-substituted chelates the first peak disappears and the other is shifted to lower field, the extent of the shift depending on the nature of the substituent, metal, and solvent [92]. When, however, a mixed complex with two rings of one type (α) and one ring of a different type (β) is examined by PMR spectroscopy, it is found that the methyl groups of the unique ring (β) are equivalent and yield only one signal. But the methyl groups of the other rings (α) exist in two different environments and so give two distinct signals. The different environments cannot be due to bonding within the individual rings, for on this basis the four methyl groups of the two α rings are equivalent. A molecular model of the compound indicates that the two methyl groups of any one α ring are directly above differently substituted rings and so are in different environments. This clearly supposes a diamagnetic shielding by the π electrons in one chelate ring of the adjacent methyl group on another ring. If the two methyl groups of an α ring are adjacent to two differently substituted rings having π clouds of unequal density, a different shielding of the two methyl groups and consequently a different chemical shift is expected. Such an effect is known in aromatic chemistry, and its appearance here seems to offer an indication of pseudoaromatic character in the acetylacetonate ring. An example of this is compound XVIII, whose methyl resonances in CCl_4 solution are: CH_3(a) = CH_3(b) = 7.38 τ; CH_3(c) = 7.59 τ; CH_3(d) = 7.56 τ. The three are in a ratio of 1:1:1 [94]. When all three rings are differently substituted, each of the six methyl groups shows its own signal.

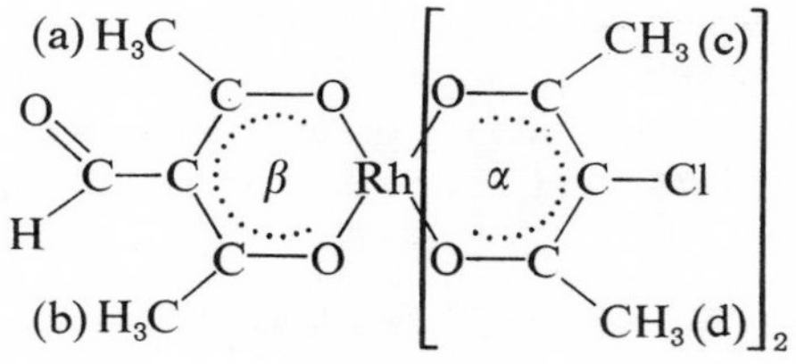

XVIII

Additional evidence for pseudoaromatic character in acetylacetonates is afforded by Hester's [169] study of the PMR spectrum of tris(acetylacetono)-silicon chloride in $CHCl_3$. The resonances and assignments are: CH_3, 7.73 τ; —CH═, 3.74 τ; −2.37 τ (probably due to HCl_2^-). So the compound should probably be formulated as $[Si(acac)_3]^+[HCl_2]^-$. The vinyl hydrogen resonance is at considerably lower field than in other diamagnetic acetylacetonates (4.46–4.75 τ) [281]. This lowering toward the value for aromatic protons (~2.65–2.85 τ) may indicate an increase in electron delocalization in the chelate ring and an increase in its aromatic character. Hester suggests that such a situation may arise by donation of the oxygen's lone pairs to the empty $d(t_{2g})$ orbitals on the silicon. This partial double bonding would give some aromatic character to the ring, and the resulting ring current could cause a deshielding of the groups attached to the chelate ring. This deshielding was observed not only for the γ proton, but also for the methyl groups, which resonate at somewhat lower field in the silicon compound (7.73 τ) than in, for example, the beryllium compound (8.02 τ) [281]. That this question is still quite an open one is clear from the recent report of Smith and Wilkins [280]. They offer an alternative explanation for the low field shift of the =CH— resonance in molecules such as $[Sn(acac)_2Cl_2]$, $[B(acac)F_2]$, and $[Si(acac)_3][HCl_2]$, in terms of electric field effects, and so feel that it is unnecessary to invoke benzenoid resonance in the latter molecule. (For an extended discussion of π-bonding in silicon compounds, see [106].)

The effect on the PMR spectrum of an acetylacetonate complex produced by substitution of phenyl or CF_3 for the terminal methyl group(s) is shown by the data of Bonati and Wilkinson [53] for $Rh(CO)_2(X)$ listed in Table 6-1.

Table 6-1 PMR Absorptions of Some Diketone Complexes[a]

Compound	Diketone=CH— (τ)	Diketone—CH_3 (τ)	Sn—CH_3 (τ)
$Rh(CO)_2(acac)$	4.48	7.89	
$Rh(CO)_2(tfac)$	3.98	7.75	
$Rh(CO)_2(bzac)$	3.82	7.84	
$Rh(CO)_2(hfac)$	3.52		
$(C_6H_5)_2Sn(acac)_2$	4.62		
$(CH_3)_2Sn(acac)_2$	4.73		9.51
$(CH_3)_2Sn(hfac)_2$	3.79		8.91
$(CH_3)_2Sn(dbm)_2$	3.25		9.07

[a] Spectra of rhodium complexes measured in CCl_4; data from [53]. Spectra of tin complexes measured in $CDCl_3$; data from [236].

The more electronegative the terminal group, the more deshielded is the ring proton. As expected, substitution has relatively little effect on the CH_3 group in the other terminal position. The shifts for the benzoylacetone compound seem also to be in part due to resonance between the phenyl group and the chelate ring. Without this resonance one would expect the PMR absorption of the =CH— in the bzac derivative to be at higher field than that in the tfac derivative. Similar data have been obtained by McCarthy and Martell [228] for some nickel(II) chelates of bis(diketone)diimine Schiff bases. The data of McGrady and Tobias [236] shown in Table 6-1 also confirm these results.

It may be of use to list the many other NMR studies of metal-β-diketone chelates that have appeared in addition to those noted above; see [61, 69, 118, 143, 166, 175, 190, 251, 254, 267, 297].

*7. **Metal-Hydrogen Bonding.*** The years since 1959 have seen the recognition of the hydride ion as a genuine ligand in coordination chemistry. Work on hydrido complexes centers about J. Chatt and his co-workers and G. Wilkinson and his group (see refs. below). Buckingham and Stephens [63, 64] have provided some of the more important theoretical discussions. In general the PMR of the hydride ion in diamagnetic transition metal complexes is quite distinctive in that it usually appears at high fields (between 10 and 50 τ) [63]. Therefore a resonance at such a high field can often be taken as an indication of the presence of a hydride ion coordinated directly with a transition metal. If the hydride complex can be isolated, then IR spectroscopy can also be used to show the presence of an M—H bond. If the metal to which the hydrogen is bonded has a spin, the hydride resonance will appear as a multiplet. Thus, for example, when *trans*-$[Rh(en)_2I_2]^+$ is treated with BH_4^-, resonances appear at 30.2 and 31.6 τ, which are assigned to the hydride in the $[Rh(en)_2IH]^+$ and $[Rh(en)_2H_2]^+$ species, respectively [250]. In both cases the resonance line is split by 27 $\pm$ 1 cps owing to coupling with Rh^{103} (spin $\frac{1}{2}$). In the same research Osborn et al. [250] gathered spectroscopic evidence which indicated that H^- should be placed in the spectrochemical series as follows:

$$I^- < Br^- < Cl^- < OH^- < H_2O < H^- \approx NCS^- < NH_3 < NO_2^-.$$

Chatt and co-workers have also reported studies with NMR data of hydride complexes of Ru(II) and Os(II) [79–81] and of Fe(II) [78], and Gillard and Wilkinson of complexes of Rh(III) containing nitrogen ligands [153]. (See also [76, 303, and pp. 251 ff. of [31].)

Various explanations have been offered for the very large, high field shift of the hydride ligand in diamagnetic transition metal complexes. These are reviewed by Buckingham and Stephens [63], who also offer a detailed general theory to explain the observations. They conclude that the principal contribution to the proton's chemical shift is the distortion by the magnetic field of the metal's partly filled *d* shell, which results in a large deshielding of the metal nucleus. They also note that the effect depends on r^{-3}, where r is the distance from the electrons to the proton, and so will depend on the M—H bond length. In a later paper [64] they applied their theory to some square-planar Pt(II) complexes, *trans*-$[PtHX(PEt_3)_2]$ ($X = NO_3^-, Cl^-, Br^-, I^-, NO_2^-, NCS^-, CN^-$). Chatt and Shaw [82] had earlier studied the PMR spectra of these same compounds, and had found that in general the value of the hydride resonance decreases with increasing *trans* effect of X (exceptions are NO_2^- and NCS^-) (see also [77]).

Miller and Prince have carried out PMR studies of aqueous [239] and DMSO [240] solutions of several metal ortho-phenanthroline complexes. They noted that the change in chemical shift of protons 2 and 9 (XIX) on chelation with a metal was a very noticeable one to higher field, and that the shift to higher field increases with a decrease in metal ion size. As the metal

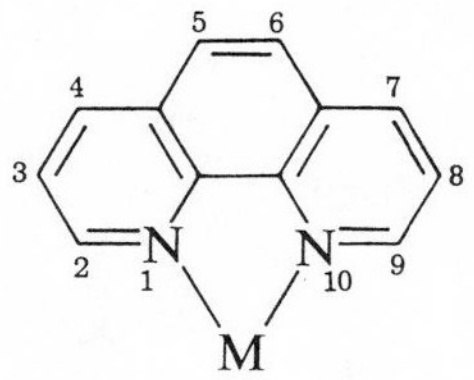

XIX

ion size decreases, so does the distance between the metal and the 2 and 9 protons; this distance is of the order of 2.8–2.9 A. Miller and Prince feel that it may be a non-bonded hydrogen-metal interaction which causes greater shielding of the protons and the consequent high field shift. This interaction would be analogous to direct M—H interaction, but much attenuated.

8. Other Structural Studies. NMR spectroscopy has also been used as an aid in elucidating the structure of several other metal chelates. For example, the PMR spectrum of tris(*o*-phenylenebisdimethylarsine)nickel(II) perchlorate indicates that it is closely similar to the Co(III) analog [54]. The complex cation therefore probably has an octahedral configuration, even though it is diamagnetic.

Clark [84] has studied the chelates formed when the tetrahalides of Ti, V, or Sn react with *o*-phenylenebisdiethylarsine. He found that, while the halides form eight-coordinate species with the methyl analog of this ligand, only six-coordinate complexes are formed with the ethyl compound. Although steric effects are probably involved, the author feels that electronic effects may be a more important factor in determining the coordination number of the metal in the complex.

Muetterties and Wright [245] report the PMR spectra of the tropolonates (XX) of the inner transition metals as well as those of Zr, Hf, Th, Nb, Ta, and several other metals. They interpret these and other data in terms of seven- and eight-coordinate structures for these chelates.

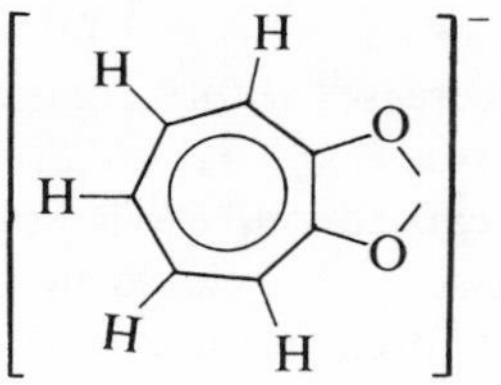

XX

F^{19} NMR has been employed along with ESR, IR, polarographic, and magnetic susceptibility measurements by Davison et al. [107] in a discussion of the structure of complexes of the type XXI. NMR data are given for complexes in which M = Ni (z = 2, 0), Pt (z = 0), Co, Fe (z = 1). They feel that metal-ligand π bonding must be invoked in order to explain all the experimental results. NMR spectroscopy has also been used by King [189] in assigning the structure of the product of the reaction of $Mo(CO)_6$ with the same ligand, bis(trifluoromethyl)dithietene.

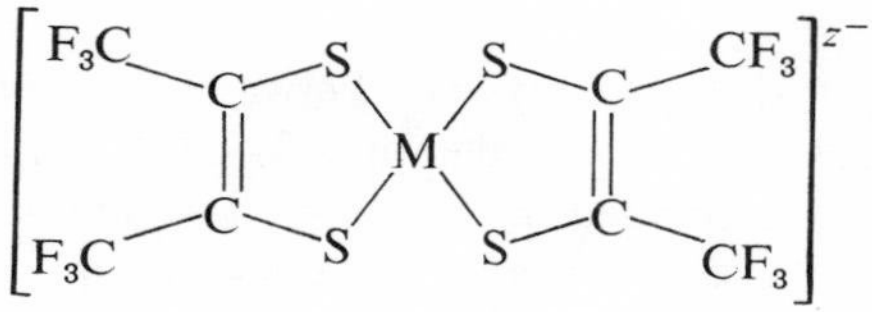

XXI

An NMR study of some Co(III)-en and/or -pn complexes of known chelate ring conformations has been carried out by Spees and co-workers [286]. The fine structure was not sufficiently well resolved, however, to permit unambiguous assignments of the different conformations of individual chelate rings. For other structural studies see [58, 66, 145, 188, 249, 276, 278].

9. NMR in Liquid Crystals. A particularly arresting development in NMR spectroscopy which may in time be the source of accurate molecular parameters is the recent discovery that molecules dissolved in the nematic mesophase [158] of a liquid crystal give highly resolved, but very complex, NMR spectra [273]. As Snyder [282] points out, the molecules of the nematic mesophase of a liquid crystal are usually rod-like in shape: "A nematic mesophase is thought to consist of domains of about 10^6 molecules. In a domain the rod-like molecules are thought to have their axis parallel, and thus a domain is anisotropic. When a magnetic field is applied to a nematic mesophase a domain tends to line up so that its axis of minimum magnetic susceptibility is parallel to the applied field. Thus in a magnetic field a nematic phase provides an anisotropic environment for dissolved molecules."

When molecules are dissolved in such a mesophase, the "NMR lines are relatively sharp because the dissolved molecules are sufficiently mobile to average dipole-dipole interactions with the solvent. The anisotropy of the molecular motion leads to large intramolecular magnetic dipole-dipole interactions between nuclei, and to corresponding structure in the NMR spectrum" [282]. In ordinary isotropic solvents these dipole-dipole interactions average to zero.

Snyder and Anderson [283, 284] have used a complex spin Hamiltonian, containing these dipole-dipole interactions as well as the ordinary terms, to analyze and simulate by computer two complex PMR spectra. They did this for an approximately 15 mole % solution of benzene in the nematic phase of *p,p'*-di-*n*-hexyloxyazoxybenzene at 79° [283] and for an approximately 40 mole % solution of C_6F_6 at 58° in the same liquid crystal [284]. Snyder [282] has also applied the technique to the PMR and F^{19} NMR spectra of monofluorobenzene in the same liquid crystal. (He notes in the latter study that this nematic crystal has been found to be a mixture of *n*-hexyloxy and methoxy derivatives). In the case of C_6F_6, for example, it was found [284] that the best fit between observed and calculated spectra is had if the following parameters are used: $J_{ortho} = -22$ cps, $J_{meta} = -4$ cps, $J_{para} = +6$ cps; $D_{ortho} = -1452.67$ cps, $D_{meta} = -271.56$ cps, and $D_{para} = -194.15$ cps, where J is the indirect spin-spin coupling constant and D is the direct nuclear magnetic dipole-dipole interaction between two nuclei. The accuracy of the fit is made clear when it is pointed out that the spectrum extends ± 2731 cps (at 60 mc sec^{-1}) from the origin, and that almost all bands (of which there appear to be 58) of the calculated and observed spectra agree within 1.5 cps. Snyder [282] notes that it is probable that the accuracy of the fit between calculated and observed spectra is limited by uncertainties in the presumed geometric parameters of the molecule being investigated. He suggests that, for systems of three or more spins, refinement of the nuclear coordinates to give a better NMR spectral fit may be a source of more accurate geometrical information about molecules.

Snyder and Meiboom [285] also found that the PMR spectra of the tetrahedral molecules TMS and neopentane in a nematic solvent show a 1:2:1 triplet pattern which they attribute to slight distortion of the solute by the solvent. Their calculations indicate that a change of about 0.1° in the Si—C—H bond angle could account for the observed spectrum of TMS. This technique has not yet been applied to metal chelate systems, but doubtless could be.

B. Paramagnetic Chelates: Spin Densities, Structures, and Bonding

1. Theory. The magnetic nuclei in a ligand or a diamagnetic chelate of the ligand exhibits ordinary chemical shifts due to shielding by the bonding electrons. In paramagnetic chelates there can be observed, in addition to the chemical shift, a much larger shift due to the interaction or coupling of the atom's nuclear spin with unpaired electron density originating in the paramagnetic metal ion. The latter shift has been termed an *isotropic hyperfine contact interaction shift* [233].

In a paramagnetic nickel chelate, for example, the metal atom contains two unpaired electrons, which can be partially delocalized onto the ligand by the formation of a $d\pi$-$p\pi$ bond. If the ligand is conjugated, molecular orbital theory indicates that this unpaired π-electron density will be found to varying degrees on all the carbon atoms of the conjugated system. Since the π-electron density in a σ bond from the carbon to another atom is zero, this would seem to eliminate the possibility of any coupling between the spin of the unpaired electron and the spin on the nucleus of a proton (or fluorine or other atom of non-zero nuclear spin) attached to a given carbon. McConnell [230], however, in a significant paper has shown how the unpaired π electron can interact through an atomic exchange mechanism with a σ bond to produce "an appreciable resultant electron spin polarization in *s* atomic orbitals at the aromatic protons." In this way the amplitude of the wave function of the unpaired electron has a finite value at the nucleus of the hydrogen atom. Consequently an electron spin-nuclear spin interaction can occur. Such an interaction, often called a *Fermi contact interaction*, depends on the unpaired electron spin density on the particular carbon, and is in general manifested by the magnitude of the proton (or F^{19}, etc.) NMR shift. In general, if the electron spin density at a given carbon is positive (α, i.e., with magnetic moment vector parallel to applied magnetic field), the signal for the proton moves to high field; if the electron spin density at a particular carbon is negative (β), the proton signal moves to low field. (A very clear qualitative discussion of contact shifts is afforded by Milner and Pratt [241].)

For these contact shifts to appear in the NMR spectrum, either $1/T_1$ (where T_1 is the electronic relaxation time for the paramagnetic system) or $1/T_e$ (where T_e is the characteristic exchange time between the paramagnetic molecules) or both must be much greater than a, the contact interaction constant. The mechanism of relaxation has been discussed by LaMar [204]. The contact interaction constant, a, can be related to measurable parameters by the equation ([234]; see also [118, 179]†)

$$\frac{\Delta H_i}{H} = \frac{\Delta f_i}{f} = -a_i \cdot \frac{\gamma_e}{\gamma_N} \cdot \frac{g\beta S(S+1)}{6S'kT\{[\exp(\Delta G/kT)] + 1\}}, \tag{6-4}$$

† The fundamental equation is given by McConnell and Robertson [234], but a form very commonly encountered is that in (6-4). In the event of a completely paramagnetic system in which the isotropic shifts obey the Curie law, $\Delta G \rightarrow -\infty$, and $[\exp(\Delta G/kT)] + 1 \rightarrow 1$. Equation (6-4) is valid for a spin-only system. If spin-orbit coupling is included, more complex equations are required; see, for example, those derived by Golding [155]

where H represents the magnetic field at which resonance of the ith nucleus occurs; ΔH_i is the change in field required to make that particular nucleus resonate when it is changed from the pure ligand or zinc chelate (both diamagnetic) to a paramagnetic chelate; f and Δf_i are the corresponding frequency and change in frequency; a_i is the contact interaction constant of the ith nucleus; γ_e and γ_N are the magnetogyric ratios of the electron and the nucleus, respectively; g is the spectroscopic splitting factor of the paramagnetic species; β is the Bohr magneton; S is the total electron spin (e.g., for paramagnetic nickel(II), $S = 1$); S' is the total spin of electrons involved in delocalization; k is the Boltzmann constant; T is the absolute temperature; and ΔG is the free energy for the equilibrium: diamagnetic chelate $\rightleftharpoons$ paramagnetic chelate. This equilibrium is a rather common feature of many Ni(II) chelates (see [134, 268–271]). In many of them there appears to be a diamagnetic (probably square-planar) $\rightleftharpoons$ paramagnetic (probably tetrahedral) equilibrium [121]. Accordingly nickel chelates have figured prominently in contact interaction studies. It has been noted that the rate of interconversion must be greater than 10^5 sec^{-1}, since the NMR spectra do not show resonances of the two forms, but only an "averaged" resonance. The optical spectra, on the other hand, can show bands for both species [121].

When the a_i values have been determined for the various resonating nuclei, they can be related to the spin densities in the $p\pi$ orbitals of the adjacent sp^2 hybridized carbon atoms by the equation [232]

$$a \cong Q\rho_c, \tag{6-5}$$

where ρ_c is the spin density on the carbon and Q is a proportionality or coupling constant. Q has been found to have a value of about -22.5 gauss for an aromatic $\cdot$C—H fragment [301]. The minus sign before the value of Q occurs because, if the carbon $p\pi$ spin density is positive, a negative spin density is placed on the proton. A value of $Q = +27$ gauss is often used for a $\cdot C—CH_3$ group [237]. The value stated for Q_{CH} is, however, only approximate (see, e.g., the recent work of Fessenden and Ogawa [144]), and that given for Q_{C-CH_3} can be quite variable [119].

Relative values of a_i and ρ_{c_i} can be obtained by (6-4) and (6-5) from measurements made at a single temperature. Absolute values of a_i can be determined, however, only if ΔG in (6-4) is known. The latter quantity can be

---

and their use in a study of iron(III) dithiocarbamate chelates [156]. Equation (6-4) differs from that used by Eaton, Phillips, and co-workers (in [119–123, 126, 259]) and Holm and co-workers (in [72, 74, 134, 173]). Horrocks [179] has clearly shown the form given here to be correct. He notes also that use of the incorrect expression has resulted in the ΔG values being too high by ($RT \ln 3$) and the ΔS values too low by ($R \ln 3$) = 2.2 e.u. The calculated values of a_i are not affected by this error.

determined by measurements of Δf over a broad temperature range. Eaton and co-workers [119] found that for a large number of aminotroponeimineates in the fully paramagnetic form, ρ_{c_β} (the absolute spin density on the β carbon of the seven-membered ring) is reasonably constant at a value of -0.0210. The constancy of this value enabled them to calculate the absolute spin densities on all the carbons in the molecule from room temperature data alone. This is done by the relationship

$$\frac{\rho_{c_i}}{\rho_{c_\beta}} = \frac{\Delta f_i}{\Delta f_{H_\beta}} \tag{6-6}$$

or

$$\rho_{c_i} = -0.0210 \times \frac{\Delta f_i}{\Delta f_{H_\beta}}. \tag{6-7}$$

2. An Example. The theory can best be realized by an example. It has been found that in the nickel(II) aminotroponeimineates (XXII) T_1 is sufficiently short to allow contact interaction shifts to appear in the NMR spectra of solutions of these compounds [259, 50]. In nickel(II) γ-phenylazo-N,N'-diphenylaminotroponeimineate ($R = C_6H_5$, $X = C_6H_5NN$ in XXII [119]) an unpaired electron is partially delocalized onto the nitrogen atoms by means of $d\pi$-$p\pi$ bonding with the nickel. This electron spin is further delocalized

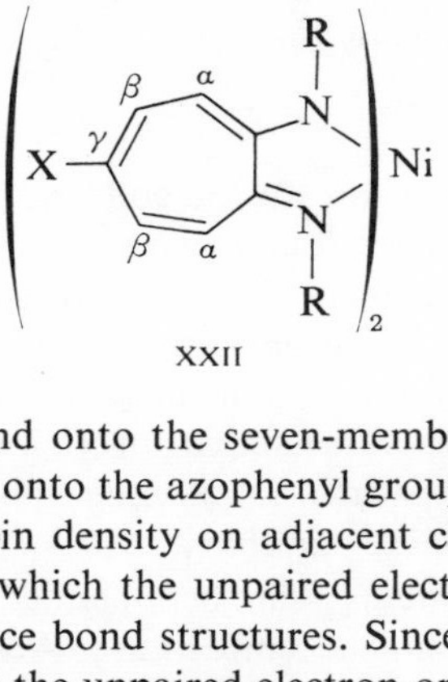

XXII

onto the phenyl groups and onto the seven-membered ring, and then from the γ position of the latter onto the azophenyl group. Spin correlation effects cause an alternation of spin density on adjacent carbons. This can be seen by noting the position in which the unpaired electron can be placed in the various contributing valence bond structures. Since valence bond structures can be written which place the unpaired electron on the α carbon and on the ortho and para carbons of the phenyl groups, we expect the $p\pi$ spin densities on these carbons to be positive. The hydrogen atoms attached to these carbons should resonate at high field. The opposite is true of the β- and meta-phenyl hydrogens.

The NMR spectrum of this nickel chelate is shown in Fig. 6-7. On the basis of their earlier studies, Eaton et al. assigned the doublets at $+927$ and -1202 cps to H_α and H_β, respectively. The azophenyl and the two equivalent

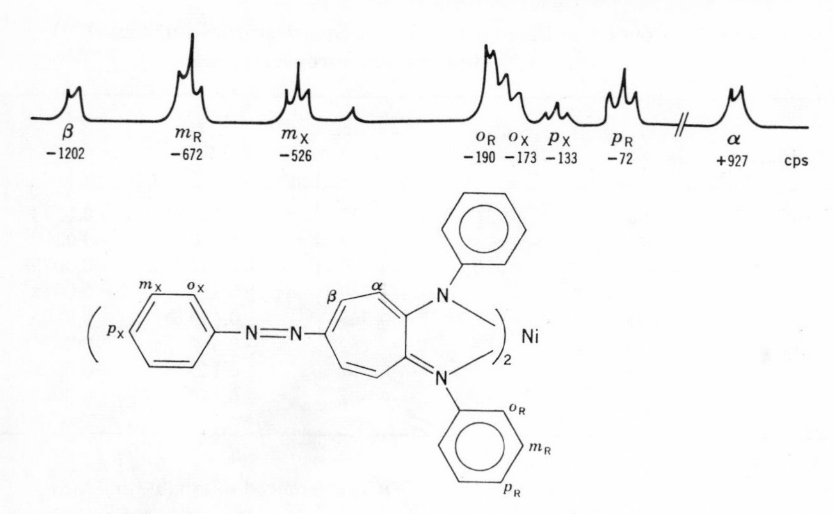

Fig. 6-7. PMR spectrum at 60 mc sec^{-1} of Ni(II) γ-phenylazo-N,N′-diphenylamino-troponeimineate in $CDCl_3$ at 25° internally referenced to TMS [119].

N-phenyl groups would be expected to show patterns which are similar, the latter having twice the intensity of the former. The para-H signals should appear as triplets and at rather high field. Accordingly the triplets at −72 and −133 cps were assigned to the p_R and p_X hydrogens, respectively. In the same way the triplets at −672 and −526 cps were assigned to the m_R and m_X hydrogens. The complex signal around −180 cps was interpreted to be two overlapping doublets and assigned to the o_R (−190 cps) and o_X (−173 cps), respectively. Using these assignments and those for the free ligand, the contact shifts and spin densities are calculated as shown in Table 6-2. In this compound about 0.1 unpaired electron is delocalized onto each ligand, but only a small part of this is delocalized onto R, the group attached to the nitrogen. When R is the 3-phenanthryl group, calculations indicate that only about $\frac{1}{47}$ unpaired electron is on each 3-phenanthryl group [121]. In general, the spin density on the seven-membered ring is relatively unaffected by the nature of the N-substituent; the two systems can therefore be treated independently.

Benson et al. [50] have also discussed the transmission of electron spin density through the groups —O—, —S—, and —NH—, and Eaton and co-workers [121] the transmission through the —SO_2— linkage. Many similar compounds containing a wide variety of substituents have also been studied by Phillips, Benson, Eaton, and co-workers [50, 119–126, 259]. Contact shifts in Fe(II) aminotroponeimineates have also been reported [229].

Table 6-2 Contact Shifts and Calculated Spin Densities for Nickel(II) γ-Phenylazo-N,N′-diphenylaminotroponeimineate[a]

Position	Shift (Free Ligand)[b]	Shift (Chelate)[b]	Contact Shift,[c] Δf	$\Delta f/\Delta f_{\mathrm{H}_\beta}$	ρ_c[d]
α	−398	+927	+1325	−1.823	+0.0383
β	−475	−1202	−727	+1.000	−0.0210
o_{R}	−435	−190	+245	−0.337	+0.00708
m_{R}	−435	−672	−237	+0.326	−0.00685
p_{R}	−435	−72	+363	−0.499	+0.0105
o_{X}	−435	−173	+262	−0.360	+0.00756
m_{X}	−435	−526	−91	+0.125	−0.00263
p_{X}	−435	−133	+302	−0.415	+0.00872

[a] From [119].
[b] Shifts are in cps at 60 mc sec^{-1} at 25° in $CDCl_3$ referenced internally to TMS.
[c] Δf = shift (chelate) − shift (free ligand).
[d] $\rho_c = -0.0210 \times \Delta f/\Delta f_{\mathrm{H}_\beta}$.

*3. **Evidence for Square-Planar⇌Tetrahedral Equilibrium.*** The paramagnetic form of the nickel(II) aminotroponeimineates is most probably tetrahedral or pseudotetrahedral, and the diamagnetic form, square-planar [126]. The conclusion is reached on the basis of several arguments. For example, the effect of R on the magnetic character of these chelates is shown in Fig. 6-8.

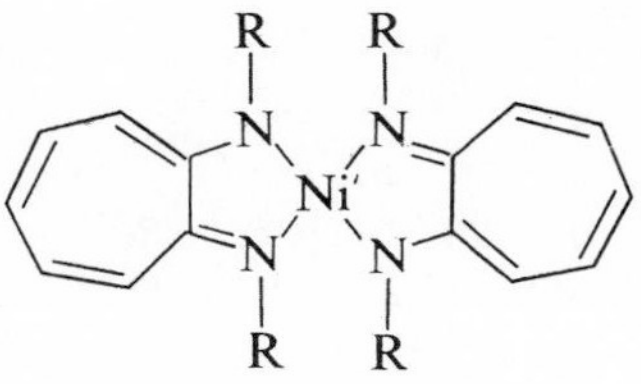

N-Substituent	magnetic character of chelate (23° in $CDCl_3$)
R = H,	diamagnetic
R = CH_3,	<1% paramagnetic
R = C_2H_5,	>99% paramagnetic

Fig. 6-8. Relation of N-substituent to magnetic character of some nickel(II) aminotroponeimineates [126].

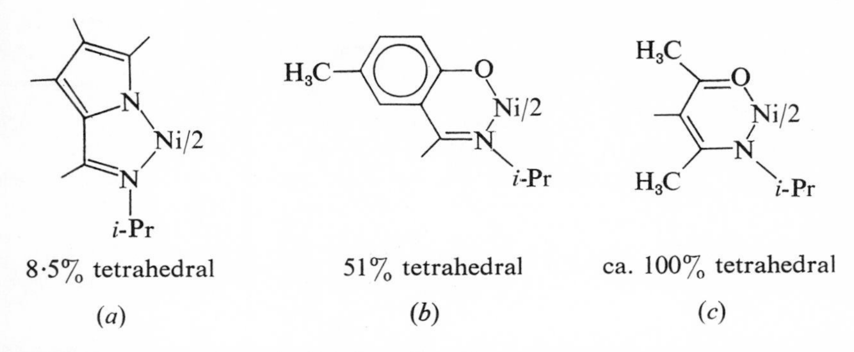

Fig. 6-9. Percentage of tetrahedral form in $CDCl_3$ solutions at 60° of (a) bis(N-isopropylpyrrole-2-aldimino)nickel(II); (b) bis(5-methyl-N-isopropylsalicylaldimino)nickel-(II); and (c) bis(4-isopropylaminopent-3-en-2-ono)nickel(II) [174].

As the alkyl group becomes larger the crowding of the groups renders the planar form more difficult to attain. This crowding can be relieved by distortion, and by the molecule's assuming a tetrahedral rather than a square-planar form. Since the paramagnetism increases with the size of the N-substituent, it can be assumed that the diamagnetic form is square-planar and the paramagnetic form, tetrahedral. With the increase of paramagnetism comes the expected increase in the observed contact shifts. For the methyl compound the shifts extend over a range of 3.75 ppm, while for the ethyl derivative the range is 202.6 ppm [121]. Additional evidence drawn from optical spectra, the existence of isomerism in nickel(II) N,N′-di-(α-naphthyl)-aminotroponeimineate, solvent effects, molecular weight determinations, etc., is offered by Eaton et al. [126]. Relative stabilization of the tetrahedral form of three other types of Ni chelates is shown in Fig. 6-9 [174]. In these compounds, also, steric effects are largely responsible for the destabilization of the planar form. For other studies of bis(β-ketoamino)-metal complexes, see [133–135, 176].

4. F^{19} Shifts; Evidence for Fluorine π Bonding. A study [122] of the H^1 and F^{19} contact interaction shifts in compound XXIII has shown that the spin density distribution in both the six- and the seven-membered rings is little affected by fluorine substitution on the phenyl groups, whether in the ortho, meta, or para positions. Q_{CF} was, however, found to be positive in sign, and not constant in absolute value for the various isomers. The fact that Q_{CF} for the chelates studied is opposite in sign to Q_{CH} must mean that a further interaction is involved in addition to the nucleus-electron interaction proceeding by way of a π-σ mechanism as in the case of aromatic C—H. Since Q for the latter interaction is negative, there must be also a

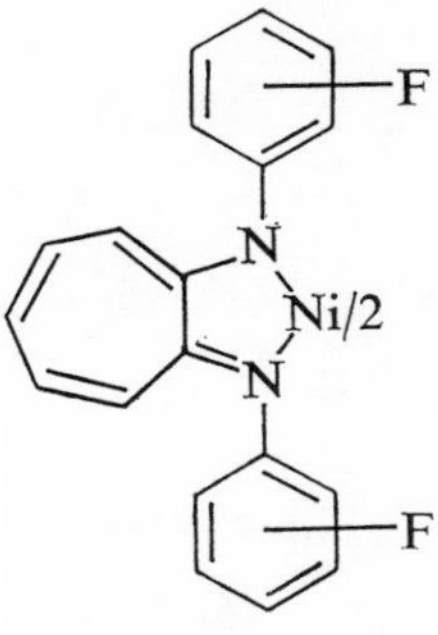

XXIII

positive component to Q_{CF}. This can be explained if there is some π bonding in the C—F bond. Thus, in addition to structures which place spin density on N, *o*-C, and *p*-C (XXIVa), structures can be written which place it on the fluorine itself (XXIVb, c). Such a situation could result from conjugation of the π system of the phenyl group with a fluorine *p* orbital containing an unshared electron pair.

A similar study was also made of the F^{19} contact shifts for the three isomers of compound XXV [119]. The results of these two studies are listed in Table 6-3. The a_F values were calculated by means of (6-4). Q_{CF} values were calculated from a_F and ρ_C values by the equation

$$a_F = Q_{CF}\,\rho_C. \tag{6-8}$$

The C—F π-bond orders for the compounds were calculated by the Hückel MO method on the fragments XXVI and XXVII. A plot of Q_{CF} against

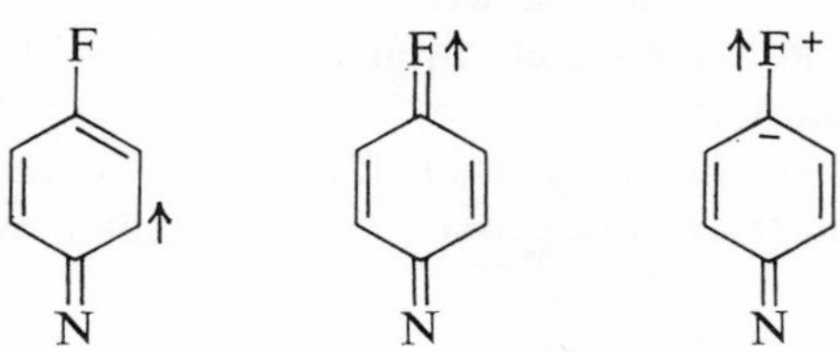

(a) (b) (c)

XXIV

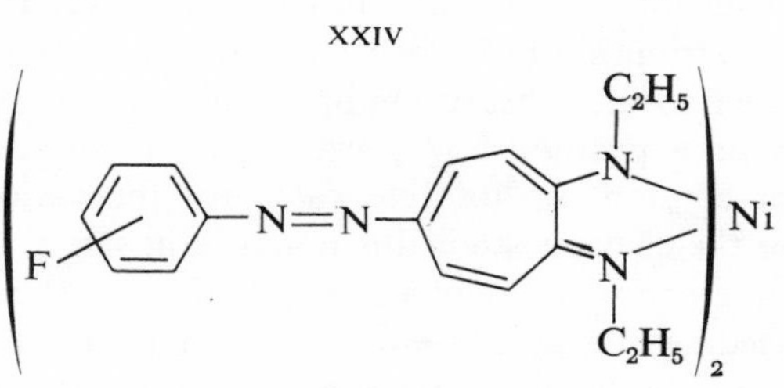

XXV

Table 6-3 Spin Densities and C—F π-Bond Orders in Some Fluorine-Substituted Nickel(II) Aminotroponeimineates

Compound	Fluorine Position	a_F (gauss)	F^{19} Contact Shift	ρ_C	Q_{CF} (gauss)	C—F π-Bond Order (p_{CF})
Azo	ortho	—[b]	−2112[c]	+0.00880[d]	+38.5	0.2067
derivative[a]	meta	—[b]	+1148[c]	−0.00250[d]	+73.8	0.2777
(XXV)	para	—[b]	−3227[c]	+0.00960[d]	+52.6	0.2173
N-Phenyl	ortho	+0.270	—[b]	+0.00657[f]	+41.1	0.212
derivative[e]	meta	−0.0325	—[b]	−0.00813[f]	+4.0	0.178
(XXIII)	para	+0.506	−863[g]	+0.0107[h]	+47.2	0.229

[a] From [119].
[b] Data not given.
[c] Shift (chelate) − shift (ligand) in cps at 56.4 mc sec^{-1} in CS_2 at 25°; internal reference: 1,2-difluorotetrachloroethane.
[d] Averaged values taken from unsubstituted positions.
[e] From [122].
[f] Values from corresponding unsubstituted positions.
[g] Shift (chelate) − shift (ligand) in cps at 40.0 mc sec^{-1} in $CDCl_3$ at 26°; internal reference: 1,2-difluorotetrachloroethane.
[h] Values from unsubstituted phenyl compound.

π-bond order (p_{CF}) for the six compounds gives a reasonably good straight line. The authors of these studies [119, 122] then postulate that Q_{CF} is made up of two parts: (1) $Q_{CF}{}^{C}$, a measure of the π-σ polarization of the ·C—F fragment; it is similar to that for the ·C—H fragment, and is, like the latter, negative in sign; and (2) $Q_{FC}{}^{F}$, a positive contribution to Q_{CF} due to spin density on the fluorine brought about by conjugation. An equation containing these contributions can be written

$$Q_{CF} = \frac{a_F}{\rho_C} = Q_{CF}{}^{C} + Q_{FC}{}^{F}(Ap_{CF}), \tag{6-9}$$

where A is a constant of order unity. From the plot of Q_{CF} against p_{CF} (on the assumption that $A = 1$), the slope, $Q_{FC}{}^{F}$, is calculated to be +580 gauss and the intercept, $Q_{CF}{}^{C}$, to be −83 gauss.

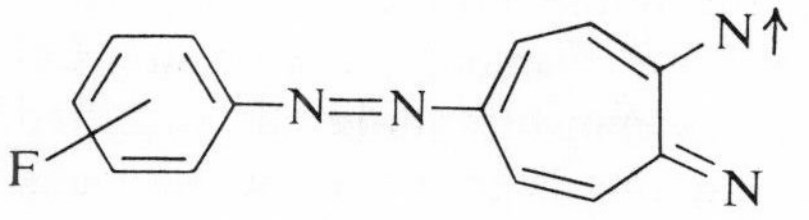

XXVI

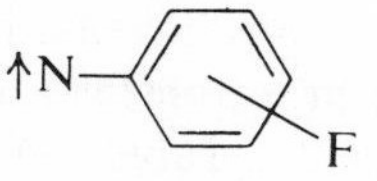

XXVII

The consistency of this treatment seems to offer a significant indication of extensive bonding in fluoroaromatic systems between the fluorine and the phenyl π system. It is also clear that, if π bonding makes significant and varying contributions to bonds, the values of Q for these bonds should not be expected to be identical. A great deal more F^{19} contact interaction data have been reported by Eaton, Josey, and Sheppard [123], as well as an extensive study of fluorine p-π interactions by Sheppard [275]. An analogous study [119] of tolyl-substituted aminotroponeimineates has offered some evidence for the existence of hyperconjugation in these systems.

*5. **Pseudocontact Interactions.*** Paramagnetic bis(acetylacetono)nickel(II), $Ni(acac)_2$, exists in the solid state as a trimer [65], and has been postulated to have the same form in solvents of low coordinating ability [138, 157]. Fackler's [136] study of the visible spectra of this compound in various benzene-pyridine mixtures has indicated that, as more pyridine is added to the mixture, two pyridine-containing compounds are formed: at low pyridine concentration, $[Ni(acac)_2]_2 \cdot py$, and at higher pyridine concentrations, $[Ni(acac)_2] \cdot 2py$. If there is any transfer of unpaired electron spin from the nickel to the pyridine, this should be seen in changes in the proton magnetic resonance spectra of pyridine coordinated to $Ni(acac)_2$. Happe and Ward [164] have studied the interaction of pyridine-type ligands with $Ni(acac)_2$ and $Co(acac)_2$. They used solutions of these ligands in $CDCl_3$ to which were added varying amounts of $Ni(acac)_2$ or $Co(acac)_2$. The pyridine-type ligands exhibit only an "average" spectrum, and therefore ligand interchange on the bonding sites must be more rapid than the difference in resonating frequencies of the protons in the coordinated and non-coordinated species. The change in a proton's resonating frequency in the complexed pyridine from that in the uncoordinated pyridine (its "contact shift") might then be expected to be proportional to the amount of paramagnetic nickel- or cobalt-acetylacetonate present. This was found to be so. The *relative* contact shifts for the various ligand protons were found to be independent of concentration of the metal chelate, and so Happe and Ward report their findings in this manner. Some of their data are reproduced in Table 6-4.

The alternation of spin density on the ligands in nickel aminotroponeimineates has been interpreted as an indication that the mechanism of electron transfer is by means of the π-electron system. If the principal mechanism for the transfer were through the σ-bonding system, the proton shifts would be expected to attenuate rapidly as the number of atoms between the metal and the protons increased. Furthermore, Horrocks, Taylor, and LaMar [181] note that in σ transfer the proton resonances should all be shifted to lower applied fields; Table 6-4 shows that this is true for the $Ni(acac)_2$-pyridine complexes. Therefore σ transfer is probably the principal spin transfer

Table 6-4 Relative Isotropic Contact Shifts[a] for Ligands Coordinated with $Ni(acac)_2$

Ligand	Proton	Isotropic Shift[b]	Ligand	Proton	Isotropic Shift[c]
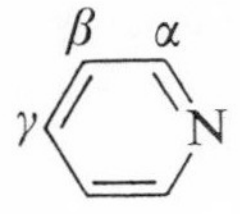 Pyridine (β α, γ, N)	α	−10.00	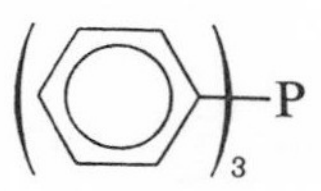 $(C_6H_5)_3P$	ortho	+10.00
	β	−2.94		meta	−7.29
	γ	−0.77		para	+7.88
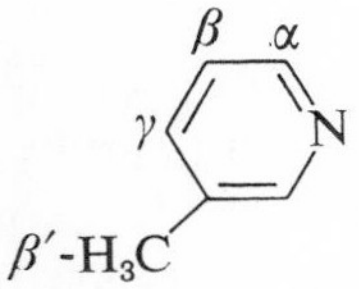 β′-H_3C-pyridine (β α, γ, N)	α	−10.00	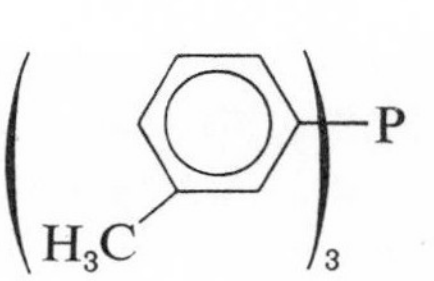 $(m\text{-}H_3CC_6H_4)_3P$	ortho	+10.00
	β	−2.96		meta	−7.86
	γ	−0.63		para	+8.36
	β′-CH_3	−0.55		m'-CH_3	+3.22
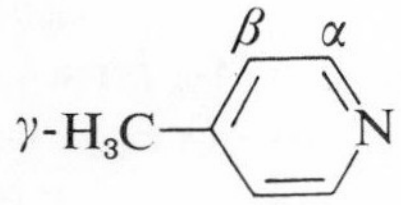 γ-H_3C-pyridine (β α, N)	α	−10.00	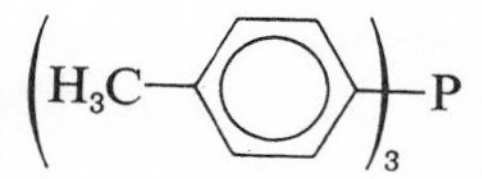 $(p\text{-}H_3CC_6H_4)_3P$	ortho	+10.00
	β	−2.78		meta	−9.02
	γ-CH_3	+0.90		para-CH_3	−12.77

[a] Contact shifts are: $(\Delta H/H) = (\Delta H/H)_{\text{ligand in complex}} - (\Delta H/H)_{\text{pure ligand}}$.
[b] From [164]. These data are from spectra of the ligands in $CDCl_3$ containing varying amounts of $Ni(acac)_2$ (TMS, internal reference). The shifts are normalized to a value of −10.00 for the α proton.
[c] From [181]. These data are taken from shifts at a concentration of 0.118 M $Ni(acac)_2$ in $CDCl_3$ (TMS, internal reference), and normalized to a value of +10.00 for the ortho proton.

mechanism. The fact that the methyl shift in γ-picóline is positive indicates that there is some alternation of the sign of the shift and therefore presumably some π transfer also in these compounds. The PMR spectra of isonitrile or triphenylphosphine ligands coordinated with $Ni(acac)_2$, on the other hand, show the alternation of contact shift and consequent alternation of spin density usually found in π systems (see Table 6-4; more extensive data are given by Horrocks et al. [181]). One must conclude then that the πd orbitals of the nickel contain some unpaired spin.

The polymeric nature of $Co(acac)_2$ in the solid state and in various solvents has been extensively studied by Cotton and co-workers [101, 102, 104]. And Fackler [137] has shown that this compound reacts with pyridine (py) in benzene to yield $Co(acac)_2 \cdot py$, $[Co(acac)_2]_2 \cdot py$, and $Co(acac)_2 \cdot 2py$, depending on pyridine concentration. One might expect that, when $Co(acac)_2$ is substituted for $Ni(acac)_2$ with the above-mentioned ligands, the total contact shifts for the corresponding protons would be similar in magnitude and sign. This was found *not* to be so. It seems probable that, in the $Co(acac)_2$ complexes, pseudocontact interactions play an important role, while they play a relatively unimportant one in the corresponding nickel complexes.

The "'*pseudocontact' interaction* . . . results from the failure of the dipolar interaction between the proton and the net electron spin magnetization on the metal to average to zero, due to the metal possessing an anisotropic **g** tensor" [205]. For a spin-only system the pseudocontact shift for proton i is given by [234]

$$\frac{\Delta H_i}{H} = \frac{-\beta^2 S(S + 1)(3\cos^2\psi_i - 1)}{27 r_i^3 kT}(g_{\parallel} + 2g_{\perp})(g_{\parallel} - g_{\perp}). \tag{6-10}$$

In this equation ψ_i is the angle between the crystal field axis of the complex and the vector r_i which joins the proton i with the metal atom; $g_{\parallel}$ and $g_{\perp}$ are the electronic g values parallel and perpendicular to the crystal field axis; the rest of the symbols have their usual meaning. The equation was derived by McConnell and Robertson [234] for the case of an axially symmetrical **g** tensor. It has been extended by LaMar et al. [208] to the case of **g** tensors of $\mathbf{C}_{2v}$ symmetry. The latter extended equation is, however, also applicable to many other symmetries when the angles are properly defined; as it is reported, it is valid for all systems possessing at least a C_2 axis [202]. Equation (6–10) is valid for the condition $T_1 \gg \tau_c$, where T_1 is the electronic relaxation time and τ_c is the molecular correlation or tumbling time. McConnell and Robertson suggested that this would be a more likely situation than its opposite, $\tau_c \gg T_1$. This equation has accordingly been used in many reported studies of pseudocontact interaction [122, 124–126, 164, 181, 203, 208, 221]. LaMar [206] has pointed out, however, that his recent study [207] of line widths in some paramagnetic arylphosphine complexes of cobalt(II) and nickel(II) halides indicates that T_1 is of the order 10^{-13} sec, while τ_c may be estimated from the Debye equation to be ca. 10^{-11} sec. He suggests that the situation $\tau_c \gg T_1$ is probably the valid one for the many systems investigated to date. The correct equation for this situation (for an axially symmetrical **g** tensor) is [234]

$$\frac{\Delta H_i}{H} = \frac{-\beta^2 S(S + 1)(3\cos^2\psi_i - 1)}{45 r_i^3 kT}(3g_{\parallel} + 4g_{\perp})(g_{\parallel} - g_{\perp}). \tag{6-11}$$

(It has been pointed out by Eaton [118] that in the original report of McConnell and Robertson [234] there is an error of signs in this equation; the equation given here is the corrected form.) LaMar [206] outlines the changes that the use of (6-11) instead of (6-10) will cause in the results of his own reports (see [203, 208]).

When the pseudocontact shifts are calculated by (6-10) for the pyridine protons of $Co(acac)_2$-pyridine and the values subtracted from the total proton shift values, the remainders (the contact shifts) should resemble those of the corresponding protons in the analogous Ni(acac)$_2$-pyridine complex, if the difference is actually due to pseudocontact interactions. Table 6-5 shows this to be true for pyridine-type ligands.

Table 6-5 Relative Contact and Pseudocontact Shifts for Co(acac)$_2$ Complexes and Comparison with Contact Shifts for Ni(acac)$_2$ Complexes[a]

		Contribution to Isotopic Shift in Co(acac)$_2$ Complex		Normalized Contact Shift of Complexes[b]	
Ligand	Proton	Pseudocontact	Contact	Co(acac)$_2$	Ni(acac)$_2$
Pyridine	α-CH	+7.82	−17.82	−10.00	−10.00
	β-CH	+4.40	−5.71	−3.21	−2.94
	γ-CH	+3.96	−1.66	−0.93	−0.77
β-Picoline	α-CH	+7.82	−17.82	−10.00	−10.00
	β-CH	+4.40	−5.11	−2.87	−2.96
	γ-CH	+3.96	−0.57	−0.32	−0.63
	β'-CH$_3$	+2.35	−0.83	−0.47	−0.55
γ-Picoline	α-CH	+7.82	−17.82	−10.00	−10.00
	β-CH	+4.40	−5.28	−2.97	−2.78
	γ-CH$_3$	+2.52	+2.08	+1.17	+0.90

[a] From [164].
[b] All shifts normalized to −10.00 for the α proton.

Horrocks et al. [181] have also noted that, for their nickel complexes, contact interactions predominate, while both contact and pseudocontact interactions are important in the cobalt chelates. The difference in behavior of the nickel and cobalt chelates arises from the fact that the ground state of Co(II) in an octahedral field is an orbital triplet, $^4T_1$. Under low symmetry distortion the orbital degeneracy of this state is reduced, and spin-orbit coupling between the split members of the ground state introduces anisotropy in the g tensor. Ni(II) in an octahedral field has an orbitally non-degenerate ground state, $^3A_2$, and spin-orbit coupling between the ground and excited states is small compared to the spin-orbit coupling in Co(II) [202]. Accordingly for Ni(II) the g tensor is reasonably isotropic, and there is no observable pseudocontact interaction shift. When $g_{\parallel} < g_{\perp}$, there is a finite, and in most cases, positive, value for the pseudocontact shift. For similar contact shift studies, see [180, 193–195, 298].

Forman, Murrell, and Orgel [147] studied the PMR spectra of Co(acac)$_3$, V(acac)$_3$, and V(3-methylacac)$_3$. Their results are probably best interpreted on the basis of π transfer of electron spin density and subsequent delocalization in the π system of the chelate. Eaton [118] notes that this is more expected in acetylacetonates than electron transfer through σ orbitals, for the latter orbitals are expected to be either strongly bonding or strongly antibonding. As such they differ greatly in energy from the metal d orbitals which are basically non-bonding. In this study Eaton has correlated the NMR shifts of many paramagnetic metal acetylacetonates with the types of metal-ligand π bonding: metal-to-ligand electron transfer and the reverse process. He also shows with reasonable certainty that pseudocontact interactions play a negligible role in the shifts observed for the first row metal tris-acetylacetonates. The data he presents indicate, however, that these shifts can no longer be ignored in the second transition series. In this series they do not, however, cause the whole contact shift. It is probable, on the other hand, that they actually do predominate in the inner transition series. Eaton [118] has also shown how these contact shifts can be related to ease of oxidation of the metal atoms.

6. Optically Active Chelates. The stereochemical behavior and conformational equilibria of variously substituted bis(salicylaldimino)nickel(II)

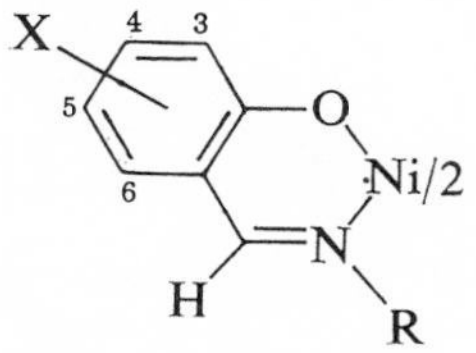

XXVIII

complexes (XXVIII) have been the subject of many investigations, and have been reviewed recently by Holm, Everett, and Chakravorty [176]. PMR spectroscopy has been used as a help in several of these investigations [71–74, 172, 173, 201, 295, 296]. Of particular interest is its use in a study of compounds in which R in XXVIII is an optically active group. These compounds show two signals for each proton or methyl group attached to the aromatic ring [173]. The two signals in any one pair have been shown by proton double resonance to be independent of each other and accordingly due to distinct molecular species [172]. Since the amines used to make the Schiff bases were mixtures of isomers and accordingly contained approximately equal amounts of a racemic [(+, +) and (−, −)] and a meso [(+, −)] form, the Schiff bases and their chelates are mixtures of diastereoisomers. Each of these isomers has its own PMR spectrum, as is clear from Fig. 6-10. From a comparison of these two spectra it can be concluded that the meso isomers have the smaller contact shifts. Since each of these isomers exists in a state of equilibrium, square-planar (diamagnetic) $\rightleftharpoons$ tetrahedral (paramagnetic), the

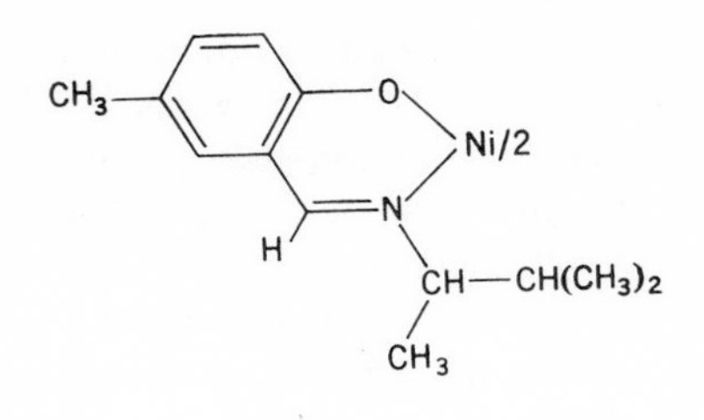

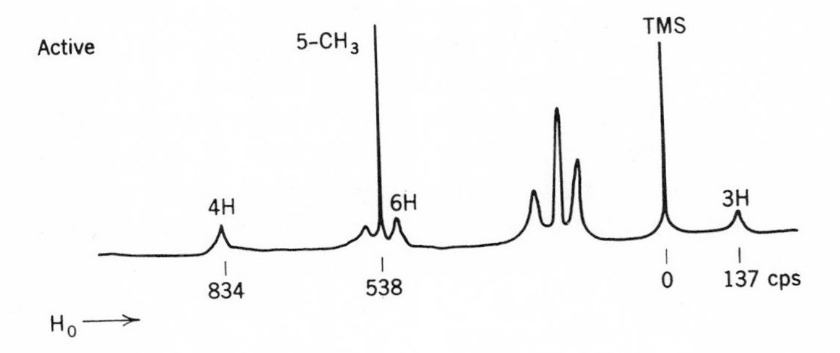

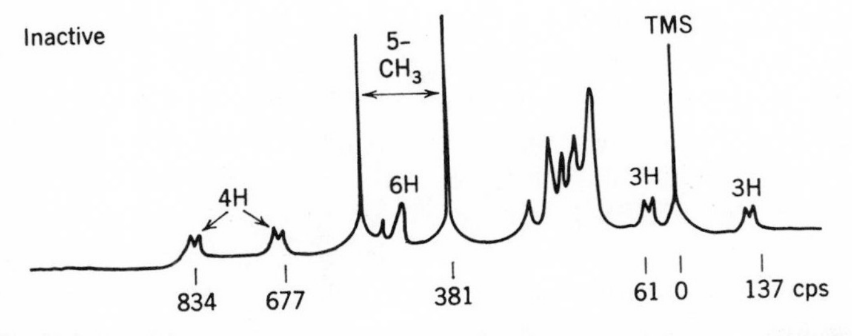

Fig. 6-10. PMR spectrum at 60 mc sec^{-1} of bis(5-methyl-N-2-(3-methyl)-butylsalicylaldimino)nickel(II) in $CHCl_3$ at 30° internally referenced to TMS. Active form (−, −); inactive meso (+, −) and racemic [(+, +) and (−, −)] forms [173].

smaller contact shifts for the meso compound mean that the equilibrium is shifted less to the right than for the *d* or *l* form.

From an analysis of the temperature dependence of the contact shifts the authors obtained thermodynamic values for the planar ⇌ tetrahedral interconversions of the various isomers. In $CDCl_3$, toluene-d_8, and CS_2 they found, for example, that $\Delta G^{60°}$ (meso) is greater than $\Delta G^{60°}$ (active). At present they can offer no satisfactory explanation for the thermodynamic data. But they offer reasons for rejecting any variation in the value of a_i and/or g for the different isomers as an explanation of the difference in contact shifts. They also noted that the ΔG and contact shift values are not linear with temperature over the whole temperature range studied. In several cases at the lower temperatures the contact shift increases rapidly with a decrease in temperature [173]; clearly this behavior does not follow (6-4). It can be explained by assuming that some molecular association takes place at lower temperatures. The associated molecular species is expected to be fully paramagnetic. Accordingly even a small amount of association could cause a significant increase in the contact interaction shifts. Such increase in paramagnetism through intermolecular association is now quite well established for nickel complexes [138, 177, 178, 271].

A study of some substituted bis(salicylaldimino)nickel(II) complexes (XXVIII) in which R is a group containing a donor site capable of coordination with the metal atom, namely, $CH(CH_3)CH_2OCH_3$, has led Chakravorty et al. [71] to postulate in this case the formation of paramagnetic, octahedral species by *intra*molecular chelate ring formation. Optical isomers were also found for these octahedral complexes, since the N-substituent contains an asymmetric carbon. The difference in contact shifts for the active and meso forms was here interpreted as being due to slightly different values of a_i for the different forms.

*7. **Mixed-Ligand Chelates.*** Studies of mixed-ligand chelates [74, 120, 125] have cast some light on the factors which influence delocalization of electron spin from a metal onto a ligand. Eaton and Phillips [125] found that mixed-ligand chelates formed within a few minutes of adding a free aminotropone-imine to a solution of the Ni(II) chelate of another aminotroponeimine. Careful comparison enabled them to assign the contact shifts of the various protons in the mixed-ligand chelate. Calculation of the spin densities indicated that the values for the mixed chelate are different from those for the non-mixed chelates, but that a reduction in spin density on one of the ligands is approximately balanced by a proportionate gain on the other ligand (Fig. 6-11).

Eaton and Phillips [125] have compared a set of variously substituted nickel(II) N-phenyl aminotroponeimineates (XXII) with the N-ethyl analogs. They define a parameter

$$\zeta_\gamma \equiv \frac{\Delta f_{\mathrm{Et}} - \Delta f_\phi}{\frac{1}{2}(\Delta f_{\mathrm{Et}} + \Delta f_\phi)} = \frac{\rho_{\mathrm{Et}} - \rho_\phi}{\rho_{\mathrm{mean}}}, \tag{6-12}$$

where Δf_{Et} is the contact shift of the γ proton of the N-ethyl ligand and Δf_ϕ that of the γ proton of the N-phenyl ligand in the mixed chelate. They calculated ζ_γ for a whole series of mixed chelates with various phenyl substituents and found that $\log \zeta_\gamma$ is an excellent linear function of σ, the Hammett parameter for the phenyl substituent. Analogous results are had when the α- or β-proton shifts are used. These results suggest that the electron-with-

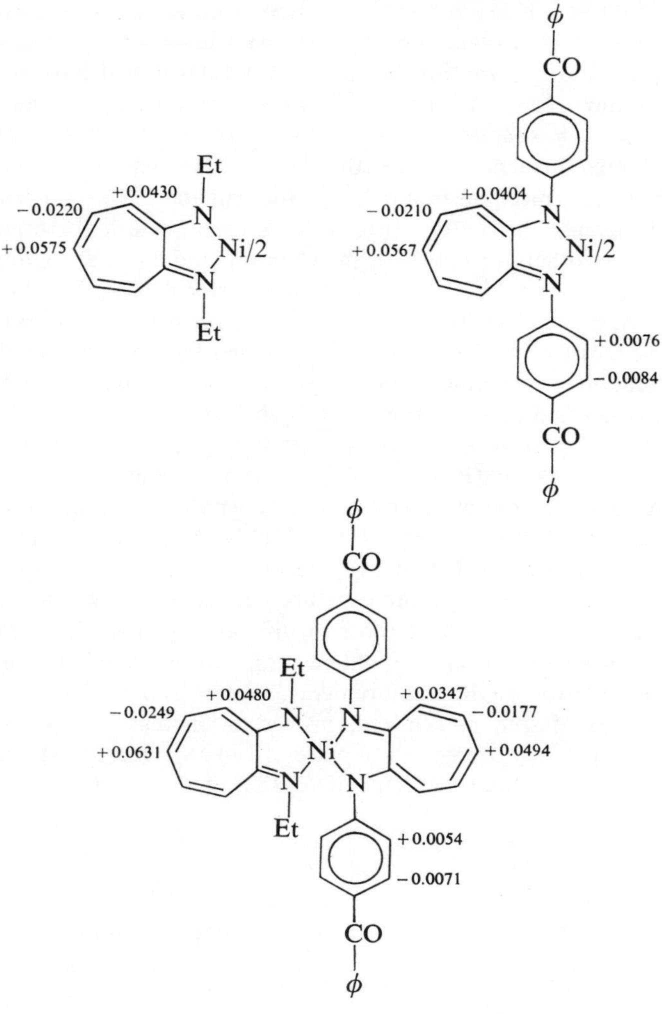

Fig. 6-11. Spin densities in symmetric and unsymmetric chelates [125].

drawing power of the substituent determines the effectiveness of a particular ligand in withdrawing spin density from the metal atom.

Two theories have been offered to explain these results. Lin and Orgel [217] have noted that in a tetrahedral mixed-ligand chelate the symmetry is reduced to $\mathbf{C}_{2v}$ and the t_2 orbitals could be split in either of the ways shown in Fig. 6-12. The order of b_1 and b_2 is not important; the authors assume for convenience that b_2 is below b_1. The important point is that b_1 will overlap with one ligand (in the xz plane) and b_2 with the other (in the yz plane). "Then, in case (A), since, at any instant, more molecules would have the lower b_2 orbital doubly occupied, and the upper b_1 orbital singly occupied, there should be more molecules with spin delocalized on the ligand in the yz plane. The rapid interconversion between the planar and the tetrahedral configuration then provides a mechanism for averaging out the effect and results in an observed increase in spin densities on one ligand with a corresponding decrease on the other. In case (B), in which b_1 and b_2 orbitals are similarly occupied, there should be no preferential delocalization" [217, p. 134]). Lin and Orgel conclude that the uneven spin density distribution in mixed-ligand chelates can probably best be explained by the fact that the ground state is doubly degenerate (^3E) in symmetrical complexes, and that this degeneracy is lifted in the unsymmetrical chelates.

More recently, however, Eaton and Phillips [125] have presented reasons for preferring situation (B) in Fig. 6-12. They point out, for example, that in $\mathbf{C}_{2v}$ symmetry the opposite situation (A) would result in two low-lying states of slightly different energy: (1) b_2 doubly occupied, b_1 singly occupied; and (2) b_2 singly occupied, b_1 doubly occupied. The population of these two states would accordingly be temperature dependent, as would the ratio of spin delocalized onto each of the two different ligands. Their calculations show that the change in ratio should be readily observable in the temperature range available for PMR measurements. Over a range of 120° they found that the ratio of spin delocalized, instead of increasing with decrease in temperature, actually decreased. In place of Lin and Orgel's explanation they offer an alternative one based on Jaffe's [182] idea of competitive π bonding

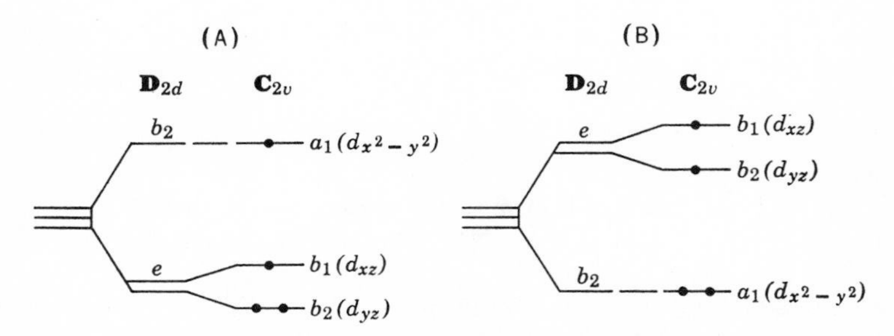

Fig. 6-12. Two possible splitting schemes for d orbitals when field is reduced from $\mathbf{D}_{2d}$ to C_{2v} symmetry [217]. According to the usual practice a coordinate system has been used for C_{2v}, which is different from that used for $\mathbf{D}_{2d}$. If the same coordinate system were retained, d_{xy} would transform as a_1 and $(1/\sqrt{2})(d_{xz} + d_{yz})$ and $(1/\sqrt{2})(d_{xz} - d_{yz})$ as b_1 and b_2, respectively.

in tetrahedral complexes. They assume situation (B) in Fig. 6-12 with b_1 and b_2 levels equally occcupied. Each of these can form a π bond with one ligand, but the extent of the π-bond formation and consequent spin density transfer to each ligand will depend on the nature of the ligand. In general there will be a competition between the two ligands, and the molecule will assume that orientation which allows maximum orbital overlap and π bond formation. They note also that this interpretation fits nicely with the excellent correlation found between the Hammett σ parameter and their own ζ parameter (6-12).

8. Study of the Outer Coordination Sphere. Stengle and Langford [289] have made use of transverse relaxation times, T_2, calculated from NMR line widths, in their study of the *outer* coordination sphere of a metal ion. They observed the F^{19} resonance of F^- and PF_6^- ions in aqueous solution and noted the effect that the presence of various paramagnetic chromium(III) complexes would have on the width of these resonances. They chose the well-defined, non-labile complexes $[Cr(ox)_3]^{3-}$, *cis*-$[Cr(en)_2Cl_2]^+$, $[Cr(en)_3]^{3+}$, and $[Cr(pn)_3]^{3+}$. Since it may be presumed that the effect of these paramagnetic ions on the F^{19} nuclei is short-ranged, the line broadening they cause will be due largely to the presence of fluorine nuclei in the second coordination sphere of the metal ion. To a rough approximation it was found that the extent of line broadening is proportional to the fraction of fluorine in the second coordination sphere of the complex ion. The authors noted also that there is an increase in the F^{19} line width as the size of the Cr(III) complex ion increases. This effect is the opposite of that expected, and must be explained in terms of ion association. As the radius of the metal complex is increased through the addition of large organic groups, solvation by water is reduced. The water molecules in the outer coordination sphere thus become easier to replace, and association between the complex ion and F^- increases. The method seems to offer a straightforward way of measuring the relative degree of outer sphere complexing for a group of complexes of similar electronic structure.

A discussion of the theory connecting proton relaxation and hydration in aqueous solutions of paramagnetic ions has been given by Pfeifer [258].

C. Structure and Bonding in Materials of Biological Interest

NMR has been used in many studies to cast light on the structure and bonding in complex biological substances, many of which are metal chelates.

1. ADP, ATP, and Related Compounds. Cohn and Hughes [88] measured the P^{31} NMR of adenosine triphosphate (ATP) (XXIX), identified the α-, β-, and γ-phosphorus peaks, and determined the chemical shift of each phosphorus in ATP and ADP (adenosine diphosphate) as a function of pH.

They further correlated the difference in shift of the β-phosphorus in ATP (as compared with those of the α-, and γ-phosphorus atoms) with the difference of reactivity of the β-phosphorus in enzyme reactions. They also studied

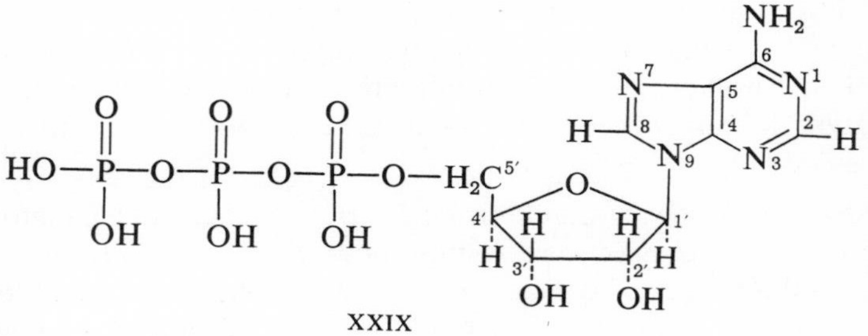

XXIX

the reaction of ATP and ADP with metal ions [89] by means of H^1 and P^{31} NMR, and found that Mg(II), Ca(II), and Zn(II) complex with the β and γ groups of ATP and with the α and β groups of ADP. The reaction sites with divalent Cu, Mn, and Co are also discussed in their paper. Hammes, Maciel, and Waugh [161] had earlier noted that the PMR of the protons H_8, H_2, and $H_{1'}$ of ATP (0.2*M* in D_2O) are not shifted by the addition of Mg^{2+} or Ca^{2+} ions. They concluded that these two ions are probably not associated with the adenine portion of ATP. Cohn and Hughes [89] noted the same for Mg^{2+} and Ca^{2+} ions, but found that Zn^{2+} ion causes a shift to lower field of the C_8 proton resonance. They interpret this as an indication of Zn-adenine bonding. Further confirmation of these results was made by Happe and Morales [163], who measured the N^{15} NMR of aqueous solutions (pH 7.0–9.6) of enriched ATP, both alone and in the presence of Mg^{2+} and Zn^{2+} ions. The presence of Mg^{2+} ion causes no significant shift in any of the nitrogen resonances. Zn^{2+} ion, on the other hand, causes significant shifts in the N_7 and NH_2 resonances and a smaller shift in that of N_9. The results are consistent with a model in which Zn^{2+} ion interacts with both N_7 and NH_2.

The work of Cohn and Hughes has been extended by Sternlicht, Shulman, and Anderson [290], who made a P^{31} and an H^1 NMR study from 0° to ca. 90° of the metal-ion binding in ATP. The P^{31} results indicate, for example, that Mn^{2+} ion is simultaneously bound to oxygens of all three phosphate groups and is approximately the same distance from all three P^{31} nuclei. These spectra along with the PMR spectra indicate that Mn^{2+} ion is bound to the adenine ring all the time it is bound to the phosphate groups. Thus at 298°K the average time the Mn^{2+} ion adheres to the phosphate groups and to the adenine group was calculated to be the same, $(5 \pm 1) \times 10^{-6}$ sec. The position of the Mn^{2+} ion in the complex was determined from the spin-lattice relaxation times, T_1, of protons H_8, H_2, and $H_{1'}$. Under certain conditions $1/T_1$ is proportional to $(\mu/r^3)^2\ \tau_c$, where μ is the magnetic moment of the ion at a distance r from the proton in question, and τ_c is the correlation time of the metal ion with respect to the proton. They were able to measure μ, τ_c, and T_1, and so to determine the distance of each of the three protons from the Mn^{2+} ion. Triangulation placed the metal ion between N_7 and the amino group on C_6 in the adenine ring. Similar conclusions were drawn for the Ni(II) and Co(II) complexes. A complete review (to 1960) of the problem of the structure of metal complexes of ADP and ATP is given by Atkinson and Morton [49].

Cohn and Leigh [90] have also employed the longitudinal relaxation time (T_1) of water to study the characteristics of manganous ion in various enzyme-substrate complexes.

2. Nucleic Acids and Related Compounds. In an effort to understand better the biochemical reactions of nucleic acids in cells, Eisinger and coworkers [127–129] have carried out several investigations involving aqueous solutions containing metal ions and deoxyribonucleic acid (DNA), ribonucleic acid (RNA), or a similar molecule. When a paramagnetic ion in water is bound to some ligand, the hydration sphere of the ion rotates less freely and the metal ion becomes more efficient in causing the relaxation of the protons in H_2O, that is, in reducing the proton T_1. They define a relaxation enhancement parameter, ε_1 [127],

$$\varepsilon_1 = \frac{R_1^*(\mathrm{M}) - R_1^*(0)}{R_1(\mathrm{M}) - R_1(0)}. \tag{6-13}$$

Here $R_1(\mathrm{M})$ and $R_1^*(\mathrm{M})$ are the proton relaxation rates ($R = 1/T_1$) at a definite divalent metal ion concentration in water, and in a solution containing the coordinating molecules, respectively. $\mathrm{R}_1(0)$ and $\mathrm{R}_1^*(0)$ are the corresponding relaxation rates when the M ions are absent. It was noted that in general, if $\varepsilon_1 > 1$, the metal ion is characteristically bonded to exterior sites, such as the phosphate groups of DNA. For $\varepsilon_1 < 1$, chelation occurs at interior sites, those which water molecules cannot reach. If $\varepsilon_1 = 1$, M-ligand bonding is absent. From the dependence of these enhancement factors on metal ion concentration, both the binding equilibrium constants and the concentration of binding sites can be calculated. In one report Eisinger, Shulman, and Blumberg [128] concluded that Cu(II), Mn(II), and Cr(III) are bound to DNA at exterior sites, while Fe(III) is bound at an interior site. The very short relaxation times observed for divalent Ni, Fe, and Co could be interpreted as an indication of either no bonding or bonding to an exterior site. The number of available absorption sites per milligram of DNA was calculated to be 7×10^{17} for Mn(II) and Cr(III) and 8×10^{17} for Fe(III)

[129]. The value for Cu(II) was not very well defined. Other studies of the bonding of Mn(II) to nucleic acids [127], and bovine serum albumin [238] have been made using the same technique. Likewise P^{31} NMR has been employed in a study [277] of the bonding between the metal ions, Mn(II) and Co(II), and RNA and adenosine monophosphate. See also [222].

3. Chlorophyll, Cytochrome, and Related Compounds. NMR has also been used by Closs, Katz, and co-workers [87] with IR and molecular weight determination [186] to study chlorophyll *a* and *b* and methyl chlorophyllide *a* and *b* (all magnesium chelates) as well as the analogous compounds containing no metal. They were able to assign the various resonance peaks to the protons of these very complex molecules (the schematic structure is shown on p. 3810 of [87]). In the difficult task of the assignment they were aided by considerations of the ring current in the macrocyclic ring and of the neighboring group anisotropies. Deuterium substitution and double resonance were also employed. In general their experiments show that the NMR spectra are very sensitive to concentration and to the degree of aggregation of the molecules. The intermolecular forces which cause aggregation are more pronounced in the magnesium chelates and seem to be due to carbonyl-magnesium interaction. They interpret the dissociation of aggregates in methanol-containing solutions on the basis of solvent(methanol)-solute interaction taking the place of solute-solute interaction. Direct molecular weight determinations and IR spectra confirm this view. Accordingly the chlorophylls seem to be predominantly monomeric in polar or basic solvents, but to exist in aggregates in non-polar solvents. A similar conclusion has recently been reached for bacteriochlorophyll [272]. Dougherty, Norman, and Katz [115] have also applied deuteron magnetic resonance to the study of chlorophyll a–d_{72}, and have reported two NMR studies [114] of plant biosynthesis concerned specifically with bacteriochlorophyll.

Several studies of the NMR spectroscopy of metal porphyrin complexes and similar molecules have appeared (see [52, 67, 70, 83, 86, 113, 124, 142, 165, 170, 291]). There are many pertinent references to the NMR of porphyrins containing no metal; see [70, 40, 185].

Kowalsky [196] has used PMR spectroscopy in research on ferri- and ferrocytochrome *c* and derivatives. Wishnia [304] has measured the T_1 of dilute solutions of Fe(III), iron conalbumin, and ferrihemoglobin. And Lumry and co-workers [219] have also studied the structure and denaturation of heme proteins by the molar relaxivities toward the protons in water of the ferric and ferrous forms of some heme complexes, and of myoglobin and hemoglobin.

4. Amino Acids and Peptides. In a series of papers, Li, Mathur, and co-workers have studied the effect of added metal ions on the PMR spectra of several amino acids and peptides [216], amino acid esters [224], glycyl peptides [225], glycylglycine and glycine amide [215], and N-methylacetamide and some thiols [226]. This is an example of their method of argumentation and of their results [224]: The PMR spectrum of ethylglycinate hydrochloride,

$$H_2NCH_2COOCH_2CH_3 \cdot HCl,$$

was taken at 60 mc sec^{-1} in D_2O with benzene as an external reference (Fig. 6-13a). (The chemical shifts here reported are in ppm upfield from benzene.) The spectrum consists of a quartet at 2.20 ppm (the ethyl—CH_2), a singlet at 2.76 ppm (the glycine—CH_2—), and a triplet at 5.22 ppm (the ethyl—CH_3). When the solution is made 0.5*M* in $CdCl_2$, complexing occurs (Fig. 6-13b). Since the —CH_2— of glycine lies between the bonding sites (—NH_2 and C═O), it shows the greatest chemical shift downfield (2.76→2.58 ppm) as the bonding site electrons are shared with the cadmium ion. The other two peaks shift from 2.20 to 2.08 ppm and from 5.22 to 5.13 ppm. The CH_3 signal shows, as expected, the least shift downfield. When the ethylglycinate solution is made 10^{-4} *M* in Cu(II) ion, the two ethyl signals are shifted very slightly, but the glycine—CH_2— signal is so broadened by the proximity of the paramagnetic Cu^{2+} ion that it is no longer observable (Fig. 6-13c).

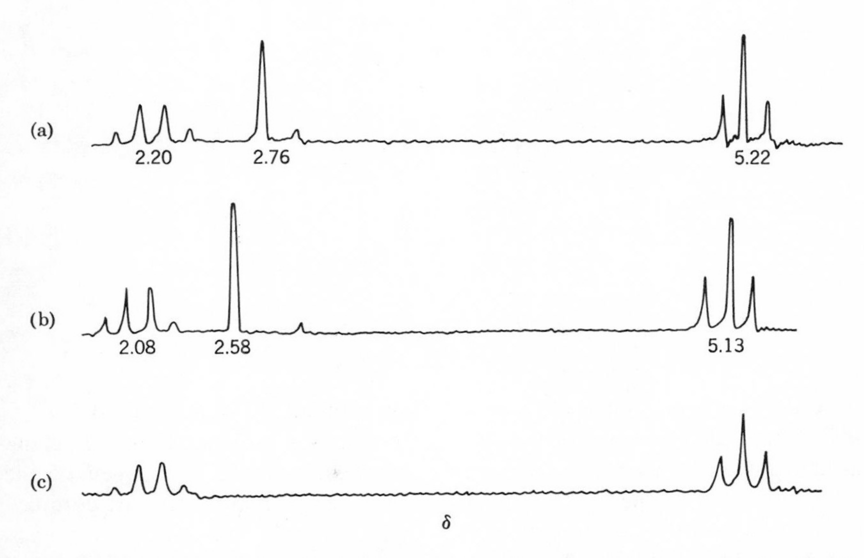

Fig. 6-13. PMR spectra in D_2O at 60 mc sec^{-1} of (a) 0.5*M* ethylglycinate hydrochloride plus 0.25*M* NaOD; (b) solution (a) plus 0.5*M* $CdCl_2$; (c) solution (a) plus $< 10^{-4}$ *M* $CuCl_2$. δ values are in ppm upfield from C_6H_6 as external reference [224].

The technique outlined here has been extended in the other papers to more complex systems. It can be a great help in defining the bonding sites when there could be ambiguity. For example, a similar analysis [226] has indicated that in N-methylacetamide($CH_3CONHCH_3$) the carbonyl oxygen, rather than the nitrogen, is the site of coordination with a metal. Likewise the PMR results for glycylserine, glycylproline, and glycyl-L-alanine [216] indicate that in chelation with a metal ion the coordination sites are the terminal amino group and the adjacent amide group. This type of analysis has also been used by Tang and Li [294] in a study of cobalt and manganese glycylglycinate in aqueous solution, both before and after oxygenation. They found the attachment of the oxygen to the metal complex to be reversible in the case of Co(II), but not in the case of Mn(II). When glycylproline is substituted for glycylglycine, no oxygenated complex is formed. This, they feel, is linked with the absence of an amide hydrogen in glycylproline. (See also [235] for a brief discussion of the effects of oxygen on Co(II)-histidine complexes.)

Use has been made of contact interaction shifts in a very interesting paper by McDonald and Phillips [235] on the structure of Co(II)-histidine complexes. Analysis of their data enabled them to identify four different complexes which form in different pH ranges. At low pH (1.0–3.5) one histidine (XXX) coordinated with Co(II) by its carboxyl group replaces one of the

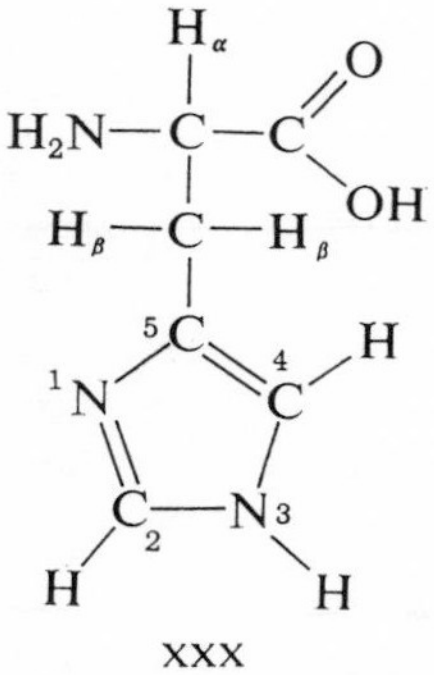

XXX

six waters from the metal ion's coordination sphere. At higher pH (4.5–10.5) they found two complexes in which histidine is a tridentate ligand, being attached via the carboxyl, the amine nitrogen, and the 1-nitrogen of the imidazole ring. In the lower part of this range a 1:1 metal-histidine complex is favored, and a 1:2 complex in the higher part. At high pH (above 11.5) a 1:2 metal-histidine tetrahedral complex forms. In it each ligand is bonded only through the nitrogens. The authors were aided in making these assignments by the paramagnetic susceptibilities of the solutions determined by the method of Evans [131]. Two other interesting results of their research are noted. They observed larger chemical shifts for a 1:2 metal-histidine complex containing one D-histidine and one L-histidine than for one containing two L- or two D-histidines. This they interpreted as an indication of stronger bonding in the former case. And their calculations indicate that the mixed complex is about 0.7 kcal $mole^{-1}$ more stable than complexes containing either two D- or two L-histidines. From a study of the pH dependence of the various proton resonances in histidine they were also able to determine unambiguously which of the protons is released at the various pH's.

NMR spectroscopy has also been used in studies of the complexes of Zn(II) with glycinate and/or pyruvate ions and related systems [213], of Mo(V) and Mo(VI) with L-histidine [287], of Zn(II) and Cd(II) with imidazole, α-alanine, and L-histidine [68], of Ni(II) and Zn(II) with cysteine and histidine [223], of Co(II) and Ni(II) with L-proline, histidine, and other ligands [241], and of Zn(II) with imidazole, purine, and pyrimidine derivatives [300]. Denning and Piper [110] have also employed PMR spectroscopy in their study of optical activity and absolute configuration of the chelates of Co(III) with L-alanine, L-leucine, and L-proline.

D. Reaction Rates

The theory of the determination of reaction rates by the broadening of NMR absorption signals has been presented by Gutowsky, McCall, and Slichter [160] and by McConnell [231]. It has recently been reviewed by Delpuech [109] and discussed in detail by Allerhand and co-workers [45]. Pearson and Anderson [255] have also summarized the principles of the method in their review of the exchange rates of ligands in complex ions. Forsén and Hoffman [148] have presented a method for studying moderately rapid exchange rates by nuclear magnetic double resonance. For an extensive review of reaction rates of transition metal complexes, see [292].

Pearson and Lanier [256] have used the NMR line broadening technique to study the rate for the reaction

$$ML_n + nL^* \rightleftharpoons ML_n^* + nL,$$

where L is a mono- or polydentate ligand. They used the divalent metals Mn, Fe, Co, Ni, and Cu and the ligands NH_3, en, glycine, sarcosine (N-methylglycine), and N,N-dimethylglycine, all in aqueous solution, and the same metals with anhydrous ethylene glycol. In all cases they found the order of reactivity of the metals to be Mn(II) > Fe(II) > Co(II) > Ni(II) ≪ Cu(II). Their results for Cu(II) with some of these ligands calculated at 27° are given in Table 6-6. The same authors also discuss in detail probable mechanisms of the exchange reactions. For example, from the data for Cu(II) complexes

Table 6-6 Rate Limits for Ligand Exchange[a]

Complex	$k_{OH}(M^{-1}\ sec^{-1})^b$	$k_{CH_2}(M^{-1}\ sec^{-1})^b$
$[Cu(en)_2]^{2+}$	2.6×10^6	1.8×10^6
$[Cu(gly)_2]$	8.2×10^6	2.6×10^6
$[Cu(sarc)_2]^c$	6×10^5	
$[Cu(dmgly)_2]^c$		1.3×10^4

[a] From [256].
[b] k_{OH} and k_{CH_2} refer to the constants calculated from line broadening of the OH and CH_2 signals, respectively.
[c] sarc = sarcosine; dmgly = N,N-dimethylglycine.

given in Table 6-6 they conclude that the rate sequence, glycine > sarcosine > dimethylglycine, is that expected for an S_N2 mechanism. Steric hindrance to the approach of the nucleophile appears to be the dominant factor. For further discussion see also the review article by Pearson and Anderson [255]. Exchange data on some chromium chelates, as for example, $[Cr(en)_3](ClO_4)_3$, are given by Pearson et al. [257].

The line-broadening technique has been used also in studies of ligand exchange in Zr, Hf, and Th-β-diketone complexes [43], of electron transfer in Fe-phenanthroline and Os-bipyridyl complexes [112, 209, 210], and of exchange in reactions involving the Cd-EDTA complex [293].

A different method was employed by Fay and Piper [141] in a study of the kinetics and mechanism of the stereochemical rearrangement of the trifluoroacetylacetonates of Al, Ga, In, Co, and Rh. They heated a $CDCl_3$ solution of *cis*-$[Co(tfac)_3]$ at 79.1° for a given time and then plunged it into a dry ice-acetone bath at −40° to quench the reaction. The F^{19} NMR spectrum of the resulting *cis-trans* mixture showed separate signals for each isomer. The areas under the curves gave a measure of the amount of each present. The procedure was repeated using several heating times. From these data they were able to calculate the first order rate constant for *cis* ⇌ *trans* interconversion to be $(1.52 \pm 0.04) \times 10^{-4}$ sec^{-1}. The same was done at other temperatures, and a plot of the rate constants versus $1/T$ (°K) gave a straight line from which the Arrhenius activation energy of 30.7 ± 0.6 kcal mole^{-1} and a frequency factor of exp (15.19 ± 0.37) sec^{-1} were calculated. On the assumption that the Eyring equation holds, an entropy of activation of 8.6 ± 1.7 e.u. was calculated from these data. They also noted that the F^{19} NMR spectral lines corresponding to the *cis* and *trans* forms tended to coalesce as the temperature was raised. For $Ga(tfac)_3$ the coalescence temperature is 61.5°, and for $Al(tfac)_3$ it is 103°. Gutowsky and Holm [159] have shown that at the temperature of coalescence the rate constant for interconversion is $(\delta\omega/2\sqrt{2})$, where $\delta\omega$ is the frequency separation in radians per second between the two lines when exchange is absent. Using this relation, Fay and Piper have estimated the first order rate constant for *cis* → *trans* conversion for the two complexes to be 38 and 34 sec^{-1}, respectively. In the same report [141] they outline in detail possible mechanisms for the observed stereochemical re-arrangements, and conclude that in all cases studied it probably proceeds via a rupture of one bond to give a symmetrical five-coordinate intermediate.

A similar study of mixed ligand chelates, $[M(tfac)_n(acac)_{3-n}]$, where M = Co, Cr, and $n = 0 - 3$, has been reported [253]. Likewise Gasser and Richards [151] have investigated the slow exchange: $[Co(NH_3)_6]^{3+} + 3en \rightleftharpoons [Co(en)_3]^{3+} + 6NH_3$ by means of Co^{59} NMR spectroscopy. And PMR spectroscopy has been employed in studying proton exchange in cobalt-ammine complexes [167].

Those who plan to use NMR spectroscopy for rate studies may wish to consult the article of Evans, Miller, and Kreevoy [132] in which this method is compared with thermal maximum and scavenging techniques.

E. Equilibrium Studies

The concentrations of the various components of an equilibrium mixture can be determined from NMR spectroscopy if each component yields a distinct signal which is adequately separated from other resonance absorptions. Measurement of the relative areas of the peaks corresponding to the different components may then allow calculation of the equilibrium constant for the reaction. For example, Pinnavaia and Fay [260] have studied ligand exchange in the complexes $[M(acac)_n(tfac)_{4-n}]$, where M is quadrivalent Zr, Hf, Ce, or Th, and n has values from 0 to 4. They used the areas under the F^{19} NMR absorption peaks to determine the concentrations of the various complexes in solution. In a 1:1 mixture of $[Zr(acac)_4]$ and $[Zr(tfac)_4]$ in benzene, equilibrium was established in less than 30 sec, and the F^{19} NMR spectrum showed four fluorine resonances which could be assigned to the $[Zr(acac)_{4-n}(tfac)_n]$ (n = 1, 2, 3, 4) (Fig. 6-14a). Measurement of the relative areas of the absorption peaks permitted the calculation of the concentrations of the various species and the values for the equilibrium quotients for the reactions

$$\begin{aligned}
&(1)\quad Zr(acac)_4 + Zr(acac)_2(tfac)_2 \rightleftharpoons 2Zr(acac)_3(tfac)\\
&(2)\quad Zr(acac)_3(tfac) + Zr(acac)(tfac)_3 \rightleftharpoons 2Zr(acac)_2(tfac)_2\\
&(3)\quad Zr(acac)_2(tfac)_2 + Zr(tfac)_4 \rightleftharpoons 2Zr(acac)(tfac)_3
\end{aligned}$$

(The concentration of $[Zr(acac)_4]$ in (1) was determined by difference, because it shows no F^{19} NMR absorption.) The values of the three constants in benzene at 31° were found to be independent of total solute concentration and of ligand composition (i.e., they are independent of f_{tfac}, the molar

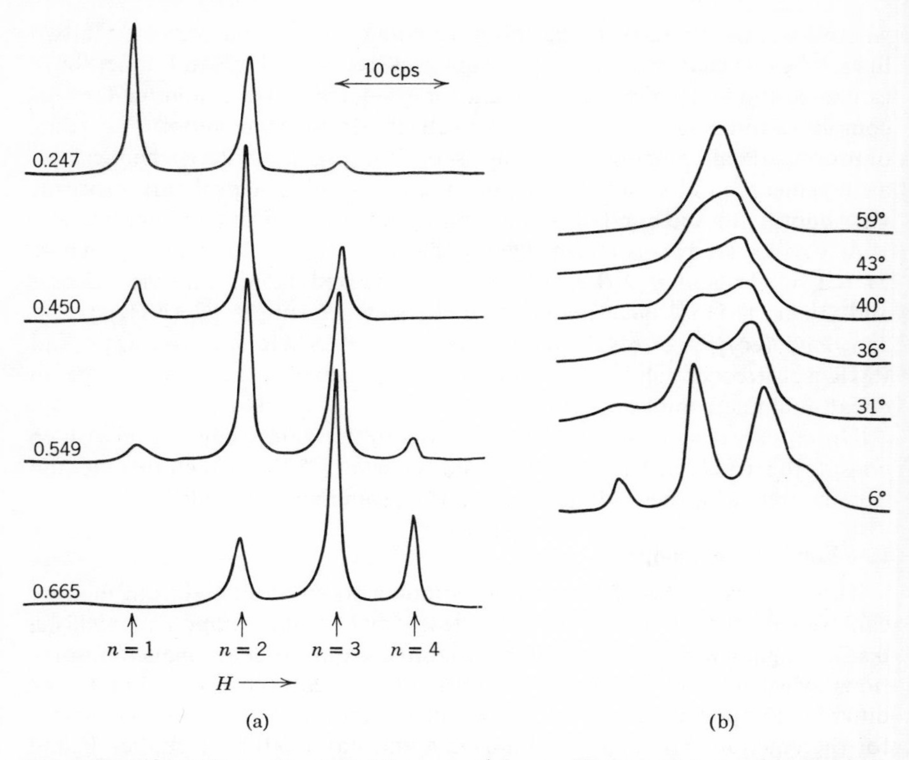

Fig. 6-14. F^{19} NMR spectra at 56.444 mc sec^{-1} for equilibrium mixtures of (a) $Zr(tfac)_4$ and $Zr(acac)_4$ in C_6H_6 solution at 31°; total solute molarity is 0.25M; numbers at left give fraction of total ligand present as tfac; (b) $Th(tfac)_4$ and $Th(acac)_4$ in C_6H_6 solution at several temperatures near the coalescence region; total solute molarity = 0.25M, tfac fraction = 0.550 [260].

fraction of total ligand present as tfac). One set of constants determined with values of f_{tfac} varying from 0.149 to 0.803 is: $K_1 = 6.09 \pm 0.21$, $K_2 = 6.08 \pm 0.20$, $K_3 = 7.23 \pm 0.17$ (errors estimated at the 95% confidence level). These values contrast with those calculated on the basis of statistical scrambling of ligands: $K_1 = K_3 = 2.67$, $K_2 = 2.25$. A study of the variation of the K's with temperature indicates that the enthalpy of the three reactions is zero within the limits of experimental error. Accordingly $R \ln K = \Delta S$. The entropy (ΔS) for the three reactions calculated from this are 3.4 ± 1.8, 3.5 ± 0.7, and 3.9 ± 0.9 e.u., respectively. The statistical ΔS values are: 1.95, 1.61, and 1.95 e.u., respectively. Similar data were obtained for CCl_4 solutions. The observed entropy changes in these two solvents are all 1–3 e.u. in excess of those calculated for random mixing of the ligands. The authors suggest that a difference in the solvation entropies of reactants and products may be the reason.

The F^{19} NMR spectra of the systems with M = Hf, Ce, and Th showed smaller chemical shifts than did the Zr mixture. From the relative intensities, however, it could be concluded that the equilibrium concentrations were rather similar in all four cases. The thorium spectrum is strongly temperature dependent; the four F^{19} lines coalesce when the temperature is raised to 43° (Fig. 6-14b). The other mixtures have higher coalescence temperatures, which could not, however, be accurately determined because of formation of decomposition products. Coalescence of the peaks indicates that all the CF_3 groups are in equivalent magnetic environments. The authors attribute this to intermolecular ligand exchange which is rapid enough to average the environments of all the CF_3 groups present. This interpretation would indicate that ligand exchange occurs more rapidly in the thorium system than in the other three.

A complementary PMR study of the mixed ligand chelates of Zr and Hf with the same two ligands has been reported by Adams and Larsen [41, 42]. See also [103] for a similar report. Fay and Piper [140] have also studied by NMR spectroscopy the *trans* $\rightleftharpoons$ *cis* equilibria in [M(tfac)$_3$] (M = Al, Ga, and Co). Walker and Li [299] have used P^{31} NMR to determine the equilibrium constant for the reaction between bis(thenoyltrifluoroacetono)zinc(II) and tri-*n*-octyl phosphine oxide; K_{eq} is about 10.1 per mole in anhydrous CCl_4. And Kula [199] has made use of PMR spectroscopy in studying solution equilibria involving Zn-EDTA complexes.

F. Stability of Chelates

The relative stability of a series of chelates can sometimes be determined from NMR spectroscopy. For example, Kula, Sawyer, Chan, and Finley [200] have recorded the PMR spectra of EDTA complexes of the alkali and alkaline earth metals, and of Zn(II), Pb(II), Hg(II), and Al(III). The chemical shift data were found to agree qualitatively with the order of the stability constants. The stabilities decrease in the order: Ca > Sr > Ba > Li > Na > K $\approx$ Cs. The position of Mg could not be determined because of precipitation below pH 5.5. PMR results indicated that Pb, Al, Zn, and Hg all form more stable chelates than the alkali and alkaline earth metals. Kula [199] has also used PMR spectroscopy to study the formation constant of a $Zn(OH)EDTA^{3-}$ complex. See also the PMR study of Erickson and Alberty [130] of alkali metal-malate ion interaction and that of Bramley et al. [57] on the uranyl-citrate system.

An interesting method of determining stability of weak complexes is based on the fact that Na^{23} has a spin of $\frac{3}{2}$ and so can give an NMR signal. Since it has an appreciable electric quadrupole moment, there is the possibility of significant line broadening. Although the electric field surrounding an isolated Na^+ is expected to be spherically symmetrical, it can be distorted when Na^+ interacts with some negative ion or dipole. This distortion will cause the nuclear quadrupole to interact with the gradient of the unsymmetrical electric field around the nucleus, which will in turn cause the NMR signal to be broadened. It is logical to expect that the greater the interaction between the Na^+ and the negative ion (or dipole), the greater will be the line broadening. The interaction can appear as a line broadening, however, only if the Na^+-negative ion (or dipole) orientation is of longer duration than that between Na^+ and H_2O. Jardetsky and Wertz [183, 302] first made this analysis when they studied the reaction of sodium ion with aqueous solutions of hydroxy and keto acids, alcohols, and phosphates. The line broadening indicated weak complex formation, while the lack of line broadening in the case of solutions of Na^+ with chloride, acetate, formate, or benzoate ions was interpreted to indicate no complex formation. Rechnitz and Zamochnick [263] have determined the formation constants of seven complexes of Na^+

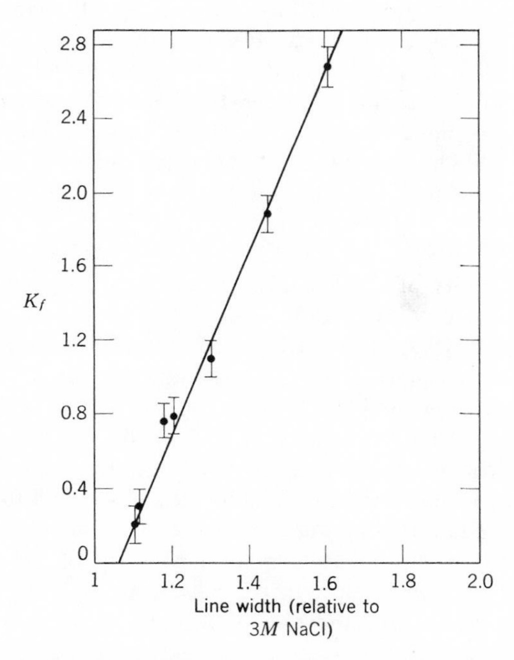

Fig. 6-15. Na^{23} NMR line width versus formation constant for complexes of Na^+ with pyruvic acid (2.7), malic acid (1.9), lactic acid (1.1), *m*-hydroxybenzoic acid (0.8_1), *o*-hydroxybenzoic acid (0.7_8), *l*-leucine (0.3), *d,l*-α-alanine (0.2) ($K_f \pm 0.1$ values in parentheses) [263].

with organic acids at pH ~8.4 by a potentiometric method, and have measured the line width of the Na^{23} NMR for these solutions. When the formation constants were plotted against the line widths of the Na^{23} NMR line (relative to 3*M* NaCl), a linear relationship was found (Fig. 6-15). It is probable that this relationship can validly be used to obtain the formation constants of other weak sodium complexes from a measurement of the Na^{23} NMR line width alone, provided conditions identical to those mentioned above are maintained.

In a group of papers Rivkind and Kozyrev [197, 198, 265, 266] have discussed a method of using relaxation times to measure formation constants of paramagnetic chelates. The results obtained often agree quite well with those obtained by the more conventional methods.

Popel and Grazhdannikov [261] have also studied complex formation by vanadyl with pyrocatechol and citrate ions by measuring relaxation times of the solvent protons. NMR studies of the complexes of glycerol with Fe(III), Cu(II), and Mn(II) [218], of diglycol borates with sodium ion [264], of Cu(II) with en [105, 242], and of V (IV) with malonic and oxalic acids [305] have also been reported.

REFERENCES

Books, Catalogs of Data

[1] *Catalog of NMR Spectral Data*, American Petroleum Institute Research Project 44, B. J. Zwolinski, Director, Chemical Thermodynamic Properties Center. Texas A and M University, College Station, Texas.

[2] *A Catalog of the NMR Spectra of Hydrogen in Hydrocarbons and Their Derivatives*, Humble Oil and Refining Co., Research and Development Division, Baytown, Texas, 1959 and 1961, 700 pp.

[3] Anderson, J. M., *Studies in Nuclear Magnetic Double Resonance*. Cambridge, Mass.: Harvard University Press, 1963, 222 pp.

[4] Bhacca, N. S., L. F. Johnson, and J. N. Shoolery, *High Resolution NMR Spectra Catalog*. Vol. I. Palo Alto, Calif.: Varian Associates, Vol. 1, 1962.

[5] Bhacca, N. S., D. P. Hollis, L. F. Johnson, and E. A. Pier, *High Resolution NMR Spectra Catalog*. Palo Alto, Calif.: Varian Associates, Vol. 2, 1963.

[6] Bible, R. H., Jr., *Interpretation of NMR Spectra: An Empirical Approach*. New York: Plenum Press, 1965, 150 pp.

[7] Chamberlain, N. F., F. C. Stehling, K. W. Bartz, and J. J. R. Reed, *NMR Data for Hydrogen*—1. Esso Research and Engineering Co., Baytown Research and Development Division, Baytown, Texas, 1965.

[8] Emsley, J. W., J. Feeney, and L. H. Sutcliffe, *High Resolution Nuclear Magnetic Resonance Spectroscopy*. New York: Pergamon, 1965, two vols., 1154 + cii pp.

[9] Fluck, E., *Die Kernmagnetische Resonanz in der Anorganischen Chemie*. Berlin: Springer-Verlag, 1963, 290 pp.

[10] Howell, M. G., A. S. Kende, and J. S. Webb, eds., *Formula Index to NMR Literature Data*. New York: Plenum Press, 1964, Vol. 1, 215 pp., Vol. 2, 400 pp.

[11] Pople, J. A., W. G. Schneider, and H. J. Bernstein, *High Resolution Nuclear Magnetic Resonance*. New York: McGraw-Hill, 1959, 513 pp.
[12] Roberts, J. D., *An Introduction to the Analysis of Spin-Spin Splitting in High Resolution Nuclear Magnetic Resonance Spectra*. New York: Benjamin, 1961, 116 pp.
[13] Slichter, C. P., *Principles of Magnetic Resonance*. New York: Harper and Row, 1963, 246 pp.
[14] Strehlow, H., *Magnetische Kernresonanz und Chemische Struktur*, Bd. VII, *von Fortschritte der physikalischen Chemie*. Darmstadt: Steinkopff, 1961, 220 pp.
[15] Varian Associates Staff, *NMR and EPR Spectroscopy*. New York: Pergamon, 1960, 288 pp.
[16] Wiberg, K. B., and B. J. Nist, *The Interpretation of NMR Spectra*. New York: Benjamin, 1962, 593 pp.

Review Articles

[17] *Ann. Rev. Phys. Chem.*, **5** (1954) to **16** (1965). Each volume contains a review of NMR (or some closely allied topic) for the year.
[18] *Bull. Soc. Chim. France*, 2695–2718 (1964). Reviews of NMR of F^{19}, of organometallics, and of kinetic studies by NMR.
[19] *Discussions Faraday Soc.*, No. 34 (1962). High Resolution NMR.
[20] Baldeschwieler, J. D., and E. W. Randall, *Chem. Rev.*, **63**, 81–110 (1963). Chemical applications of nuclear magnetic double resonance.
[21] Bishop, E. O., *Ann. Rept. Progr. Chem.* (*Chem. Soc. London*), **58**, 55–78 (1961). Nuclear magnetic resonance.
[22] Brownstein, S., *Chem. Rev.*, **59**, 463–96 (1959). High resolution NMR and molecular structure.
[23] Conroy, H., "NMR in organic structural elucidation," *Advan. Org. Chem.*, **2**, 265–328 (1960).
[24] Corio, P. L., *Chem. Rev.*, **60**, 363–429 (1960). Analysis of NMR spectra.
[25] Eaton, D. R., and W. D. Phillips, "NMR of Paramagnetic Molecules," in *Advances in Magnetic Research*, J. S. Waugh, ed. New York: Academic, Vol. 1, 1965, pp. 103–148.
[26] Gillespie, R. J., and R. F. M. White, "NMR and stereochemistry," *Progr. Stereochem.*, **3**, 53–94 (1962).
[27] Hardeman, G. E. G., *Philips Tech. Rev.*, **24** (7) 206–220 (1962–63). Double magnetic resonance.
[28] Jardetzky, O., and C. D. Jardetzky, "Introduction to magnetic resonance spectroscopy. Methods and biochemical applications," *Methods Biochem. Anal.*, **9**, 235–410 (1962).
[29] Jones, R. A. Y., and A. R. Katritzky, *Angew. Chem. Intern. Ed. Engl.*, **1**, 32–41 (1962). Phosphorus NMR spectroscopy.
[30] Mavel, G., "Résonance magnétique nucléaire (RMN) et interactions moléculaires en phase liquide," *J. Chim. Phys.*, **61** (1, 2) 182–194 (1964).
[31] Muetterties, E. L., and W. D. Phillips, "The use of NMR in inorganic chemistry," *Advan. Inorg. Chem. Radiochem.*, **4**, 231–292 (1962).
[32] Nachod, F. C., and W. D. Phillips, eds., *Det. Org. Struct. Phys. Methods*, **2** (1962). Several articles on NMR.
[33] Parello, J., *Bull. Soc. Chim. France*, 2033–62 (1964). Coupling between nuclear spins in NMR and the method of double resonance.

[34] Richards, R. E., "Nuclear magnetic resonance," *Advan. Spectry.*, **2**, 101–188 (1961).
[35] Smith, J. A. S., *Quart. Rev. (London)*, **7**, 279–306 (1953). NMR absorption.
[36] Stothers, J. B., *Quart. Rev. (London)*, **19**, 144–167 (1965). C^{13} NMR spectroscopy.
[37] Wertz, J., *Chem. Rev.*, **55**, 829–956 (1955). Nuclear and electronic spin magnetic resonance.
[38] Wheatley, P. J., "Nuclear Magnetic Resonance," in *The Determination of Molecular Structure*. Oxford: Clarendon Press, 1959, pp. 233–55.

Other References

[39] Abel, E. W., R. P. Bush, F. J. Hopton, and C. R. Jenkins, *Chem. Commun.*, 58 (1966).
[40] Abraham, R. J., A. H. Jackson, G. W. Kenner, and D. Warburton, *J. Chem. Soc.* 853 (1963).
[41] Adams, A. C., and E. M. Larsen, *J. Am. Chem. Soc.*, **85**, 3508 (1963).
[42] Adams, A. C., and E. M. Larsen, *Inorg. Chem.*, **5**, 228 (1966).
[43] Adams, A. C., and E. M. Larsen, *Inorg. Chem.*, **5**, 814 (1966).
[44] Allen, G., J. Lewis, R. F. Long, and C. Oldham, *Nature*, **202**, 589 (1964).
[45] Allerhand, A., H. S. Gutowsky, J. Jonas, and R. A. Meinzer, *J. Am. Chem. Soc.*, **88**, 3185 (1966).
[46] Aochi, Y. O. and D. T. Sawyer, *Inorg. Chem.*, **5**, 2085 (1966).
[47] Archer, R. D., and B. P. Cotsoradis, *Inorg. Chem.*, **4**, 1584 (1965).
[48] Asperger, R. G., and Chui Fan Liu, *Inorg. Chem.*, **4**, 1395 (1965).
[49] Atkinson, M. R., and R. K. Morton, *Comp. Biochem.*, **2**, 1 (1960).
[50] Benson, R. E., D. R. Eaton, A. D. Josey, and W. D. Phillips, *J. Am. Chem. Soc.*, **83**, 3714 (1961).
[51] Berry, R. S., *J. Chem. Phys.*, **32**, 933 (1960).
[52] Bertele, E., et al, *Angew. Chem.*, **76**, 393 (1964).
[53] Bonati, F., and G. Wilkinson, *J. Chem. Soc.*, 3156 (1964).
[54] Bosnich, B., R. Bramley, R. S. Nyholm, and M. L. Tobe, *J. Am. Chem. Soc.*, **88**, 3926 (1966).
[55] Boucher, L. J. and J. C. Bailar, Jr., *J. Inorg. Nucl. Chem.*, **27**, 1093 (1965).
[56] Bradley, D. C., and C. E. Holloway, *Chem. Commun.*, 284 (1965).
[57] Bramley, R., W. F. Reynolds, and I. Feldman, *J. Am. Chem. Soc.*, **87**, 3329 (1965).
[58] Brannan, J. R., and D. T. Sawyer, *Inorg. Chem.*, **4**, 1070 (1965).
[59] Brinckman, F. E., H. S. Haiss, and R. A. Robb, *Inorg. Chem.*, **4**, 936 (1965).
[60] Brockway, L. O., and J. Y. Beach, *J. Am. Chem. Soc.*, **60**, 1836 (1938).
[61] Brown, N. M. D., and P. Bladon, *Chem. Commun.*, 304 (1966).
[62] Bryson, A., and G. H. Nancollas, *Chem. Ind. (London)*, 654 (1965).
[63] Buckingham, A. D., and P. J. Stephens, *J. Chem. Soc.*, 2747 (1964).
[64] Buckingham, A. D., and P. J. Stephens, *J. Chem. Soc.*, 4583 (1964).
[65] Bullen, G. J., R. Mason, and P. Pauling, *Inorg. Chem.*, **4**, 456 (1965).
[66] Burdett, J. L., and L. L. Burger, *Can. J. Chem.*, **44**, 111 (1966).
[67] Busch, D. H., J. H. Weber, D. H. Williams, and N. J. Rose, *J. Am. Chem. Soc.*, **86**, 5161 (1964).
[68] Carlson, R. H., and T. L. Brown, *Inorg. Chem.*, **5**, 268 (1966).
[69] Carty, A. J., D. G. Tuck, and E. Bullock, *Can. J. Chem.*, **43**, 2559 (1965).
[70] Caughey, W. S., and W. S. Koski, *Biochemistry*, **1**, 923 (1962).
[71] Chakravorty, A., J. P. Fennessey, and R. H. Holm, *Inorg. Chem.*, **4**, 26 (1965).
[72] Chakravorty, A., and R. H. Holm, *Inorg. Chem.*, **3**, 1010 (1964).

[73] Chakravorty, A., and R. H. Holm, *Inorg. Chem.*, **3**, 1521 (1964).
[74] Chakravorty, A., and R. H. Holm, *J. Am. Chem. Soc.*, **86**, 3999 (1964).
[75] Chan, S. I., R. J. Kula, and D. T. Sawyer, *J. Am. Chem. Soc.*, **86**, 377 (1964).
[76] Chatt, J., *Proc. Chem. Soc.*, 318 (1962).
[77] Chatt, J., L. A. Duncanson, and B. L. Shaw, *Proc. Chem. Soc.*, 343 (1957).
[78] Chatt, J., F. A. Hart, and D. T. Rosevear, *J. Chem. Soc.*, 5504 (1961).
[79] Chatt, J., and R. G. Hayter, *Proc. Chem. Soc.*, 153 (1959).
[80] Chatt, J., and R. G. Hayter, *J. Chem. Soc.*, 2605 (1961).
[81] Chatt, J., and R. G. Hayter, *J. Chem. Soc.*, 6017 (1963).
[82] Chatt, J., and B. L. Shaw, *J. Chem. Soc.*, 5075 (1962).
[83] Clarke, D. A., R. Grigg, and A. W. Johnson, *Chem. Commun.*, 208 (1966).
[84] Clark, R. J. H., *J. Chem. Soc.*, 5699 (1965).
[85] Clifton, P., and L. Pratt, *Proc. Chem. Soc.*, 339 (1963).
[86] Closs, G. L., and L. E. Closs, *J. Am. Chem. Soc.*, **85**, 818 (1963).
[87] Closs, G. L., J. J. Katz, F. C. Pennington, M. R. Thomas, and H. H. Strain, *J. Am. Chem. Soc.*, **85**, 3809 (1963).
[88] Cohn, M., and T. R. Hughes, *J. Biol. Chem.*, **235**, 3250 (1960).
[89] Cohn, M., and T. R. Hughes, *J. Biol. Chem.*, **237**, 176 (1962).
[90] Cohn, M., and J. S. Leigh, *Nature*, **193**, 1037 (1962).
[91] Collman, J. P., *Advan. Chem. Ser.*, **37**, 78 (1963).
[92] Collman, J. P., *Angew. Chem. Intern. Ed. Engl.*, **4**, 132 (1965).
[93] Collman, J. P., and E. T. Kittleman, *J. Am. Chem. Soc.*, **83**, 3529 (1961).
[94] Collman, J. P., R. L. Marshall, and W. L. Young, *7th I.C.C.C. Rept.*, No. 1 G 1, Stockholm, June, 1962.
[95] Collman, J. P., R. L. Marshall, and W. L. Young, *Chem. Ind.* (*London*), 1380 (1962).
[96] Collman, J. P., R. L. Marshall, W. L. Young, and S. D. Goldby, *Inorg. Chem.*, **1**, 704 (1962).
[97] Collman, J. P., R. L. Marshall, W. L. Young, and C. T. Sears, *J. Org. Chem.*, **28**, 1449 (1963).
[98] Collman, J. P., R. A. Moss, H. Maltz, and C. C. Heindel, *J. Am. Chem. Soc.*, **83**, 531 (1961).
[99] Collman, J. P., and Jui-Yuan Sun, *Inorg. Chem.*, **4**, 1273 (1965).
[100] Cooke, D. W., *Inorg. Chem.*, **5**, 1141 (1966).
[101] Cotton, F. A., and R. C. Elder, *J. Am. Chem. Soc.*, **86**, 2294 (1964).
[102] Cotton, F. A., and R. H. Holm, *J. Am. Chem. Soc.*, **82**, 2979 (1960).
[103] Cotton, F. A., P. Legzdins, and S. J. Lippard, *J. Chem. Phys.*, **45**, 3461 (1966).
[104] Cotton, F. A., and R. H. Soderberg, *Inorg. Chem.*, **3**, 1 (1964).
[105] Cox, P. F., and L. O. Morgan, *J. Am. Chem. Soc.*, **81**, 6409 (1959).
[106] Cruickshank, D. W. J., *J. Chem. Soc.*, 5486 (1961).
[107] Davison, A., N. Edelstein, R. H. Holm, and A. H. Maki, *Inorg. Chem.*, **3**, 814 (1964).
[108] Day, R. J., and C. N. Reilley, *Anal. Chem.*, **36**, 1073 (1964).
[109] Delpuech, J. J., *Bull. Soc. Chim. France*, 2697 (1964).
[110] Denning, R. G., and T. S. Piper, *Inorg. Chem.*, **5**, 1056 (1966).
[111] Dewar, D. H., J. E. Fergusson, P. R. Hentschel, C. J. Wilkins, and P. P. Williams, *J. Chem. Soc.*, 688 (1964).
[112] Dietrich, M. W., and A. C. Wahl, *J. Chem. Phys.*, **38**, 1591 (1963).
[113] Dolphin, D., R. L. N. Harris, A. W. Johnson, and I. T. Kay, *Proc. Chem. Soc.*, 359 (1964).

[114] Dougherty, R. C., H. L. Crespi, H. H. Strain, and J. J. Katz, *J. Am. Chem. Soc.*, **88**, 2854, 2856 (1966).
[115] Dougherty, R. C., G. D. Norman, and J. J. Katz, *J. Am. Chem. Soc.*, **87**, 5801 (1965).
[116] Douglass, D. C., and A. Fratiello, *J. Chem. Phys.*, **39**, 3161 (1963).
[117] Dudek, E. P., and M. L. Snow, *Inorg. Chem.*, **5**, 395 (1966).
[118] Eaton, D. R., *J. Am. Chem. Soc.*, **87**, 3097 (1965).
[119] Eaton, D. R., A. D. Josey, R. E. Benson, W. D. Phillips, and T. L. Cains, *J. Am. Chem. Soc.*, **84**, 4100 (1962).
[120] Eaton, D. R., A. D. Josey, W. D. Phillips, and R. E. Benson, *Discussions Faraday Soc.*, **34**, 77 (1962).
[121] Eaton, D. R., A. D. Josey, W. D. Phillips, and R. E. Benson, *J. Chem. Phys.*, **37**, 347 (1962).
[122] Eaton, D. R., A. D. Josey, W. D. Phillips, and R. E. Benson, *Mol. Phys.*, **5**, 407 (1962).
[123] Eaton, D. R., A. D. Josey, and W. A. Sheppard, *J. Am. Chem. Soc.*, **85**, 2689 (1963).
[124] Eaton, D. R., and E. A. LaLancette, *J. Chem. Phys.*, **41**, 3534 (1964).
[125] Eaton, D. R., and W. D. Phillips, *J. Chem. Phys.*, **43**, 392 (1965).
[126] Eaton, D. R., W. D. Phillips, and D. J. Caldwell, *J. Am. Chem. Soc.*, **85**, 397 (1963).
[127] Eisinger, J., F. Fawaz-Estrup, and R. G. Shulman, *J. Chem. Phys.*, **42**, 43 (1965).
[128] Eisinger, J., R. G. Shulman, and W. E. Blumberg, *Nature*, **192**, 963 (1961).
[129] Eisinger, J., R. G. Shulman, and B. M. Szymanski, *J. Chem. Phys.*, **36**, 1721 (1962).
[130] Erickson, L. E., and R. A. Alberty, *J. Phys. Chem.*, **66**, 1702 (1962).
[131] Evans, D. F., *J. Chem. Soc.*, 2003 (1959).
[132] Evans, P. G., G. R. Miller, and M. M. Kreevoy, *J. Phys. Chem.*, **69**, 4325 (1965).
[133] Everett, G. W., Jr., and R. H. Holm, *Proc. Chem. Soc.*, 238 (1964).
[134] Everett, G. W., Jr., and R. H. Holm, *J. Am. Chem. Soc.*, **87**, 2117 (1965).
[135] Everett, G. W., Jr., and R. H. Holm, *J. Am. Chem. Soc.*, **87**, 5266 (1965); **88**, 2442 (1966).
[136] Fackler, J. P., Jr., *J. Am. Chem. Soc.*, **84**, 24 (1962).
[137] Fackler, J. P., Jr., *Inorg. Chem.*, **2**, 266 (1963).
[138] Fackler, J. P., Jr., and F. A. Cotton, *J. Am. Chem. Soc.*, **83**, 2818, 3775 (1961).
[139] Fay, R. C., and T. S. Piper, *J. Am. Chem. Soc.*, **84**, 2303 (1962).
[140] Fay, R. C., and T. S. Piper, *J. Am. Chem. Soc.*, **85**, 500 (1963).
[141] Fay, R. C., and T. S. Piper, *Inorg. Chem.*, **3**, 348 (1964).
[142] Ferguson, J., and B. O. West, *J. Chem. Soc.* (*A*), 1565, 1569 (1966).
[143] Ferraro, J. R., *J. Inorg. Nucl. Chem.*, **26**, 225 (1964).
[144] Fessenden, R. W., and S. Ogawa, *J. Am. Chem. Soc.*, **86**, 3591 (1964).
[145] Fetter, N. R., and D. W. Moore, *Can. J. Chem.*, **42**, 885 (1964).
[146] Figgis, B. N., J. Lewis, R. F. Long, R. Mason, R. S. Nyholm, P. J. Pauling, and G. B. Robertson, *Nature*, **195**, 1278 (1962).
[147] Forman, A., J. N. Murrell, and L. E. Orgel, *J. Chem. Phys.*, **31**, 1129 (1959).
[148] Forsén, S., and R. A. Hoffman, *J. Chem. Phys.*, **39**, 2892 (1963).
[149] Freeman, R., and N. S. Bhacca, *J. Chem. Phys.*, **45**, 3795 (1966).
[150] Fritz, H. P., I. R. Gordon, K. E. Schwarzhans, and L. M. Venanzi, *J. Chem. Soc.*, 5210 (1965).
[151] Gasser, R. P. H., and R. E. Richards, *Mol. Phys.*, **3**, 163 (1960).

[152] Gibson, D., J. Lewis, and C. Oldham, *J. Chem. Soc.* (*A*), 1453 (1966).
[153] Gillard, R. D., and G. Wilkinson, *J. Chem. Soc.*, 3594 (1963).
[154] Gillard, R. D., and G. Wilkinson, *J. Chem. Soc.*, 6041 (1963).
[155] Golding, R. M., *Mol. Phys.*, **8**, 561 (1964).
[156] Golding, R. M., W. C. Tennant, C. R. Kanekar, R. L. Martin, and A. H. White, *J. Chem. Phys.*, **45**, 2688 (1966).
[157] Graddon, D., and E. Watton, *Nature*, **190**, 906 (1961).
[158] Gray, G. W., *Molecular Structure and the Properties of Liquid Crystals.* New York: Academic, 1962, pp. 31–38.
[159] Gutowsky, H. S., and C. H. Holm, *J. Chem. Phys.*, **25**, 1228 (1956).
[160] Gutowsky, H. S., D. W. McCall, and C. P. Slichter, *J. Chem. Phys.*, **21**, 279 (1953).
[161] Hammes, G. G., G. E. Maciel, and J. S. Waugh, *J. Am. Chem. Soc.*, **83**, 2394 (1961).
[162] Hammond, G. S., D. C. Nonhebel, and C. S. Wu, *Inorg. Chem.*, **2**, 73 (1963).
[163] Happe, J. A., and M. Morales, *J. Am. Chem. Soc.*, **88**, 2077 (1966).
[164] Happe, J. A., and R. L. Ward, *J. Chem. Phys.*, **39**, 1211 (1963).
[165] Harris, R. L. N., A. W. Johnson, and I. T. Kay, *Chem. Commun.*, 355 (1965).
[166] Hawthorne, M. F., and M. Reintjes, *J. Am. Chem. Soc.*, **86**, 5016 (1964).
[167] Hayashi, S., *Nippon Kagaku Zasshi*, **85**, 256 (1964); **86**, 364 (1965).
[168] Henney, R. C., H. F. Holtzclaw, Jr., and R. C. Larson, *Inorg. Chem.*, **5**, 940 (1966).
[169] Hester, R. E., *Chem. Ind.* (*London*), 1397 (1963).
[170] Hill, H. A. O., J. M. Pratt, and R. J. P. Williams, *J. Chem. Soc.*, 2859 (1965).
[171] Holloway, C. E., R. R. Luongo, and R. M. Pike, *J. Am. Chem. Soc.*, **88**, 2060 (1966).
[172] Holm, R. H., A. Chakravorty, and G. O. Dudek, *J. Am. Chem. Soc.*, **85**, 821 (1963).
[173] Holm, R. H., A. Chakravorty, and G. O. Dudek, *J. Am. Chem. Soc.*, **86**, 379 (1964).
[174] Holm, R. H., A. Chakravorty, and L. J. Theriot, *Inorg. Chem.*, **5**, 625 (1966).
[175] Holm, R. H., and F. A. Cotton, *J. Am. Chem. Soc.*, **80**, 5658 (1958).
[176] Holm, R. H., G. W. Everett, Jr., and A. Chakravorty, *Progr. Inorg. Chem.*, **7**, 83 (1966).
[177] Holm, R. H., and T. McKinney, *J. Am. Chem. Soc.*, **82**, 5506 (1960).
[178] Holm, R. H., and K. Swaminathan, *Inorg. Chem.*, **1**, 599 (1962).
[179] Horrocks, W. D., Jr., *J. Am. Chem. Soc.*, **87**, 3779 (1965).
[180] Horrocks, W. D., Jr., R. H. Fischer, J. R. Hutchison, and G. N. LaMar, *J. Am. Chem. Soc.*, **88**, 2436 (1966).
[181] Horrocks, W. D., Jr., R. C. Taylor, and G. N. LaMar, *J. Am. Chem. Soc.*, **86**, 3031 (1964).
[182] Jaffe, H. H., *J. Phys. Chem.*, **58**, 185 (1954).
[183] Jardetzky, O., and J. E. Wertz, *J. Am. Chem. Soc.*, **82**, 318 (1960).
[184] Jensen, R. E., and R. T. Pflaum, *Anal. Chim. Acta*, **32**, 235 (1965).
[185] Johnson, A. W., and D. Oldfield, *J. Chem. Soc.* (*C*), 794 (1966).
[186] Katz, J. J., G. L. Closs, F. C. Pennington, M. R. Thomas, and H. H. Strain, *J. Am. Chem. Soc.*, **85**, 3801 (1963).
[187] Kawasaki, Y., and T. Tanaka, *J. Chem. Phys.*, **43**, 3396 (1965).
[188] King, J. P., B. P. Block, and I. C. Popoff, *Inorg. Chem.*, **4**, 198 (1965).
[189] King, R. B., *Inorg. Chem.*, **2**, 641 (1963).
[190] Klose, G., H. Mueller, and E. Uhlemann, *Z. Naturforsch.*, **19b**, 952 (1964).
[191] Kluiber, R. W., *J. Am. Chem. Soc.*, **82**, 4839 (1960).

[192] Kluiber, R. W., *Inorg. Chem.*, **4**, 829 (1965).
[193] Kluiber, R. W., and W. D. Horrocks, Jr., *J. Am. Chem. Soc.*, **88**, 1399 (1966).
[194] Kluiber, R. W., and W. D. Horrocks, Jr., *Inorg. Chem.*, **5**, 152 (1966).
[195] Kluiber, R. W., and W. D. Horrocks, Jr., *J. Am. Chem. Soc.*, **87**, 5350 (1965).
[196] Kowalsky, A., *Biochemistry*, **4**, 2382 (1965).
[197] Kozyrev, B. M., *Discussions Faraday Soc.*, **19**, 135 (1955).
[198] Kozyrev, B. M., and A. I. Rivkind, *Dokl. Akad. Nauk SSSR*, **98**, 97 (1954); *Chem. Abstr.*, **50**, 2286e (1956).
[199] Kula, R. J., *Anal. Chem.*, **37**, 989 (1965).
[200] Kula, R. J., D. T. Sawyer, S. I. Chan, and C. M. Finley, *J. Am. Chem. Soc.*, **85**, 2930 (1963).
[201] LaLancette, E. A., D. R. Eaton, R. E. Benson, and W. D. Phillips, *J. Am. Chem. Soc.*, **84**, 3968 (1962).
[202] LaMar, G. N., private communication.
[203] LaMar, G. N., *J. Chem. Phys.*, **41**, 2992 (1964).
[204] LaMar, G. N., *J. Am. Chem. Soc.*, **87**, 3567 (1965).
[205] LaMar, G. N., *J. Chem. Phys.*, **43**, 235 (1965).
[206] LaMar, G. N., *J. Chem. Phys.*, **43**, 1085 (1965).
[207] LaMar, G. N., *J. Phys. Chem.*, **69**, 3212 (1965).
[208] LaMar, G. N., W. D. Horrocks, Jr., and L. C. Allen, *J. Chem. Phys.*, **41**, 2126 (1964).
[209] Larsen, D. W., and A. C. Wahl, *J. Chem. Phys.*, **41**, 908 (1964).
[210] Larsen, D. W., and A. C. Wahl, *J. Chem. Phys.*, **43**, 3765 (1965).
[211] Legg, J. I., and D. W. Cooke, *Inorg. Chem.*, **4**, 1576 (1965).
[212] Legg, J. I., and D. W. Cooke, *Inorg. Chem.*, **5**, 594 (1966).
[213] Leussing, D., and C. K. Stanfield, *J. Am. Chem. Soc.*, **86**, 2805 (1964).
[214] Lewis, J., R. F. Long, and C. Oldham, *J. Chem. Soc.*, 6740 (1965).
[215] Li, N. C., L. F. Johnson, and J. N. Shoolery, *J. Phys. Chem.*, **65**, 1902 (1961).
[216] Li, N. C., R. L. Scruggs, and E. D. Becker, *J. Am. Chem. Soc.*, **84**, 4650 (1962).
[217] Lin, W. C., and L. E. Orgel, *Mol. Phys.*, **7**, 131 (1963–4).
[218] Lohmann, W., C. F. Fowler, A. J. Moss, Jr., and W. H. Perkins, *Experientia*, **21**, 31 (1965).
[219] Lumry, R., H. Matsumiya, F. A. Bovey, and A. Kowalsky, *J. Phys. Chem.*, **65**, 837 (1961).
[220] Lussan, C., *J. Chim. Phys.*, **61**, 462 (1964).
[221] Luz, Z., and S. Meiboom, *J. Chem. Phys.*, **40**, 1058 (1964).
[222] Luz, Z., and R. G. Shulman, *J. Chem. Phys.*, **43**, 3750 (1965).
[223] Martin, R. B., and R. Mathur, *J. Am. Chem. Soc.*, **87**, 1065 (1965).
[224] Mathur, R., and N. C. Li, *J. Am. Chem. Soc.*, **86**, 1289 (1964).
[225] Mathur, R., and R. B. Martin, *J. Phys. Chem.*, **69**, 668 (1965).
[226] Mathur, R., S. M. Wang, and N. C. Li, *J. Phys. Chem.*, **68**, 2140 (1964).
[227] McCarthy, P. J., and G. H. Bidell, unpublished results.
[228] McCarthy, P. J., and A. E. Martell, *Inorg. Chem.*, **6**, 781 (1967).
[229] McClellan, W. R., and R. E. Benson, *J. Am. Chem. Soc.*, **88**, 5165 (1966).
[230] McConnell, H. M., *J. Chem. Phys.*, **24**, 764 (1956).
[231] McConnell, H. M., *J. Chem. Phys.*, **28**, 430 (1958).
[232] McConnell, H. M., and D. B. Chesnut, *J. Chem. Phys.*, **28**, 107 (1958).
[233] McConnell, H. M., and C. H. Holm, *J. Chem. Phys.*, **27**, 314 (1957); **28**, 749 (1958).
[234] McConnell, H. M., and R. E. Robertson, *J. Chem. Phys.*, **29**, 1361 (1958).

[235] McDonald, C. C., and W. D. Phillips, *J. Am. Chem. Soc.*, **85**, 3736 (1963).
[236] McGrady, M. M., and R. S. Tobias, *J. Am. Chem. Soc.*, **87**, 1909 (1965).
[237] McLachlan, A. D., *Mol. Phys.*, **1**, 233 (1958).
[238] Mildvan, A. S., and M. Cohn, *Biochemistry*, **2**, 910 (1963).
[239] Miller, J. D., and R. H. Prince, *J. Chem. Soc.*, 3185 (1965).
[240] Miller, J. D., and R. H. Prince, *J. Chem. Soc.*, 4706 (1965).
[241] Milner, R. S., and L. Pratt, *Discussions Faraday Soc.*, **34**, 88 (1962).
[242] Morgan, L. O., J. Murphy, and P. F. Cox, *J. Am. Chem. Soc.*, **81**, 5043 (1959).
[243] Muetterties, E. L., *Inorg. Chem.*, **4**, 769 (1965).
[244] Muetterties, E. L., and W. D. Phillips, *J. Am. Chem. Soc.*, **79**, 322 (1957).
[245] Muetterties, E. L., and C. M. Wright, *J. Am. Chem. Soc.*, **87**, 4706 (1965).
[246] Muller, J.-C., *Bull. Soc. Chim. France*, 2027 (1964).
[247] Musher, J. I., and E. J. Corey, *Tetrahedron*, **18**, 791 (1962).
[248] Nukada, K., O. Yamamoto, T. Suzuki, M. Takeuchi, and M. Ohnishi, *Anal. Chem.*, **35**, 1892 (1963).
[249] Nyholm, R. S., M. R. Snow, M. H. B. Stiddard, *J. Chem. Soc.*, 6564, 6570 (1965).
[250] Osborn, J. A., R. D. Gillard, and G. Wilkinson, *J. Chem. Soc.*, 3168 (1964).
[251] Paciorek, K. L., and R. H. Kratzer, *Inorg. Chem.*, **5**, 538 (1966).
[252] Palmer, J. W., and F. Basolo, *J. Phys. Chem.*, **64**, 778 (1960).
[253] Palmer, R. A., R. C. Fay, and T. S. Piper, *Inorg. Chem.*, **3**, 875 (1964).
[254] Parshall, G. W., and F. N. Jones, *J. Am. Chem. Soc.*, **87**, 5356 (1965).
[255] Pearson, R. G., and M. M. Anderson, *Angew. Chem. Intern. Ed. Engl.*, **4**, 281 (1965).
[256] Pearson, R. G., and R. D. Lanier, *J. Am. Chem. Soc.*, **86**, 765 (1964).
[257] Pearson, R. G., J. Palmer, M. M. Anderson, and A. L. Allred, *Z. Elektrochem.*, **64**, 110 (1960).
[258] Pfeifer, H., *Arch. Sci. (Geneva)*, **14**, 357 (1961).
[259] Phillips, W. D., and R. E. Benson, *J. Chem. Phys.*, **33**, 607 (1960).
[260] Pinnavaia, T. J., and R. C. Fay, *Inorg. Chem.*, **5**, 233 (1966).
[261] Popel, A. A., and E. D. Grazhdannikov, *Khim. Redkikh Elementov, Leningr. Gos. Univ.*, 147 (1964); *Chem. Abstr.*, **61**, 6616g (1964).
[262] Ramsey, N. F., and E. M. Purcell, *Phys. Rev.*, **85**, 143 (1952).
[263] Rechnitz, G. A., and S. B. Zamochnick, *J. Am. Chem. Soc.*, **86**, 2953 (1964).
[264] Ritchey, W., 140th ACS Meeting, September, 1961, *Reports*, p. 35 N.
[265] Rivkind, A. I., *Dokl. Akad. Nauk SSSR*, **100**, 933 (1955); *Chem. Abstr.*, **50**, 1392b (1956).
[266] Rivkind, A. I., *Russ. J. Inorg. Chem. (English Transl.)*, **2**, 77 (1957).
[267] Robinson, S. D., and B. L. Shaw, *J. Chem. Soc.*, 4806 (1963).
[268] Sacconi, L., *Experientia Suppl.*, **9**, 148 (1964).
[269] Sacconi, L., A. D. 602990, 1964, 32 pp., Avail. O.T.S.; *Chem. Abstr.*, **62**, 1545c (1965).
[270] Sacconi, L., M. Ciampolini, and N. Nardi, *J. Am. Chem. Soc.*, **86**, 819 (1964).
[271] Sacconi, L., P. Paoletti, and M. Ciampolini, *J. Am. Chem. Soc.*, **85**, 411 (1963).
[272] Sauer, K., J. R. L. Smith, and A. J. Schultz, *J. Am. Chem. Soc.*, **88**, 2681 (1966).
[273] Saupe, A., and G. Englert, *Z. Naturforsch.*, **19a**, 172 (1964); *Phys. Rev. Letters*, **11**, 462 (1963).
[274] Sawyer, D. T., and J. R. Brannan, *Inorg. Chem.*, **5**, 65 (1966).
[275] Sheppard, W. A., *J. Am. Chem. Soc.*, **87**, 2410 (1965).
[276] Shiner, V. J., Jr., and D. Whittaker, *J. Am. Chem. Soc.*, **87**, 843 (1965).

[277] Shulman, R. G., H. Sternlicht, and B. J. Wyluda, *J. Chem. Phys.*, **43**, 3116 (1965).
[278] Siddall, T. H., and C. A. Prohaska, *Inorg. Chem.*, **4**, 783 (1965).
[279] Smith, J. A. S., and E. J. Wilkins, *Chem. Commun.*, 381 (1965).
[280] Smith, J. A. S., and E. J. Wilkins, *J. Chem. Soc.* (A), 1749 (1966).
[281] Smith, J., and J. Thwaites, *Discussions Faraday Soc.*, **34**, 143 (1962).
[282] Snyder, L. C., *J. Chem. Phys.*, **43**, 4041 (1965).
[283] Snyder, L. C., and E. W. Anderson, *J. Am. Chem. Soc.*, **86**, 5023 (1964).
[284] Snyder, L. C., and E. W. Anderson, *J. Chem. Phys.*, **42**, 3336 (1965).
[285] Snyder, L. C., and S. Meiboom, *J. Chem. Phys.*, **44**, 4057 (1966).
[286] Spees, S. T., Jr., L. J. Durham, and A. M. Sargeson, *Inorg. Chem.*, **5**, 2103 (1966).
[287] Spence, J. T., and J. Y. Lee, *Inorg. Chem.*, **4**, 385 (1965).
[288] Stehling, F. C., *Anal. Chem.*, **35**, 773 (1963).
[289] Stengle, T. R., and C. H. Langford, *J. Phys. Chem.*, **69**, 3299 (1965).
[290] Sternlicht, H., R. G. Shulman, and E. W. Anderson, *J. Chem. Phys.*, **43**, 3123 (1965).
[291] Storm, C. B., and A. H. Corwin, *J. Org. Chem.*, **29**, 3700 (1964).
[292] Stranks, D. R., "The Reaction Rates of Transitional Metal Complexes," *Mod. Coord. Chem.*, 78–173 (1960).
[293] Sudmeier, J. L., and C. N. Reilley, *Inorg. Chem.*, **5**, 1047 (1966).
[294] Tang, P., and N. C. Li, *J. Am. Chem. Soc.*, **86**, 1293 (1964).
[295] Thwaites, J. D., and L. Sacconi, *Inorg. Chem.*, **5**, 1029 (1966).
[296] Thwaites, J. D., I. Bertini, and L. Sacconi, *Inorg. Chem.*, **5**, 1036 (1966).
[297] Toporcer, L. H., R. E. Dessy, and S. I. E. Green, *Inorg. Chem.*, **4**, 1649 (1965).
[298] Van Hecke, G. R., and W. D. Horrocks, Jr., *Inorg. Chem.*, **5**, 1968 (1966).
[299] Walker, W. R., and N. C. Li, *J. Inorg. Nucl. Chem.*, **27**, 411 (1965).
[300] Wang, S. M., and N. C. Li, *J. Am. Chem. Soc.*, **88**, 4592 (1966).
[301] Weissman, S. I., T. R. Tuttle, and E. deBoer, *J. Phys. Chem.*, **61**, 28 (1957).
[302] Wertz, J. E., and O. Jardetzky, *Archiv. Biochem. Biophys.*, **65**, 569 (1956).
[303] Wilkinson, G., in *Advances in the Chemistry of Coordination Compounds*, S. Kirshner, ed. New York: Macmillian, 1961, p. 50.
[304] Wishnia, A., *J. Chem. Phys.*, **32**, 871 (1960).
[305] Zimmerman, J., *J. Chem. Phys.*, **21**, 1605 (1953).

ABBREVIATIONS

Me	methyl, CH_3
Et	ethyl, C_2H_5
Pr	propyl, C_3H_7
Bu	butyl, C_4H_9
Ph	phenyl, C_6H_5
py	pyridine, C_5H_5N
bipy	2,2′-bipyridine, $C_{10}H_8N_2$
terpy	2,2′,2″-terpyridine, $C_{15}H_{11}N_3$
phen	1,10-phenanthroline, $C_{12}H_8N_2$
en	ethylenediamine, $H_2NCH_2CH_2NH_2$
pn	propylenediamine, $H_2NCH_2CH(CH_3)NH_2$
tn	trimethylenediamine, $H_2NCH_2CH_2CH_2NH_2$
i-bn	isobutanediamine, $H_2NC(CH_3)_2CH_2NH_2$
bdn	1,3-diaminobutane, $H_2N(CH_2)_2CH(NH_2)CH_3$
cptn	1,2-diaminocyclopentane, $H_2N(C_5H_8)NH_2$
chxn	1,2-diaminocyclohexane, $H_2N(C_6H_{10})NH_2$
stien	stilbenediamine, $H_2NCH(C_6H_5)CH(C_6H_5)NH_2$
den	diethylenetriamine, $H_2N(CH_2)_2NH(CH_2)_2NH_2$
tren	triaminotriethylamine, $N(CH_2CH_2NH_2)_3$
trien	triethylenetetramine, $H_2N(CH_2)_2NH(CH_2)_2NH(CH_2)_2NH_2$
(ac)H	acetic acid, CH_3COOH
(ox)H_2	oxalic acid, $(COOH)_2$
(mal)H_2	malonic acid, $CH_2(COOH)_2$
(tart)H_2	tartaric acid, HOOCCH(OH)CH(OH)COOH
(IDA)H_2	iminodiacetic acid, $HN(CH_2COOH)_2$
(NTA)H_3	nitrilotriacetic acid, $N(CH_2COOH)_3$
(EDTA)H_4	ethylenediaminetetraacetic acid, $(HOOCCH_2)_2N(CH_2)_2N(CH_2COOH)_2$

$(PDTA)H_4$	propylenediaminetetraacetic acid, $(HOOCCH_2)_2NCH_2CH(CH_3)N(CH_2COOH)_2$
(gly)H	glycine, H_2NCH_2COOH
(α-ala)H	α-alanine, $CH_3CH(NH_2)COOH$
(β-ala)H	β-alanine, $H_2NCH_2CH_2COOH$
(leu)H	leucine, $H_2NCH(COOH)CH_2CH(CH_3)_2$
(thr)H	threonine, $CH_3CH(OH)CH(NH_2)COOH$
(prol)H	proline, C_4H_8NCOOH
(acac)H	acetylacetone, $CH_3COCH_2COCH_3$
(tfac)H	trifluoroacetylacetone, $CF_3COCH_2COCH_3$
(hfac)H	hexafluoroacetylacetone, $CF_3COCH_2COCF_3$
(bzac)H	benzoylacetone, $C_6H_5COCH_2COCH_3$
(dbm)H	dibenzoylmethane, $C_6H_5COCH_2COC_6H_5$
(DPM)H	dipivaloylmethane, $(CH_3)_3CCOCH_2COC(CH_3)_3$
bu	biuret, $H_2NCONHCONH_2$
TMG	tetramethylguanidine, $[(CH_3)_2N]_2CNH$
(dmg)H	dimethylglyoxime, $HONC(CH_3)C(CH_3)NOH$
daes	2,2′-di(aminoethyl)sulfide, $(H_2NCH_2CH_2)_2S$
(dtp)H	diethyldithiophosphoric acid, $(C_2H_5O)_2PS_2H$
(dtc)H	diethyldithiocarbamic acid, $(C_2H_5)_2NCS_2H$
(dsep)H	diethyldiselenophosphoric acid, $(C_2H_5O)_2PSe_2H$

INDEX